History of Space Shuttle Rendezvous

Mission Operations Directorate
Flight Dynamics Division

October 2011

National Aeronautics and
Space Administration

Lyndon B. Johnson Space Center
Houston, Texas

Space Shuttle Rendezvous and Proximity Operations Missions (For Missions To ISS See Inside Back Cover)

Mission	Flight	Year	Profile	Target	Comments
Rendezvous or Prox Ops Demo	7	1983	Deploy/Rtrv	SPAS-01	Proximity operations only.
	41B	1984	Deploy/Rndz	IRT	No rendezvous due to IRT balloon failure.
	51G	1985	SK	none	Station-keeping test of proximity operations autopilot.
	61B	1985	Deploy/SK	radar reflector	Station-keeping test of proximity operations autopilot.
	37	1991	Deploy/Rndzs	GRO	GRO used as target for star tracker navigation test.
	63	1995	Ground-Up	Mir	Approached to within 37 feet of Mir on +V Bar.
	134	2011	Ground-Up	ISS	STORRM DTO (after undock) on an Orion rndz profile.
Satellite Servicing	41C	1984	Ground-Up	Solar Max	Retrieved and repaired after second rendezvous.
	51A	1984	Ground-Up	1. Palapa-B2 2. Westar-VI	Both maneuvered to meet downrange and planar constraints and retrieved by an astronaut flying the MMU.
	51D	1985	Deploy/Rndz	SYNCOM IV-3	Contingency rendezvous after deployment and activation failure.
	51I	1985	Ground-Up	SYNCOM IV-3	Rendezvous & EVA planned in four months. Elliptical orbit.
	49	1992	Ground-Up	INTELSAT VI (F-3)	Hybrid Control Box, 3 rendezvous. Lambert targeting problem.
	61	1993	Ground-Up	Hubble	Servicing Mission 1
	82	1997	Ground-Up	Hubble	Servicing Mission 2
	103	1999	Ground-Up	Hubble	Servicing Mission 3A
	109	2002	Ground-Up	Hubble	Servicing Mission 3B
	125	2009	Ground-Up	Hubble	Servicing Mission 4
Deployment & Retrieval of Scientific Satellite	51G	1985	Deploy/Rtrv	SPARTAN-101	Incorrect SPARTAN attitude at retrieval.
	51F	1985	Deploy/Rtrv	PDP	On-board targeted proximity operations.
	39	1991	Deploy/Rtrv	IBSS-SPAS II	Most complex deploy/retrieve profile flown.
	56	1993	Deploy/Rtrv	SPARTAN-201-01	Laser range and range rate sensor test.
	51	1993	Deploy/Rtrv	ORFEUS-SPAS 1	Long range, in-front and behind station-keeping.
	60	1994	Deploy/Rtrv	WSF-1	WSF-1 problems prevented deployment.
	64	1994	Deploy/Rtrv	SPARTAN-201-02	First successful test of Trajectory Control Sensor laser.
	66	1994	Deploy/Rtrv	CRISTA-SPAS 1	Football for data collection. +R Bar Mir approach corridor test.
	63	1995	Deploy/Rtrv	SPARTAN-204	Deploy day after Mir rendezvous. Traj. designed to avoid Mir.
	69	1995	Deploy/Rtrv Deploy/Rtrv	1. SPARTAN-201-03 2. WSF-2	1. Incorrect SPARTAN attitude at retrieval. 2. Long range, in-front station-keeping.
	72	1996	Deploy/Rtrv	OAST-Flyer	Also called SPARTAN-206
	77	1996	Deploy/Rtrv Deploy/Rndz	1. SPARTAN-207-IAE 2. PAMS-STU	1. Inflatable Antenna Experiment 2. Three rendezvous and station-keeping (650 m on -V Bar).
	80	1996	Deploy/Rtrv Deploy/Rtrv	1. ORFEUS-SPAS 2 2. WSF-3	1. Relative GPS test for ISS ESA ATV. 2. Long range, in-front station-keeping.
	85	1997	Deploy/Rtrv	CRISTA-SPAS 2	Tested ISS TORVA & +V Bar approach using keel camera.
	87	1997	Deploy/Rtrv	SPARTAN-201-04	SPARTAN activation failure, EVA retrieval. VGS test.
	95	1998	Deploy/Rtrv	SPARTAN-201-05	VGS test
Retrieval & Return to Earth	32	1990	Ground-Up	LDEF	Hot final approach due to radar procedure issue.
	57	1993	Ground-Up	EURECA (ESA)	Solar array latch failure, corrected during EVA.
	72	1996	Ground-Up	SFU (Japan)	Hybrid control box. Solar array retraction failure & jettison.
Mir	71	1995	Ground-Up	Mir	Docked to Buran port on Kristall Module. Crew exchange.
	74	1995	Ground-Up	Mir	Installed Shuttle Docking Module on Kristall.
	76	1996	Ground-Up	Mir	Resupply & U.S. crew delivery.
	79	1996	Ground-Up	Mir	Resupply & U.S. crew exchange.
	81	1997	Ground-Up	Mir	Resupply & U.S. crew exchange.
	84	1997	Ground-Up	Mir	Resupply & U.S. crew exchange. GPS & laser test for ATV.
	86	1997	Ground-Up	Mir	Resupply & U.S. crew exchange. GPS test for ATV.
	89	1998	Ground-Up	Mir	Resupply & U.S. crew exchange.
	91	1998	Ground-Up	Mir	Resupply & U.S. crew return.

ATV = Automated Transfer Vehicle, **CRISTA** = Cryogenic Infrared Spectrometers and Telescopes for the Atmospheric, **EURECA** = European Retrievable Carrier, **EVA** = Extra Vehicular Activity, **IBSS** = Infrared Background Signature Survey, **GPS** = Global Positioning System, **GRO** = Gamma Ray Observatory, **INTELSAT** = International Telecommunications Satellite, **IRT** = Integrated Rendezvous Target, **LDEF** = Long Duration Exposure Facility, **MMU** = Manned Maneuvering Unit, **OAST** = Office of Aeronautics and Space Technology, **ORBT** = Optimized R-Bar Targeted Rendezvous, **ORFEUS** = Orbiting and Retrievable Far and Extreme Ultraviolet Spectrometer, **PAMS-STU** = Passive Aerodynamically Stabilized Magnetically Damped Satellite-Satellite Test Unit, **PDP** = Plasma Diagnostics Package, **Rndz** = Rendezvous, **Rtrv** = Retrieve, **SFU** = Space Flyer Unit, **SK** = Station-Keeping, **SPARTAN** = Shuttle Pointed Autonomous Tool For Astronomy, **SPAS** = Shuttle Pallet Satellite, **STORRM DTO** = Sensor Test for Orion Relnav Risk Mitigation Detailed Test Objective, **SYNCOM** = Synchronous Communication, **TORRA** = Twice Orbital Rate R Bar Approach, **TORVA** = Twice Orbital Rate V Bar Approach, **VGS** = Video Guidance Sensor, **WSF** = Wake Shield Facility

History of Space Shuttle Rendezvous

October 2011

By:

Original Signed By:

John L. Goodman

Flight Design and Dynamics
United Space Alliance, LLC

Approved By:

Original Signed By:

Christine M. Reichert

Chief
Orbit Dynamics Branch
Flight Dynamics Division
Mission Operations Directorate
NASA Johnson Space Center

This page intentionally left blank.

TABLE OF CONTENTS

This page intentionally left blank.

PREFACE

This technical history is intended to provide a technical audience with an introduction to the rendezvous and proximity operations history of the Space Shuttle Program. It details the programmatic constraints and technical challenges encountered during shuttle development in the 1970s and over thirty years of shuttle missions. An overview of rendezvous and proximity operations on many shuttle missions is provided, as well as how some shuttle rendezvous and proximity operations systems and flight techniques evolved to meet new programmatic objectives.

Since the publication of the first edition in October of 2006 additional historical information has been collected. This revised edition provides additional information on Mercury, Gemini, Apollo, Skylab, and Apollo/Soyuz. Some chapters on the Space Shuttle have been updated and expanded.

Four special focus chapters have been added to provide more detailed information on shuttle rendezvous. A chapter on the STS-39 mission of April/May 1991 describes the most complex deploy/retrieve mission flown by the shuttle. Another chapter focuses on the Hubble Space Telescope servicing missions. A third chapter gives the reader a detailed look at the February 2010 STS-130 mission to the International Space Station. The fourth chapter answers the question why rendezvous was not completely automated on the Gemini, Apollo, and Space Shuttle vehicles.

Unfortunately, the brief coverage given in this history to these missions, and the Mir and ISS missions in particular, does not do justice to the tremendous amount of work required to overcome challenges and successfully fly. The diligence of the NASA civil servant and contractor team was the key to the success of shuttle rendezvous and proximity operations. Now that the Program is ending I have wondered how factors behind this success could be described to personnel working on future flight programs, either human or robotic.

In November of 1998 Space Shuttle Program Manager Tommy W. Holloway published a memo titled "The Future."* The first shuttle mission in support of the ISS was to be launched in about two weeks and the flight manifest was challenging. Holloway mentioned several close calls within the Shuttle Program and referred to several space and aviation incidents not connected with human spaceflight. Holloway went on to detail 8 points that were required to ensure flight safety and mission success. The first time I read this list I recognized it as a description of the flight safety culture I had seen demonstrated by rendezvous and proximity operations personnel in my 25 year career with the Shuttle Program.

The technical achievements portrayed in the pages of this history are due to the following:

1. Each member of the Space Shuttle Team is accountable and responsible for his/her task, function, or project. The Program, Safety, and Mission Assurance, or the phantom "They," etc., are not. We, individually, are responsible.

2. Individual Space Shuttle Team member skills and expertise must be continually pursued and honed. Thinking we know it all and complacency are enemies.

3. Adequate and thorough analysis is mandatory. Understanding the limitation of the analysis is just as important. Using "similarity" and "gut feeling" or "extrapolation" is dangerous. The Mission Evaluation Room's motto, "In God we trust, all others bring data," will serve us well.

4. Adequate and thorough testing in the best possible environment is mandatory. Understanding the limitations of the test is as important as understanding the results. Bad tests are worse than no tests; they mislead you.

5. Individual rigor and discipline to do it right are mandatory. Lackadaisical attitude, lack of attention to detail, and not implementing procedures correctly, etc., are precursors to failure.

6. Take time to do it right, to ensure there will be a tomorrow. Cutting corners and hurrying to do a job are sure ways to fail. If you don't think you have the time to do it right, take time out!

7. Communication and sharing of data, concepts, and ideas across the Space Shuttle Team are the checks and balances that keep us on track. Not having data is bad; not sharing is worse.

8. Learn from close calls. We should not only investigate the specific close call but review like areas in other systems, processes, and designs.

Mr. Holloway's instructions are timeless.

John L. Goodman

---

* Holloway, Tommy W., "The Future," MA-98-068, Space Shuttle Program, NASA Johnson Space Center, November 20, 1998.

This page intentionally left blank.

INTRODUCTION

In the late 1950s rendezvous and docking was recognized as necessary for building space stations and assembling vehicles in low Earth orbit to perform exploration missions. Rendezvous and docking was a key component of the Apollo lunar orbit rendezvous architecture adopted by NASA in 1962. During Project Gemini rendezvous and docking technology and mission techniques were developed and successfully demonstrated in preparation for the Apollo lunar missions. Nominal rendezvous and docking was successfully performed on the Apollo missions. Lunar missions also required extensive development of contingency mission techniques to ensure crew safety and successful return to Earth. Knowledge and experience gained from the Gemini and Apollo missions was later successfully applied to the Skylab and Apollo/Soyuz missions.

In 1972, at the time of the Shuttle Program contract awards, rendezvous and docking technology and flight techniques were considered to be mature and the challenges well understood. Space Shuttle software and hardware sub-system design introduced more automation into rendezvous and lowered crew work-load compared to Gemini and Apollo vehicles. Space Shuttle missions were intended to require less pre-flight preparation than Gemini and Apollo missions to achieve low life cycle costs. This included simplified and standardized mission planning and training, lower number of mission support personnel, high flight rates, elimination of extensive flight-to-flight analysis, no computation of flight specific trajectory data, and no generation of customized onboard charts for each mission.

However, the differences between rendezvous and docking as flown with Gemini and Apollo vehicles and the proposed Space Shuttle mission requirements were only gradually understood as the 1970s proceeded, after much Space Shuttle hardware design had been completed. While the Space Shuttle was a far more capable and flexible spacecraft than Gemini and Apollo, these differences required extensive mission specific procedure and trajectory development. The challenges presented by these differences included:

1) Space Shuttle rendezvous as an optional service (secondary to satellite deployment) that had a lower priority for spacecraft and ground system development funds and resources.

2) Rendezvous as a mission success objective as opposed to safety critical rendezvous and docking (i.e. return the lunar exploration crew to Earth).

3) Rendezvous with spacecraft not originally designed and equipped to support Space Shuttle rendezvous.

4) Use of the Space Shuttle to perform space station construction, large scale space station re-supply, satellite retrieval, and satellite repair.

5) Rendezvous with spacecraft that did not possess cooperative relative navigation aids (i.e. no radio-frequency transponders or lights).

6) Capture and berthing of a spacecraft with a robotic arm rather than docking.

7) Adaptation of piloting and relative motion trajectory design techniques to ensure safe spacecraft deployment, including deployment of spacecraft with high energy upper stages.

8) Plume impingement on target spacecraft caused by Space Shuttle Reaction Control System (RCS) jets that were designed to support atmospheric entry and orbital flight without rendezvous considerations.

9) Integration of rendezvous operations with Extra-Vehicular Activity (EVA, i.e. space walks).

10) Execution of rendezvous activities in parallel with multiple secondary payload objectives (all having equal priority) by a single team or round the clock teams of astronauts.

11) Use of the Space Shuttle to perform space station orbit raising and debris avoidance burns while docked.

12) Integration of Space Shuttle thermal protection inspections by the shuttle and International Space Station (ISS) crews into the rendezvous procedures and flight techniques.

As these challenges emerged they were primarily overcome through the development of crew and ground personnel procedural work-arounds (nominal and contingency). Development of new piloting techniques and procedures was preferred over hardware and software development to meet cost and schedule constraints. A new flight phase, called proximity operations, was defined by March of 1977 due to the considerable mission technique development required to successfully overcome plume impingement and propellant budget challenges associated with capturing target spacecraft. In contrast, Gemini and Apollo final approach and docking was relatively straightforward and did not require mission specific procedure and trajectory development of later shuttle missions.

There were two notable cases of hardware and software development performed to enable the Space

Approved for public release via STI DAA 24483. See statement on title page.

9

Shuttle to meet new mission objectives associated with proximity operations, docking, and docked operations with a space station (Mir and ISS). The shuttle on-orbit Digital Autopilot (DAP) was modified to permit use of 25 pound thrust vernier RCS jets to be used for orbit raising burns while the shuttle was docked to Mir and the ISS. Other DAP modifications were made to enable the shuttle to effectively control the attitude the mated stack.

The second case concerned precise relative navigation sensors during final approach and docking. Gemini and Apollo vehicles did not posses relative navigation sensors to support manually piloted final approach (starting at approximately 100 feet) through docking due to the large docking capture envelopes. In the 1970s shuttle rendezvous sensors were designed to support rendezvous burns at long ranges (tens of miles and thousands of feet). Simulations showed that manual piloting to completely stop relative motion to facilitate satellite capture with the robotic arm did not require precise, close-in relative navigation sensors.

By the late 1980s studies of shuttle docking with Space Station Freedom revealed that the shuttle rendezvous radar was not adequate in terms of minimum useable range and accuracy to support final approach and docking while meeting plume impingement and docking hardware capture envelope constraints. This led to the development of crew hand-held and shuttle payload bay laser sensors and a laptop computer application that greatly enhanced crew situational awareness of relative motion. These new proximity operations tools reduced proximity operations propellant consumption and eased the manual piloting task.

However, budget constraints forced the use of some off-the-shelf hardware and software (laptop computer hardware and operating system, and the hand-held laser). To meet cost and schedule constraints, certification of the new tools was at a lower level than software and hardware used for ascent and entry, resulting in a requirement that the crews be able to accomplish final approach and docking without them. This development philosophy did result in proximity operations procedures and tools that were effective at enabling the crew to achieve mission objectives and meet safety and mission success requirements. However, the resulting system was labor intensive and required close coordination between multiple crew members and ground support personnel.

In spite of unforeseen technical, requirements, budget, and schedule challenges the Space Shuttle was successful at meeting mission objectives associated with rendezvous, proximity operations, and docking. From June 1983 to July 2011, 78 Space Shuttle missions had at least one rendezvous or proximity operations objective. These missions were successful in achieving the relative position and velocity required for grapple of or docking with the target spacecraft (in spite of on-board systems failures on several missions). Several flights required multiple attempts at target spacecraft capture, but ultimately succeeded in accomplishing mission objectives. An understanding of factors behind the success of shuttle rendezvous and proximity operations, and how programmatic and technical challenges shaped vehicle operation and mission design is essential for mitigating cost, schedule, and technical risk in current and future programs.[1,2]

Reference

1. Goodman, J. L., *Knowledge Capture and Management for Space Flight Systems*, NASA Contractor Report NASA/CR-2005-213692, NASA Johnson Space Center, October 2005. See the NASA Technical Reports server at http://ntrs.nasa.gov/, or the Johnson Technical Reports server at http://ston.jsc.nasa.gov/collections/TRS/.

2. *Aerospace Safety Advisory Panel Annual Report for 2010*, NASA, Washington, DC, January 2011. See http://oiir.hq.nasa.gov/asap/index.html (accessed January 15, 2011).

Space Shuttle Rendezvous and Proximity Operations Missions

Mission	Flight	Year	Profile	Target	Comments
Rendezvous or Prox Ops Demo	7	1983	Deploy/Rtrv	SPAS-01	Proximity operations only.
	41B	1984	Deploy/Rndz	IRT	No rendezvous due to IRT balloon failure.
	51G	1985	SK	none	Station-keeping test of proximity operations autopilot.
	61B	1985	Deploy/SK	radar reflector	Station-keeping test of proximity operations autopilot.
	37	1991	Deploy/Rndzs	GRO	GRO used as target for star tracker navigation test.
	63	1995	Ground-Up	Mir	Approached to within 37 feet of Mir on +V Bar.
	134	2011	Ground-Up	ISS	STORRM DTO (after undock) on an Orion rndz profile.
Satellite Servicing	41C	1984	Ground-Up	Solar Max	Retrieved and repaired after second rendezvous.
	51A	1984	Ground-Up	1. Palapa-B2 2. Westar-VI	Both maneuvered to meet downrange and planar constraints and retrieved by an astronaut flying the MMU.
	51D	1985	Deploy/Rndz	SYNCOM IV-3	Contingency rendezvous after deployment and activation failure.
	51I	1985	Ground-Up	SYNCOM IV-3	Rendezvous & EVA planned in four months. Elliptical orbit.
	49	1992	Ground-Up	INTELSAT VI (F-3)	Hybrid Control Box, 3 rendezvous. Lambert targeting problem.
	61	1993	Ground-Up	Hubble	Servicing Mission 1
	82	1997	Ground-Up	Hubble	Servicing Mission 2
	103	1999	Ground-Up	Hubble	Servicing Mission 3A
	109	2002	Ground-Up	Hubble	Servicing Mission 3B
	125	2009	Ground-Up	Hubble	Servicing Mission 4
Deployment & Retrieval of Scientific Satellite	51G	1985	Deploy/Rtrv	SPARTAN-101	Incorrect SPARTAN attitude at retrieval.
	51F	1985	Deploy/Rtrv	PDP	On-board targeted proximity operations.
	39	1991	Deploy/Rtrv	IBSS-SPAS II	Most complex deploy/retrieve profile flown.
	56	1993	Deploy/Rtrv	SPARTAN-201-01	Laser range and range rate sensor test.
	51	1993	Deploy/Rtrv	ORFEUS-SPAS 1	Long range, in-front and behind station-keeping.
	60	1994	Deploy/Rtrv	WSF-1	WSF-1 problems prevented deployment.
	64	1994	Deploy/Rtrv	SPARTAN-201-02	First successful test of Trajectory Control Sensor laser.
	66	1994	Deploy/Rtrv	CRISTA-SPAS 1	Football for data collection. +R Bar Mir approach corridor test.
	63	1995	Deploy/Rtrv	SPARTAN-204	Deploy day after Mir rendezvous. Traj. designed to avoid Mir.
	69	1995	Deploy/Rtrv	1. SPARTAN-201-03	1. Incorrect SPARTAN attitude at retrieval.
			Deploy/Rtrv	2. WSF-2	2. Long range, in-front station-keeping.
	72	1996	Deploy/Rtrv	OAST-Flyer	Also called SPARTAN-206
	77	1996	Deploy/Rtrv	1. SPARTAN-207-IAE	1. Inflatable Antenna Experiment
			Deploy/Rndz	2. PAMS-STU	2. Three rendezvous and station-keeping (650 m on -V Bar).
	80	1996	Deploy/Rtrv	1. ORFEUS-SPAS 2	1. Relative GPS test for ISS ESA ATV.
			Deploy/Rtrv	2. WSF-3	2. Long range, in-front station-keeping.
	85	1997	Deploy/Rtrv	CRISTA-SPAS 2	Tested ISS TORVA & +V Bar approach using keel camera.
	87	1997	Deploy/Rtrv	SPARTAN-201-04	SPARTAN activation failure, EVA retrieval. VGS test.
	95	1998	Deploy/Rtrv	SPARTAN-201-05	VGS test
Retrieval & Return to Earth	32	1990	Ground-Up	LDEF	Hot final approach due to radar procedure issue.
	57	1993	Ground-Up	EURECA (ESA)	Solar array latch failure, corrected during EVA.
	72	1996	Ground-Up	SFU (Japan)	Hybrid control box. Solar array retraction failure & jettison.
Mir	71	1995	Ground-Up	Mir	Docked to Buran port on Kristall Module. Crew exchange.
	74	1995	Ground-Up	Mir	Installed Shuttle Docking Module on Kristall.
	76	1996	Ground-Up	Mir	Resupply & U.S. crew delivery.
	79	1996	Ground-Up	Mir	Resupply & U.S. crew exchange.
	81	1997	Ground-Up	Mir	Resupply & U.S. crew exchange.
	84	1997	Ground-Up	Mir	Resupply & U.S. crew exchange. GPS & laser test for ATV.
	86	1997	Ground-Up	Mir	Resupply & U.S. crew exchange. GPS test for ATV.
	89	1998	Ground-Up	Mir	Resupply & U.S. crew exchange.
	91	1998	Ground-Up	Mir	Resupply & U.S. crew return.

ATV = Automated Transfer Vehicle, **CRISTA** = Cryogenic Infrared Spectrometers and Telescopes for the Atmospheric, **EURECA** = European Retrievable Carrier, **EVA** = Extra Vehicular Activity, **IBSS** = Infrared Background Signature Survey, **GPS** = Global Positioning System, **GRO** = Gamma Ray Observatory, **INTELSAT** = International Telecommunications Satellite, **IRT** = Integrated Rendezvous Target, **LDEF** = Long Duration Exposure Facility, **MMU** = Manned Maneuvering Unit, **OAST** = Office of Aeronautics and Space Technology, **ORBT** = Optimized R-Bar Targeted Rendezvous, **ORFEUS** = Orbiting and Retrievable Far and Extreme Ultraviolet Spectrometer, **PAMS-STU** = Passive Aerodynamically Stabilized Magnetically Damped Satellite-Satellite Test Unit, **PDP** = Plasma Diagnostics Package, **Rndz** = Rendezvous, **Rtrv** = Retrieve, **SFU** = Space Flyer Unit, **SK** = Station-Keeping, **SPARTAN** = Shuttle Pointed Autonomous Tool For Astronomy, **SPAS** = Shuttle Pallet Satellite, **STORRM DTO** = Sensor Test for Orion Relnav Risk Mitigation Detailed Test Objective, **SYNCOM** = Synchronous Communication, **TORRA** = Twice Orbital Rate R Bar Approach, **TORVA** = Twice Orbital Rate V Bar Approach, **VGS** = Video Guidance Sensor, **WSF** = Wake Shield Facility

Space Shuttle Rendezvous and Proximity Operations Missions to the ISS

Mission	Flight	Year	Profile	Target	Comments
ISS Assembly and Supply	88 (2A)	1998	Ground-Up	ISS	Captured Zarya with RMS, attached Unity Node with PMA 1 & 2.
	96 (2A.1)	1999	Ground-Up	ISS	First docking with ISS. ISS resupply and outfitting.
	101 (2A.2a)	2000	Ground-Up	ISS	ISS resupply and outfitting.
	106 (2A.2b)	2000	Ground-Up	ISS	ISS resupply and outfitting.
	92 (3A)	2000	Ground-Up	ISS	Radar failure. Z1 Truss, PMA 3, Ku comm, & CMGs installed.
	97 (4A)	2000	Ground-Up	ISS	Delivered P6 truss (with solar arrays & radiators).
	98 (5A)	2001	Ground-Up	ISS	Delivered Destiny lab.
	102 (5A.1)	2001	Ground-Up	ISS	Tail forward approach. MPLM resupply. Crew exchange.
	100 (6A)	2001	Ground-Up	ISS	Tail forward approach. Installed robotic arm. MPLM resupply.
	104 (7A)	2001	Ground-Up	ISS	Delivered Quest Airlock (installed with ISS robotic arm).
	105 (7A.1)	2001	Ground-Up	ISS	MPLM resupply. Crew exchange.
	108 (UF-1)	2001	Ground-Up	ISS	MPLM resupply. Crew exchange.
	110 (8A)	2002	Ground-Up	ISS	Delivered S0 truss and Mobile Transporter.
	111 (UF-2)	2002	Ground-Up	ISS	MPLM resupply. Mobile base installation. Crew exchange.
	112 (9A)	2002	Ground-Up	ISS	Delivered S1 truss, radiators & CETA cart A.
	113 (11A)	2002	Ground-Up	ISS	Delivered P1 truss, radiators & CETA cart B. Crew exchange.
	114 (LF-1)	2005	Ground-Up	ISS	MPLM Resupply. CMG replacement. First RPM.
	121 (ULF-1.1)	2006	Ground-Up	ISS	MPLM Resupply. Add third ISS crewmember.
	115 (12A)	2006	Ground-Up	ISS	P3/P4 truss.
	116 (12A.1)	2006	Ground-Up	ISS	P5 Truss, SPACEHAB
	117 (13A)	2007	Ground-Up	ISS	S3/S4 Truss
	118 (13A.1)	2007	Ground-Up	ISS	S5 Truss
	120 (10A)	2007	Ground-Up	ISS	U.S. Node 2, first flight of Lambert guidance upgrade.
	122 (1E)	2008	Ground-Up	ISS	Columbus Laboratory
	123 (1J/A)	2008	Ground-Up	ISS	Kibo Logistics Module, Dextre Robotics System
	124 (1J)	2008	Ground-Up	ISS	Kibo Pressurized Module, Japanese Remote Manipulator System
	126 (ULF2)	2008	Ground-Up	ISS	MPLM
	119 (15A)	2009	Ground-Up	ISS	S6 truss segment
	127 92J/A)	2009	Ground-Up	ISS	Kibo JEM EF, Kibo Japanese ELM-ES
	128 (17A)	2009	Ground-Up	ISS	Leonardo MPLM, LMPESSC, Vernier RCS failure.
	129 (ULF3)	2009	Ground-Up	ISS	ELC1, ELC2
	STS-130 (20A)	2010	Ground-Up	ISS	Tranquility Node 3, Cupola. TCS failure during approach.
	STS-131 (19A)	2010	Ground-Up	ISS	Leonardo MPLM, radar fail.
	STS-132 (ULF4)	2010	Ground-Up	ISS	ICC, MRM1, COAS bulb replacement.
	STS-133 (ULF5)	2011	Ground-Up	ISS	ELC4, Leonardo PMM
	STS-134 (ULF6)	2011	Ground-Up	ISS	ELC3, AMS-2, STORRM DTO during rndz & docking.
	STS-135 (ULF7)	2011	Ground-Up	ISS	Raffaello MPLM, LMC, return to Earth of failed ammonia pump. ISS yaw maneuver after orbiter undocking to facilitate engineering photos during orbiter half-lap fly-around.

A = Assembly, **AMS** = Alpha Magnetic Spectrometer, **ATV** = Automated Transfer Vehicle, **CETA** = Crew and Equipment Translation Aid, **CMG** = Control Moment Gyro, **COAS** = Crew Optical Alignment Sight, **ELC** = EXPRESS Logistics Carrier, **ELM-ES** = Experiment Logistics Module - Exposed Section, **EVA** = Extra Vehicular Activity, **ICC** = Integrated Cargo Carrier, **ISS** = International Space Station, **JEM EF** = Japanese Experiment Module Exposed Facility, **LF** = Logistics Flight, **LMC** = Lightweight Multi-purpose Carrier, **LMPESSC** = Lightweight Multi-Purpose Experiment Support Structure Carrier, **MPLM** = Multi-Purpose Logistics Module, **MRM** = Mini Research Module, **ORBT** = Optimized R-Bar Targeted Rendezvous, **PMA** = Pressurized Mating Adapter, **PMM** = Permanent Multi-Purpose Module, **RCS** = Reaction Control System, **RMS** = Remote Manipulator System, **Rndz** = Rendezvous, **RPM** = R Bar Pitch Maneuver, **STORRM DTO** = Sensor Test for Orion Relnav Risk Mitigation Detailed Test Objective, **TCS** = Trajectory Control Sensor, **TORRA** = Twice Orbital Rate R Bar Approach, **TORVA** = Twice Orbital Rate V Bar Approach, **UF** = Utilization Flight, **ULF** = Utilization & Logistics Flight

PART I – BEFORE THE SPACE SHUTTLE

This page intentionally left blank.

CHAPTER 1 - EARLY STUDIES

Early Studies

The pace of theoretical work on rendezvous in the United States picked up in the late 1950s.[1-4] Of particular importance were studies conducted by NASA Langley Research Center into manual and automatic rendezvous techniques. Early Langley research was a key factor behind development, advocacy of (Figure 1.1), and acceptance of the Lunar Orbit Rendezvous mission profile for Apollo.[5-6]

Figure 1.1 John C. Houbolt discussing Lunar Orbital Rendezvous (LOR).

Aldrin Dissertation

In 1959, Air Force Major Edwin E. Aldrin arrived at MIT to pursue a Masters, and eventually a PhD degree in astronautical engineering. Aldrin chose rendezvous as his dissertation topic, hoping to work for the Air Force or NASA. Many early rendezvous studies were written from a theoretical perspective. Aldrin brought an aviator's perspective to rendezvous techniques, having flown the F-86 in Korea (two MiG-15 kills) and later flew the F-100. His dissertation and later work as an astronaut was to influence development of piloting techniques for Gemini and Apollo, and in particular the development of back-up techniques in the event of computer or other system failures.[7-9] Of particular importance was Aldrin's understanding of how relative motion of spacecraft differed from flying airplanes.

Satellite Inspector (SAINT)

SAINT (later known as Program 621A or 720) was a United States Air Force sponsored satellite inspection program that grew out of studies conducted in 1959.[10] The SAINT profile was similar to the "first apogee rendezvous" technique frequently depicted in early 1960s era papers on rendezvous. It has long been suspected that the support given by NASA Associate Administrator Robert C. Seamans for the Lunar Orbit Rendezvous profile was based on his experience at RCA with SAINT.[11,12] In early 1962, NASA was interested in applying SAINT experience to the Gemini and Apollo programs. A briefing on SAINT was given to Manned Spacecraft Center (MSC) personnel in Houston in April of 1962. However, as details of Gemini rendezvous were the subject of much debate and not finalized until late 1964, it is not clear how much influence SAINT had on Gemini rendezvous techniques at the time of the briefing. It is unlikely that SAINT, which involved rendezvous of an unmanned inspector spacecraft with an "uncooperative" target, took into account manual piloting concerns and safety of flight issues that needed to be addressed for Gemini and Apollo. SAINT was canceled in December 1962.

MORAD

By early 1961 studies had indicated that rendezvous was technically feasible and would be useful for space missions. Some proposed missions could be flown using existing boosters to rendezvous and assemble spacecraft in orbit, rather than having to develop larger boosters. The Manned Orbital Rendezvous and Docking (MORAD) project was proposed by NASA Langley rendezvous investigators as an extension to Mercury in early 1961. MORAD was intended to provide early proof of the feasibility of manned rendezvous and establish confidence in rendezvous techniques in the areas of control of the closure maneuver, handling of the docking phase, and manned operation in orbit. It would have demonstrated rendezvous and docking using a modified single-seat Mercury spacecraft (passive) and an active target vehicle. Two targets were discussed, one that needed to be developed and would be launched by a Scout booster. The other was the SAINT vehicle.

A Scout would place the target on an intercept course with the Mercury spacecraft. The astronaut was expected to acquire the target vehicle flashing light at a range of from 50 to 100 miles. The target would maneuver during the terminal phase based on control inputs from the Mercury astronaut (Figure 1.2). The Mercury periscope was to be replaced by equipment that enabled the astronaut to control the braking and docking of the target spacecraft via a radio command link (Figure 1.3). Target spacecraft telemetry (such as attitude and range) would be displayed to the astronaut.

The piloting task involved yawing and pitching the Mercury so that the astronaut field of view was along the line-of-sight to the target, after which the Mercury automatic stabilization mode would then be engaged. The target would be maneuvered so that it achieved and maintained an appropriate attitude with respect to the line-of-sight to the Mercury. The astronaut would then detect and null line-of-sight motion by watching the

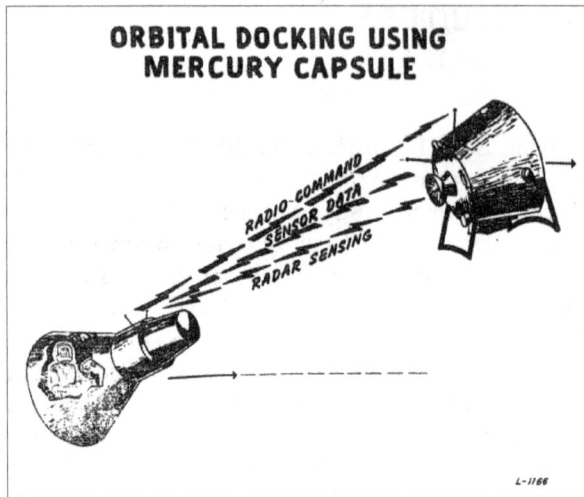

Figure 1.2 Sketch of a Mercury spacecraft and a notional target vehicle (1961).

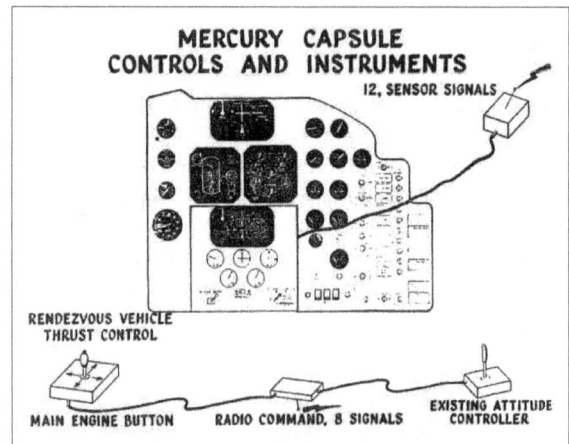

Figure 1.3 Controls for remote piloting of the target vehicle from the Mercury spacecraft (1961).

target against a star field background. A braking schedule would be executed to attain the appropriate relative position and velocity for docking.

The MORAD rendezvous demonstration was later deemed too dangerous for a one-man spacecraft. MORAD was dropped in favor of rendezvous demonstrations in the Mercury Mark II program, later renamed Project Gemini.[12]

References

1. Clohessy, W. H., and R. S. Wiltshire, "Terminal Guidance System For Satellite Rendezvous," *Journal of the Aerospace Sciences,* Volume 27, Number 9, September 1960, pp. 653-659.

2. Houbolt, J. C., "Problems And Potentialities Of Space Rendezvous," *Astronautica ACTA VII,* 1961, pp. 406-429.

3. Pennington, J. E., and R. F. Brissenden, "Visual Capability in Rendezvous," *Astronautics and Aerospace Engineering,* AIAA, Volume 1, Number 1, February 1963, pp. 96-99.

4. Vogeley, A. W., and R. F. Brissenden, "Survey of Rendezvous Progress," *Proceedings of the AIAA Guidance and Control Conference,* Cambridge, MA, August 12-14, 1963, in Guidance and Control II, Academic Press, New York, NY, 1964.

5. Hansen, J. R., *Enchanted Rendezvous, NASA Monographs in Aerospace History Series #4,* NASA, Washington, D. C., 1995.

6. Houbolt, J. C., "Lunar Rendezvous," *International Science and Technology,* February 1963, pp. 62-70.

7. Aldrin, E. E., *Line-Of-Sight Guidance Techniques For Manned Orbital Rendezvous,* PhD Dissertation, Massachusetts Institute of Technology, Cambridge, MA, January 1963.

8. Kramer, P. C., E. E. Aldrin and W. E. Hayes, "Onboard Operations For Rendezvous," *Gemini Summary Conference,* NASA SP-138, Manned Spacecraft Center, Houston, TX, February 1-2, 1967, pp. 27-40.

9. Aldrin, E. E., *Men From Earth,* Bantam Books, New York, NY, 1989.

10. Chun, Clayton K. S., "A Falling Star: SAINT, America's First Antisatellite System," *Quest,* Volume 6, Number 2, Summer 1998, pp. 44-48.

11. Seamans, Robert C., *Aiming At Targets,* NASA SP-4106, NASA, Washington, D.C., 1996.

12. Grimwood, James M., Barton C. Hacker and Peter J. Vorzimmer, *Project Gemini Technology and Operations – A Chronology,* NASA SP-4002, National Aeronautics and Space Administration, Washington, D.C., 1969.

CHAPTER 2 - MERCURY

The ability of an astronaut to spot a target against a star or Earth background, judge distances, or estimate target spacecraft attitude with the naked eye had to be determined. Several experiments were conducted during Project Mercury to gauge the ability of an astronaut to perform these tasks.[1,2] Test results were used in planning for Gemini.

After separation of the Mercury spacecraft from the Atlas booster, John Glenn, Scott Carpenter, Wally Schirra and Gordon Cooper all maneuvered their spacecraft to visually acquire the booster through the large window (Figure 2.1).[1-5] Glenn estimated the initial range as 200 feet, which was later confirmed with ground analysis. He tracked the booster for six or seven minutes, and estimated the range at the end of the tracking to be two miles behind and one mile below. Carpenter observed and photographed the booster for about 8.5 minutes and was able to discern a tumbling motion. Schirra maintained a track attitude using the Mercury fly-by-wire control system on the sunlit side of the Earth. The booster appeared to him to be black, rather than the silver color observed by Glenn and Carpenter. Booster relative motion appeared to be as predicted preflight, which Schirra took as verification that the Mercury gyros and horizon scanners were operating properly.[4] Cooper observed the Atlas for about 8 minutes, and was able to discern hardware details and propellant vapor emissions from the sustainer engine.[1-5]

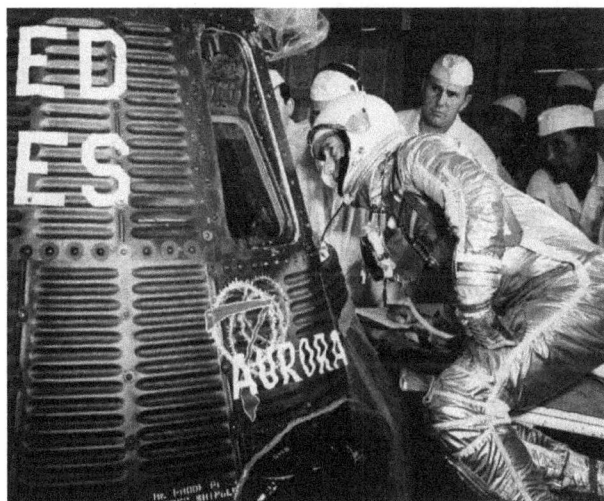

Figure 2.1 Scott Carpenter looks into his Aurora 7 spacecraft before entering the spacecraft for the launch (May 24, 1962).

During the flights of Aurora 7 (MA-7, Carpenter) and Faith 7 (MA-9, Cooper), a 20-inch, multi colored balloon was to be ejected from the antenna canister and inflated at the end of a 100-foot tether from the antenna canister. Atmospheric drag data was to be measured over one

orbit, and the ability of the astronaut to observe the balloon would be determined. On Aurora 7 the balloon was deployed but failed to fully inflate due to a seam failure. Although the balloon failed to inflate properly, Carpenter was able to discern the various colors (Figure 2.1).[2,3,5] The balloon experiment underwent a thorough testing process before the flight of Faith 7, but failed to deploy from the antenna canister.[5]

A flashing light experiment was conducted on Faith 7. A 5.75-inch diameter sphere equipped with two xenon gas discharge lamps that flashed at a rate of about once per second was ejected towards the Earth from the Faith 7 spacecraft. This set up relative motion that would place the sphere against both an Earth and space background with increasing range from the spacecraft. Cooper was unable to spot the light during the first daylight and night periods after deployment. At the end of the second daylight period just before sunset, he spotted it as a continuous (non-flashing) light just below the horizon, apparently due to reflected sunlight (it was against a black background). Flashing was observed at sunset, and the strobes remained visible through the night period. A flashing light was easier to spot than a non-flashing light. It was estimated that the brightness at a range of about 15 miles was about +2 magnitude. He was not able to acquire the light during the next daylight period, but did spot it again halfway through the following night period, although it was dim.[2,5]

References

1. Swenson, L. S., Grimwood, J. M., and Alexander, C. C., *This New Ocean – A History of Project Mercury*, NASA SP-4201, National Aeronautics and Space Administration, Washington, D.C., 1966.

2. McDivitt, J. A., "Rendezvous, Rescue And Recovery Aspects Of Space Operations," *Proceedings of the Symposium on Space Rendezvous, Rescue and Recovery*, American Astronautical Society, Edwards Air Force Base, CA, September 10-12, 1963, in Volume 16, Part 2, *Advances In The Astronautical Sciences*, pp. 12-19.

3. *Results Of The Second U.S. Manned Orbital Space Flight*, May 24, 1962, NASA SP-6, NASA, Washington, D.C., 1962.

4. *Results Of The Third U.S. Manned Orbital Space Flight*, October 3, 1962, NASA SP-12, NASA, Washington, D.C., December 1962.

5. *Mercury Project Summary - Including Results Of The Fourth Manned Orbital Flight*, May 15-16, 1963, NASA SP-45, Washington, D.C., Oct, 1963.

Approved for public release via STI DAA
24483. See statement on title page.

17

This page intentionally left blank.

CHAPTER 3 - GEMINI

Introduction

President Kennedy's goal of sending Americans to and from the moon by the end of the sixties necessitated the development of rendezvous. Gemini served as a technology demonstration and risk reduction program in preparation for the lunar landings, and established an experience base of operational techniques and mission planning for Apollo. An overall goal of Gemini was to prove that a human crew could manually accomplish activities in zero-gravity that were similar to those required for the Apollo Program.

Objectives included successful demonstration of rendezvous and docking, long-duration flight, guided (piloted or automated) lifting entry to a targeted splashdown point, and Extra Vehicular Activity (EVA). Technical challenges associated with rendezvous and docking included closed loop execution of using manual piloting techniques, crew training and simulation, mission planning for nominal and off-nominal conditions, launch windows that minimized out-of-plane insertion error, coordinated launch of target and chaser

spacecraft on the same day, and the development of contingency procedures (both pre-mission and in near real time) in response to vehicle performance problems. Ten rendezvous profiles and nine different dockings were successfully completed. Gemini successfully demonstrated a number of rendezvous techniques that are listed in (Tables 3.1 and 3.2). Many of these techniques were later used in the Apollo and Space Shuttle Programs.[1-10]

Table 3.1 Gemini Rendezvous Accomplishments

- Coelliptic rendezvous from above and below
- Stable orbit, direct ascent and equal period (football) rendezvous
- Rendezvous during both orbital night and day
- Use of only optical measurements (no radar)
- Station-keeping and docking
- Simultaneous countdown of chaser and target launch vehicles
- Launch during a narrow launch window
- Real time maneuver targeting using data from ground based or onboard navigation sensors
- Conducting multiple rendezvous operations in a single mission within a propellant budget
- Extra-Vehicular Activity while docked

Table 3.2 Gemini Rendezvous and Station-Keeping Missions

Flight	Year	Chaser	Target	Profile	Comments
Gemini IV	1965	Gemini IV	2nd Stage	Station-keeping, separate & rendezvous	Station-Keeping attempt resulted in separation. Rendezvous canceled.
Gemini V	1965	Gemini V	REP		REP deployed but rendezvous canceled due to fuel cell power down. Successful radar test with other REP hardware at Kennedy Space Center.
Gemini VII	1965	Gemini VII	2nd Stage	Station-Keeping	Approached to within 15 meters.
Gemini VIA	1965	Gemini VIA	Gemini VII	Ground-Up (M=4)[1]	GATV-5002 destroyed during launch, Gemini VII used instead. First successful space rendezvous.
Gemini VIII	1966	Gemini VIII	GATV-5003	Ground-Up (M=4)[1]	First successful space docking. Failed on Gemini thruster forced undocking and activation of re-entry RCS system and contingency splash-down in the Pacific.
Gemini IX	1966	Gemini IX	1. ATDA 2. ATDA 3. ATDA	1. Ground-Up (M=3)[1] 2. Football re-rendezvous 3. Rendezvous from above	GATV-5004 destroyed during launch, ATDA launched instead. No docking as ATDA shroud failed to separate. Three rendezvous profiles flown.
Gemini X	1966	Gemini X	1. GATV-5005 2. GATV-5003	1. Ground-Up (M=4)[1] 2. Phase from in front, above & behind to set up coelliptic	Two rendezvous profiles flown. Second rendezvous without radar and strobes due to Agena power loss.
Gemini XI	1966	Gemini XI	1. GATV-5006 2. GATV-5006	1. Ground-Up (M=1) 2. Stable Orbit	Two rendezvous profiles flown. First stable orbit rendezvous.
Gemini XII	1966	Gemini XII	GATV-5001	Ground Up (M=3)[1]	Radar failure, angles only rendezvous.

ATDA = Augmented Target Docking Adapter, **GATV** = Gemini Agena Target Vehicle, **M** = revolution that rendezvous was completed on, **REP** = Radar Evaluation Pod, **2nd Stage** = Titan II Booster Second Stage
[1] Ground-Up rendezvous profiles other than Gemini XI had a coelliptic phase before the TPI maneuver.

Gemini VI Profile Selection

In 1962, some Langley rendezvous specialists moved with the Space Task Group to the newly formed Manned Spacecraft Center (MSC) in Houston. NASA and contractor personnel from various disciplines at MSC, and the MSC Mission Planning and Analysis Division (MPAD) in particular, turned rendezvous theory into reality during Gemini.[1] By the spring of 1964, three candidate rendezvous profiles (Figures 3.1, 3.2, and 3.3) had been identified and were under consideration by the Trajectories and Orbits Panel (chaired by Bill Tindall, Table 3.3) for Gemini VI, the first planned Gemini rendezvous and docking with an Agena. Discussions were conducted in May and June of 1964.[11-16]

Tangential Orbit (Mission Plan 1) - This profile involved launching the Gemini spacecraft into an elliptical orbit tangential to the Agena Target Vehicle (ATV) orbit (Figure 3.1). Rendezvous would occur near the apogee of the fourth Gemini orbit. However, this technique did not guarantee proper lighting conditions or consistent relative dynamics in the terminal phase under dispersed conditions.

Concentric Flight Profile (Mission Plan 2) - This used the same maneuver plan as the ground targeted phase in the tangential orbit profile, but had a different terminal phase. Rather than ground targeting placing the spacecraft on an intercept trajectory, it placed the Gemini in a co-elliptic orbit with respect to the target spacecraft (Figure 3.2). The intercept maneuver, Terminal Phase Initiation (TPI), was executed while the chaser vehicle was on an orbit coelliptic with the Agena. The length of the co-elliptic phase could be controlled to ensure appropriate lighting during the terminal phase and adequate coverage by ground tracking. The terminal phase would begin once a trajectory criterion was met.

First Apogee or Direct Rendezvous (Mission Plan 3) - The Titan II booster would place the Gemini spacecraft on an intercept trajectory with the Agena (Figure 3.3). Gemini would achieve radar lock on the target soon after orbit insertion. However, the short amount of time for the crew to conduct on-orbit checkout of Gemini systems and rendezvous procedures made the timeline impractical. Furthermore, the trajectory was highly sensitive to ascent dispersions and liftoff delays. Trajectory dispersions would have to be corrected by the on-board system, without help from ground tracking. In case of a dispersed trajectory that made rendezvous impossible, a backup rendezvous profile was needed.

After an extensive trade study, a coelliptic rendezvous profile was chosen for execution on Gemini VI at a meeting on June 15, 1964 (Figure 3.2).[13] The length of the coelliptic phase permitted control over terminal phase lighting, and provided a terminal phase that was less sensitive to trajectory dispersions than the direct rendezvous and tangential orbit profiles. Of the

Table 3.3 First Gemini/Agena Mission Rendezvous Profile Selection Panel

Name	Organization
M. Czarnik	McDonnell
W. B. Evans	Gemini Program Office
R. R. Carley	Gemini Program Office
E. M. See, Jr.	Astronaut Office
E. E. Aldrin, Jr.	Astronaut Office
J. B. Jones	Flight Crew Support Division
G. S. Lunney	Flight Control Division
A. D. Aldrich	Flight Control Division
L. C. Dunseith	Mission Planning and Analysis Division
E. C. Lineberry	Mission Planning and Analysis Division
R. P. Parten	Mission Planning and Analysis Division
H. W. Tindall, Jr.	Mission Planning and Analysis Division

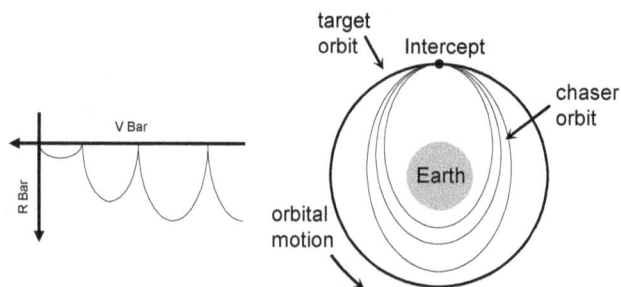

Figure 3.1 Tangential Orbit

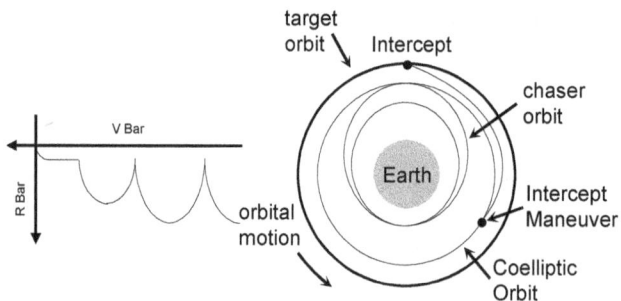

Figure 3.2 Coelliptic

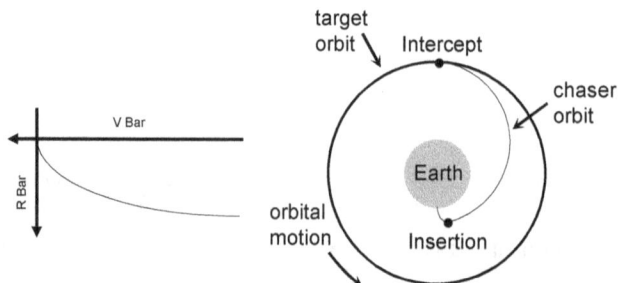

Figure 3.3 Direct Intercept (First Apogee)

three concentric provided the most flexibility, had a terminal phase that was the least sensitive to dispersions and facilitated easier definition of backup procedures. This helped ensure standardized crew procedures and training, even with mission-to-mission variations in the pre-terminal phase rendezvous profile. Furthermore, compared to the direct rendezvous profile, the crew did not have to conduct rendezvous activities during the first orbit, as it was preferred to spend the first orbit conducting spacecraft systems checks. The coelliptic approach also facilitated the use of manual backup guidance techniques in the event of system failures (sensor failure, computer failure, or loss of communications with Mission Control).[3] Two mid-course correction maneuvers followed TPI, allowing correction of dispersions before final braking. Since most of the mission planning and Mission Control software development had concerned the tangential orbit profile, significant changes were required to support the concentric profile.

The profile from orbital insertion to radar acquisition was designed to use ground targeted burns to control phasing and out-of-plane dispersions. Gemini terminal phase rendezvous profiles (Figure 3.4) were designed to maximize the success of manually piloted rendezvous in the presence of trajectory dispersions and systems problems. A terminal phase initiated from a coelliptic orbit was selected that allowed control over terminal phase lighting, provided less sensitivity to trajectory dispersions, ensured standardized crew procedures and training, provided sufficient time for crew execution of nominal and backup procedures, and facilitated the use of manual backup guidance techniques in the event of system failures. The crew flew an inertial final approach by controlling closing rate and the rotation rate of the inertial line-of-sight to the target.

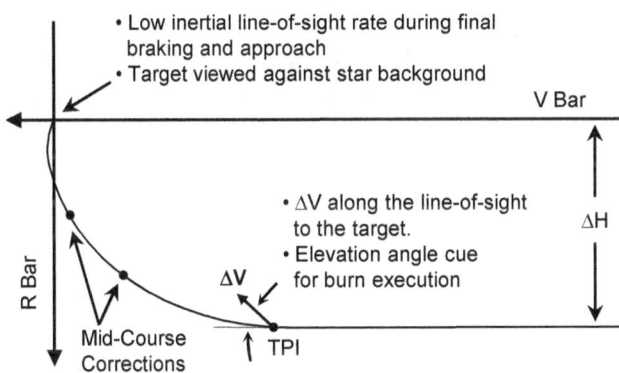

Figure 3.4 Terminal phase for coelliptic rendezvous.

Three different parameters had to be chosen for the terminal phase (Figure 3.4): 1) the ΔH for the co-elliptic orbit, 2) the angular transfer from Terminal Phase Initiation to intercept, and 3) a thrusting methodology that permitted burn execution in the event of spacecraft system failures. For the relative geometry cue, elevation angle

of the target with respect to the Gemini local horizontal was chosen over range, relative radial velocity, time and minimum ΔV. In the presence of dispersions, range, relative radial velocity and elevation angle were equally insensitive. Elevation angle determination was the least vulnerable to equipment failures and easy to incorporate into backup procedures. Elevation angle was also a convenient attitude reference for the crew.

A transfer angle of between ~130 to ~140 degrees also ensured an approach from below, which enabled the crew to use the star field as a reference during the approach along a constant inertial line-of-sight vector. Transfer angles in this range were the least sensitive to TPI Time of Ignition (TIG) and ΔV dispersions. ΔH was chosen as a trade-off between visual sighting requirements, minimizing the impact of dispersions in the terminal phase, minimizing closure rate when approaching target, and minimizing ΔV. Lower values of ΔH were favored. It was decided that the TPI burn should have a ΔV that was along the line-of-sight to the target. The advantage of this burn over a horizontal ΔV was that it was easier to execute with backup procedures.

Rendezvous Target Vehicles

The docking target for Gemini was a modification of the Agena upper stage and was launched by an Atlas. Agena was equipped with an L-Band radar transponder and strobes to facilitate relative navigation. L-Band antennas were arranged to support radar tracking from any direction. Orbital adjustments could be performed using the Agena propulsion system based on either ground command before the docking phase or crew command after docking. The crew could also command Agena systems before docking through the radar. Docking was performed with the Agena powered up and under stable attitude control using its own attitude control system. The attitude control system of either the Gemini or the Agena could be used to control the attitude of the docked vehicles.

After the launch failure of the Agena for Gemini VI, the Gemini VII vehicle was modified with a radar transponder to serve as a target and was launched before Gemini VI (only station-keeping was performed, there was no capability for Gemini to Gemini docking). Later in the program a back-up target, the Augmented Target Docking Adapter (ATDA) was built and launched by an Atlas to serve as the Gemini VIII rendezvous target. The ATDA consisted of the cone hardware attached to a Gemini spacecraft entry RCS system module.

Docking Hardware

The Agena used cone and latch hardware to capture three fittings on the nose of the Gemini. Docking hardware design was driven by a need for high reliability and short development schedule. The crew could view

Approved for public release via STI DAA 24483. See statement on title page.

21

the docking hardware and an Agena mounted status display panel during docking. The cone hardware was carried by the Agena and ATDA vehicles, and was built by the Gemini manufacturer. Docking occurred at a relative approach velocity of approximately 0.75 feet/second.

Crew Interface, Procedures, and On-Board Computers

Gemini was the first human spacecraft to carry a digital computer. For rendezvous the computer was used to compute maneuver solutions once radar data was available. Since Gemini had an on-board computer and IMU, both were used to provide a back-up ascent guidance capability in the event of a Titan II radio guidance failure. The crew could switch to the back-up guidance mode based on pre-defined criteria. For entry, the computer could fly the vehicle automatically or provide the crew with cues for manual piloting. The crew interface is shown in Figure 3.5.

Spacecraft and avionics design was simplified through the use of manual sequencing and systems management, taking advantage of the crew's ability to diagnose failures and take the appropriate corrective action. Rendezvous crew procedures were developed to cover nominal systems performance, IMU failure, radar failure, and computer failure.[18]

On-Board Navigation and Ground Navigation

The Mercury ground tracking network was upgraded to support Gemini. Gemini required ground based orbit determination of both the chaser and target vehicles for computation of chaser orbital adjustments outside of radar range.[16] Limitations in ground tracking accuracy drove the development of closed loop rendezvous techniques for the terminal phase. A rendezvous radar and an associated L-Band transponder on the target spacecraft (another Gemini, Agena, or the Augmented

Target Docking Adapter) provided measurements of range, range rate, and line-of-sight angles.[17] Line-of-sight angles could also be obtained from the IMU by bore sighting the Crew Optical Alignment Sight (COAS) on the target spacecraft. While on-orbit the IMU was aligned using horizon sensors with respect to a local level reference frame. Rendezvous procedures enabled the crew to successfully rendezvous and dock in the presence of an IMU, computer, or radar failure.

On-Board Maneuver Targeting

Targeting for the Terminal Phase Initiation and Mid-Course Correction maneuvers was performed using the Clohessy-Wiltshire equations in the digital computer, based on radar measurements. Maneuver charts served as a back-up to the computer for terminal phase burn computations based on range and elevation to the target at specific times before the maneuvers. Chart solutions were compared with on-board and Mission Control burn solutions.[18]

Gemini IV

Gemini IV (June 1965) had an ambitious mission plan of station-keeping with the Titan II second stage, an Extravehicular Activity (EVA, or spacewalk) to the second stage, followed by a separation and rendezvous with the second stage. Gemini IV was not equipped with radar, nor were the maneuvers to be targeted on-board. Two flashing lights were placed midway up the Titan second stage, 180 degrees apart, to assist the crew in tracking it. If successful, these would have been the first station-keeping, EVA, and rendezvous activities conducted by the U.S. space program.

After completion of the spacewalk and station-keeping activities, a separation and rendezvous sequence was planned (Figure 3.6). After nulling the relative velocity on the +V Bar, a posigrade maneuver would be performed to initiate the separation. 90 minutes later, at a

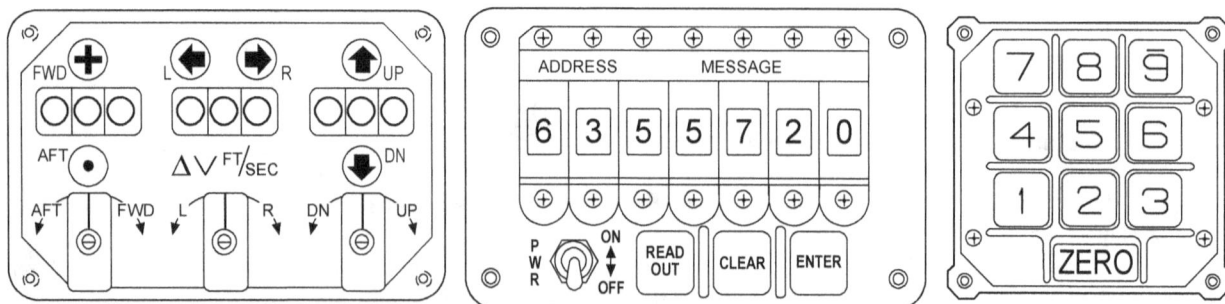

Figure 3.5 Gemini Computer Keyboard and Displays

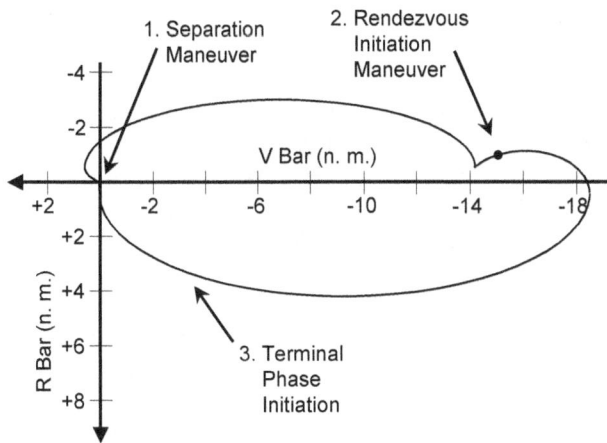

Figure 3.6 Proposed Gemini IV rendezvous with Titan II second stage (June 1965).

range of about 15 nautical miles behind the second stage, a maneuver would be executed placing the Gemini on an intercept trajectory over an orbital transfer of 309 degrees. At a range of about 4.6 nautical miles (Terminal Phase Initiation in Figure 3.6) the crew would begin trajectory control to null the rotation rate of the inertial-line-of-sight to the second stage. This was to be accomplished by observing the two flashing lights against the star background. If the star background were not available data from the Gemini inertial platform would be used. The final phase of the rendezvous was to be conducted in daylight. After another period of station-keeping, the Gemini was to execute another posigrade separation maneuver and the crew would move on to other mission activities.

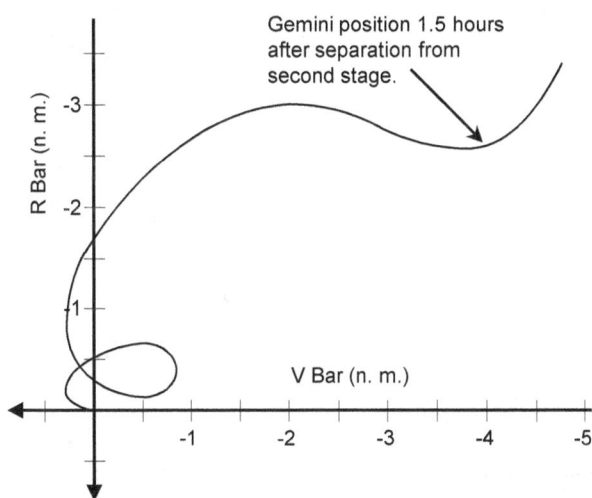

Figure 3.7 Gemini IV relative motion during station keeping attempt with Titan II second stage (June 1965).

The crew performed thruster firings to separate from the second stage and set up the station-keeping phase. However, more ΔV was obtained than planned, possibly due to plume impingent effects on the second stage and Gemini. Plume impingement from OAMS and/or force from separation pyrotechnics imparted a 3 foot/second ΔV on the second stage. This complicated the station-keeping phase (Figure 3.7), along with difficult lighting conditions, unexpected second stage tumbling, loss of an aft firing RCS thruster, difficulty judging range and range rate with eye observations, second stage propellant venting (particularly when going from orbital night to daylight), insufficient pre-flight training and simulations, and lack of crew understanding of the difference between flying an aircraft in the atmosphere and flying a spacecraft in orbit to achieve desired relative motion with another vehicle. Propellant consumption and timeline concerns led to cancellation of the EVA with the second stage and the subsequent rendezvous with it.[6, 8] The EVA was successfully conducted later in the mission. After the mission changes were made to the crew training and mission planning processes to ensure that future station-keeping and rendezvous activities would be successful. Gemini IV was an important learning experience.[19-20]

Gemini V

Gemini V was the first Gemini mission to carry rendezvous radar and fuel cells. The target was the Radar Evaluation Pod (REP) that was deployed from the back of the Gemini V adapter module (Figure 3.8). It was equipped with a radar transponder and flashing lights. After deployment, about 23 minutes of radar data was obtained. Cooper commanded RCS thruster firings to attempt to null the rotation rate of the inertial line-of-sight by observing REP motion against the stars. However, the rendezvous activity was canceled after a cryogenic oxygen heater failure in a fuel cell forced the crew to shut down the radar and other systems to conserve power. Later in the mission Gemini V successfully tracked an identical Radar Evaluation Pod that was located at the Kennedy Space Center. The crew also successfully performed a rendezvous with an imaginary target after performing four maneuvers based on data from Mission Control.[6, 8]

Figure 3.8 Gemini V REP at rear of Gemini Equipment Module (Aug. 1965).

Gemini VIA

The first rendezvous and docking was scheduled for the Gemini VI mission. On October 25, 1965, the Agena GATV-5002 was launched on an Atlas. However, the Agena stage exploded at the start of a maneuver soon after separation from the Atlas. The launch of Gemini VI, which was scheduled to occur about 100 minutes after the Agena launch, was cancelled. NASA developed a revised mission plan so that Gemini VII would be launched next, and then Gemini VIA (renamed due to the new mission plan) would be launched and use Gemini VII as the rendezvous target. This would permit a rendezvous demonstration without a delay caused by the Agena failure investigation. No docking would be performed, but Gemini VII would be equipped with the same radar transponder and lights carried by the Agena.

Gemini VIA finally launched on December 15, after a delay caused by a failure to launch on December 12. The first spacecraft rendezvous was successfully flown as planned (Figure 3.9). Three revolutions of station-keeping were performed (Figure 3.10). An in-plane and out-of-plane fly-around of Gemini VII was performed, and the Gemini VIA docking light was used during night station-keeping. Gemini VIA approached no closer than one foot to Gemini VII. Gemini VIA separated and returned to Earth the next crew day.[21-24]

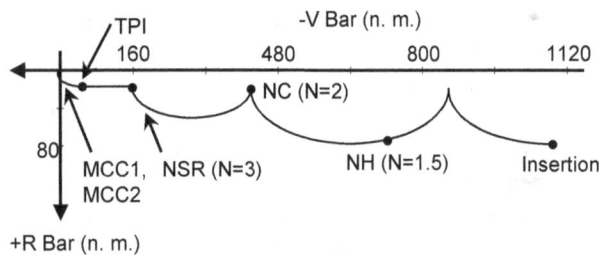

Figure 3.9 Gemini VIA M=4 ground-up rendezvous profile.

Figure 3.10 Gemini VII photographed by the Gemini VIA crew after the first space rendezvous (Dec. 1965).

Gemini VII

Gemini VII was launched on December 4, 1966. Before the launch of Gemini VIA radar at the Kennedy Space Center was used to test the Gemini VII transponder as it flew near the Cape. Gemini VII served as the target vehicle for Gemini VIA (Figure 3.10).

One objective of the Gemini VII mission was to conduct the station-keeping exercise with the Titan II second stage that was attempted, but unsuccessful, on Gemini IV. The Gemini VII station-keeping plan took advantage of the Gemini IV experience by applying lessons learned to mission planning, crew procedures, and crew training. The station-keeping and relative motion activities were conducted immediately after orbit insertion and 11 days before the launch of Gemini VIA.

After separation from the Titan II second stage the crew was to move ahead of the stage by about 100 feet, then turn around and approach the stage and perform station-keeping. Establishment of station-keeping was to be performed more rapidly than on Gemini IV to avoid undesirable relative motion. After station-keeping the crew was to perform a radial down burn to establish a relative motion football about 5 miles long (Figure 3.11).[25, 26] This football would bring Gemini VII back to the vicinity of the Titan II second stage after one revolution of the Earth. The Titan II second stage was equipped with strobes to assist the crew in locating it visually. However, no radar measurements were to be taken since Gemini VII was not equipped with the rendezvous radar (it did have the radar transponder to support the Gemini VIA rendezvous later in the flight), nor were there any on-board targeted burns to ensure good relative trajectory performance. After 2.5 revolutions of Earth the crew was to perform a posigrade perigee raise burn. This burn would raise the Gemini VII orbital altitude and end the football relative motion.

However, the actual Gemini VII trajectory performance did not go as planned.* In 1995 James Oberg published an analysis of the station-keeping and relative motion activity that was verified by the Gemini VII crew.† [25, 26] After orbit insertion the crew performed the standard manual 2 foot/second posigrade burn to

---

* More detail is provided on Gemini VII and the Apollo 16 brute force rendezvous (see the next chapter) than on other missions since both are examples of trajectory dispersions and undesirable relative motion that could have presented a crew safety hazard or placed mission success at risk.

† James Oberg documented his Gemini VII analysis in part to transmit lessons about trajectory control and mission planning to Shuttle Program rendezvous personnel. Another purpose of his analysis was to debunk claims that the Gemini VII crew had observed an Unidentified Flying Object (UFO) during this phase of the mission. These claims were made based on misinterpretation of crew comments and a misunderstanding of Gemini VII relative motion with respect to the Titan II second stage.

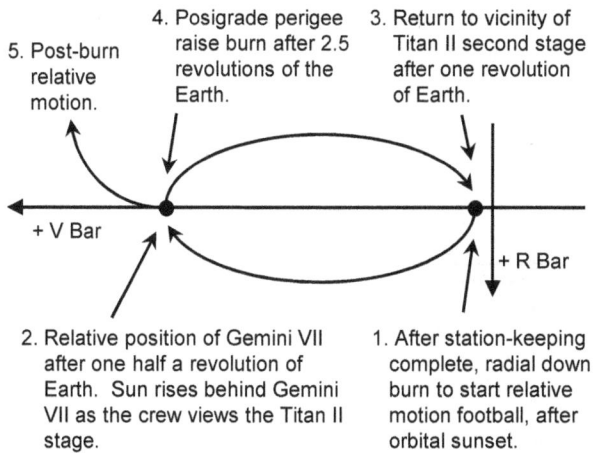

Figure 3.11 Planned relative motion of Gemini VII with respect to the Titan II second stage. Figure based on a 1995 analysis by James Oberg. [25]

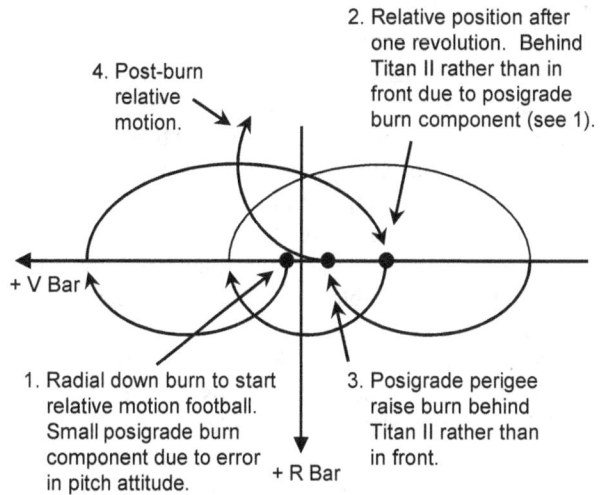

Figure 3.12 Likely relative motion of Gemini VII with respect to the Titan II second stage. Figure based on a 1995 analysis by James Oberg. [25]

separate from the Titan II second stage. The crew then yawed Gemini VII 180 degrees so that they could observe the stage and film it with a 16 mm motion picture camera. The stage was venting propellant and was surrounded by small debris. Additional RCS jet firings were performed to close in on the upper stage and establish station-keeping at a range of approximately 50 to 60 feet. [25, 26]

At this point the upper stage was backlit by the sun making crew observation difficult. To remove the Sun from the line-of-sight the crew fired RCS jets to move Gemini VII out-of-plane in the orbital north direction. This set up cyclic out-of-plane motion that caused Gemini VII to periodically cross the orbital plane of the upper stage and re-encounter the associated debris cloud. Propellant venting caused the stage to tumble and induced translation. The crew performed additional RCS jet firings to maintain station-keeping in response to the translation caused by the venting. The crew later stated that the venting made station-keeping with the upper stage more difficult than the later station-keeping with the controlled Gemini VIA spacecraft. [25, 26]

Approximately 15 minutes after separation from the upper stage the crew performed the radial down burn of approximately 9 feet/second to establish the football relative motion (Figure 3.11). The desired spacecraft attitude for the burn was -25 degrees in pitch and zero degrees yaw. However, the actual attitude was -41.5 degrees in pitch and 49 degrees in yaw. The large value for yaw was due to the earlier out-of-plane maneuver to remove the Sun from the crew line-of-sight. The pitch attitude error added a small posigrade component to the burn. This posigrade component resulted in a dispersed relative motion trajectory (Figure 3.12). [25, 26]

Gemini VII returned to the general vicinity of the Titan II second stage after one revolution (Figure 3.12). However, according to Oberg's analysis the dispersion

introduced by the posigrade burn component (in turn due to the burn attitude error) resulted in Gemini VII being behind the second stage (on the minus V Bar, Figure 3.12), rather than ahead of it (on the plus V Bar, Figure 3.11) as was planned. The crew reported that the stage was tumbling and surrounded by a cloud of particles. In addition, particles with out-of-plane motion were crossing the path of Gemini VII. These were likely ice crystals from vented Titan II propellant or from Gemini VII RCS jet firings conducted to move the spacecraft out-of-plane to resolve the orbital lighting problem. One revolution later the crew reported that the stage was ahead of them (Figure 3.12). [25, 26]

About half a revolution later the crew performed the posigrade perigee raise burn of 59 feet/second. This burn was supposed to have been performed ahead (on the plus V Bar, Figure 3.11) of the Titan II upper stage but due to the dispersed trajectory was performed behind it (on the minus V Bar, Figure 3.12). During the burn Gemini VII encountered more particles from the upper stage. [25, 26]

The dispersed trajectory resulted in Gemini VII moving in the direction of the upper stage (Figure 3.12), rather than away from it in the pre-mission plan (Figure 3.11). It is not known exactly how close Gemini VII came to the Titan II second stage, but if contact had occurred the relative velocity would have been high enough to result in loss of the crew and the Gemini VII spacecraft. [25, 26]

Oberg estimated that translation from upper stage venting likely had little impact on the overall relative motion, but the posigrade football burn component (caused by the burn attitude error) and differential atmospheric drag caused the dispersed relative motion trajectory (Figure 3.12). [25, 26]

Gemini VII conducted the first successful proximity operations with another spacecraft (station-keeping with the Titan II second stage). It was also the first use of football relative motion.

Several lessons can be drawn from the Gemini VII experience. Pre-planned canned burns simplify mission planning, crew procedures, and Mission Control procedures. However, these burns assume that the spacecraft is in the correct attitude (or in some cases a particular position relative to the other spacecraft). Attitude errors (and in some cases relative position errors) can result in undesirable and potentially unsafe relative motion.

A chaser spacecraft should approach a target spacecraft only if it has some means of quantifying and verifying relative motion through relative sensor data. In addition, a means of adjusting the relative motion based on sensor data is required. Without relative navigation (radar) the crew was not able to verify desired relative motion, nor was it able to correct undesirable motion by performing on-board targeted burns. Insight into dispersed relative motion would have permitted Mission Control to recognize the risk to safety of executing the pre-mission planned perigee raise burn behind the Titan II second stage, rather than ahead of it. Had the safety risk been recognized a new burn could have been planned that could have met mission requirements while ensuring safe relative motion.

There is some risk (increased propellant consumption, collision, optical sensor contamination, etc.) to station-keeping with target spacecraft whose attitude is not controlled and that are venting propellants. Furthermore, mission design should ensure that orbital lighting is appropriate for accomplishing mission objectives without requiring unplanned and un-analyzed rotational and translational maneuvers. Such maneuvers can result in increased propellant consumption and undesirable relative motion that can present a safety hazard or complicate subsequent mission activities.

Gemini VIII

During the investigation of the GATV-5002 failure from Gemini VI, McDonnell proposed building a backup rendezvous target consisting of off-the-shelf hardware. An Agena Target Docking Adapter (ATDA) would be bolted to a Gemini re-entry RCS system module. The backup target would allow the Gemini Program schedule for rendezvous missions to proceed in the event the Agena problem could not be quickly resolved. The ATDA was delivered to KSC in February of 1966. Gemini VIII mission planning proceeded under the assumption that either an Agena or the ATDA would be used.

The Gemini Agena (GATV-5003) was successfully placed in orbit on March 16, 1966, and was followed by Gemini VIII. The Gemini VIA rendezvous profile (Figure 3.13) was also used for this mission. Radar measurements were obtained at a range of 180 nm, and visual

acquisition occurred at 76 nm. In the terminal phase, as Gemini VIII moved into daylight, the crew turned off the Agena lights by radio command. A fly-around of the Agena was performed, and the crew performed the first docking in space at a rate of 6 inches/second at sunset. No electrical discharge occurred when the spacecraft touched.

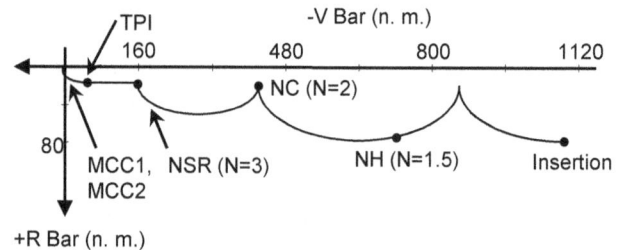

Figure 3.13 Gemini VIII M=4 ground-up rendezvous profile.

The crew tested the ability of the Agena Attitude Control System (ACS) to maneuver the docked spacecraft. An uncommanded left roll occurred as the ACS was firing and the crew undocked and performed a separation maneuver thinking that there was a problem with the Agena ACS. After undocking, the Gemini roll and yaw rates increased due to a failed on Gemini RCS thruster. The rolling motion interrupted communications with the ground-tracking network. The crew shut off the Gemini Orbit Attitude and Maneuvering System (OAMS), engaged the re-entry RCS system and stabilized the vehicle. Later the OAMS was re-activated and the crew used the circuit breakers to determine which thruster was malfunctioning. A flight rule called for a contingency return if the re-entry RCS was activated on-orbit. Splashdown occurred in the western Pacific about 10 hours and 42 minutes after launch. The Agena was parked in a higher orbit for possible use on a future mission.

Gemini IX

A shorter rendezvous, on the third revolution (M=3) was planned as the concept for lunar rendezvous also involved an M=3 profile (Figure 3.14). Burns from the M=4 profile were combined to support the shorter timeline while preserving ground tracking opportunities. An Insertion Velocity Adjustment Routine (IVAR) burn was to be performed soon after separation from the Titan II second stage to correct in-plane insertion velocity errors. A phasing maneuver would be performed at the first apogee, and a corrective combination would be performed at the start of revolution three to adjust phasing, height, and wedge angle. The co-elliptic maneuver would be executed after 90 degrees of orbital travel, and the TPI burn would be performed later on the third revolution.

On May 17, 1966, the Atlas carrying the Agena target (GATV-5004) failed and the Gemini IX launch was scrubbed. The program decided to take the ATDA out of storage and launch it rather than use an Agena. This would ensure that all subsequent Gemini missions had an Agena target.

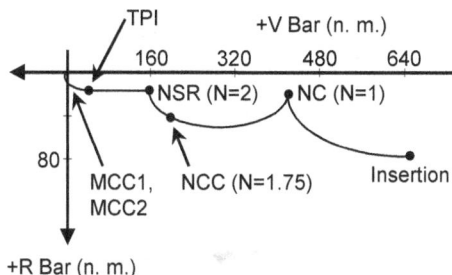

Figure 3.14 Gemini IX M=3 ground-up rendezvous.

The ATDA was successfully placed in orbit on June 1. Telemetry indicated that confirmation of the aerodynamic shroud separation from the ATDA had not occurred and that the ATDA RCS system could not stabilize the attitude dynamics. The Gemini IX liftoff was scrubbed when a ground equipment problem prevented an update of launch targeting data in time to make the launch window. Gemini IX was successfully launched on June 3.

Radar lock was intermittent due to the shroud still on the ATDA. Solid lock and measurements were eventually obtained. During daylight the ATDA was visible at long range due to the white color of the shroud. At night the ATDA lights were not continuously visible, due to the shroud. The shroud had partially opened, but had not separated from the ATDA (Figure 3.15). The crew performed a fly-around of the slowly tumbling target and closed to within 3 feet to describe the shroud and associated separation hardware in detail to Mission Control. The crew backed away from the ATDA before the ground cycled the ATDA docking cone through a rigidize/derigidize sequence in an attempt to free the shroud. The attempt failed, and suggestions for the Gemini to nudge the shroud or perform an EVA to cut the shroud lose were turned down for vehicle and crew safety reasons.

About 45 minutes after intercept, the crew performed a radial burn to separate and start the equi-period football re-rendezvous sequence (Figure 3.16). A sextant was used to measure the angle of the line-of-sight to the ATDA above the horizon and radar was not used. The TPI maneuver involved an 80 degree transfer to intercept. The second rendezvous was successful and the crew separated from the ATDA to initiate phasing for a third rendezvous the next day.

The third rendezvous was to end with a final approach from above during daylight to evaluate proposed techniques for Command/Service Module active rendezvous with the Lunar Module in lunar orbit. After phasing in front of the ATDA, Gemini IX

Figure 3.15 ATDA with shroud photographed by the Gemini IX crew (June 1966).

performed a height adjustment and flew a co-elliptic profile above and in front of the ATDA (Figure 3.17). Radar was used for the rendezvous. The crew had some difficulty visually observing the ATDA in daylight until

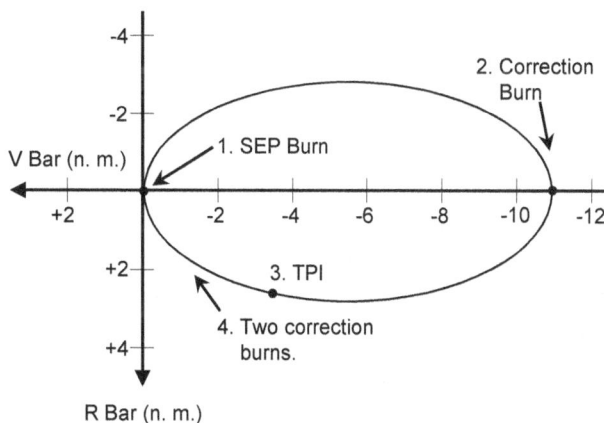

Figure 3.16 Gemini IX-A football re-rendezvous.

the terminal phase, but radar performance was good and the rendezvous was successful. Additional attempts to free the shroud by moving the docking cone and firing the ATDA RCS thrusters were not successful.

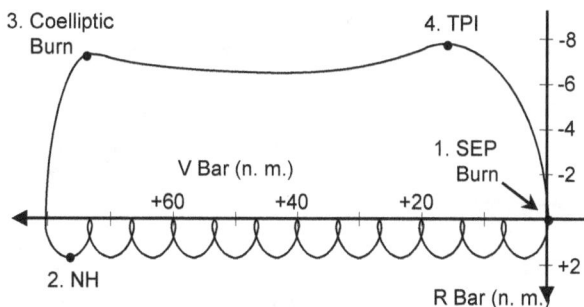

Figure 3.17 Gemini IX-A rendezvous from in front and above.

Gemini X

The plan for Gemini X was to perform an M=4 docking (Figure 3.18) with the Gemini X Agena (GATV-5005). The Gemini X Agena would then be used perform some of the rendezvous burns to enable Gemini X to rendezvous with the Gemini VIII Agena (GATV-5003) that was in a higher parking orbit. If the Gemini X Agena had been lost during launch, Gemini X would have flown a M=16 profile to rendezvous with the Gemini VIII Agena. Accommodating both rendezvous plans resulted in a launch window of 35 seconds.[4]

Both the Agena (GATV-5005) and Gemini X were launched on July 18, 1966. A larger capacity computer enabled more maneuvers to be computed on-board. The final braking was performed in darkness. An IMU misalignment resulted in an out-of-plane error at intercept of about half a mile.[1] More propellant was used than planned due to the correction of the trajectory dispersion.

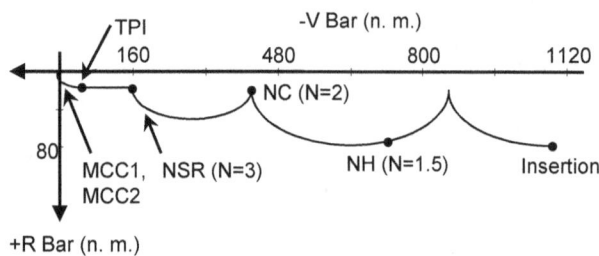

Figure 3.18   Gemini X M=4 ground-up rendezvous profile.

The docking was successful. Mission Control decided to use the Agena for longer than originally planned to make up for the lower Gemini X propellant quantity. Gemini X later undocked from the Agena (GATV-5005) in preparation for the rendezvous with the Gemini VIII Agena (GATV-5003).

The second rendezvous, with the Gemini VIII Agena (GATV-5003), was performed without radar or strobes since the GATV-5003 batteries were dead. There was no power for the Agena radar transponder. The high apogee while phasing for the second rendezvous required mission planners to avoid crew radiation exposure in the South Atlantic Anomaly. The Gemini VIII Agena, which no longer was capable of RCS attitude control, was in a gravity gradient attitude (engine to the Earth, docking cone to space). A docking was not performed. Extended station-keeping was performed while a micrometeoroid package was retrieved from the Agena during an EVA.

Gemini XI

The Agena target (GATV-5006) and Gemini XI were launched on September 12, 1966. The primary objective was a demonstration of first orbit (M=1) rendezvous (Figure 3.19). The IVAR maneuver would

correct for phasing, height, and wedge angle errors. IVAR would set up a nodal crossing after 90 degrees of orbital travel, at which point another maneuver would correct the out-of-plane error. The TPI maneuver, designed to be near apogee, was to set up an intercept after 130 degrees of orbit travel.

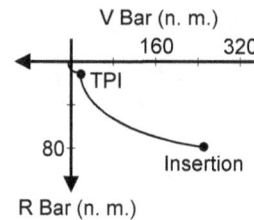

Figure 3.19 Gemini XI M=1 ground-up rendezvous.

During the rendezvous it was noted that radar signal strength fluctuated. Rendezvous and docking was successful. Gemini XI then undocked and performed a fly-around as part of a charged particle experiment. Another docking was performed, this time by the pilot. While docked an EVA was performed to attach a tether between the Agena and Gemini XI. Later in the mission the spacecraft undocked and the tether experiment was conducted.

Due to the available propellant remaining, a second rendezvous profile, stable orbit, was test flown (Figure 3.20). The term stable orbit comes from performing long range station-keeping on the –V Bar until a transfer to intercept is performed at a time that provides appropriate lighting, rather than using a coelliptic segment to control lighting at intercept. This involved a 292 degree transfer from a point approximately 22 nm behind the Agena on the –V Bar. Radar was not used due to the problem noted during the first rendezvous.[4]

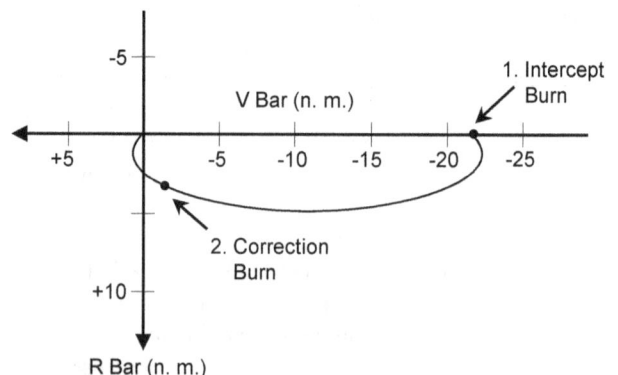

Figure 3.20 Gemini XI stable orbit rendezvous.

Gemini XII

In the spring of 1966, there was some discussion of the Gemini XII spacecraft performing a rendezvous with the Apollo AS-204 (later called Apollo 1) Command/Service Module (CSM) in the fall of 1966.

Approved for public release via STI DAA 24483. See statement on title page.

28

Since the Block 1 Apollo CSM for the flight was not capable of rendezvous, Gemini XII would be the active vehicle.

The Apollo would be launched first. If CSM system checks could be performed on the first flight day, the Gemini would launch the next day. The low Apollo orbit (lower than an Agena orbit, and driven by a requirement to be able to deorbit with the RCS thrusters in the event of a Service Propulsion System failure) would result in a lower catch-up rate and short Gemini launch windows. To provide two launch windows per day the Apollo orbital inclination would be higher than the Agena inclination on previous Gemini missions. The CSM would probably execute plane change maneuvers to lower yaw steering requirements on the Titan II booster. The standard Gemini VI rendezvous profile with rendezvous on the fourth orbit and a 15 nautical mile delta height during the coelliptic phase was suggested.

The Apollo vehicle was not equipped with a L-band transponder to enable tracking by the Gemini radar. While there were optical rendezvous navigation (no radar) procedures for Gemini, the capability was considered to be limited and placing a radar transponder on the CSM was considered mandatory.

In the end, the proposal was not approved and the AS-204 mission was delayed into 1967, beyond the end of the Gemini Program. A later proposal for Gemini XII to rendezvous with the Orbiting Astronomical Observatory was also turned down. The fire on January 27, 1967, ended the AS-204 mission.

Gemini XII was to rendezvous with the Agena on the third orbit. (Figure 3.21) During the coelliptic phase radar range rate data was spurious and appeared to have lost lock on the Agena. Rather than wait and see if the problem could be resolved Aldrin began taking sextant (optical) data to support targeting for the upcoming TPI maneuver. The crew proceeded under the assumption of a no-radar rendezvous. Although radar lock was occasionally obtained, the range rate data remained spurious. The two Mid-Course Correction maneuvers were small, and Lovell spotted the Agena lights stationary against the star background, indicating a good trajectory. Rendezvous and docking was successfully accomplished (Figure 3.22).[19, 27, 28]

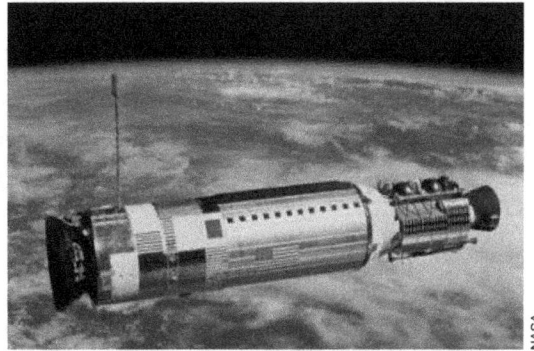

Figure 3.22 Agena target spacecraft photographed from Gemini XII (Nov. 1966).

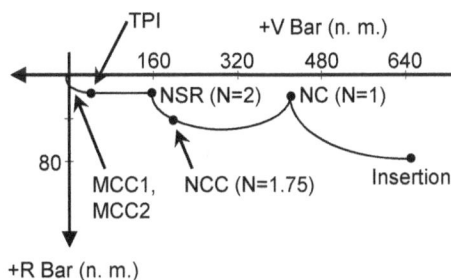

Figure 3.21 Gemini XII M=3 ground-up rendezvous.

References

1. Lunney, G. S., "Summary of Gemini Rendezvous Experience," *AIAA Flight Test, Simulation and Support Conference*, AIAA, Reston, VA, 1967.

2. Kramer, P. C., E. E. Aldrin and W. E. Hayes, "Onboard Operations For Rendezvous," *Gemini Summary Conference*, NASA SP-138, Manned Spacecraft Center, Houston, TX, February 1-2, 1967, pp. 27-40.

3. Harland, D., *How NASA Learned To Fly In Space*, Collector's Guide Publishing Inc., Burlington, Ontario, Canada, 2004.

4. Evans, W. B., and M. R. Czarnik, "Summary of Rendezvous Operations," *Gemini Summary Conference*, NASA SP-138, Manned Spacecraft Center, 1967, pp. 7-20.

5. Houbolt, J. C., "An Assessment of Rendezvous Accomplishments," *Applied Mechanics Reviews*, Vol. 20, No. 1, January 1967.

6. Hacker, B. C., and J. M. Grimwood, *On The Shoulders Of Titans*, NASA SP-4203, NASA, Washington, D.C., 1977.

7. Young, J. W., "Gemini Status Report," *10th Symposium of the Society of Experimental Test Pilots*, Beverly Hills, Calif., Sept. 23-24, 1966. *Society of Experimental Test Pilots Technical Review*, Vol. 8, No. 2, pp. 23-43.

8. McDivitt, J. A., and N. A. Armstrong, "Gemini Manned Flight Program to Date," *9th Symposium of the Society of Experimental Test Pilots*, Beverly Hills, CA, Sept. 24-25, 1965. *Society of Experimental Test Pilots Technical Review*, vol. 7, no. 4, 1965, pp. 134-166.

9. See, E. M., D. K. Slayton, A. B. Shepard, L. G. Cooper, N. A. Armstrong, and E. E. Aldrin, "Pilot's Report on Gemini/Apollo Projects," *8th Symposium of the Society of Experimental Test Pilots*, Beverly Hills, CA, Sept. 25-26, 1964. *Society of Experimental Test Pilots Technical Review*, Vol. 7, No. 2, 1964, pp. 173-191.

10. Grimwood, James M., Barton C. Hacker, and Peter J. Vorzimmer, *Project Gemini Technology and Operations – A Chronology*, NASA SP-4002, National Aeronautics and Space Administration, Washington, D.C., 1969.

11. Chamberlin, J. A., and J. T. Rose, "Gemini Rendezvous Program," *Journal of Spacecraft and Rockets*, Volume 1, Number 1, January 1964, pp. 13-18.

12. Parten, R. P., and J. P. Mayer, "Development Of The Gemini Operational Rendezvous Plan," *AIAA Journal of Spacecraft And Rockets*, Vol. 5, No. 9, Sept. 1968, pp. 1023-1028.

13. Lineberry, E. C., "Gemini VI-A Rendezvous Mission Planning," *Gemini Midprogram Conference*, NASA SP-121, Manned Spacecraft Center, 1966, pp. 277-282.

14. Burton, J. R., and W. E. Hayes, "Gemini Rendezvous," *Astrodynamics, Guidance and Control Conference*, AIAA, Reston, VA, 1964.

15. Kramer, Paul, "The Gemini Rendezvous Debate," December 7, 1991. Kramer wrote his recollections at the request of James Oberg, who was researching rendezvous history. See JSC-35055 in the A Note On Sources chapter in this document.

16. Pavelka, E. L., E. C. Lineberry, and W. J. Kennedy, "Ground Control and Monitoring of Rendezvous," *Gemini Summary Conference*, NASA SP-138, Manned Spacecraft Center, Houston, TX, February 1-2, 1967, pp. 21-26.

17. Quigley, W. W., "Gemini Rendezvous Radar," *The Microwave Journal*, Vol. VIII, No. 6, June 1965, pp. 39-45.

18. Kramer, P. C., E. E. Aldrin and W. E. Hayes, "Onboard Operations For Rendezvous," *Gemini Summary Conference*, NASA SP-138, Manned Spacecraft Center, Houston, TX, February 1-2, 1967, pp. 27-40.

19. Aldrin, E. E., *Men From Earth*, Bantam Books, New York, NY, 1989.

20. Pyron, E. E., "Hybrid Simulation Used for Developing Gemini Rendezvous Procedures and for Flight Crew Indoctrination," *AIAA Flight Test, Simulation and Support Conference*, AIAA Reston, VA, 1967.

21. *Gemini 6 - The NASA Mission Reports*, CG Publishing Inc., Burlington, Ontario, Canada, 1999.

22. Schirra, W. M., T. P. Stafford, and D. F. Grimm, "Gemini VI Rendezvous," *Society of Experimental Test Pilots Technical Review*, Vol. 8, No. 1, 1966, pp. 77-87.

23. *Gemini 7 - The NASA Mission Reports*, CG Publishing Inc., Burlington, Ontario, Canada, 2002.

24. Stafford, T. P., W. M. Schirra and D. F. Grimm, "Rendezvous of Gemini VII and Gemini VI-A," *Gemini Midprogram Conference*, NASA SP-121, Manned Spacecraft Center, Houston, TX, February 23-25, 1966, pp. 283-300.

25. Oberg, James, "30th Anniversary of the World's First Prox Ops," *Flight Design and Dynamics Newsletter Special Supplement*, Rockwell Space Operations Company, December 1995. See Gemini Rendezvous Papers (JSC-35055) in the A Note On Sources chapter.

26. Oberg, James, "Gemini-7: Lessons and Legends (30th Anniversary Revisit)," September 15, 1995. See http://jamesoberg.com. Accessed March 11, 2011.

27. *Gemini 12 - The NASA Mission Reports*, CG Publishing Inc., Burlington, Ontario, Canada, 2003.

28. Hayes, W. E., and W. W. Hinton, "Planning for a First Orbit Rendezvous Mission," *Proceedings of the 17th International Astronautical Congress*, International Astronautical Federation, October 9-15, 1966, Vol. 4, pp. 35-46.

Approved for public release via STI DAA 24483. See statement on title page.

30

CHAPTER 4 - APOLLO

Introduction

The choice of the lunar orbit rendezvous mission profile for Apollo was announced on July 11, 1962, after a trade study comparing it to Earth orbit rendezvous and direct descent to the lunar surface.[1] Rendezvous techniques developed and flight proven in the Gemini Program were applied to Apollo. Piloting and mission planning techniques from TPI through docking were essentially those developed and flight proven during the Gemini Program. However, Apollo required the development of new rendezvous profile concepts to cover both nominal and contingency rendezvous burns before TPI.[2]

Contingency Rendezvous

Nominal Apollo rendezvous involved launch of the Lunar Module (LM) ascent stage (Figure 4.1) from the lunar surface, followed by lunar orbit insertion and rendezvous with the LM as the active vehicle performing burns and terminal braking. However, in the event the

Figure 4.1 Apollo 17 LM during inspection by CSM before docking (Dec. 1972)

LM was not able to continue as the active vehicle the Command Service Module (CSM, Figure 4.2) could become the active vehicle and complete the rendezvous. Unlike Gemini, Apollo rendezvous was safety critical and docking was required for the entire crew to return to Earth (Figure 4.3).

Contingency rendezvous could also be performed after CSM/LM separation and before the lunar landing.

Table 4.1 Apollo Rendezvous and Station-Keeping Missions

Flight	Year	Chaser	Target	Profile	Comments
Apollo 7 C	1968	CSM	1. S-IVB 2. S-IVB	1. Station-keeping 2. Coelliptic	1. Aligned CSM with docking target. 2. CSM active angles only rendezvous
Apollo 8 C Prime	1968	CSM	S-IVB	Station-keeping	Observe SLA Panel jettison & lighting.
Apollo 9 D	1969	1. CSM *Gumdrop* 2. LM *Spider*	1. S-IVB/LM 2. CSM *Gumdrop**	1. Transposition 2. Coelliptic	2. First LM active rendezvous in LEO. LM tracking light failure.
Apollo 10 F	1969	1. CSM *Charlie Brown* 2. LM *Snoopy*	1. S-IVB/LM 2. CSM *Charlie Brown* [†]	1. Transposition 2. Coelliptic	2. First LM active rendezvous in lunar orbit.
Apollo 11 G	1969	1. CSM *Columbia* 2. LM *Eagle*	1. S-IVB/LM 2. CSM *Columbia* [†]	1. Transposition 2. Coelliptic	2. Docking performed with AGS due to IMU gimbal lock. CSM VHF ranging break locks.
Apollo 12 H-1	1969	1. CSM *Yankee Clipper* 2. LM *Intrepid*	1. S-IVB/LM 2. CSM *Yankee Clipper* [†]	1. Transposition 2. Coelliptic	2. LM tracking light failure. No angle marks caused CSM TPM solutions to diverge.
Apollo 13 H-2	1970	1. CSM *Odyssey* 2. LM *Aquarius*	1. S-IVB/LM 2. CSM *Odyssey* [†]	1. Transposition 2. Coelliptic	2. No rendezvous due to canceled lunar landing.
Apollo 14 H-3	1971	1. CSM *Kitty Hawk* 2. LM *Antares*	1. S-IVB/LM 2. CSM *Kitty Hawk* [†]	1. Transposition 2. Short	1. Successful hard dock on 6th attempt. 2. Hard dock on first attempt. CSM docking probe returned to Earth for analysis.
Apollo 15 J-1	1971	1. CSM *Endeavour* 2. LM *Falcon*	1. S-IVB/LM 2. CSM *Endeavour* [†]	1. Transposition 2. Short	
Apollo 16 J-2	1972	1. CSM *Casper* 2. LM *Orion*	1. S-IVB/LM 2. CSM *Casper* [†]	1. Transposition 2. Short	LM return to CSM before descent to Moon due to CSM gimbal problem.
Apollo 17 J-3	1972	1. CSM *America* 2. LM *Challenger*	1. S-IVB/LM 2. CSM *America* [†]	1. Transposition 2. Short	

* For Apollo 9, LM active for docking. [†] CSM active for docking. **AGS** = Abort Guidance System, **Coelliptic** = Coelliptic Flight Profile, **CSM** = Command/Service Module, **IMU** = Inertial Measurement Unit, **LEO** = Low Earth Orbit, **LM** = Lunar Module, **S-IVB** = Third stage of Saturn IB or Saturn V, **SLA** = Spacecraft LM Adapter, **Short** = Direct Rendezvous, **TPM** = Terminal Phase Mid-course maneuver, **Transposition** = Transposition and docking maneuver to extract LM from S-IVB, **VHF** = Very High Frequency.

Figure 4.2 Apollo 15 CSM (July 1971).

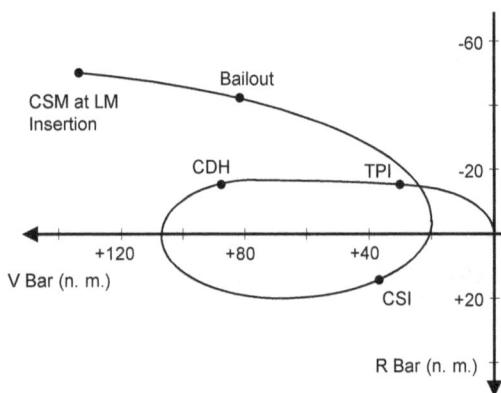

Figure 4.3 CSM active rendezvous after CSM bailout for a LM insertion underspeed of 18 feet per second.

The LM could abort the landing before or after the beginning of the powered descent.[3] For contingency rendezvous after an aborted landing attempt the LM was nominally the active vehicle. However, as with the nominal rendezvous, the CSM could become the active vehicle if LM performance issues prevented the LM from completing the rendezvous.

Contingency rendezvous scenarios were defined as follows.

• Contingency rendezvous after CSM/LM undocking and separation.

• Contingency rendezvous following abort from the LM descent orbit.

• Contingency rendezvous following a LM abort during powered descent.

• Contingency rendezvous after LM abort from the lunar surface soon after landing (the stay/no stay decision).

• Contingency rendezvous after nominal LM ascent stage orbit insertion.

• Contingency rendezvous after an anytime LM ascent stage lift-off from the lunar surface (low mission planning priority).

Only one contingency rendezvous was flown during the Apollo Program, on Apollo 16 in April of 1972. After CSM and LM separation in lunar orbit a contingency brute force re-rendezvous was successfully flown.

Rendezvous Target Vehicles

The LM or CSM could serve as the active or passive vehicles during a rendezvous mission. Both vehicles possessed relative navigation sensors, relative navigation software to process relative measurements, and software for targeting burns.[4] During nominal or contingency lunar orbit rendezvous relative navigation and burn targeting functions were exercised, systems performance permitting, regardless of whether the vehicle was active or passive. Data and status could be shared between the vehicles to aid in performance monitoring, decision making, and to provide redundancy. CSM active rendezvous required that rendezvous procedures and piloting could be successfully executed by the only crew member in the CSM, the CSM pilot.

After the Trans Lunar Injection (TLI) burn the CSM separated from the Saturn V third stage (the S-IVB) and performed the transposition and docking maneuver (Figure 4.4). The CSM then extracted the LM from the S-IVB. Although this involved relative motion the maximum separation distance was on the order of 100 feet and no relative navigation or burn targeting procedures of the type performed in lunar orbit were required.

Figure 4.4 Apollo 9 view of LM and S-IVB during the transposition maneuver (March 1969).

In lunar orbit the LM was nominally the active vehicle with rendezvous occurring after the completion of lunar surface activities. However, in case of a systems issue after CSM and LM separation and before initiation of powered descent to the lunar surface, the LM could return to the CSM. A LM performance issue during ascent from the lunar surface could cause the LM crew to perform a contingency rendezvous profile. In case the LM could not complete a rendezvous with the CSM, the CSM could go active and rendezvous with the LM.

During the Apollo 7 Earth orbit rendezvous with the S-IVB the CSM served as the active vehicle for a test of the CSM single crewman piloting rendezvous capability. During the Apollo 9 Earth orbit rendezvous test of the LM the LM served as the active vehicle throughout rendezvous and docking. However, starting with Apollo 10, the CSM was the active vehicle during the final docking maneuver with the LM in lunar orbit, at ranges typically less than 100 feet. This was due to the docking axis and CSM pilot line-of-sight axis being parallel. LM active docking required the LM commander to rotate his head 90 degrees to view the CSM through an overhead window.

LM ascent stage lifetime of the thermal control, power, and life support systems was a constraint on rendezvous. The ascent stage lifetime was 7.5 to 8 hours fully powered. This could be extended to ~14 hours with some equipment powered down. While on the lunar surface consumables were drawn from the descent stage.

Docking Hardware

An innovation over the Gemini docking hardware was that the Apollo probe and drogue permitted crew transfer between vehicles in a pressurized environment. However, probe and drogue hardware had to be removed from the docking tunnel by the crew before transfer could take place. An advantage to this design was that malfunctioning docking hardware could be returned to Earth for inspection. The probe was returned to Earth on Apollo 14 since five attempts were required to successfully engage the capture latches before the LM could be extracted from the S-IVB.[5] The probe was normally abandoned with the LM ascent stage after the lunar EVA crew returned to the CSM.

The LM originally had two docking ports. The top one was to be used when the CSM docked with the LM and extracted it from the S-IVB stage. This aligned the thrust vector of the LM descent and ascent engines with the CSM/LM center of mass in the event either LM propulsion system was needed in a contingency, as during Apollo 13.[6] The front port (which also served as the lunar EVA hatch) was to be used when the LM (as the active vehicle) docked with the CSM in lunar orbit. Later the forward docking port was eliminated to simplify the LM structural design, the forward hatch was customized for lunar EVA, and a docking window was added above the

head of the commander to support LM active docking.

The LM was equipped with a docking target mounted on the top of the LM to support a CSM active docking. To support a LM active docking, a docking target was placed in one of the CSM rendezvous windows by the CSM pilot.

Relative Navigation

The LM and CSM Apollo relative sensors had a sufficient acquisition envelope and maximum range to support relative measurement processing throughout the nominal and contingency rendezvous profiles. The ranging limit for both the CSM and LM was about 320 nautical miles. Both the CSM and LM had flashing lights to aid visual tracking. Although the LM was nominally the active vehicle during rendezvous, the requirement for CSM active contingency rendezvous required the execution of relative navigation functions by the CSM pilot during nominal LM active rendezvous.

State vectors from rendezvous navigation on both vehicles were available to Mission Control, and could be uplinked to either vehicle. Normally the LM post orbit insertion state vector was the only one uplinked to the CSM. The LM Primary Guidance and Navigation Section (PGNS), LM Abort Guidance Section (AGS), and CSM Primary Guidance, Navigation, and Control System (PGNCS) all used Kalman filters for relative measurement processing and state vector updates.[7, 8]

The primary LM relative navigation sensor was rendezvous radar providing measurements of range, range rate, shaft angle, and trunnion angle for Kalman filtering.[9] LM rendezvous navigation using the radar was active before and after each burn. The covariance matrix was reinitialized after each burn, before more radar data was processed. In addition, raw radar measurements of range, range rate, and line-of-sight angle rates were available to the LM crew for piloting cues. The radar had both cooperative (with a transponder on the CSM) and passive skin tracking modes. LM radar weight, reliability, accuracy, and thermal issues led the Apollo Program to consider replacing the radar with an optical system consisting of a hand held sextant for the LM pilot, a LM star tracker, and a xenon strobe on the CM. However, the radar was retained in 1966 due to the success of a LM weight reduction program and performance concerns with not having a direct source of range and range-rate measurements. The LM radar was available in time to support the first test of LM active rendezvous on Apollo 9 in March of 1969.

The LM was equipped with a Crew Optical Alignment Sight (COAS) for backup line-of-sight angle data. Strobes were located on the CSM and to facilitate COAS optical tracking with the human eye by the LM crew using a COAS. The LM COAS could be mounted in one of two positions. The first position was in the LM commander's window. After rendezvous but before docking, the COAS was moved to the overhead docking window (above the commander). This enabled the commander to sight on the CSM docking target in a CSM rendezvous window. However, the CSM was normally active during docking. Apollo 9 was the only Apollo mission during which the LM was active for docking. During Apollo 13 the COAS was used in the commander's window for burn attitude cues.[6]

CSM rendezvous radar was deleted in 1964 as part of a weight reduction effort. At that time it was believed that sextant line-of-sight angle measurements would be sufficient for CSM relative navigation. However, by 1967 it became apparent that angles only relative navigation was insufficient to support CSM active rendezvous. VHF ranging was added to the CSM/LM VHF communications system later to provide a CSM ranging measurement capability. VHF ranging first flew on Apollo 10 in May of 1969.[10] In case of a LM radar failure, CSM VHF ranging data could be voiced from the CSM to the LM and manually entered into the LM AGS.

For angle measurements, the CSM was equipped with a sextant and a COAS, with a corresponding strobe on the LM to support optical tracking. CSM state vectors for both vehicles were used by the PGNCS for vehicle pointing to aid in sextant tracking and pointing the rendezvous radar transponder at the LM. Sextant marks and VHF ranging marks were taken at a rate of about one per minute.

Crew Interface, Procedures, and On-Board Computers

The LM was the first U.S. human spacecraft with redundant computers and Inertial Measurement Unit (IMUs). The primary flight computer and stable member IMU in the LM PGNS were the same as was used in the CSM PGNCS. The CSM and LM primary computers had a simple display and keyboard interface (Figure 4.5). The display could show three 5-digit numbers in either decimal or octal formats. The Apollo user interface and computer capacity was a significant improvement over Gemini. Crew communication with the LM PGNS and CSM PGNCS computers used a noun and verb nomenclature that was limited to 99 of each type.

The AGS was a backup LM GNC system with a computer and strapdown IMU. It provided basic functionality to permit the LM to establish a safe orbit and rendezvous with the CSM after a LM PGNS failure. It also had a rendezvous navigation and maneuver targeting capability. The AGS was available during all LM flight phases (pre-descent orbit coast, powered descent, lunar surface, powered ascent, rendezvous and docking). However, the AGS could not support a lunar landing. AGS hardware and software were developed by different contractors than the PGNS hardware and software. The AGS concept was similar to that of the Shuttle Backup Flight System (BFS).

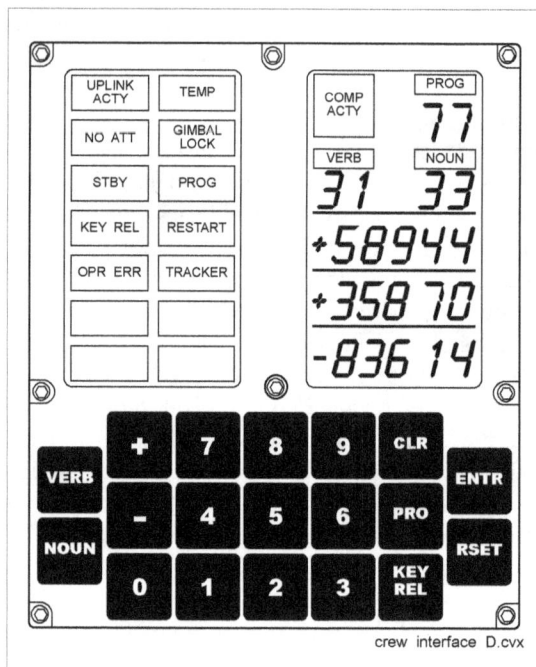

Figure 4.5 Apollo Display and Keyboard Panel (DSKY).

The mission commander executed maneuvers and managed the PGNS while the LM pilot managed the AGS. As with the LM PGNS, the CSM PGNCS had a complete rendezvous capability.

On-Board Maneuver Targeting

Both the LM (PGNS and AGS) and CSM (PGNCS) computers could compute lunar rendezvous maneuvers for either vehicle to perform Coelliptic Flight Profile, short rendezvous profile (Apollo 14 and subsequent missions), or contingency rendezvous profiles. The CSM and LM computers used a Lambert algorithm and other targeting routines designed for the lunar rendezvous profiles. Due to the safety critical nature of rendezvous, computer independent burn chart solutions were also computed on-board both the LM. The LM crew had five maneuver solutions available to them for cross checks. These included maneuver solutions from the LM PGNS, LM AGS, LM charts, CSM PGNCS, and Mission Control. On-board LM and CSM burn solutions based on relative navigation were primary, ground solutions served as backup. Stable orbit targeting software was also available in the CSM PGNCS and LM PGNS computers but was never used.[11, 12]

The delta-velocity vector burned by the LM was voiced to the CSM pilot for incorporation into the CSM navigation estimate of the LM state vector. Starting with Apollo 10 the CSM PGNCS could compute LM maneuver solutions for those burns before TPI. The CSM pilot normally computed and voiced over to the LM out-of-plane maneuver solutions. CSM sextant navigation provided more accurate estimates of out-of-plane dynamics than LM rendezvous radar navigation.

Automation

Although the CSM was nominally the passive vehicle during rendezvous, the CSM pilot performed relative navigation and targeting tasks in case a LM systems problem kept the LM from completing the rendezvous. CSM procedures for both the passive and active vehicle roles were complex and labor intensive. The first test of Apollo CSM single crew member piloting for rendezvous occurred during the Apollo 7 rendezvous with its own S-IVB stage (October 1968).

Comments made after the return of Apollo 11 by CSM pilot Michael Collins about the heavy workload (approximately 850 keystrokes) led the Apollo Program to seriously consider automating CSM rendezvous procedures in the fall of 1969. This resulted in the MINKEY program that was flown on the last three lunar missions (the J series). The level of automation in the CSM passive and active rendezvous procedures was limited by on-board computer capacity and the need for the CSM pilot to manually take sextant marks. When available, ground monitoring of CSM systems was very helpful in reducing the workload of the pilot so he could concentrate on guidance, navigation, and control procedures and performance monitoring. The addition of automation did not reduce the flexibility inherent in the CSM PGNCS rendezvous procedures.[13]

Other Relative Motion Analysis

Relative motion analysis and flight techniques developed for rendezvous were also applied to ensure that undesirable contact between Apollo spacecraft and other spacecraft components did not occur. Re-contact analysis was applied to both nominal and contingency procedures. The analysis involved the Command Module, Service Module, Launch Escape Tower, Lunar Module ascent stage, Lunar Module descent stage, Saturn S-IVB stage, Spacecraft Lunar Module Adapter Panels, Apollo lunar sub-satellite (deployed on Apollo missions 15 and 16), CSM experiment instrument booms, the Service Module Scientific Instrument Module bay door, and the Command Module docking ring and probe adapter.[14]

Development and Evolution of the Nominal Lunar Orbit Rendezvous Profile

Three types of nominal lunar rendezvous profiles were developed for Apollo. These were the direct ascent (never flown), the four burn coelliptic flight profile (CSI/CDH), and the short rendezvous. The goal throughout the development and evolution of the rendezvous profiles was a standard terminal phase that was insensitive to orbit insertion dispersions, could be easily flown manually by the crew, and did not exceed the propellant capacity of the LM ascent stage RCS.[2] This goal was achieved with the CSI/CDH and short

rendezvous profiles. Nominal profile development went through eleven phases, from early 1963 through late 1969.[15, 16]

Phase 1 – Direct Ascent

The direct ascent profile was conceived in the early 1960s before the vehicle hardware and dispersions were defined (Figure 4.6). LM ascent guidance established an intercept trajectory at orbit insertion. The transfer angle from insertion to intercept varied from 120 degrees to 300 degrees depending on when the launch occurred within the approximately 5 minute launch window. Two mid-course correction maneuvers were added before intercept so that delta-velocity requirements during the terminal phase (braking) were within the capability of the LM ascent RCS. This also enabled ascent dispersions to be corrected before the final approach.

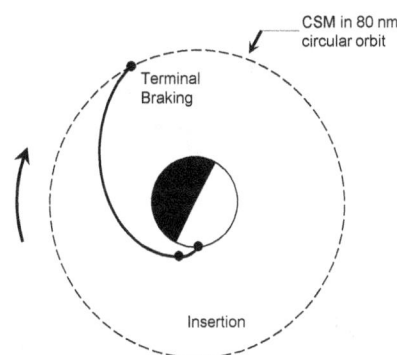

Figure 4.6 Phase 1, Direct Ascent.

However, there were three problems associated with this profile. First, the final approach angle (direction) in the terminal phase was variable, which complicated crew procedures. Secondly, much of the profile occurred behind the Moon, therefore there was no ground support available due to the lack of communication. Third, since an intercept trajectory was targeted at orbit insertion the insertion (ascent guidance) targets were a function of the lift-off time within the launch window. The launch window was sensitive to unsafe perigee after insertion.

Phase 2 – Standard Insertion Parking Orbit with Standard Direct Intercept at Variable Time

A profile using a parking or phasing orbit concept was developed that would standardize the final approach direction, permit insertion into a standard orbit, and permit communication with Mission Control during some portions of the rendezvous (Figure 4.7). The insertion orbit was standardized to an 8 to 10 nm insertion altitude with a 10 to 20 nm apolune. This eliminated the unsafe perilune problem with the previous profile. There was a standard 160 degree transfer from the Terminal Phase

Approved for public release via STI DAA 24483. See statement on title page.

35

Initiation (TPI) burn to intercept. The TPI time was variable and was a function of the lift-off time within the launch window. In addition, the profile provided a planar correction (wedge angle) before the terminal phase. TPI also occurred on the near side of the Moon, permitting Mission Control support for that burn.

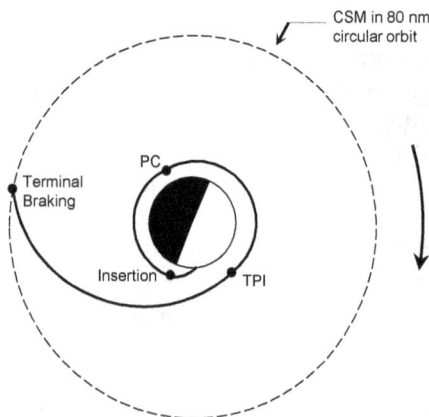

Figure 4.7 Phase 2, standard insertion parking orbit with standard direct intercept at variable time.

Phase 3 – Original Three Burn Coelliptic Profile

By early 1965 three issues became apparent that had to be addressed. First, the TPI time varied based on the lift-off time within the launch window. This in turn resulted in variable orbital lighting conditions during the terminal phase (braking, station-keeping, and docking). Second, slow and relatively constant approach rates before the TPI burn were needed to assist relative sensor tracking and crew monitoring activities. Third, the 70 nm delta-height at TPI resulted in final braking burns too large to be executed by the LM ascent stage RCS. Use of the ascent engine would require an attitude maneuver to burn attitude causing the crew to lose sight of the CSM for a period of time during the final approach. This was not desirable from a safety standpoint.

These issues were addressed in a new three burn profile that incorporated a coelliptic orbit before TPI (Figure 4.8). The insertion orbit apolune varied as a function of lift-off time. The coelliptic burn, called Constant Delta Height (CDH), was executed half a revolution after insertion. The launch window limited the coelliptic delta-height to between 15 nm and 50 nm. Braking requirements associated with the 50 nm delta-height were within the capability of the LM RCS. The variation in the insertion orbit, and the resulting coelliptic delta-height, caused TPI to be executed at a fixed time, independent of the lift-off time with in the launch window. This provided a standard 150 degree transfer to intercept.

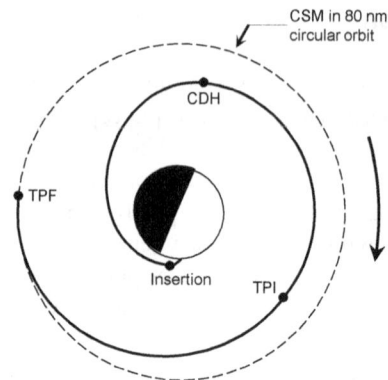

Figure 4.8 Phase 3, Direct Coelliptic Sequence.

Phase 4 – The Original CSI/CDH Profile

Dispersion analyses conducted from late 1965 to mid 1966 revealed that orbit insertion dispersions could result in large TPI time slips. The variable insertion orbit was believed to require complex crew monitoring techniques. A Coelliptic Sequence Initiation (CSI) burn was added to the profile between insertion and CDH to correct for dispersions (in conjunction with the CDH burn) and permit a standard insertion orbit (Figure 4.9).

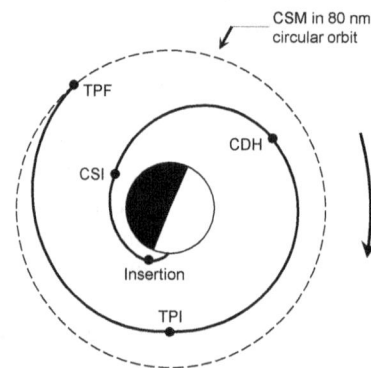

Figure 4.9 Phase 4, original CSI-CDH Coelliptic Sequence.

The new standard insertion orbit was 10 by 30 nm, with insertion occurring at perigee. The CSI burn was executed 30 minutes after insertion. It was constrained to the local horizontal to ensure that the perilune would not be lowered. CDH was executed at the next apolune following CSI to minimize propellant consumption. It was usually a horizontal burn due to the circular 80 nm CSM orbit. Delta-height at CDH varied from 15 to 50 nm and the transfer time from CSI to CDH varied from ~51 to ~28 minutes due to the variation in lift-off time within the launch window. CSI adjusted the phasing rate, or the delta-height of the post CDH orbit, to permit launch within the nominal launch window and to correct for dispersions. The TPI burn occurred approximately over the landing site, permitting Mission Control support of TPI. The TPI to intercept transfer angle was standardized at 140 degrees.

This new profile made the TPI time less sensitive to orbit insertion dispersions. More standardized range and range rate profiles permitted the development of backup burn solution charts. Desired standard insertion conditions also decreased procedural complexity.

Phase 5 – Optimized Terminal Phase

By early 1967 the profile lighting became standard with TPI occurring at a fixed lunar longitude. The priority of Mission Control support for burns before TPI was changed to be higher than support for post TPI activities. The profile was modified to provide a 130 degree TPI to intercept transfer angle to optimize the terminal braking phase. TPI was changed to occur 20 minutes before orbital sunset to provide optimum terminal phase lighting.

Phase 6 – CSM Orbit Decrease to 60 nm

By early 1968 three problems with the rendezvous profile became apparent. 1) The LM RCS and the SM RCS (for the CSM rescue of the LM case) had insufficient propulsion capability to complete a rendezvous with the large delta-heights associated with the nominal launch window. 2) Raising the apolune insertion orbit was not feasible given the small LM ascent propulsion system propellant margin. 3) In some dispersed cases the LM could be outside the 400 nm range of the LM radar during the early part of the rendezvous. These issues were resolved by lowering the CSM parking orbit from 80 nm to 60 nm. This also lowered propellant requirements for the LM ascent stage during LM aborts and the descent stage during landing. Elimination of the 4 to 5 minute launch window bounded the acceptable coelliptic delta-height to ~25 nm. The lower delta-height variation allowed the crew timeline between CSI and TPI to become more standardized.

Phase 7 – Extended CSI-CDH Profile

Shortcomings of the Phase 6 profile were insufficient time between insertion and the CSI burn for IMU alignments and CSM VHF ranging and sextant tracking of the LM. In addition, there was a need to perform a planar correction burn (wedge angle) at some point between insertion and the TPI burn. Such a burn would avoid the need for large out-of-plane burns during the final approach.

The extended CSI-CDH sequence had increased time between insertion and CSI, and CSI and CDH. Both transfer times were 50 minutes. The additional time before CSI permitted an IMU alignment and sufficient

relative tracking, both of which increased the accuracy of the CSI burn. The additional time between burns also made the crew timeline more manageable. A Plane Change (PC) burn was placed ~29 minutes before CDH. PC established a nodal crossing at CDH for the out-of-plane correction. This provided a nearly co-planar terminal phase even with out-of-plane dispersions at orbit insertion. TPI lighting was delayed to the mid-point of orbital night since CDH now occurred a few minutes before the Phase 6 TPI time.

Phase 8 – Controlling the CSI to CDH Transfer Time

By late 1968 it became apparent that the transfer time between CSI and CDH could significantly decrease under dispersed conditions. CDH was performed at the first apsis after CSI. Completion of the plane change at CDH could result in a large out-of-plane delta velocity component at CDH.

The timeline between CSI and CDH was standardized regardless of dispersions by performing CDH one half a revolution after CSI, instead of at the first apsis after CSI (Figure 4.10). CSI was performed at apolune, 55 minutes after insertion, to avoid a large radial delta-velocity component at CDH. This resulted in nearly horizontal CDH burns unless there was a large radial dispersion at orbital insertion. The nodal crossing was targeted by CSI to occur at the PC burn point. Using CSI to establish the nodal crossing at PC saved propellant since the CSI delta-velocity was normally larger than the CDH delta-velocity.

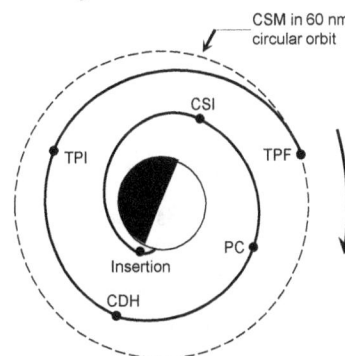

Figure 4.10 Phase 8, CSI-CDH Coelliptic Sequence with controlled CSI to CDH transfer time.

Phase 9 – Insertion Orbit Changed to 45 nm by 10 nm

Simulations involving the crew revealed that the LM was outside the CSM VHF ranging tracking range before CSI. This was resolved by inserting the LM into a 45 nm by 10 nm orbit. Under nominal conditions CSI occurred at the desired coelliptic delta-height of 15 nm, resulting in a near zero CDH delta-velocity. TPI occurred at the mid-point of orbital night, 33 minutes after the CDH burn.

Phase 10 – Apollo 11 CSI-CDH Profile

It was desired to increase the CDH to TPI transfer time by 5 minutes to accommodate possible early TPI time slips due to navigation and burn execution dispersions. To increase the transfer time between CDH and TPI by 5 minutes without delaying TPI, lunar orbit insertion targeted an upward radial component of 30 feet/second. This decreased the transfer time from insertion to CSI and CSI continued to be performed at apolune.

This profile (Figures 4.11 and 4.12) was flown on Apollo 11, Apollo 12, and was planned for Apollo 13. However, there were four unresolved issues with the profile. 1) The terminal phase lighting was not optimum. 2) Under certain dispersions CDH could have a radial delta-velocity component. 3) The LM ascent propulsion system had low margins. 4) The total time between orbit insertion and docking was too long.

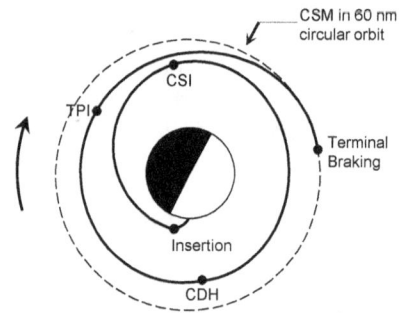

Figure 4.11 Phase 10, Coelliptic Flight Profile as flown on Apollo 11 and Apollo 12.

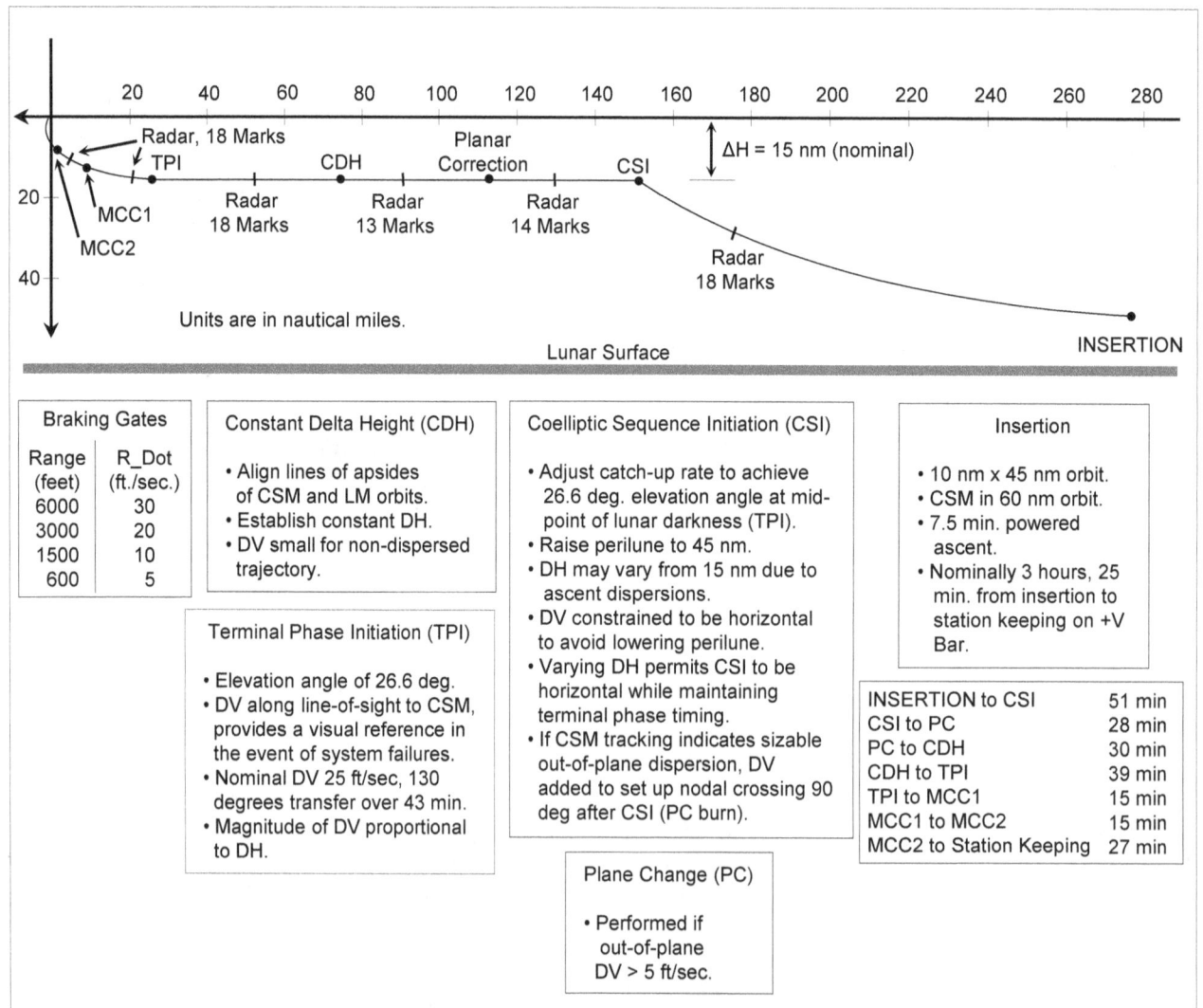

Braking Gates	
Range (feet)	R_Dot (ft./sec.)
6000	30
3000	20
1500	10
600	5

Constant Delta Height (CDH)

- Align lines of apsides of CSM and LM orbits.
- Establish constant DH.
- DV small for non-dispersed trajectory.

Terminal Phase Initiation (TPI)

- Elevation angle of 26.6 deg.
- DV along line-of-sight to CSM, provides a visual reference in the event of system failures.
- Nominal DV 25 ft/sec, 130 degrees transfer over 43 min.
- Magnitude of DV proportional to DH.

Coelliptic Sequence Initiation (CSI)

- Adjust catch-up rate to achieve 26.6 deg. elevation angle at mid-point of lunar darkness (TPI).
- Raise perilune to 45 nm.
- DH may vary from 15 nm due to ascent dispersions.
- DV constrained to be horizontal to avoid lowering perilune.
- Varying DH permits CSI to be horizontal while maintaining terminal phase timing.
- If CSM tracking indicates sizable out-of-plane dispersion, DV added to set up nodal crossing 90 deg after CSI (PC burn).

Plane Change (PC)

- Performed if out-of-plane DV > 5 ft/sec.

Insertion

- 10 nm x 45 nm orbit.
- CSM in 60 nm orbit.
- 7.5 min. powered ascent.
- Nominally 3 hours, 25 min. from insertion to station keeping on +V Bar.

INSERTION to CSI	51 min
CSI to PC	28 min
PC to CDH	30 min
CDH to TPI	39 min
TPI to MCC1	15 min
MCC1 to MCC2	15 min
MCC2 to Station Keeping	27 min

Figure 4.12 Nominal Apollo 11 (Mission G) Profile

Approved for public release via STI DAA 24483. See statement on title page.

38

Phase 11 – Short Rendezvous Profile

The CSI-CDH coelliptic profile was successfully flown on Apollo 11 (July 1969) and Apollo 12 (November 1969). However, the time from insertion to station-keeping with the CSM was approximately 3.5 hours. Planning for the Apollo 14 mission indicated a crew day of 23.5 hours. Excellent vehicle performance on Apollo 11 and Apollo 12 coupled with concerns about the length of the crew day led to development of a short rendezvous profile that saved about 2 hours. Much of the design of this profile was complete by January of 1970. The profile was similar to one flown on Gemini XI (September 1966). This profile was successfully flown on Apollo missions 14, 15, 16, and 17.

Short rendezvous was a precisely timed orbit insertion providing a TPI burn relative position that in turn yielded appropriate terminal phase lighting (Figures 4.13 and 4.14). TPI was executed on a fixed time after orbit insertion, 38 minutes for Apollo 14, 45 minutes for Apollo 15, and 47 minutes for missions 16 and 17. Short rendezvous lift-off time was about 2.5 minutes earlier than the lift-off time for the CSI-CDH profile. CSM and LM IMU platforms were aligned before lift-off. For a nominal mission no alignments were conducted between lift-off and docking due to the short timeline. The maximum range between the CSM and LM during the rendezvous was about 145 nm, well within the tracking ranges of the relative sensors.

The required insertion orbit was approximately that required for the CSI-CDH coelliptic profile. Short rendezvous nominally was a 10 nm x 48 nm insertion orbit, while the CSI-CDH coelliptic profile had a nominally 10 nm x 45 nm insertion orbit. Apolune of the insertion orbit could vary from 45 nm to 50 nm, and the transfer time from insertion to TPI varied from 38 minutes to 45 minutes. Variations were due to the CSM parking orbit and lunar stay time. A Mission Control computed tweak burn performed 2 or 3 minutes after orbit insertion could correct for nominally expected insertion dispersions.

The TPI delta-velocity was no longer along the line-of-sight vector to the CSM. TPI had to be performed with the LM ascent propulsion system, or the CSM Service Propulsion System in the event of a CSM rescue of the LM. TPI was targeted to force a nodal crossing 90 degrees later, at the second Mid-Course Correction burn, to permit correction of out-of-plane dispersions.

The terminal approach geometry was the same as the CSI-CDH rendezvous. TPI was executed on time rather than on elevation angle since on a nominal short rendezvous the desired elevation angle occurred twice (nominal TPI time and ~18 minutes after insertion). In addition, in some dispersed cases the desired TPI elevation angle did not occur in the desired time frame.

In the event of degraded systems performance on either vehicle before lift-off, the CSI-CDH profile would be flown. If degraded systems performance occurred during ascent, the crew could switch (also called a "bailout") to the CSI-CDH profile ~5 minutes after insertion (Figure 4.15). For example, if the post insertion tweak burn delta-velocity was greater than 60 feet/second or the insertion out-of-plane wedge angle was greater than 0.5 degrees a bailout to the CSI-CDH was performed. The CSI-CDH coelliptic profile was also retained for pre-descent aborts and powered descent aborts before lunar landing.

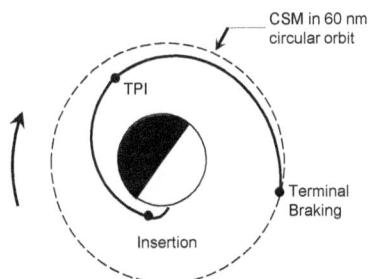

Figure 4.13 Phase 11, Short (Direct) rendezvous profile as flown on Apollo missions 14 through 17.

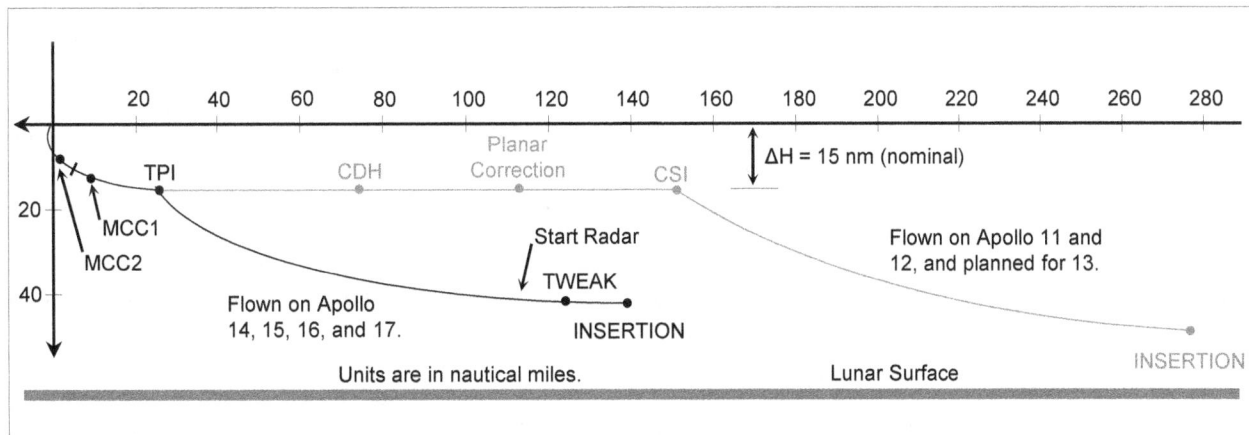

Figure 4.14 Comparison of Short and Coelliptic Rendezvous Profiles

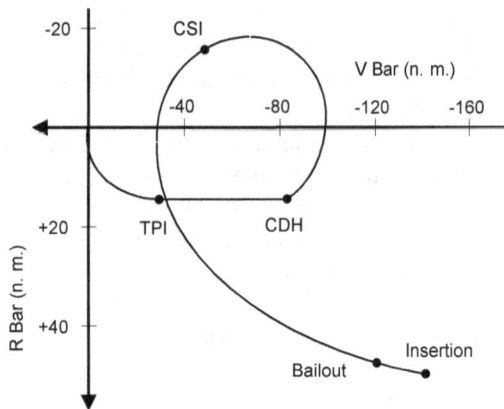

Figure 4.15 LM active rendezvous following a LM bailout for an insertion underspeed of 50 feet per second.

Apollo Missions

This section provides a summary of rendezvous activities that occurred on the Apollo missions.[17] Most of the activities described include the transposition and docking maneuver after Trans Lunar Injection, undocking in lunar orbit before LM powered descent to the surface, and rendezvous after completion of lunar surface exploration. More details may be found in the references.

AS-204 (Apollo 1, Scheduled for February 1967)

At the time of the fatal fire on January 27, 1967, the AS-204 (later known as Apollo 1) mission was scheduled for launch on Tuesday, February 21, 1967. The mission was to test Block 1 Apollo systems. The Block 1 vehicle was not equipped for rendezvous and docking, nor was AS-204 to carry a Lunar Module.

Near the end of the second orbit, the CSM was to separate from the S-IVB stage. The crew would perform the transposition maneuver that would be flown on subsequent missions before docking with and extraction of the LM (no LM was to be carried on Apollo 1). Station-keeping was to be performed while the crew photographed and filmed liquid oxygen and liquid hydrogen venting by the S-IVB. After leaving the S-IVB, the CSM would not have re-rendezvoused with it.[18]

Apollo 7 (C Mission, October 1968)

Apollo 7 (October 1968) was the first crewed flight of the Command/Service Module and performed extensive systems checks over a 12-day mission in low Earth orbit. After separation from the S-IVB the crew performed a transposition and simulated docking maneuver. A LM was not carried on the mission but a docking target was installed on the S-IVB. One S-IVB adapter panel only deployed to about 25 degrees rather than the desired 45 degrees. The crew aligned the CSM with the docking target, but did not approach too close due to the angle of the one panel. The panels would be

jettisoned on all subsequent flights after CSM separation and before docking with the LM. After completion of the station keeping and photography activities the CSM separated and phased ahead of the S-IVB.

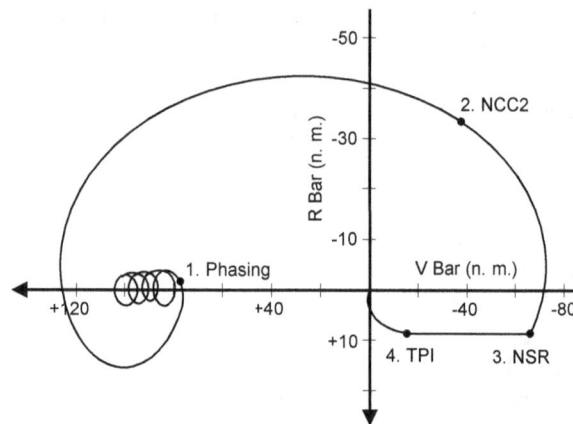

Figure 4.16 Apollo 7 rendezvous with S-IVB stage (Oct. 1968).

On the second day of the mission a rendezvous was conducted with the S-IVB (Figure 4.16). This was to evaluate the CSM contingency LM rescue rendezvous capability planned for later missions in the event of an aborted lunar landing or LM system problems.[19] The ability of one crewmember, the CSM pilot, to fly the rendezvous would also be evaluated. The rendezvous could have been delayed one day in the event of a change to the mission timeline caused by vehicle performance issues. Atmospheric drag and orbital lifetime uncertainties of the S-IVB ruled out any delay in rendezvous execution beyond one day.

About 26.5 hours into the mission, at a range of about 70 nm in front of the S-IVB, a burn was conducted to initiate the rendezvous. Sextant marks were taken during daylight periods to support computation of burns. Unlike subsequent Apollo flights, no LM navigation or burn solution data was available for cross checking CSM burn and navigation data. Mission Control computed burn solutions were available to the CSM crew for comparison. A coelliptic approach as flown on Gemini and planned for the lunar missions was conducted. The TPI delta-velocity was designed to occur along the line-of-sight to the target, but the crew could not see the S-IVB flashing lights through the CSM windows during TPI execution.

Execution of the rendezvous procedures by one crewmember proved to be challenging. The crew was able to use COAS subtended angles to estimate range to the S-IVB during braking, but would have felt more comfortable with radar data. Range derived from subtended angles could be more difficult with the much smaller LM. Starting with Apollo 10 (May 1969) VHF range measurements were available to the CSM pilot. LM rendezvous radar data (if available) could be voiced over by the LM crew during a CSM active rendezvous.

On the second and third days of the flight the S-IVB was tracked with the CSM sextant at ranges of 80, 160, and 320 nm. On the fourth day the CSM, carrying a LM radar transponder, was tracked by a LM rendezvous radar located at White Sands, New Mexico. Some 47 seconds of data over ranges from 390 to 415 nautical miles were obtained. [20, 21]

Apollo 8 (C-Prime Mission, December 1968)

Apollo 8 did not carry a Lunar Module and no rendezvous activities were performed. However, after separation from the S-IVB the CSM performed the transposition maneuver so that the crew could monitor Spacecraft LM Adapter (SLA) panel separation and orbital lighting. The panels separated from the S-IVB without any danger of re-contact and orbital lighting was adequate for docking. Since there was no LM, the crew did not take the CSM too close to the S-IVB. Formation flying with the S-IVB was accomplished without difficulty. An additional separation maneuver was performed to ensure adequate CSM separation from the S-IVB. [22, 23]

Apollo 9 (D Mission, March 1969)

Apollo 9 was the first flight of the LM. The mission involved extensive tests of CSM and LM systems in low Earth orbit. Included was a 6 hour 23 minute rendezvous activity that tested all CSM and LM systems associated with rendezvous (Figure 4.17). Testing rendezvous in low Earth orbit was desirable before rendezvous in lunar orbit was conducted. Successful rendezvous and docking was required to ensure the safe return to Earth of the LM crew. Apollo 9 performed the first safety critical space rendezvous. The profile included burns that progressively increased the separation distance between the CSM and LM as systems on both vehicles exhibited expected performance. The profile ended with the execution of a coelliptic CSI-CDH-TPI rendezvous. [24]

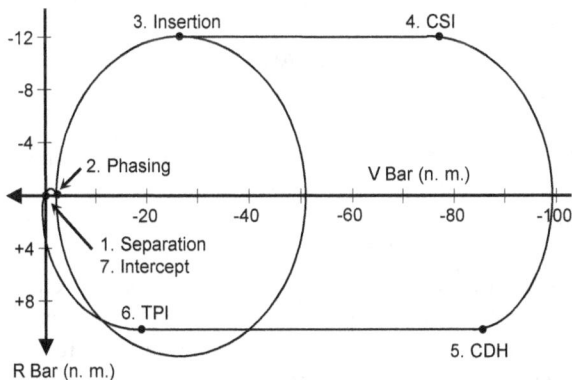

Figure 4.17 Apollo 9 LM separation and rendezvous with the CSM in Earth orbit (March 1969).

After orbit insertion the CSM separated from the S-IVB. The transposition maneuver was completed and the CSM successfully docked with the LM and extracted it from the S-IVB. The crew encountered no difficulty removing the docking probe from the docking tunnel. The probe had to be removed for the crew to enter the LM.

On the fourth flight day, the day before the undocking, separation, and rendezvous, an Extra Vehicular Activity (EVA) was conducted. The crew successfully tested a contingency procedure for EVA crew transfer from the LM to the CSM if the docking tunnel could not be used.

Undocking, rendezvous, and docking occurred on the fifth flight day. After undocking LM attitude maneuvers were performed to permit the CSM pilot to visually inspect and photograph the LM. The CSM then performed a 5 foot/second radial down separation maneuver to set up mini-football relative motion. The mini-football permitted both vehicles to perform IMU alignments while not requiring precise station-keeping piloting at the same time, provided a large enough range between the vehicles to permit LM rendezvous radar checkout, and permitted a contingency re-rendezvous without performing relative navigation and on-board burn targeting. The mini-football was the same as that planned for the lunar missions between separation and LM execution of the Descent Orbit Insertion (DOI) burn. [24]

A 90.7 foot/second phasing burn, primarily radial up, was performed by the LM Descent Propulsion System (DPS) under the control of the LM AGS at a range of about 2 nm from the CSM. This burn, computed only by Mission Control, established a larger football profile (equi-period orbit) with respect to the CSM. The delta-height of the achieved football was 12.2 nm.

From this orbit the LM could execute the nominal rendezvous plan or perform a contingency return to the CSM in the event the systems performance of either vehicle prevented continuation of the nominal rendezvous plan. In the event that a LM or CSM performance problem required cancelation of the nominal CSI-CDH-TPI rendezvous and required a sooner than planned docking, an abort TPI burn opportunity was placed in the equi-period orbit. The TPI time was defined by a 27.5 degree elevation angle of the LM to CSM line-of-sight with respect to the local horizontal. On a nominal equi-period football TPI would occur at a delta-height 10 nm below the CSM orbit. Both the LM and CSM incorporated relative sensor measurements (LM radar, CSM sextant) and computed the abort TPI burn solutions. The abort TPI burn was not executed since both vehicles were performing as expected. Good agreement between the nominal pre-mission, LM, CSM, and Mission Control abort TPI solutions indicated that systems on both vehicles were functioning well. [24, 25]

A Mission Control computed 42.7 foot/second phasing burn (item 3 on Figure 4.17), primarily posigrade, was executed to place the LM in a coelliptic orbit above that of the CSM. The burn was performed with the LM DPS and was controlled by the LM PGNS. After the burn the coelliptic delta-height was 12.2 nm. Once the burn was executed there were no other contingency return options in the event of degraded LM or CSM systems performance. Docking required execution of the entire CSI-CDH-TPI profile by either the LM (nominal) or the CSM (contingency mirror image burns).

The CSM pilot performed sextant tracking and mirror image burn targeting for the CSI, CDH, TPI, and the two mid-course correction burns. Execution of mirror image burns would have been performed in the event a CSM rescue of the LM was required (i.e. the LM could not complete the rendezvous). Mirror image burns would have to be executed within one minute of the planned LM ignition time for a burn.

CSI was a 40 foot/second retrograde (horizontal) burn performed with LM RCS. The LM descent stage was jettisoned (staged) after the start of RCS thrusting. The CSI burn was constrained to be horizontal at a fixed time-of-ignition to place the LM at the desired TPI line-of-sight elevation angle at a desired time. It was computed by the LM, CSM, and Mission Control. After CSI was performed a maximum range of 98 nm between the vehicles was reached.

After CSI the CSM pilot noted that the LM tracking light had failed. The LM crew did not see any flashes from the tracking light reflected by the LM RCS quads. The failure of the LM tracking light during the CSI burn limited CSM sextant tracking to orbital daylight. No sextant marks were taken between CSI and CDH, but sextant marks during orbital daylight between CDH and TPI were obtained. The LM was visible during daylight out to the maximum range in orbital daylight of 70 nm. When operating the LM tracking light was also visible during orbital day. CSM VHF ranging was not available until the Apollo 10 mission (May 1969). [25]

The CDH burn occurred at the first apsidal crossing after CSI. It was performed with the LM Ascent Propulsion System and consisted of a 39.2 feet/second retrograde (horizontal) component and a 13.7 radial up component. It was designed to align the semi-major axes of the LM and CSM orbits, and establish equal differential altitudes at apogee and perigee. The nominal differential altitude of 10 nm was achieved.

After CDH the LM maneuvered to the TPI burn attitude. The LM crew noted a decrease in the radar signal strength. It was later determined that this was due to a CSM maneuver to an attitude that placed the line-of-sight to the LM 20 degrees above the CSM +X body axis. In this attitude the signal strength of the CSM radar transponder was reduced.

TPI was targeted to occur at the standard 27.5 degree elevation angle and 25 minutes before sunrise. CSM transfer angle from TPI time to intercept was 130 degrees. The TPI and terminal phase design was the same as that flown on some Gemini missions. TPI was executed with the LM RCS. It was a 21.7 feet/second burn along the line-of-sight to the CSM.

The two mid-course correction burns were under 2 feet/second in each axis. Due to the LM tracking light failure the CSM pilot was unable to take any sextant marks to support CSM computation of the two mid-course corrections. Braking gate execution was nominal. Only small corrections to inertial line-of-sight rate were required. Station-keeping was established within 100 feet of the CSM.

After sunrise during final approach the CSM pilot determined the range to the LM using a diastimeter mounted in a forward looking CSM window. This optical device permitted the CSM pilot to determine the range to a spacecraft of known dimensions out to a range of 3 nm. The range data was accurate and readily accessible to the CSM pilot. It was carried on the Apollo 9 CSM as a backup source of range measurements to the LM radar.

The crew elected to dock as soon as possible to preserve margin in orbital daylight in the event of docking difficulties. LM active docking was complicated and took longer than expected due to a brightly lit CSM, a poorly lighted CSM docking target, and a dim COAS reticle pattern. The commander had difficulty distinguishing the illuminated COAS reticle pattern with the mirror-like surface of the brightly lit CSM in the background. The CSM pilot talked the commander in to a range of 4 or 5 feet. At that point the COAS reticle became visible and the commander (in the LM) completed the docking.

The LM active docking was awkward from an ergonomic perspective due to position of the COAS in the LM docking window, above the head of the commander. The commander's line-of-sight through the COAS during docking was 90 degrees from his view of the LM displays and controls. For a CSM active docking the CSM displays and controls were much closer to the CSM pilot line-of-sight through the COAS mounted in the docking window. CSM active dockings were performed on all subsequent missions, with LM active docking reserved as a contingency procedure.

The LM PGNS and AGS successfully processed radar measurements. The CSM pilot managed CSM attitude to ensure proper pointing of the CSM radar transponder. PGNS and AGS burn targeting solutions closely matched those of Mission Control throughout the rendezvous. Relative navigation and mirror image burn targeting by the CSM was successful. A Mission Control up link of state vectors to either vehicle was not required between separation and docking. CSM pilot rendezvous procedures were successfully performed. However, Mission Control monitoring of telemetry and systems status not associated with rendezvous permitted the CSM pilot to focus on the rendezvous procedures. [24, 25]

Apollo 10 (F Mission, May 1969)

Apollo 10 was a lunar orbit dress rehearsal of the Apollo 11 lunar landing mission. It was the first flight of the LM in lunar orbit. The rendezvous profile was designed to exercise all CSM and LM rendezvous related systems and most procedures (Figure 4.18). [26] Apollo 10 was the first flight of CSM VHF ranging for relative navigation.

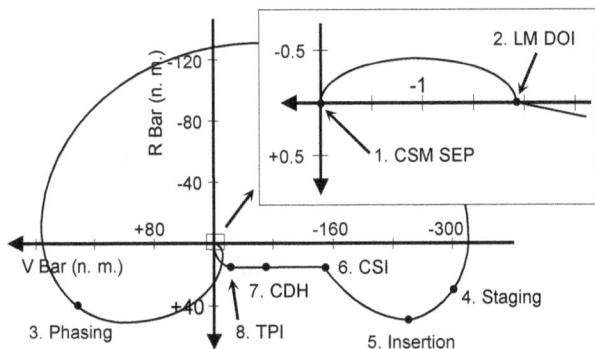

Figure 4.18 Apollo 10 LM separation from and rendezvous with the CSM in lunar orbit (May 1969).

Before the transposition and docking maneuver the mission commander and CSM pilot changed seats. The crew wore helmets and gloves during this activity and through LM pressurization. After completion of the TLI burn the CSM separated from the S-IVB and the adapter panels were jettisoned. Maximum separation from the S-IVB/LM was ~150 feet, about 100 feet further than planned. A delta-velocity of about 1.2 feet/second was executed to close on the S-IVB. The COAS reticle pattern was washed out due to the brightness of the LM, but the pattern became visible as the range during the approach decreased. Minimal lateral and vertical translations were required to align the COAS with the LM docking target. Sunlight was not a problem during docking. Closing rate at docking was estimated at ~0.2 feet/second. [27, 28]

The 8 hour 10 minute lunar orbit rendezvous activity was planned for the 5th flight day. The nominal mission plan was as follows. On the fourth flight day, in lunar orbit, the CSM/LM orbit was to be circularized at 60 nm. Undocking was to be followed by 25 minutes of station-keeping for CSM pilot inspection of the LM. A 2.5 foot/second radically down separation burn would be performed to set up mini-football relative motion with a nominal maximum relative separation of 2 nm. At the mid-point of the mini-football the LM DPS would perform a 71 foot/second retro-grade burn, called Descent Orbit Insertion (DOI), to lower the LM perilune to 8 nm, resulting in an 8 x 60 nm orbit.

Later, with the LM below and phasing ahead of the CSM, the LM DPS would perform a 195 foot/second phasing burn to place the LM in a dwell orbit (8 x 194 nm). This orbit would take the LM ahead of, above, and then behind and below the CSM. This burn would not be performed on a lunar landing mission. After LM descent staging the LM ascent stage would perform a 207 foot/second insertion burn (8 x 43.6 nm orbit) to set up the trajectory for the CSI burn, the first burn of the CSI-CDH-TPI coelliptic rendezvous sequence. The insertion burn design was to establish trajectory conditions similar to those after lunar lift-off and lunar orbit insertion. [26]

The rest of the rendezvous sequence was similar to that flown on Apollo 9, except that CSI was below the CSM orbit rather than above it. The maximum spacecraft separation during the rendezvous was planned to be 350 nm. All burns from CSI through establishment of station-keeping would be performed with the LM RCS. CSI would place the LM in a 42.9 x 46.2 nm orbit. CDH would establish a constant delta-height of 15 nm and TPI would establish an intercept trajectory. [26]

Actual rendezvous details were as follows. The CSM and LM undocked on the 12th lunar revolution. The 2.5 foot/second separation burn had a retrograde 0.2 feet/second component, rather than being purely radial down as planned. This resulted in a 0.4 nm greater range at DOI than planned. After separation LM radar and CSM VHF ranging tracking was initiated. DOI successfully lowered perilune to 8.5 nm. After DOI the CSM pilot tracked the LM with the sextant and VHF ranging. The LM appeared as a bright star against the lunar surface until a range of ~125 nm, when it disappeared. LM radar tracking of the CSM was also performed after DOI. [28]

The phasing burn placed the LM in a 12 x 190 nm phasing orbit. After the phasing burn the CSM pilot resumed sextant and VHF ranging relative navigation. The LM was tracked with the sextant during orbital night at ranges exceeding 230 nm, and in daylight out to 275 nm. VHF ranging marks were incorporated out to ~275 nm, but higher range measurements were observed. [27]

Ten minutes before the insertion burn the LM descent stage was jettisoned while the LM was under AGS control. The LM crew donned helmets and gloves for staging. The insertion burn placed the LM in a 11.0 x 46.5 nm orbit. The CSM was prepared to execute a mirror image insertion burn in case the LM could not execute the burn. [27]

Both the LM and CSM began relative navigation to support targeting for the subsequent CSI burn. On-board and Mission Control CSI burn solutions were in agreement. CSI was a 45.3 foot/second posigrade burn. The CDH burn was performed with the LM RCS under AGS control. CDH components were 0.1 feet/second posigrade and 3 feet/second radial down. CDH established a near nominal delta-height of 14.9 nm. [27]

TPI time-of-ignition slip was ~2 minutes late. TPI was along the line-of-sight vector with components of 21.7 feet/second posigrade, -5.7 feet/second out-of-plane, and 9.6 feet/second radial up. Both mid-course correction burns were less than 2 feet/second. Braking gates and

43

line-of-sight control was performed behind the Moon and out of communications with Earth. When the vehicles resumed communications with Mission Control they were station-keeping. The LM established station-keeping at about 20 feet. The LM was placed in an AGS attitude hold and the CSM was active for docking. [27]

The maximum VHF ranging measurement was 320 nm, well above the maximum specified operating range of 200 nm. At ranges from 3,000 to 300 feet VHF ranging and LM radar range agreed to within ~100 feet.[27]

Burn solutions computed by the LM (radar relative navigation) and CSM (sextant and VHF ranging relative navigation) were in agreement throughout the rendezvous. Propellant consumption by both the CSM and LM was below the pre-flight predictions. Maintaining minimum attitude rates and efficient docking execution was a factor behind lower than anticipated CSM propellant consumption. [27]

Apollo 11 (G Mission, July 1969)

The transposition maneuver began as scheduled 20 seconds after CSM separation from the S-IVB to provide at least 70 feet of separation between the spacecraft. Per procedure the CSM pilot used the delta-velocity counter on the Entry Monitoring System (EMS) display to determine how much RCS jet thrusting was required to return to the LM/S-IVB stack. However, the EMS data did not make sense to the CSM pilot. The CSM approach to and docking with the LM/S-IVB stack was nominal. The CSM docking light was not required. The COAS reticle pattern was dim but became more visible once the CSM was very close to the LM. The approach rate at docking was estimated to be 0.1 feet/second. [29, 30]

In lunar orbit after undocking the CSM maintained the undocking rate until the range between the vehicles was ~40 feet. At that point the CSM pilot nulled the relative velocity based on visual viewing of the LM. The LM performed a 360 degree yaw attitude maneuver to permit the CSM pilot to visually verify that the LM landing legs had properly deployed. After separation the LM tracked the CSM with the rendezvous radar, and the CSM tracked the LM with VHF ranging. Both sensors were in agreement.[30] The CSM pilot updated the state vector using both VHF ranging and sextant measurements. This was not part of the crew procedure between the DOI burn and the Powered Descent Initiation (PDI) burns but it gave the CSM pilot confidence that the relative navigation hardware and software was working before the rendezvous the next day. [29]

After the lunar exploration activities were complete, lunar orbit insertion and a LM IMU alignment was performed, followed by initiation of rendezvous radar tracking. The LM PGNCS, LM AGS, Mission Control, and CSM PGNCS CSI burns solutions agreed to within 0.2 feet/second in each axis. Due to the higher than expected elliptical nature of the CSM orbit (63.2 x 56.8

nm) the range rate was outside the envelope of range rate values needed for the CSI and CDH back-up chart burn solutions. Therefore backup chart burn solutions for CSI and CDH were not available. The LM insertion and overall rendezvous profile were nominal, including a nominal 15 nm delta-height (Figure 4.19). [29]

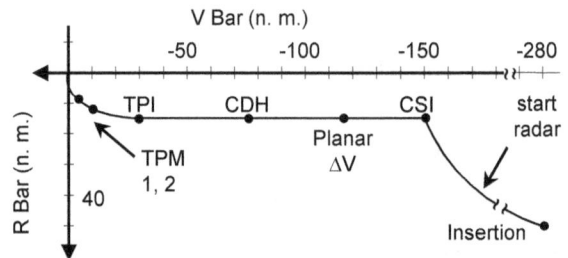

Figure 4.19 Coelliptic Flight Plan, Apollo missions 11, 12, and planned for 13 (1969-1970).

The post-TPI intercept trajectory was nominal and line-of-sight rates during braking were low and easily controlled. The commander used the line-of-sight rate needles driven by the radar for piloting cues. After the LM began station-keeping with the CSM it maneuvered to the docking attitude. The LM PGNS IMU went into gimbal lock as the crew maneuvered to the docking attitude while trying to avoid placing the sun in the fields of view of the LM windows. As a result the AGS was used to maintain LM attitude during docking. The CSM active docking was successful. [29, 30]

Apollo 12 (H-1 Mission, November 1969)

Transposition and docking after the Trans Lunar Injection burn were nominal. The CSM docking light was not used. Due to a problem with the Entry Monitoring System (EMS) velocity counter an accurate measurement of delta-velocity executed during the maneuver could not be made. Propellant consumption and maximum separation distance were higher than expected. The crew recommended that RCS thrusting be based on time rather than EMS delta-velocity measurement to simplify the procedures. [31, 32]

After separation in lunar orbit the CSM pilot observed the LM DOI burn through the sextant. Range at that point was ~3.5 nm. After DOI the CSM pilot performed sextant and VHF tracking of the LM. [31]

The CSM pilot was not able to track the LM ascent stage during powered ascent. Near the end of lunar ascent a late switch throw resulted in a LM Ascent Propulsion System over-burn at insertion of ~30 feet/second. This was quickly corrected with a RCS trim burn.

After insertion the LM crew performed an in-flight alignment and radar tracking of the CSM was begun.

Radar lock-on occurred at a range of about 235 nm.[31] No out-of-plane corrections were required before the TPI burn. All CSM and LM burn solutions during the rendezvous were in agreement. Once the LM entered orbital darkness the CSM pilot observed that the LM tracking light was not working. It apparently failed at some point after the CSI burn. No further CSM sextant data was obtained and only VHF ranging measurements were processed by the CSM. This led to expected inaccuracies in the CSM computed solutions for the two mid-course correction burns.[32] The LM pilot commented that the AGS and chart solutions had a heavy work load with many opportunities for procedural error. A more automated form of backup systems was needed. [31]

During the terminal phase line-of-sight rate corrections were not required until a range of about 1,000 feet, and the required corrections were small. The docking was nominal and the closure rate was estimated to be ~0.2 or ~0.3 feet/second. [32]

Apollo 13 (H-2 Mission, April 1970)

After TLI the S-IVB maneuvered the stack to the transposition and docking attitude. Once this attitude was achieved the S-IVB maintained an inertial attitude hold. Maximum spacecraft separation during the CSM transposition and docking maneuver was about 80 feet, with a CSM pitch rate during the maneuver of about 1.5 degrees/second. The CM pilot reported that sunlight on the LM docking target washed out the COAS. The COAS was therefore set at maximum brightness, making it difficult for the CM pilot to see the LM docking target. Just before docking the CSM shadowed the LM docking target and target visibility was improved. Closing rate at docking was about 0.2 feet/second. The docking and spring ejection of the LM/CSM from the S-IVB was nominal. [6, 33, 34]

No rendezvous was performed since the lunar landing was canceled. Had the oxygen tank combustion and rupture not occurred, and the lunar landing had been performed, Apollo 13 would have flown the same coelliptic flight profile flown on Apollo 11 and Apollo 12.

Just before re-entry, the Service Module was jettisoned from the Command Module before the Lunar Module was jettisoned. The commander (flying from the LM) pitched the CM/LM stack so that the departing SM could be photographed for the accident investigation.[6, 33]

Apollo 14 (H-3 Mission, February 1971)

Lighting was not a problem during transposition and docking, the COAS was always visible, and the docking light was not needed when the LM docking target was in shadow. However, the crew did have difficulty achieving a hard dock. Hard docking was achieved on the fifth attempt. The crew verified that all 12 docking hardware latches were locked. [5, 35, 36]

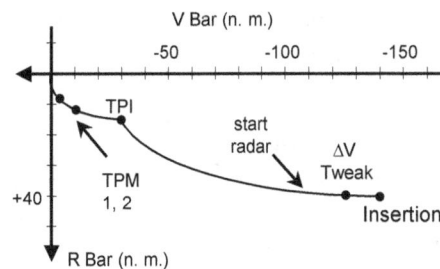

Figure 4.20 Short rendezvous profile, Apollo missions 14, 15, 16, and 17 (1971-1972)

Apollo 14 was the first mission to fly the short rendezvous profile (Figure 4.20). The concentric flight plan flown on earlier missions was a back-up procedure in case of performance or system problems. Ascent was to insert the LM into a 51x9 nm orbit with TPI occurring 38 minutes after orbit insertion. The nominal liftoff time was ~2.5 minutes before the nominal liftoff time for the coelliptic flight profile. The liftoff window duration was 30 seconds to keep the orbit insertion perilune above 8 nm. If required a Mission Control computed tweak burn using the LM RCS could be performed after insertion to correct for off-nominal insertion conditions. The total time from insertion to rendezvous was ~85 minutes. [37]

After the tweak burn a manual attitude maneuver to the radar tracking attitude was performed. Radar measurements were successfully incorporated into the LM PGNS and AGS. However, due to a VHF ranging problem only sextant measurements were processed by the CSM PGNCS Kalman filter. TPI was successfully executed with the LM Ascent Propulsion System. [35]

After TPI the CSM pilot obtained VHF ranging data. Both VHF range and sextant angle measurements were processed. The two mid-course correction burns and the braking gates were nominal. After the second mid-course correction burn the AGS failed and was not recovered. After the LM established station-keeping the CSM performed a 360 degree pitch maneuver for LM crew inspection of the CSM. The LM then maneuvered to place the LM docking target in the field of view of the CSM pilot. The CSM active docking was successful with no difficulties encountered. [35]

Apollo 15 (J-1 Mission, July-August 1971)

After TLI, the transposition and docking maneuver was nominal. The LM was illuminated by the Sun, there were no problems with shadows and the CSM docking light was not used. The closing rate before docking was ~0.1 feet/second. At contact there was no indication of probe capture latch engagement. The crew applied one to two seconds of forward thrusting and the latch capture was indicated. After docking it was noted that one latch was not locked onto the docking ring. The crew manually re-cocked and latched it. [38]

In lunar orbit undocking was performed 25 minutes late since the crew had to re-plug a loose docking probe umbilical plug in the docking tunnel.

Before rendezvous all Scientific Instrumentation Module (SIM) bay experiments were deactivated. This included retracting booms and closing covers of cameras and experiments. CSM VHF ranging locked on at a range of about 136 nm (closer than normally seen in the simulators), during powered ascent and before orbit insertion. [39] The CSM pilot voiced the VHF range to the LM. The LM crew verified that the measured range agreed with that computed by the PGNS and the AGS. Mission Control told the LM crew that the post-insertion tweak burn would not be performed. Mission Control also informed the LM crew that due to the CSM orbit TPI would be off nominal and the final approach to the CSM would be near horizontal. [40]

Apollo 15 was the first flight of the CSM MINKEY (minimum keystroke) rendezvous program and it worked as advertised, cycling through the burns and tracking periods. [13, 40] The CSM pilot was not able to sight the LM through the sextant until after the LM passed into orbital darkness. The CSM tracking lights were not visible to the LM crew until about 40 minutes after sunset at a range of 18 nm. CSM burn solutions were compared to the LM solutions and were within limits. The LM solutions were used for all burns. [38]

After the second mid-course correction burn the CSM was maneuvered to the COAS tracking attitude. As the LM approached the 1500 foot braking gate the radar line-of-sight rate pointers were not providing data to the commander for line-of-sight control. The commander controlled the line-of-sight rates by keeping the CSM centered in the COAS. The CSM pilot verified line-of-sight rates by observing the LM through the CSM COAS. [39]

The LM began station-keeping with the CSM at a range of about 100 feet. As a result of the COAS back-up line-of-sight control procedure the LM was out-of-plane by about 20 degrees. The CSM pilot maneuvered to an attitude that permitted the LM crew to photograph the SIM bay. The CSM then maneuvered back to the docking attitude. The CSM performed the docking maneuver at a closing rate of ~0.1 feet/second. After docking the SIM bay experiments were re-activated. [40]

Apollo 16 (J-2 Mission, April 1972)

After Trans Lunar Injection the transposition and docking maneuver was performed. The CSM pilot reported that at transposition maneuver completion the COAS cross hairs were almost exactly aligned with the docking target on the LM. After contact additional RCS firings were required to center the vehicle. Color

television was transmitted to Earth during transposition and docking.

In lunar orbit, just after undocking, on the 12th revolution of the Moon, the CSM was to perform a 1 foot/second radial down SM RCS burn for separation. This burn was supposed to place the CSM on a mini-football that would return the CSM to the vicinity of the LM after one lunar orbit if neither vehicle performed any subsequent burns (Figure 4.21). About 1.5 hours after undocking, near the end of revolution 12, the CSM would perform a circularization burn. CSM circularization had to be performed before the LM performed PDI.

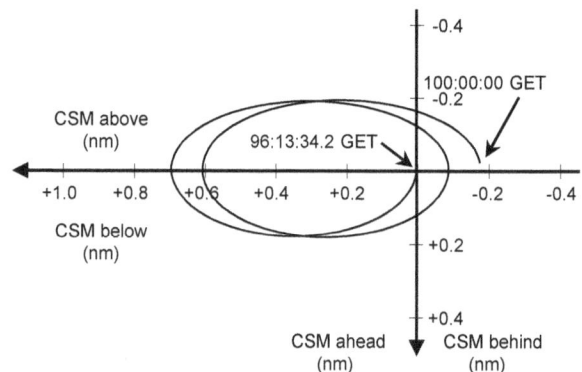

Figure 4.21 Planned CSM relative motion, from Apollo 16 post flight report by Allan DuPont.[45]

CSM and LM undocking occurred as scheduled on the 12th revolution. Some LM inspection photographs were taken but the LM did not do any attitude maneuvers to facilitate photography by the CSM pilot. At acquisition of signal on revolution 13 the CSM pilot reported to Mission Control that the circularization burn had not been executed as planned. During the pre-ignition checklist a Service Propulsion System (SPS) secondary yaw gimbal check indicated a problem with the yaw gimbal drive servo. A flight rule stated that four servo loops had to be operative for the circularization burn to be executed.[41] If the servo problem could not be resolved the lunar landing attempt would be canceled, the spacecraft would have to dock, and the crew would perform a Trans Earth Injection (TEI) burn using the LM descent engine to return to Earth. About 10 hours (5 revolutions) were available for the issue to be resolved and still perform a lunar landing. After 10 hours the LM ground-track would pass far enough away from the landing site that the LM descent stage could not fly the LM to the site. [42]

The CSM pilot maintained visual contact with the LM. At one point a maneuver to an optimum communications attitude was performed to permit Mission Control to monitor two gimbal tests in real time. Mission Control decided that the two spacecraft should re-establish station-keeping while the servo loop issue was worked by ground personnel. This involved a CSM re-rendezvous with the LM. The CSM was not performing relative navigation of the LM using the

Kalman filter, per procedure. The recommended re-rendezvous procedure was to perform a CSM active brute-force rendezvous. A delta-velocity (5 feet/second) from an empirical formula would be executed by the CSM at the predicted point of closest approach to the LM on revolution 14, a range of about 2,000 feet (Figure 4.21, GET 100 hours). This procedure had been tested in a pre-flight simulation. [41, 42, 43, 44]

Both spacecraft were behind the Moon and out of contact with Mission Control at the predicted time of closest approach. The CSM pilot began VHF ranging to obtain range data to the LM. The LM performed an attitude maneuver, based on instructions from the CSM pilot, so that the LM radar would be pointed at the CSM to facilitate radar acquisition and tracking.

The CSM pilot began to execute the 5 foot/second burn to establish a closing rate, which was a retrograde burn. However, he noted that the CSM continued to have an increasing opening rate with respect to the LM. The CSM pilot then stopped the burn after 3 feet/second had been executed. He then took out the 3 feet/second executed thus far due to a low perilune concern. The retrograde attempt to establish a closing rate began near apolune, which would lower the perilune altitude. A low perilune altitude could represent a safety hazard. The CSM pilot asked the LM if they had a burn chart for re-rendezvous. The LM crew did not.[42] Figure 4.22 is a post-flight simulation plot showing relative motion resulting from this burn if it had been executed to completion.

After acquisition of signal on revolution 14 it became apparent to Mission Control that the CSM pilot had not initiated the re-rendezvous and that the post

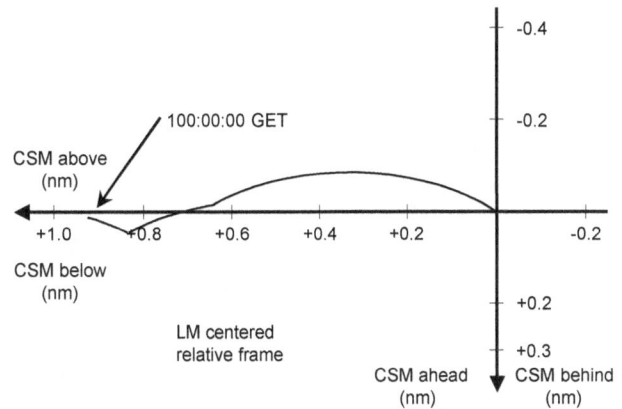

Figure 4.22 Post-flight simulation of the brute force rendezvous using the original recommended 5 foot/second rendezvous start burn at GET 100 hours, created by Allan DuPont.[45] This plot does not reflect actual relative motion during the mission.

separation relative motion was not as predicted (i.e. the CSM was not on a mini-football relative trajectory that would return it to the vicinity of the LM). The CSM was ahead of the LM rather than behind it and phasing away (Figure 4.23). The on-board CSM state indicated a perilune of 6.3 nm but Mission Control ground tracking indicated a perilune of 9 nm. This led the CSM pilot to voice a concern about the quality of the CSM on-board state vector. The CSM pilot requested that Mission Control perform short arc tracking to verify the integrity of the CSM state vector.[44]

Just before perilune the brute force rendezvous was re-initiated (Figure 4.24). The LM crew provided the CSM pilot with piloting cues based on LM radar range,

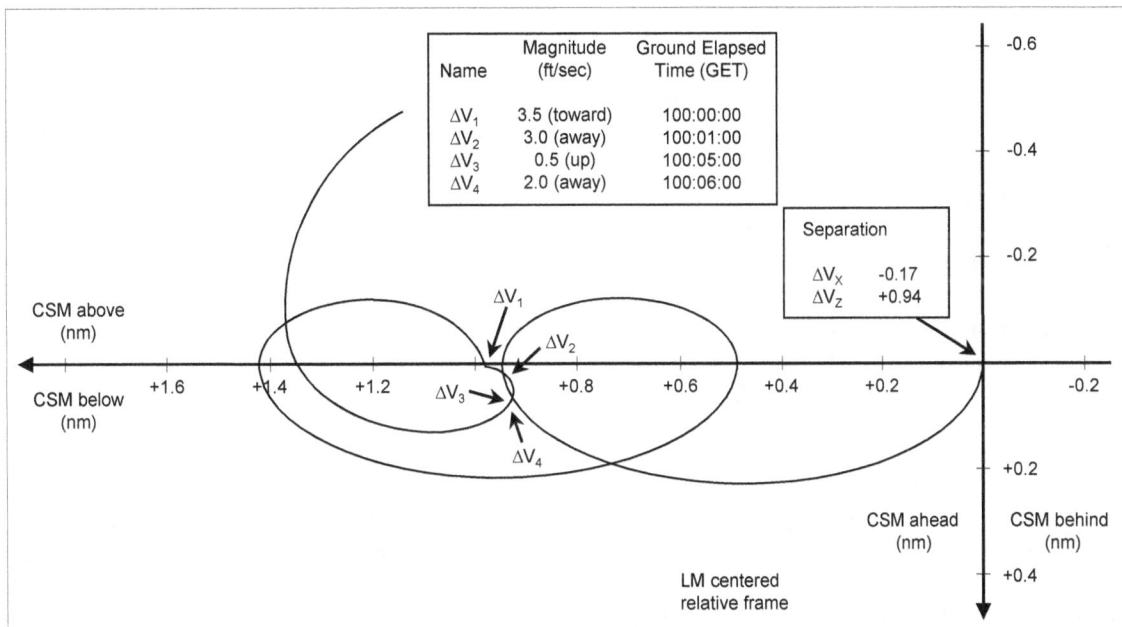

Figure 4.23 Post-flight reconstruction of Apollo 16 CSM relative motion with respect to the LM created by Allan DuPont.[45] The first rendezvous burn began at a GET of 100 hours. The brute force rendezvous is not shown.

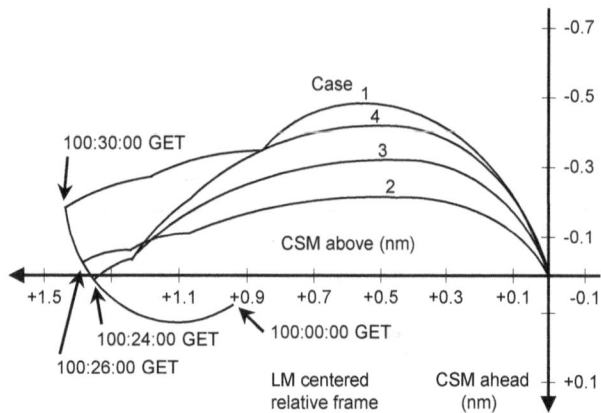

Figure 4.24 Post-flight simulations of the brute force rendezvous for various start times, created by Allan DuPont.[45] These are not exact reconstructions of the actual relative motion.

range rate, and inertial line-of-sight rate measurements. The cues were in terms of thrusting north or south (out-of-plane), or towards the Moon or away from it (in-plane). At times translating the line-of-sight rate measurement needle data into CSM pilot thrusting cues was difficult. The CSM pilot kept the inertial line-of-sight rate measurements nulled and maintained a closing rate. [42]

The CSM VHF ranging and LM radar range were in close agreement throughout the rendezvous. The CSM pilot could have computed range rate using VHF ranging data and a stopwatch, but it was simpler to rely on the LM range rate voiced by the LM crew. [43]

At a range of about 4,000 feet the LM tracking light was turned on due to orbital night. LM tracking light was visible to the CSM pilot. At a range of about 2,400 feet Mission Control asked that the LM radar and tracking light be turned off as soon as was feasible to conserve power. At a range of about 2,000 feet the CSM pilot could observe the LM due to Earth-shine. Later the LM crew asked the CSM pilot to turn on the CSM lights. The CSM spotlight was turned on at 500 feet. By 300 feet Earth-shine had disappeared and the LM was visible using the CSM spotlight. The LM radar and tracking light were turned off to conserve power. [42]

The CSM COAS was useful for verifying line-of-sight control, particularly inside the 500 foot range where it was more useful than radar measurements of inertial line-of-sight rate. The CSM pilot had to mentally calibrate the COAS to discern between line-of-sight rate and attitude dead-banding. Range rate was difficult to assess visually, even at close range. [43]

Eventually the crew was given a go to execute PDI on revolution 16 and a second separation burn was performed by the CSM. The incident resulted in a 5.75 hour delay (approximately three lunar revolutions) in execution of the CSM circularization burn and LM descent initiation. The delay also required Mission Control to re-plan all subsequent mission activities in the

crew timeline. The circularization burn was successfully performed using the primary servo loop. The secondary loop was not needed. [41]

Post flight analysis later indicated that the first separation burn had a small retrograde component, rather than being purely radial down.[45] This resulted in the CSM separating below and ahead of the LM, rather than the CSM being on a mini-football (Figure 4.22).

As a result of this incident the Apollo 17 crew carried a mini-football re-rendezvous chart. This chart had been developed earlier in the program and was carried on the initial lunar missions. It provided the LM crew with required range rate and inertial line-of-sight rate data to be achieved by the CSM pilot. This would facilitate a CSM active short range rendezvous without resorting to a CSM bailout and execution of a CSI-CDH rendezvous profile. These required values would be used by the LM crew along with radar measurements of range rate and inertial line-of-sight rate to determine piloting cues to be voiced to the CSM pilot. Using these piloting cues the CSM pilot could initiate a rendezvous from any point on a mini-football for various transfer times.

After lunar exploration activities were completed the ascent stage of the LM was inserted into a 40.2 nm x 7.9 nm orbit at a range of about 170 nm from the CSM. The range at insertion was about 33,000 feet too close and a 10 foot/second tweak burn was executed to correct for the dispersion. Two or three minutes after insertion the crew performed a pitch-up maneuver and radar tracking of the CSM was begun. The crew also visually spotted the CSM at this time due to sunlight reflected from the CSM. The crew maintained visual sight of the CSM until orbital sunset. Radar tracking was maintained until the maneuver to the TPI burn attitude. The CSM pilot spotted the LM through the sextant at ~100 nm, and observed the LM flashing light in the scanning telescope at ~70 nm. [41, 43]

The TPI burn was 78 feet/second. Once station-keeping was established the LM performed a 360 degree yaw maneuver for photography and LM inspection by the CSM pilot. Then the CSM pilot performed a pitch-over to permit the LM crew to photograph bubbles on the surface of the CSM. Docking was successful. The LM crew left their helmets and gloves on until docking due to the amount of lunar dust in the LM. [43]

Apollo 17 (J-3 Mission, December 1972)

The crew reported that the S-IVB was steady as a rock during transposition and docking. Closing rate at docking was ~0.1 feet/second. Once docking was achieved a talkback barber pole indicated a possible docking ring latch malfunction. During hard docking the COAS indicated a one degree right yaw error with respect to the LM docking target. After LM pressurization, hatch removal, and inspection it was discovered that the handles for latches 7, 9, and 10 were not locked. The crew locked the handle for latch 10 by pushing on it. Latches 7 and 9 were locked and manually fired to lock the handles. After hatch replacement the CSM/LM stack separated from the S-IVB about 45 minutes after CSM docking with the LM. During a subsequent LM activity it was discovered that docking latch 4 was not properly latched. Troubleshooting was delayed until later in the mission. Later the crew successfully cocked the latch. [46, 47]

In lunar orbit undocking and separation were nominal. The LM crew visually tracked the CSM as it maneuvered for landmark tracking over the landing site.[46]

After the lunar surface activities were complete the LM ascent stage was inserted into lunar orbit. CSM VHF tracking of the LM began soon after insertion at a range of 155 nm. A 10 foot/second tweak burn was performed soon after insertion. This placed the LM on a nominal trajectory for TPI. VHF ranging broke lock during the tweak burn but was re-established after the burn. The LM crew spotted the CSM during orbital daylight at a range of about 110 nm.[46] After sunset the CSM was not visually observed again until the tracking lights were spotted at about 40 nm. The CSM docking light was not discernable until well within 40 nm. The LM crew stated that the rendezvous was nominal. The TPI burn was 53.8 feet/second.

The CSM pilot did not detect the LM flashing light using the scanning telescope or sextant until orbital sunset, when the LM was at a range of 80 nm. The sun was not in the field of view during the daylight attempts to acquire and track the LM. The CSM TPI burn solution agreed with the LM solution, but the CSM second Mid-Course Correction solutions did not agree. The CSM pilot mounted a television monitor on a strut beside the commander's couch and kept the LM in the television field of view during the terminal phase of the rendezvous.

The LM did a fly-around of the CSM for inspection, particularly of the SIM bay. The LM maneuvered to the docking attitude and transferred station-keeping responsibility to the CSM. After the CSM maneuvered to the docking attitude the LM performed pitch and yaw maneuvers, and then stood by for the CSM active docking. The CSM pilot noted that the LM moved during docking more than the S-IVB/LM stack. The first docking attempt, at a rate of probably less than 0.1 feet/second, likely failed due to the slow closing rate. The CSM backed away about 3 feet and re-initiated the approach. Docking was successfully accomplished.[46-47]

Summary

Nominal Apollo rendezvous procedures from TPI through docking relied heavily on piloting and trajectory design techniques developed and flight proven during the Gemini Program. However, Apollo rendezvous was far more complex. Both the LM and CSM were capable of being either the active or passive vehicle throughout the rendezvous. Between 1963 and 1969 nominal rendezvous profile development went through ten iterations, finally resulting in the CSI-CDH-TPI profile flown on Apollo 11 (July 1969). By late 1969 the confidence in LM and CSM systems performance led MPAD personnel to design a short rendezvous profile to shorten the crew day. The short profile was first flown on Apollo 14 (February 1971). The CSI-CDH profile was retained as a contingency profile to be flown in the event of off-nominal systems performance. The complexity of the CSM pilot task during rendezvous resulted in automation of many rendezvous procedures starting with Apollo 15 (July-August 1971). However, the most significant challenge of Apollo rendezvous was the development of contingency rendezvous flight techniques to be flown in the event of a LM abort before or after Powered Descent Initiation.

References

1. Hansen, J. R., *Enchanted Rendezvous*, NASA Monographs in Aerospace History Series #4, NASA, Washington, D. C., 1995.

2. Young, K. D., and J. D. Alexander, "Apollo Lunar Rendezvous," *AIAA Journal of Spacecraft*, Volume 7, Number 9, AIAA, Washington, D.C., September 1970, pp. 1083 - 1086.

3. Bennett, F. V., *Apollo Experience Report: Mission Planning For Lunar Module Descent and Ascent*, NASA-TN-D-6846, NASA Manned Spacecraft Center, 1972.

4. Saponaro, J. A., and S. L. Copps, "Operations and Functions of the Apollo Guidance Computer During Rendezvous," *Proceedings of the Third International Conference on Automatic Control in Space*, Toulouse, France, March 2-6, 1970, International Federation of Automatic Control.

5. Langley, R. D, *Apollo Experience Report: The Docking System*, NASA-TN-D-6854, NASA Manned Spacecraft Center, June 1972.

6. Goodman, John L. "Apollo 13 Guidance, Navigation, and Control Challenges," *AIAA Space 2009 Conference & Exposition*, September 14-17, 2009, Pasadena, CA.

7. Phillips, R., and P. Kachmar, "Apollo Rendezvous Navigation," *Guidance and Control 2004, Advances in the Astronautical Sciences*, Volume 118, Univelt, Inc, San Diego, CA, 2004.

8. Chen, P. P., "Real-Time Kalman Filtering of Apollo LM/AGS Rendezvous Radar Data," *Proceedings of the AIAA Guidance, Control and Flight Mechanics Conference*, AIAA, Reston, VA, 1970.

9. Rozas, P., and A. R. Cunningham, *Apollo Experience Report: Lunar Module Landing Radar And Rendezvous Radar*, NASA-TN-D-6849, NASA Manned Spacecraft Center, 1972.

10. Panter, W. C., and P. W. Shores, *Apollo Experience Report: Very High Frequency Ranging System*, NASA-TN-D-6851, NASA Manned Spacecraft Center, 1972.

11. Johnson, Madeline S., and Donald R. Giller, *MIT's Role in Project Apollo, Volume 5: The Software Effort*, R-700, Charles Stark Draper Laboratory, Massachusetts Institute of Technology, March 1971.

12. Regelbrugge, R. R., Apollo Experience Report: Apollo Spacecraft and Ground Software Development for Rendezvous, MSC Internal Note No. 70-FM-153, MSC-02676, NASA Manned Spacecraft Center, September 28, 1970.

13. Copps, S. L., and J. H. Goode, "Operations and Functions of the MINKEY Rendezvous Computer Program in the Apollo Command Module Computer," *1971 National Space Meeting Proceedings - Space Shuttle/Space Station/Nuclear Shuttle Navigation*, Institute of Navigation, Fairfax, VA, 1971.

14. McAdams, R. E., and C. J. Gott, Williamson, M. L., *Apollo Experience Report: Spacecraft Relative Motion and Recontact Analyses*, NASA-TN-D-7920, NASA Johnson Space Center, 1975.

15. *The Apollo 11 Adventure*, MSC Internal Note No. 70-FM-20, MSC-01562, Mission Planning and Analysis Division, NASA/MSC, February 5, 1970.

16. Alexander, J. D., and R. W. Becker, *Apollo Experience Report: Evolution of the Rendezvous-Maneuver Plan for the Lunar-Landing Missions*, NASA-TN-D-7388, NASA Johnson Space Center, 1973.

17. *Apollo Program Summary Report*, NASA JSC-09423, NASA Johnson Space Center, April 1975.

18. Shayler, D. J., *Apollo - The Lost and Forgotten Missions*, Springer, Berlin, Germany, 2002.

19. Callihan, J. C., and P. M. Dugge, "Rendezvous Navigation for the Apollo VII Mission," *5th American Institute of Aeronautics and Astronautics Annual Meeting and Technical Display*, AIAA, Reston, VA, 1968.

20. Apollo VII Technical Debriefing, NASA Manned Spacecraft Center, October 27, 1968, in *Apollo 7 - The NASA Mission Reports*, CG Publishing Inc., Burlington, Ontario, Canada, 2000.

21. *Apollo 7 Mission Report*, MSC-PA-R-68-15, NASA Manned Spacecraft Center, Houston, TX, December 1968.

22. Apollo 8 Technical Debriefing, NASA Manned Spacecraft Center, January 2, 1969, in *Apollo 8 - The NASA Mission Reports*, CG Publishing Inc., Burlington, Ontario, Canada, 1999.

23. *Apollo 8 Mission Report*, MSC-PA-R-69-1, NASA Manned Spacecraft Center, Houston, TX, February 1969.

24. *Apollo 9 - The NASA Mission Reports*, CG Publishing Inc., Burlington, Ontario, Canada, 1999.

25. *Apollo 9 Mission Report*, MSC-PA-R-69-2, NASA Manned Spacecraft Center, Houston, TX, May 1969.

26. Apollo 10 Press Kit, Release No. 69-68, NASA, May 7, 1969, in *Apollo 10 - The NASA Mission Reports*, CG Publishing Inc., Burlington, Ontario, Canada, 2000.

27. *Apollo 10 Mission Report*, MSC-00126, NASA Manned Spacecraft Center, Houston, TX, August 1969.

28. Apollo 10 Technical Crew Debriefing, NASA Manned Spacecraft Center, June 2, 1969, in *Apollo 10 - The NASA Mission Reports*, CG Publishing Inc., Burlington, Ontario, Canada, 2000.

29. Apollo 11 Technical Crew Debriefing, NASA Manned Spacecraft Center, July 31, 1969, in *Apollo 11 - The NASA Mission Reports*, CG Publishing Inc., Burlington, Ontario, Canada, 1999.

30. *Apollo 11 Mission Report*, MSC-00171, NASA Manned Spacecraft Center, Houston, TX, November 1969.

31. Apollo 12 Technical Crew Debriefing, NASA Manned Spacecraft Center, December 1, 1969, in *Apollo 12 - The NASA Mission Reports*, CG Publishing Inc., Burlington, Ontario, Canada, 1999.

32. *Apollo 12 Mission Report*, MSC-01855, NASA Manned Spacecraft Center, Houston, TX, March 1970.

33. Apollo 13 Technical Crew Debriefing, NASA Manned Spacecraft Center, April 24, 1970, in *Apollo 13 - The NASA Mission Reports,* CG Publishing Inc., Burlington, Ontario, Canada, 2000.

34. *Apollo 13 Mission Report*, MSC-02680, NASA Manned Spacecraft Center, Houston, TX, September 1970.

35. *Apollo 14 Mission Report*, MSC-04112, NASA Manned Spacecraft Center, May 1971.

36. Shepard, A. B., Jr., "Apollo 14 Mission Report," *15th Society of Experimental Test Pilots Symposium, Beverly Hills*, Calif., Sept. 16-18, 1971. Society of Experimental Test Pilots, Technical Review, vol. 10, no. 4, 1972, pp. 109-114.

37. Apollo 14 Technical Crew Debriefing, NASA Manned Spacecraft Center, February 17, 1971, in *Apollo 14 - The NASA Mission Reports*, CG Publishing Inc., Burlington, Ontario, Canada, 2000.

38. Scott, D. R., and R. F. Gordon, "Apollo 15 Mission Report," *15th Society of Experimental Test Pilots Symposium*, Beverly Hills, CA., Sept. 16-18, 1971. Society of Experimental Test Pilots, Technical Review, vol. 10, no. 4, 1972, pp. 115-129.

39. Apollo 15 Technical Crew Debriefing, MSC-04561, NASA Manned Spacecraft Center, August 14, 1971, in *Apollo 15 - The NASA Mission Reports*, CG Publishing Inc., Burlington, Ontario, Canada, 2001.

40. *Apollo 15 Mission Report*, MSC-05161, NASA Manned Spacecraft Center, December 1971.

41. *Apollo 16 Mission Report*, MSC-07230, NASA Manned Spacecraft Center, August 1972.

42. *Apollo 16 Spacecraft Commentary*, April 16 - April 27, 1972, NASA Manned Spacecraft Center.

43. Apollo 16 Technical Crew Debriefing, MSC-06805, NASA Manned Spacecraft Center, May 5, 1972, in *Apollo 16 - The NASA Mission Reports*, CG Publishing Inc., Burlington, Ontario, Canada, 2002.

44. Mattingly, T. K., "Results of Apollo 16," *16th Society of Experimental Test Pilots Symposium, Beverly Hills*, Calif., Sept. 28-30, 1972. Society of Experimental Test Pilots, Technical Review, vol. 11, no. 2, 1973, pp. 159-170.

45. DuPont, Al, *Post-Flight Report on the Brute-Force Rendezvous Performed During the Apollo 16 Mission*, FM32 (72-189), Mission Analysis Branch, Mission Planning and Analysis Division, NASA/MSC, October 4, 1972. See JSC-35056, Volume 1, in the A Note on Sources chapter in this document.

46. *Apollo 17 Mission Report*, JSC-07904, NASA Johnson Space Center, March 1973.

47. Apollo 17 Technical Crew Debriefing, MSC-07631, NASA Manned Spacecraft Center, January 4, 1973, in *Apollo 17 - The NASA Mission Reports*, CG Publishing Inc., Burlington, Ontario, Canada, 2002.

This page intentionally left blank.

CHAPTER 5 - SKYLAB

Programmatic and Mission Objectives

Skylab was the United States first space station (Figure 5.1).[1-3] Three missions, Skylab (SL) 2, 3, and 4, were flown to perform medical evaluation of long duration human flight, solar astronomy, and Earth surveys. The Apollo CSM was modified to serve as a crew and limited cargo transport vehicle to and from the Skylab station. Like most Gemini missions rendezvous and docking was completed on flight day one. This requirement was driven by a need to place flight day one medical samples in the Skylab freezer within 24 hours after they were obtained. Orbital inclination was 50 degrees and the nominal Skylab workshop insertion orbit was 234 nm circular.

All three Skylab missions successfully performed rendezvous and docking. An additional objective was Skylab 2 CSM station-keeping on flight day one for a stand-up EVA in an attempt to free a partially deployed solar array. This foreshadowed later Space Shuttle missions that combined proximity operations and EVA for satellite repair.

Figure 5.1 Skylab viewed by Skylab 4 crew during the final fly-around before returning to Earth (Feb. 1974).

Early Concepts – The Apollo Applications Program

The initial design of America's first space station involved the on-orbit conversion by astronauts of a spent S-IVB stage, launched by a Saturn IB, into an orbital workshop. This was known as the wet workshop.[4-6] The Apollo Telescope Mount (ATM, a scientific package mounted on a LM ascent stage) would be launched on an unmanned Saturn IB. A subsequently launched Apollo CSM was to rendezvous with the ATM, dock with it, and bring it to the workshop. Two crewman would transfer to the ATM and dock to a workshop radial port, while the third crewman docked the CSM to an axial port. There were Saturn IB/CSM payload capability concerns due to propellant needed to support the double rendezvous, and there were safety concerns with splitting the crew between the ATM and CSM during docking.

Studies were performed in 1968 of an automated or remotely controlled (from the workshop) docking of the ATM, rather than using a crewed Apollo CSM. This would eliminate payload and safety concerns. The workshop crew was to monitor (both via transmitted data and looking out a window) the terminal phase of either an automated or remote controlled rendezvous. Lighting conditions required for crew visual monitoring dictated that the ATM (chaser) approach the workshop (target) from ahead and above, rather than from behind and below as was done when the chaser spacecraft was crewed. The ATM would phase from below and behind, pass the workshop, then an NH burn would have been performed to arrive at a co-elliptic (CDH) start point above and ahead of the workshop. The TPI and mid-course correction burns followed.

On July 18, 1969, the wet workshop was eliminated in favor of the dry S-IVB workshop assembled on Earth, launched by the first two stages of a Saturn V. The dry configuration did not require on-orbit conversion. The

Table 5.1 Skylab

Flight	Year	Chaser	Target	Profile	Comments
Skylab 2 (SL-2)	1973	CSM	1. Skylab 2. Skylab 3. Skylab	1. Coelliptic M=5, FA 2. FA, Station-keeping 3. FA	1. Inspection, then soft docking to prepare for EVA (2.). 2. Stand-up EVA attempt failed to free solar panel. Redocking difficulty, IFM required CM depressurization, followed by successful hard dock. 3. Before final separation and deorbit.
Skylab 3 (SL-3)	1973	CSM	1. S-IVB 2. Skylab	1. Station-keeping 2. Coelliptic M=5, FA	2. Plume impingement observed on thermal parasol during pre-docking fly-around. Fly-around after undocking canceled due to RCS Quad leaks (some RCS jets inhibited).
Skylab 4 (SL-4)	1973/ 1974	CSM	1. Skylab 2. Skylab	1. Coelliptic M=5 2. FA	1. No fly-around before docking. Hard dock achieved on the third attempt. 2. Final inspection before deorbit.

Coelliptic = Coellptic Flight Profile, **CM** = Command Module, **CSM** = Command/Service Module, **EVA** = Extra-Vehicular Activity, **FA** = Fly Around, **IFM** = In Flight Maintenance, **M** = docking on the Mth revolution, **RCS** = Reaction Control System.

ATM was to be integral with the dry workshop, rather than launched separately and brought to the workshop by the LM. This eliminated the need for complicated rendezvous operations to assemble the cluster. The program was renamed Skylab in February of 1970.[4, 5]

Rendezvous Target Vehicle

Skylab was the only rendezvous target vehicle in the program (Figure 5.1). It carried strobes to facilitate sextant tracking out to a range of 300 nm during orbital night. The strobes were turned off by Mission Control after the second Mid-Course Correction burn, just after orbital sunrise, so that the crew did not have to view the bright strobes during final approach. Eight running lights were placed around Skylab, and were color coded to aid in attitude determination. Four smaller lights were mounted on the ends of antennas. Skylab was also equipped with a VHF transponder to support CSM VHF ranging.

Skylab was the first U.S. spacecraft equipped with two docking ports. The Apollo probe and drogue hardware was used. The axial port was the primary port and was used for all dockings. The radial port served as a back-up. If a rescue mission was required the unusable CSM would be undocked from the axial port and disposed of without a crew onboard, enabling the rescue CSM to use the axial port. If the unusable CSM could not be undocked from the axial port the radial port would have been used by the rescue CSM.

The normal Skylab attitude was solar inertial to facilitate power generation by the solar arrays. During rendezvous a different attitude was maintained to optimize VHF ranging performance. Skylab was maneuvered back to the solar inertial attitude for the nominal axial docking case. Docking to the axial port occurred just after orbital noon, at which time the axial port was facing generally in the direction of flight. For a rescue mission using the backup radial docking port, a different attitude was flown to ensure proper lighting at docking.

The Skylab attitude control system was the primary means of control with a CSM docked to the axial port. Momentum dumping was inhibited during docking. For the case of two CSMs docked or one CSM docked to the radial port, the attitude control system of the radial port CSM was to provide the primary means of attitude control.

Crew Interface, Procedures, and On-Board Computers

Extensive use was made of piloting procedures used in the Apollo Program. The rendezvous procedure book for the nominal mission also served as the rendezvous book for the rescue mission. The Minimum Keystroke (MINKEY) program developed for the CSM and first flown on Apollo 15 was used to reduce crew workload.

Independent and complete rendezvous procedures for the nominal, computer failure, and IMU failure cases were in the rendezvous procedures book. Additional procedures were created for optics failures (obscured visibility, frozen optics), no VHF ranging, a blank Entry Monitoring System Delta-Velocity/Range display, mark button inoperable, and Skylab tracking light failures.

On-Board Navigation

Skylab was equipped with strobes to enhance visual acquisition and a VHF transponder to support VHF ranging. Before TPI only sextant tracking was used to update the CSM state vector using a Kalman filter during both orbital day and night periods. Data from VHF ranging was used for the chart maneuver solutions. Both sextant and VHF ranging measurements were used by the Kalman filter to update the state vector after TPI. A change from the Apollo missions was that range rate could be determined from the VHF tracking and displayed to the crew. However, VHF range rate measurements were not processed by the Kalman filter in the PGNCS. In the rescue case with a disabled CSM at the axial port VHF ranging may not have been acquired until short range. COAS subtended angles were a backup method of range determination.

On-Board Maneuver Targeting

Since the CSM flew a different rendezvous profile than the lunar missions, new targeting algorithms were written, tested, and certified for the Skylark CSM PGNCS software. As with the lunar missions a chart solution was computed for comparison with the PGNCS and ground solutions. A chart solution was only used if the onboard computer and Mission Control burn solutions were not available.

Skylab Rescue (SL-R)

In the event of loss of CSM return to Earth capability, or the crew could not enter the CSM from Skylab, astronauts Brand and Lind were to fly a rescue mission using the next mission's Saturn IB/CSM. An additional CSM and Saturn IB launch vehicle was available to rescue the Skylab 4 crew if required.[3]

Response time varied from 10 to 45.5 days, depending on where the next vehicle was in the launch flow.[4] Trajectory planning was to be the same as the nominal Skylab mission, with a duration of five days. Before rendezvous, a spring loaded device would be used by the Skylab crew to separate the malfunctioning CSM from the station. In the event the CSM could not be separated, the rescue CSM would use the radial docking port. In the event of a communications problem during a rescue mission, hand, spotlight, and Skylab running light signals were defined to allow the Skylab crew to signal the rescue CSM crew.

The rescue CSM would be modified before launch to accommodate five crewmembers. Two would be launched and perform rendezvous and docking. No more than 40 hours after docking the rescue CSM would return to Earth with the three Skylab crew and two rescue CSM crew members.[5]

The Skylab 4 CSM and Saturn IB launch vehicle was prepared for a rescue of the Skylab 3 crew due to two malfunctioning SM RCS quads. However, the Skylab 3 CSM was judged capable of returning the crew to Earth and a rescue mission was not flown. Brand and Lind also trained for a proposed Skylab deorbit mission (SL-5), which was later dropped due to crew safety concerns.[5]

Nominal Rendezvous Profile

The nominal profile (Figure 5.2) supported a docking on the fifth revolution (M=5).[7] This was the earliest docking that could be supported by ground tracking coverage. The first mission, SL-2 (May 1973) had a requirement for a flight day 1 docking so that medical samples taken could be placed in the Skylab freezer within 24 hours of collection. Rendezvous profiles for SL-3 (July 1973), SL-4 (November 1973), and the Skylab rescue mission (not flown) were essentially the same. Nominal launch time for the M=5 rendezvous was the near the midpoint of the 16 minute launch window. M=6, 7, and 8 docking opportunities also existed.

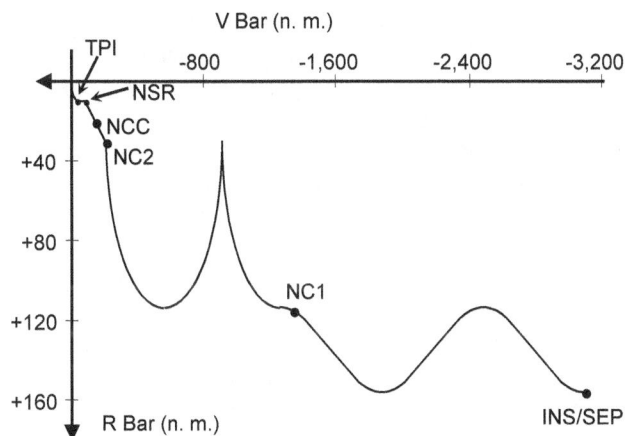

Figure 5.2   Relative motion for nominal Skylab 2 rendezvous on the fifth orbit (May 1973).

The NC1-NC2-NCC-NSR-TPI-TPM1-TPM2 profile was designed to provide the following: 1) Attain a range of 300 nm or less 36 minutes before the NCC burn to facilitate VHF tracking acquisition, 2) Maximize ground tracking coverage, 3) Permit the same basic rendezvous plan to be flown for M=5 through M=8 dockings, and

4) Conserve RCS propellant by using the SPS for all burns before the mid-course burns (TPM1 and TPM2).

The crew timeline from launch through NC1 was the same for the M=5, 6, 7, and 8 docking opportunities. For an increase in M number beyond 5 the number of revolutions between NC1 and NC2 was increased by 1. For example, for M=6 there were 2.5 revolutions between NC1 and NC2, and 3.5 revolutions for M=7.

The CSM separated from the S-IVB with a 3 foot/second posigrade burn with the S-IVB maintaining a local horizontal attitude hold. NC1 was a horizontal phasing maneuver performed 1.5 revolutions after insertion, at apogee, with the CSM about 12 nm behind and slightly above the S-IVB. A 1.5 revolution transfer was to permit sufficient time for ground tracking and Mission Control computation of the NC1 burn solution. The NC1 (phasing adjustment) and NC2 (targeted as a height adjustment) combination was designed to place the CSM within 300 nm of Skylab 36 minutes before NCC to permit VHF and sextant tracking. NC1 was targeted on-board but the ground solution was considered primary since no tracking data was incorporated into the on-board navigation state before NC1.

If an NPC were required NC1 would be targeted (zero out-of-plane velocity) so that NPC would occur at the second nodal crossing after NC1, ~69 minutes after NC1 and before NC2.

Between NC1 and NC2 Mission Control uplinked attitude commands and maneuver times to Skylab. Skylab maneuvered to a Z body axis/local vertical attitude to facilitate CSM acquisition of the Skylab VHF transponder. This attitude also facilitated sextant tracking by the crew. The Skylab maneuver from solar inertial to the tracking attitude was completed by orbital midnight. The station maneuvered back to solar inertial two revolutions later, at orbital midnight, just after the TPI burn.

After NPC (if required) and before NC2 the crew optically tracked Skylab with the sextant and incorporated measurements into the navigation state. For the nominal M=5 docking NC2 occurred 1.5 revolutions after NC1. The on-board solution was primary for NC2. NC2 was a horizontal height correction.

Sextant tracking was resumed after NC2 to update the navigation state in the CSM computer. VHF tracking began before NCC. VHF data was used for NCC, NSR, and TPI back-up chart planar (no out-of-plane component) targeting solutions. It was not processed by the Kalman filter before TPI.

NCC was executed 0.5 revolutions after NC2. It provided phasing, height, and planar control to clean up trajectory dispersions. NCC and NSR were targeted as a pair since they had compensating errors that reduced dispersions at TPI. If the NCC/NSR on-board solution was good the NSR solution was recorded by the crew for later execution.

Sextant and VHF tracking was resumed after NCC. However, the NSR solution computed before the NCC burn was not updated based on this tracking data. NSR was performed 37 minutes after NCC to place the CSM in a coelliptic orbit 10 nm below Skylab. Sextant and VHF tracking resumed after NSR.

NC1, NC2, NCC, and NSR were executed in a heads-down attitude to minimize the change in attitude between the tracking attitude and the burn attitude. No sextant or VHF tracking was performed before NC1 but the NC1 burn attitude was heads-down for consistency. The minimum key-stroke (MINKEY) program that automatically sequenced the computer between navigation, IMU alignment, burn targeting, and burn execution was used.

TPI occurred 2 minutes before orbital midnight at an elevation angle of 27 degrees and a range of ~22 nm. The maximum TIG slip for TPI was +/-10 minutes. After TPI both sextant and VHF ranging data were processed by the Kalman filter to update the CSM computer navigation state. VHF data continued to be used for back-up planar burn targeting solutions.

TPI was followed by two mid-course correction burns, TPM1 and TPM2. After TPM2 the crew established line-of-sight rate control and controlled range rate at procedure specified ranges. Before docking Skylab momentum dumping was inhibited. After docking momentum dumping was enabled once stable attitude control was established.

SL-1 (Workshop Launch, May 1973)

The Skylab workshop was launched on May 14, 1973, by a Saturn V. About a minute after launch, during the period of maximum dynamic pressure, telemetry indicated that a micrometeoroid shield had prematurely deployed. Once on-orbit telemetry indicated that one solar array wing had released for deployment but was not fully extended. Temperature telemetry indicated that both solar array wings were missing, along with an absence of voltage readings from both wings.[1] The solar arrays attached to the Apollo Telescope Mount successfully deployed and were producing power. If the two wings were gone the ability of Skylab to produce power was cut in half. In addition, temperatures in the workshop were rising, suggesting that the micrometeoroid shield was gone.

The first Skylab crew (SL-2) was scheduled for launch on the next day, May 15, 1973. However, the launch was delayed until May 25 to provide NASA personnel at the Johnson and Marshall Space Centers time to develop a sunshade that would be erected by the SL-2 crew. In addition, extensive crew training was conducted for new procedures. The Command Module (CM) was re-stowed with the sunshades and other items to replace those in Skylab that may have been damaged by the high temperatures.[3]

SL-2 (First Visit, May - June 1973)

The first Skylab visit crew was launched on May 25, 1973. Rendezvous on the fifth orbit (M=5) was required since an M=6, 7, or 8 rendezvous would result in limited television coverage of the Skylab inspection and limited passes over the United States after rendezvous.[8]

About seven hours before liftoff two sunshades (also called parasols) arrived at launch pad 39B and were packed into the Command Module.[3] Launch and orbit insertion were nominal. After CSM separation from the S-IVB the CSM performed an automatic pitch around to allow the crew to observe the S-IVB and the launch adapter panels. The panels were fully deployed at 45 degrees. They were not jettisoned as on lunar missions to reduce orbital debris. Four revolutions later excess propellant was vented through the J-2 engine bell to deorbit the S-IVB into the Pacific Ocean north of Hawaii.

Some differences in crew positioning while executing rendezvous procedures were required to accommodate the re-stowage in preparation for the Stand-up Extra Vehicular Activity (SEVA) to be performed after the first docking. The Science Pilot couch was stowed under the Commander's couch and the Science Pilot performed rendezvous burn procedures while sitting on center stowage boxes. The crew remain suited during the rendezvous but with helmets and gloves off.

Sextant tracking was not performed before NC-2 due to the special Skylab attitude for power generation and thermal control. The Skylab tracking lights were not visible to the CSM crew in this attitude. Sextant marks were obtained before NCC. The NCC/NSR matched pair on-board targeting solution agreed with the Mission Control burn solution.

A chart solution was not available for the NCC burn due to the lack of VHF ranging measurements before NCC. A chart solution for NSR was not computed since the NCC/NSR matched pair solution was deemed acceptable. Both NCC and NSR were successfully executed with burn residuals under 0.2 ft/sec in each axis.

VHF ranging was initiated at a range of 117.08 nm. Sextant tracking was performed after NSR. The on-board TPI burn solution was in agreement with the Mission Control TPI solution. The mid-course correction burn solutions were low. Line-of-sight rates during braking were also low.[9]

The crew established station-keeping, then performed a fly-around of the workshop. The crew inspection included verbal descriptions of the damaged workshop, still photography, and 15 minutes of television transmitted to Mission Control.[8] The crew reported that solar array wing one was only partially deployed due to debris, solar array two was missing, and the micrometeoroid shield was gone. During braking and the fly-around the crew noted that Service Module RCS jet firings disturbed the attitude of Skylab, resulting in Skylab workshop RCS firings.[9]

The crew then soft docked, ate a meal, and prepared for the SEVA. A suit integrity check was completed before undocking. After undocking the CSM was piloted close to the end of solar array wing one. The cabin was vented to vacuum, but thrust generated by oxygen venting through the open hatch was apparent to the commander as a translation. However, it did not prevent the commander from maintaining station-keeping.[9] Attempts by the EVA crew member to deploy the solar array wing resulted in unwanted CSM translation. CSM jet firings to maintain station-keeping resulted in CSM RCS jet plume impingement on Skylab, which in turn perturbed the Skylab attitude and caused the Skylab RCS jets to fire. Several attempts to deploy the array failed and the stand-up EVA was called off due to approaching orbital darkness.

The crew was unable to soft dock since the three capture latches would not capture the probe. After executing three backup procedures the crew depressurized the CSM, opened the docking hatch, and removed probe. If hard docking could not be achieved Mission Control provided the crew with a burn pad to set up overnight phasing away from Skylab. The hatch was closed and hard dock was achieved without a soft dock, as had been done on Apollo 14.[3] Eight docking attempts had been performed. The length of the crew day was 22 hours.[8]

At the end of the mission undocking was performed with an initial separation rate of 0.4 feet/second. The crew was suited for undocking. At a range of 300 feet the crew began a fly-around. The Skylab inspection included still photography, television, and 16 mm motion picture. After 90 degrees of fly-around the crew noted that the vehicle was drifting away from Skylab. Translational RCS firings caused the attitude to drift in pitch and yaw. RCS firings to correct the attitude impinged on the solar parasol, Apollo Telescope Mount solar arrays, and the Skylab discone antenna. The impingement led to a decision to discontinue the fly-around. The separation burn was executed in all three axes since the fly-around was not in-plane.

SL-3 (July - September 1973)

After orbit insertion on July 28, 1973, the CSM separated from the S-IVB and the CSM crew performed an S-IVB observation activity. The crew estimated the distance to the S-IVB to be 30 to 90 meters. Range determination was not possible due to crew unfamiliarity with the exact size of the S-IVB and the lack of reference objects. The crew reported that maintaining station-keeping was a simple exercise.[10]

Before the NC-1 burn the crew reported fireflies passing by window #5. Mission Control told the crew to isolate SM RCS Quad B due to a propellant leak. VHF ranging lockon occurred at the predicted range, and the Skylab flashing beacon was sighted about 5.5 hours after lift-off, also at the predicted range.[10]

VHF range and range rate data indicated that the first braking gate at 6,000 feet was successfully executed to 30 feet/second. Subsequent braking was almost continuous since only two RCS quads were available to support −X body axis braking. The near continuous thrusting prevented accurate VHF range rate determination need by the crew for the braking phase. The reduced control authority made it difficult to control line-of-sight rates, primarily in the vertical direction.[10]

Station-keeping was begun approximately eight hours after lift-off. The crew found it easy to note when relative motion went to zero, but it was impossible to determine the exact range to Skylab. Due to the difficulty in estimating range the crew had to reduce the range rate to zero before they approached close enough to visually estimate range rate.[10] A fly-around inspection was performed with live television transmission to Mission Control. The CSM flew too near to the thermal parasol due to the difficulty of piloting the CSM with two SM RCS quads isolated.[11] Mission Control personnel observed via live television that the parasol erected by the SL-2 crew was flapping in response to SM RCS plume impingement. The CSM was immediately flown away from the parasol.

Docking occurred about 30 minutes after arrival with a closing rate estimated to be less than 1 foot/second. The crew opened the hatch and moved into Skylab about 10 hours after launch. The crew suffered from space adaptation syndrome during flight days one, two, and three.[10]

On flight day 6 SM RCS Quad D was isolated due to another propellant leak. Mission Control and supporting personnel developed alternate flight control and deorbit procedures in response the loss of two RCS quads. The CSM provided orbital adjustment capability for Skylab, but orbit trim burns scheduled for flight days 5, 31, and 53 were cancelled due to the isolated RCS quads.[3] In case the SM RCS problems prevented the crew from returning, the SL-4 Saturn IB and CSM were prepared for a rescue mission flown by two astronauts (Brand and Lind). The CM would be modified to carry a total of five crew members and would dock to the back-up, or radial docking port. In the end the rescue flight was not required.[3]

Skylab systems problems led NASA to consider launching the SL-4 crew before the SL-3 crew returned to Earth. This would avoid a period when the station would be unmanned. The SL-4 CSM could dock to the back-up radial port and SL-3 could undock and return the next day. A second option was for the SL-4 CSM to station-keep with Skylab while the SL-3 CSM undocked. The SL-4 CSM would then dock to the primary axial port while the SL-3 crew returned to Earth. However, neither of these options was exercised.[3]

After undocking on September 25, 1973, when the range had increased to over 30 meters, several CSM flight control systems tests were successfully conducted.[10] A fly-around for Skylab inspection was not performed since SM RCS jets on Quads B and D were not available. Due to the SM RCS problem the two burn deorbit sequence was changed to one SPS burn executed when the CSM was ~5,900 feet ahead of and below Skylab.[3]

SL-4 (November 1973 - February 1974)

The SL-4 crew was launched on November 16, 1973. The launch and rendezvous were nominal. Station-keeping began about 7.5 hours after lift-off. Soft capture and hard docking occurred about 30 minutes later after two unsuccessful attempts. The failures were attributed to a low approach rate. The approach rate for the successful docking was estimated to be between 0.8 and 1 foot/second.[12] The crew entered the station on flight day two. Two crew members suffered from space adaptation syndrome on flight days one, two, and three.

Undocking occurred on February 8, 1974. A fly-around was performed for inspection and to photograph Skylab.

The crew later reported that station-keeping and the fly-around inspection were easier to perform than was expected from training. Intermittent ground station coverage resulted in excessive voice traffic during the infrequent ground station passes. The crew used a handheld computer (a Hewlett Packard HP-35 calculator) and recommended that the new hand held computers with program cards be used for back-up rendezvous burn computations in place of the charts.[12-13] This was later done on Apollo/Soyuz (July 1975), when the CSM crew used a Hewlett Packard HP-65 calculator to compute back-up rendezvous burn solutions.

References

1. Compton, W. David, and Charles D. Benson, *Living and Working in Space – A History of Skylab*, NASA, Washington, DC, 1983.

2. Cooper, Henry S. F., *A House in Space*, Holt, Rinehart, and Winston, New York, NY, 1976.

3. Baker, David, *The History of Manned Spaceflight*, Crown Publishers, Inc., New York, NY, 1981.

4. Newkirk, Roland W., Ivan D. Ertel, and Courtney G. Brooks, *Skylab: A Chronology*, NASA SP-4011, NASA, Washington, DC, 1977.

5. Shayler, David J., *Skylab: America's Space Station*, Springer Praxis Books, 2001.

6. Gatland, Kenneth, *The Pocket Encyclopedia of Spaceflight in Color*, The MacMillian Company, New York, NY, 1967.

7. *Final Skylab Prelaunch Targeting and Rendezvous Mission Techniques*, MSC-07479, NASA Manned Spacecraft Center, Houston, TX, December 22, 1972.

8. *Skylab Mission Report, First Visit*, JSC-08414, NASA Johnson Space Center, Houston, TX, August 28, 1973.

9. Conrad, Charles, and Dr. Joseph P. Kerwin, "Skylab 1 and 2: Pilot's Report," *Proceedings of the 17th Symposium of the Society of Experimental Test Pilots*, Beverly Hills, CA, September 26-29, 1973, pages 211-218.

10. *Skylab Mission Report, Second Visit*, JSC-08662, NASA Johnson Space Center, Houston, TX, January 28, 1974.

11. SL-3 commander Alan Bean explanation to John Goodman, July 10, 2009.

12. *Skylab Mission Report, Third Visit*, JSC-08963, NASA Johnson Space Center, Houston, TX, July 28, 1974.

13. Hitt, David, Owen K. Garriott, and Joe P. Kerwin, *Homesteading Space: The Skylab Story*, University of Nebraska Press, 2008.

CHAPTER 6 - APOLLO/SOYUZ TEST PROJECT

Introduction

Talks were held in 1970 between Soviet and American space officials to explore the possibility of a joint space flight and development of space rescue techniques. Discussions expanded over the next two years and led to the Apollo-Soyuz (or Soyuz-Apollo) Test Project (ASTP). ASTP became a part of the process of détente. Programmatic challenges of ASTP included cultural differences, language barriers, use of different atmospheres in the spacecraft, and the development of androgynous docking hardware that would permit space rescue by vehicles of the same or different countries.[1, 2, 3]

In June of 1971, just after the successful docking of Soyuz 11 with the Salyut space station, the Soviets proposed an Apollo docking with a Salyut space station. A Soyuz docking with Skylab was also proposed. However, the decision for an Apollo docking with a Soyuz was made official in April of 1972 (Figure 6.1).[1, 4]

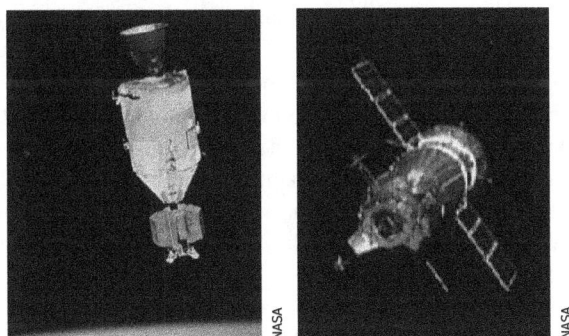

Figure 6.1 Apollo CSM and Docking Module (DM) (left) as seen from Soyuz (left) and Soyuz as seen from Apollo (right) (July 1975).

The probe and drogue docking mechanisms used by Apollo and Soyuz vehicles required different hardware on the vehicles to be docked. An androgynous docking system that permitted two vehicles with identical docking hardware, called APAS-75, was developed.

Apollo/Skylab CSM hardware, software and the coelliptic rendezvous technique were successfully adapted and flown to support rendezvous and docking with Soyuz. Apollo Command Service Module (CSM) 111, originally built for a lunar mission, was modified for ASTP. In the event of a Soyuz failure the Soyuz crew could return in the Apollo CSM with the Apollo crew. Additional foot restraints, restraint harnesses, and helmets were carried.[5, 6]

Two unmanned Soyuz test flights were flown in support of Apollo/Soyuz. These were Kosmos 638 (April 1974) and Kosmos 672 (August 1974). A crewed mission, Soyuz 16, was a six day flight flown in December of 1974. The two man crew tested the APAS-75 docking unit. The NASA ground tracking network also tested American communications equipment installed on Soyuz 16.[4, 7]

The launches of both the Soyuz and Apollo vehicles were successful. Docking and joint activities were conducted as per the pre-mission nominal timeline. After the final Apollo-Soyuz undocking, challenging piloting in close proximity to the Soyuz was successfully performed in support of scientific experiments. All mission objectives were met (Table 6.1).[1, 6, 8]

ASTP was the first mission on which a geostationary satellite was used to expand communications beyond that provided by ground tracking sites. The NASA ATS-6 communications satellite was launched on May 30, 1974. To support ASTP it was located at 35 degrees east longitude over the east coast of Africa. It expanded continuous communications with the CSM up to approximately 50 minutes per revolution.[5, 6]

Rendezvous Target Vehicles

A number of mission concepts were explored from 1970 to 1972 that involved different chaser and target vehicles. One scenario involved a Soyuz docking with either the primary or backup Skylab (Skylab B). However, Skylab mission planning and hardware development was too advanced to permit Skylab modification to accept a Soyuz vehicle. In addition

Table 6.1 Apollo/Soyuz

Flight	Year	Chaser	Target	Profile	Comments
Apollo/ Soyuz	1975	1. CSM 2. CSM/DM 3. CSM/DM 4. CSM/DM	1. S-IVB/DM 2. Soyuz 3. Soyuz 4. Soyuz	1. Transposition & docking 2. Coelliptic M=29 3. Prox Ops & re-docking 4. Prox Ops & final sep.	2. Based on Skylab profile. 3. Solar occultation experiment. 4. Test active mode of Soyuz APAS docking unit at sep. Atomic O_2/N_2 measurements.

APAS = Androgynous Peripheral Assembly System, **Coelliptic** = Coelliptic Flight Profile, **CSM** = Command/Service Module, **DM** = Docking Module, **M** = docking on the Mth revolution, **Prox Ops** = Proximity Operations, **SEP** = Separation

launch of the Skylab B was questionable due to budget concerns. Another scenario used the Skylab backup CSM to conduct a cooperative docking with a Salyut station, followed by a fourth visit to Skylab. This mission would have occurred approximately 18 months after the launch of Skylab.[9] An Apollo/Salyut mission was planned, but the Soviets eventually indicated that it would be too costly to modify a Salyut with a second docking port that could support Apollo. An Apollo/Soyuz docking was selected in April of 1972 for cost and schedule reasons.[1]

The Soyuz spacecraft was a modified version of the 7K-T used in the Salyut program (Figure 6.1). The power and life support systems were upgraded to support longer autonomous flight. A VHF communications system was added to support voice communications with the CSM and VHF ranging by the CSM relative navigation function. The Igla rendezvous system was removed from Soyuz. However, the Soyuz crew was trained to fly a manual final approach and docking from a range of 300 meters.[4]

The Soviets reduced their O2/N2 14 psia cabin atmosphere to 10 psia, a pressure level they had never flown. Apollo used the 5 psia O2 atmosphere flown on the lunar missions and Skylab.[2]

The CSM was the active vehicle and the Soyuz was passive for the initial docking and a final docking on the last day of the joint flight. However, for the final docking the Soyuz APAS-75 hardware was in the active role.

Docking Hardware

Technical discussions concerning rendezvous and docking were conducted in Moscow in October of 1970 during a visit by Bob Gilruth, Caldwell Johnson and Glynn Lunney of the Manned Spacecraft Center in Houston.[1] Johnson gave an overview of his double ring and cone or androgynous docking mechanism, which he had been working on since May of 1962. The device had been proposed for the Apollo lunar spacecraft, but the drogue and probe was selected November of 1963.[10] Johnson also proposed the same concept for the Apollo Applications Program in 1967. NASA recommended a peripheral androgynous docking system for future missions. It was peripheral since the docking hardware was on the periphery of the docking tunnel. It was androgynous in that both spacecraft had identical docking hardware that could function in either an active or passive role during docking. Soviets began working on an androgynous docking concept in 1968.[1, 4]

Consideration was given to using the legacy probe and cone concept flown on Apollo and Soyuz vehicles, but this was discarded in favor of new technology development.[4] The probe and drogue design used by Apollo had to be disassembled to be removed and permit crew transfer. A spacecraft equipped with a probe could only dock with a drogue equipped spacecraft. The limitations of the probe and drogue led to the joint development of the Androgynous Peripheral Assembly System 75, or APAS-75. Both vehicles would fly with identical docking hardware that could function in either an active or passive role. The passive APAS could perform a contingency separation from the active APAS. The APAS-75 allowed crew transfer without requiring removal of the docking hardware by the crew.[4]

The ASTP Soyuz was modified to include an androgynous device. The APAS was too large to be placed on the Apollo Command Module and was on the Docking Module instead.[4] The legacy Apollo probe and drogue were used for CSM docking with the DM.

Docking Module

A Docking Module (DM) was designed as an airlock that would enable the crews to safely acclimatize before transferring between the vehicles since different atmospheres were used (Figure 6.1). The DM performed several other functions as well. These were 1) Serve as a structural adapter between the Apollo legacy probe and drogue docking mechanism and new androgynous docking mechanism (APAS-75), 2) Carry communications gear compatible with Soyuz frequencies, and 3) House Earth resources survey equipment for use after the joint part of the international mission ended. One end of the DM was equipped with a Soyuz compatible androgynous docking device (APAS-75), while the other had an Apollo compatible drogue. The DM was carried on top of the S-IVB stage of the Saturn IB launcher, in the same manner as the LM was carried in the Saturn V. DM also had redundant VHF/FM simplex transceivers operating at the Soyuz frequency of 121.75 MHz.[6]

Docking Alignment Tools

A 1971 proposal called for a centerline camera and window in the DM hatch to support docking. One target would be in center of Salyut hatch, while another would be aligned with the CSM COAS. This was canceled in October of 1972 for technical and budget reasons. Docking targets visible to the CSM COAS in the CM rendezvous window were judged to be adequate.

The Soyuz carried primary and backup docking targets to support alignment just before docking. The primary Apollo docking target was erectable and attached to the Soyuz docking assembly. The CSM crew member flying the docking sighted on the target using the COAS. Soyuz was also equipped with a fixed backup docking target in the event that the primary target failed to deploy. Soyuz was also equipped with two beacons and four orientation lights to aid the CSM crew during rendezvous.

A Soyuz docking target was placed on the DM for use by the Soyuz crew. It would be observed through the Soyuz periscope.

On-Board Relative Navigation

Apollo CSM relative sensors were the same as for Skylab. A sextant provided line-of-sight angle measurements. The Soyuz was equipped with a VHF transponder to support VHF ranging. VHF range rate was also available but was not processed in the Apollo computer Kalman filter.

On-Board Maneuver Targeting

The Apollo CSM computer used the same targeting and relative navigation algorithms as were used for Skylab rendezvous. However, the paper chart solutions were replaced by targeting algorithms on a Hewlett-Packard HP-65 calculator with a magnetic card reader. Back-up targeting could be performed for the coelliptic (NSR), Terminal Phase Initiation and Mid-Course Correction maneuvers.

Plume Impingement

Soviet concerns about Apollo CSM plume impingement on the Soyuz solar arrays were triggered by film of CSM RCS plume impingement effects on the Skylab parasol. The four Apollo SM -X (forward firing) RCS jets were inhibited within 2 seconds of contact. Only two of the four CSM roll jets were used while docked due to loading concerns with the Soyuz solar arrays.[1, 11]

Communications Testing

Enhanced communications and ground tracking tests were conducted through the ATS-6 satellite. Use of ATS-6 extended communications from 15 minutes to 49 minutes per orbit. This foreshadowed the use during the Shuttle Program of the Tracking and Data Relay Satellite System (TDRSS).[6]

Nominal Rendezvous Profile

The nominal Soyuz rendezvous orbit was 121.5 nm circular inclined 51.8 degrees to the equator. Apollo launch opportunities occurred on five consecutive days. The first three opportunities provided a docking on the 29th Apollo revolution (M=29, flight day 3), the fourth opportunity on M=14 (flight day 2), and the fifth opportunity on M=13 (flight day 2). The first through the fourth Apollo launch opportunities occurred on Soyuz daily orbit number 4 while the fifth Apollo launch opportunity occurred on Soyuz daily orbit number 3. Soyuz daily orbit 1 began when the ascending ground-track crossed 20 degrees east longitude. Nominal CSM insertion was into an 81 x 90 nm orbit with the nominal launch time in the planar window about 3 minutes before the minimum yaw steering launch time. This provided a

total launch window of 8 minutes while providing propellant margin for engine failures by conserving Saturn S-IVB propellant allocated for yaw steering.[12]

After orbit insertion the Apollo CSM was to separate from the S-IVB and perform a transposition and docking maneuver to remove the DM from the S-IVB. DM extraction was followed by a separation burn. At third apogee a burn was performed to circularize the CSM/DM orbit at 90 nm.[12]

The CSM/DM flew a nominal coelliptic profile based on the Skylab NC1-NC2-NCC-NSR-TPI sequence (Figures 6.2 and 6.3). However, unlike Skylab, which performed flight day one dockings, the nominal ASTP docking was on flight day 3. Alternate ground targeted phase profiles were developed to support different launch opportunities (M=14 and M=13 flight day 2 dockings), in part due to the high 51.8 degree inclination of the Soyuz orbit.[12]

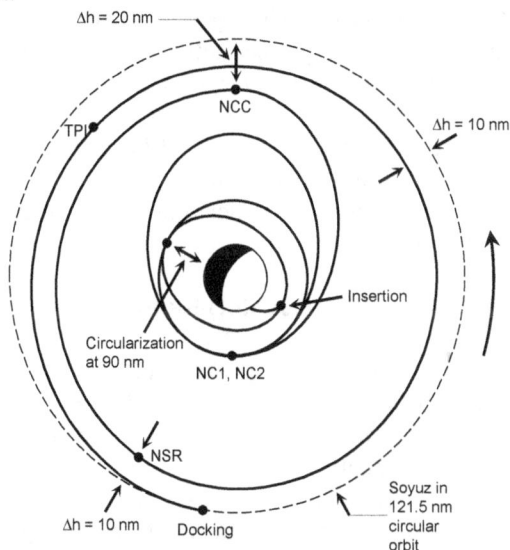

Figure 6.2 Rendezvous profile for the first Apollo launch opportunity (July 1975).

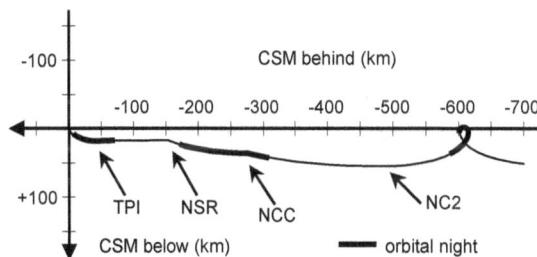

Figure 6.3 Apollo/Soyuz relative motion for the first Apollo launch opportunity.

After the circularization burn ground tracking data would be acquired and orbit determination results used to compute the NC1 burn. This burn was executed on flight day 1. Since the Soyuz circularization burn was not to be executed until after NC1 the desired Soyuz orbit was used for NC1 targeting. If required, a plane change burn (NPC) could be executed 90 or 270 degrees after NC1.

NC1 would be used to create the common node for the NPC. NPC was the last burn opportunity planned for flight day 1.[12]

The Soyuz circularization burn was scheduled for execution while the Apollo crew was executing a scheduled sleep period. On flight day 2 an Apollo Phasing Correction Maneuver (PCM) opportunity existed to permit phasing errors resulting from the NC1 and Soyuz circularization burns to be corrected.[12]

The remaining burns (NC2-NCC-NSR-TPI) plus two mid-course corrections after TPI were executed on flight day 3. NC2 was a horizontal burn that controlled relative altitude to ensure an appropriate range to Soyuz for relative measurement acquisition. NCC controlled relative altitude, phasing, and wedge angle at the NSR (coelliptic) burn point. NSR established a nominal coelliptic delta-height of 10 nm. TPI was executed on a line-of-sight elevation angle of ~27 degrees, nominally 2 minutes before orbital midnight. Two mid-course correction burns were planned after TPI. Zero line-of-sight rate control began at a range of approximately one nautical mile.[12]

After the final undocking the CSM flew a challenging relative motion profile (what would become known two years later as proximity operations) in support of several scientific experiments.[12]

Contingency S-IVB (Booster) Rendezvous

Problems with the upper stage of the Saturn IB booster, the S-IVB, could have required the CSM to leave the vicinity of the S-IVB before the DM could be extracted. This contingency could occur anytime within the first hour after orbit insertion, but was most likely to occur right after orbit insertion. Once the ground executed procedures to safe the S-IVB (taking from 3 to 6 hours) the CSM could return for DM extraction. The low altitude at orbit insertion (80 to 90 nautical miles), a flight rule limiting CSM perigee altitude to greater than 70 nautical miles, and the possibility of S-IVB re-entry as soon as 12 hours after insertion prevented use of the standard coelliptic rendezvous profile.[13]

The emergency separation procedure involved the crew applying a 20 foot/second CSM RCS ΔV in whatever direction the CSM happened to be pointed. Separation burn ΔV was to be voiced to Mission Control, or relayed digitally through the ATS-6 communications satellite (if available). A one rev transfer would be performed out to a point trailing the S-IVB of 53 nautical miles (Figure 6.4). If required as a part of the safing procedure, S-IVB propellant venting would be performed just after the NC maneuver. The venting could impart ~15 feet/second of ΔV to the S-IVB, and CSM maneuver targeting (NC, SOR1, SOR2) performed by Mission Control would be altered if propellant venting was performed. NC would initiate phasing to return to the S-IVB.

The only ground update of the CSM state vectors for the CSM and S-IVB would be performed after NC. Ground tracking at other times during the rendezvous was not accurate enough to support rendezvous. Optical measurements using the sextant would be taken during all daylight periods. VHF ranging was not available for S-IVB rendezvous, nor was the S-IVB equipped with lights to aid optical navigation (as it was on Apollo 7). The lack of an atmospheric drag model in the CSM computer further complicated relative navigation. The S-IVB attitude was designed to minimize atmospheric drag and provide attitude stability, while the CSM attitude was defined to maximize atmospheric drag while permitting sextant tracking of the S-IVB. The attitudes of both vehicles minimized differential drag effects on relative motion.

The NC/SOR1 sequence was to place the CSM at a point on the V Bar 6 nautical miles behind the S-IVB. The SOR2 maneuver executed at this point would establish a coelleptic orbit (stable orbit rendezvous profile). Later a TPI maneuver with a transfer angle of

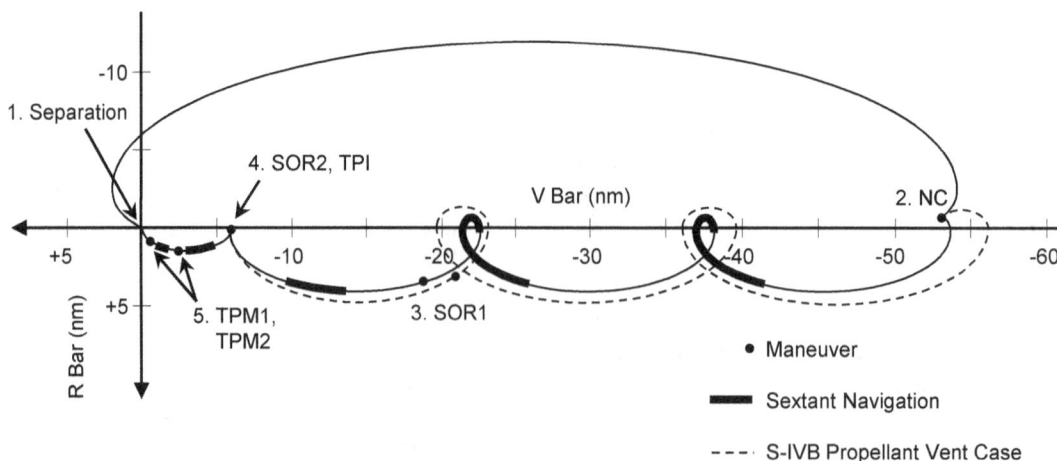

Figure 6.4 Apollo-Soyuz contingency S-IVB stable orbit rendezvous, not executed (July 1975).

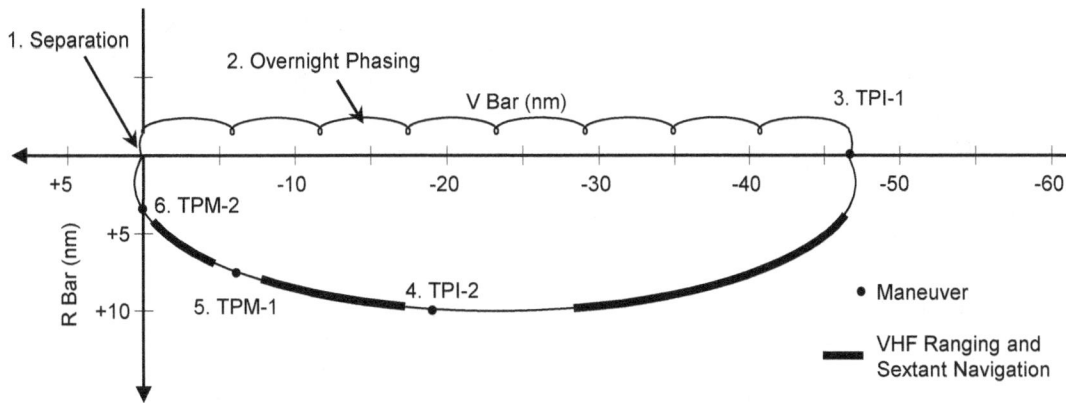

Figure 6.5 Apollo stable orbit contingency re-rendezvous with Soyuz, not executed (July 1975).

200 degrees to intercept would be performed. The TPI and both TPM maneuvers were computed by the crew. COAS subtended angles were to be used for range estimation, and range rate was to be estimated by the Entry Monitoring System using an initial value for range from the CSM computer.

Contingency Re-rendezvous With Soyuz

In the event that Apollo attempts to dock with Soyuz on the planned day of rendezvous were not successful, an alternate plan was developed that would enable more attempts to be made on the following flight day (Figure 6.5).[14] Apollo was to separate from Soyuz and phase behind it overnight. A 2 foot/second posigrade separation maneuver and differential drag effects would place the CSM ~47.6 nautical miles behind Soyuz after 8 revolutions. At this point, the re-rendezvous would be initiated by a Mission Control computed TPI-1 maneuver which placed the CSM on a 290 degree intercept trajectory. The profile was a stable orbit instead of a coelliptic. The TPI-1 maneuver was designed so that at the TPI-2 point (a nominally zero ΔV maneuver), the same relative trajectory conditions (delta height, transfer angle, elevation angle) would exist as at the TPI maneuver on a coelliptic profile. This profile design also provided terminal phase lighting and geometry conditions identical with the terminal phase of the coelliptic profile. The TPI-2, TPM-1, and TPM-2 maneuvers were computed using the CSM computer.

Launch, Rendezvous, and Docking Performance (Flight Days 1 through 3)

After orbital insertion the S-IVB propellant and cold gas dumps were successfully accomplished. The separation and stable orbit contingency re-rendezvous plan was not needed (Figure 6.4).

After separation from the DM/S-IVB, the CSM turned around at a range of 80 to 90 feet. After transposition, as the CSM approached the DM/S-IVB, the COAS reticle was washed out by reflected sunlight from the Earth background. The commander placed his hand behind the COAS glass to block the sunlight and confirmed that the green reticle was properly illuminated. The washout made it difficult to achieve a proper roll alignment. The commander maintained station-keeping until the DM/S-IVB approached the horizon and proceeded with the docking. There was some risk with delaying docking since orbital night was approaching and the S-IVB was not equipped with lighting. Docking was successful. Separation of the CSM/DM from the S-IVB was nominal.[8, 15] The NC-1 burn was successfully performed.

Docking probe removal was delayed until flight day 2 due to a problem with the probe. The crew also successfully extended and retracted the APAS guide ring as a test.[15]

NC-2, the first burn executed on flight day 3, was nominal. VHF ranging acquisition occurred at the expected range of approximately 120 nm. The NCC burn was nominal. The crew observed the Soyuz flashing beacon before the NSR burn. Although sextant tracking was not scheduled before NSR, it could have been performed. The crew chose to follow procedure and process only VHF marks before NSR.

At a range of about 50 nm to 60 nm the crew noticed VHF interference from ground stations in the United States, Europe, and the Soviet Union. TPI time of ignition slip was approximately three minutes. The delta-velocity magnitudes for the two Mid-Course Correction burns after TPI were 0.2 feet/second and 0.4 feet/second. Line-of-sight rates were null and no inertial motion of the Soyuz was observed by the crew at a range of 0.25 nm. Some inertial line-of-sight motion occurred at close range but it was corrected by the CSM commander.[8]

The Soyuz maintained an LVLH attitude hold with the blunt end of the spacecraft pointed in the direction of motion. The CSM flew over the top of the Soyuz to achieve the proper position for final approach. The Soyuz

Approved for public release via STI DAA 24483. See statement on title page.

63

performed a small pitch maneuver, followed by a 60 degree roll maneuver. This permitted the CSM to keep the high gain antenna pointed at the ATS-6 communications satellite.

At final approach the Soyuz was below the horizon as viewed by the Apollo crew. Reflected sunlight from the Earth background washed out the COAS reticle. Once the Soyuz reached the appropriate attitude it was near the horizon and the COAS reticle pattern was faintly visible. Approach rate at docking was estimated to be about 0.4 feet/second.[8, 15] The docking was televised live using the ATS-6 communications satellite. Later on flight day 3 the crew checked the COAS alignment. It was centered on the Soyuz docking target. Apollo was able to dock with Soyuz on the planned day of rendezvous, and the separation and stable orbit re-rendezvous contingency plan was not needed (Figure 6.5).

Undocking and Astronomy Experiments on Flight Day 5

The first undocking was performed on flight day 5, the final day of the joint flight. After the Apollo active undocking an artificial solar eclipse experiment was performed. The CSM flew a proximity operations profile (Figure 6.6) out to a range of approximately 200 meters in front of the Soyuz. At this point the CSM subtended an angle of about two solar diameters as seen from the Soyuz. The Soyuz crew then performed solar photography.[6, 12, 16]

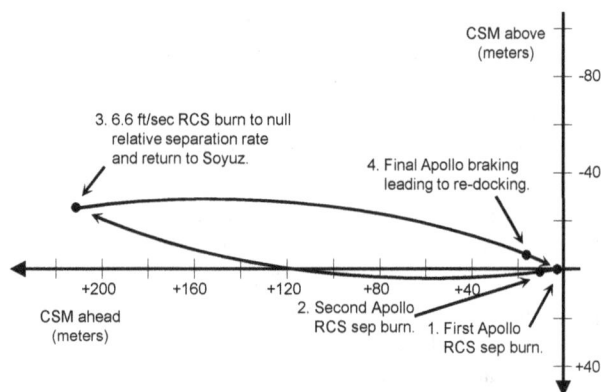

Figure 6.6 Apollo motion relative to Soyuz for the artificial solar eclipse experiment.

The delta-velocity count on the EMS appeared to drift during the final approach. The EMS provided the only indication, other than visual observation of the Soyuz, of the approach rate. During the approach for the second docking the Soyuz was against an Earth background as observed from the CSM. The CSM COAS reticle was washed out by sunlight reflected from the Earth, as had occurred during transposition and docking with the DM (flight day 1) and the docking with Soyuz (flight day 3). This made it difficult to see the COAS

reticle pattern when the pattern was superimposed on the Soyuz or the Earth. Once the stand-off cross docking target became visible it provided a good reference for the CSM crew.[15] The second docking was nominal, but after capture a yaw and pitch rotation was noticeable to both crews. The APAS units were able to withstand the large structural loads. The docking was completed successfully.

After the mission the second docking was investigated. During final approach the Soyuz was in a somewhat different attitude than planned. Additional maneuvering by the CSM was required. Sunlight in the field of view of the CSM crew member flying the vehicle complicated the approach. Due to a CSM sideways RCS firing after contact the mated stack was rotating. The CSM IMU was approaching gimbal lock and another CSM RCS firing was performed to avoid gimbal lock. CSM RCS jets were not supposed to be fired during docking mechanism operation.[4]

During the second and final undocking the Soyuz was active. The Ultraviolet Absorption (UVA) experiment involved three proximity operations profiles, two out of plane and one in-plane.[6, 16] Ultraviolet light was produced by lamps and collimating mirrors on the Docking Module. The light was reflected by an array of corner reflectors on the Soyuz. An instrument on the Docking Module collected the reflected ultraviolet light. To prevent Doppler effects caused by spacecraft orbital velocity from shifting the lamp frequency away from the atmospheric absorption frequency the CSM had to be maneuvered so that the light beams were perpendicular to the orbital velocity. Three data takes were performed.

The first was an out-of-plane orbital south profile out to a distance of 150 meters (Figure 6.7). The second was an out-of-plane orbital north profile out to a distance of 500 meters (Figure 6.8). The third data take was in-plane as the CSM performed the final separation from Soyuz (Figure 6.9). Data was obtained at a range of 1500 meters as the CSM passed above the Soyuz.[6, 12, 16] The profile flown required more accurate CSM pointing than the CSM auto-pilot had been designed to provide.[8] Data could not be obtained on the 150 meter profile. This required changing to the aft Soyuz retro-reflector for the 500 meter profile and modification of the planned Soyuz maneuvering to support the 500 meter profile. Data was successfully obtained during the 500 meter and 1500 meter profiles.[5] VHF ranging data was available to the crew during the flight day 5 proximity operations activities. However, range biases made the data inaccurate (measured range much smaller than actual range) and the crew did not consider it reliable at the data take ranges.[15]

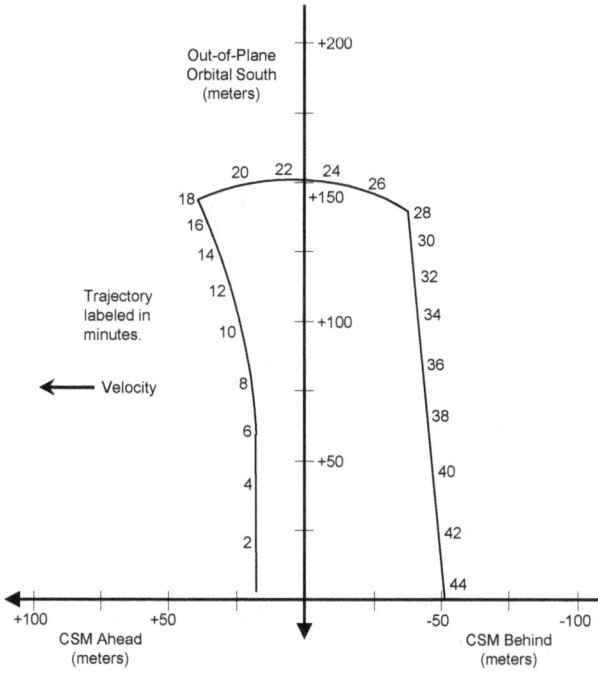

Figure 6.7 Relative motion during 150 meter out-of-plane UVA data take.

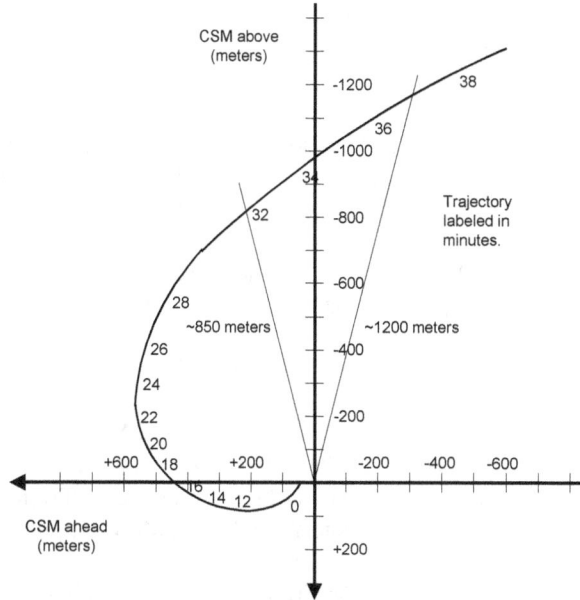

Figure 6.9 Relative motion during final Apollo separation and 1000 meter in-plane UVA data take.

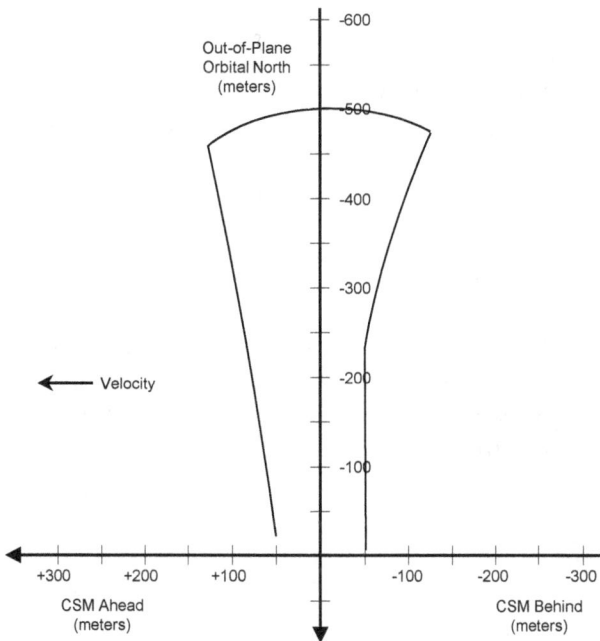

Figure 6.8 Relative motion during 500 meter out-of-plane UVA data take.

References

1. Ezell, E. C., and L. N. Ezell, *The Partnership: A History of the Apollo-Soyuz Test Project*, NASA Special Publication-4209, NASA History Series, 1978.

2. Slayton, D. K., "Apollo-Soyuz Test Project," *18th Society of Experimental Test Pilots Symposium*, Beverly Hills, CA, September 25-28, 1974.

3. Faget, Dr. Maxime A., "Background and Planning for the Apollo/Soyuz Mission," *Proceedings of the 6th Control in Space Conference*, International Federation of Automatic Control, Yerevan, Armenia, August 26-31, 1974.

4. Syromiatnikov, Vladimir, *100 Stories About Docking and Other Adventures in Space*, Volume 1, Twenty Years Back, Universitetskaya Kniga, Moscow, 2005.

5. *Apollo Soyuz Mission Evaluation Report*, JSC-10607, NASA Johnson Space Center, December 1975.

6. Hardee, S. N., "Mission Description," *Apollo-Soyuz Test Project Summary Science Report*, Volume I, NASA SP-412, NASA, Washington, DC, 1977.

7. Filipchenko, A., "The Soyuz – Ready For Flight," "Soyuz k polëtu gotov," *Aviatsiya I Kosmonavtika*, No. 3, March 1975, pages 38-39, NASA Technical Translation (NASA TT F-16369), NASA, Washington, DC, June 1975.

8. Bean, A. L., and R. E. Evans, "Apollo-Soyuz Joint Mission Results," *Proceedings of the Nineteenth Symposium of the Society of Experimental Test Pilots*, Beverly Hills, CA, September 24-27, 1975, pp. 173-178.

9. Newkirk, Roland W., Ertel, Ivan D., and Brooks, Courtney G., *Skylab: A Chronology*, SP-4011, NASA, Washington, DC, 1977.

10. Langley, R. D., *Apollo Experience Report: The Docking System*, NASA-TN-D-6854, NASA Manned Spacecraft Center, June 1972.

11. Penchuk, A., J. F. Turnbull, R. W. Schlundt, and J. E. Jones, "Digital Attitude Control of the Skylab and Apollo/Soyuz Orbital Assemblies," *Proceedings of the 6th Control in Space Conference*, International Federation of Automatic Control, Yerevan, Armenia, August 26-31, 1974.

12. *Apollo-Soyuz Test Project Operational Trajectory*, JSC Internal Note No. 75-FM-18, JSC-09532, Mission Planning and Analysis Division, NASA Johnson Space Center, March 19, 1975.

13. Wilson, S. W., *Dispersion Analysis of the ASTP Booster Rendezvous Procedure*, TRW, June 5, 1975. See JSC-35056, Volume 2, in the A Note on Sources chapter in this document.

14. Wilson, S. W., *Dispersion Analysis of the Soyuz Re-Rendezvous Procedure*, TRW, June 30, 1975. See JSC-35056, Volume 2, in the A Note on Sources chapter in this document.

15. *ASTP Technical Crew Debriefing*, JSC-09823, Crew Training and Procedures Division, NASA Johnson Space Center, August 8, 1975.

16. Giuli, T. R., "Apollo-Soyuz Science Results," Paper IAF-76-149, *27th Congress of the International Astronautical Federation*, Anaheim, CA, October 10-16, 1976.

PART II – THE SPACE SHUTTLE

This page intentionally left blank.

CHAPTER 7 - SPACE SHUTTLE – A NEW DIRECTION IN MISSION ACTIVITIES

Space Shuttle rendezvous and proximity operations represented a significant departure from Gemini and Apollo.[1] Rendezvous was considered a secondary service, while the primary service was payload deployment. As a result rendezvous was not given as high a priority as it was in the Gemini and Apollo Programs.

There were also important technical differences between the shuttle and the previous two programs. Most rendezvous targets would not possess active navigation aids (transponders or lights), nor were many of them originally designed to support rendezvous, retrieval, and on-orbit servicing. Shuttle rendezvous missions also involved deploy and retrieval of the same or different spacecraft on the same mission, and on some missions more than one rendezvous.

Relative chaser and target spacecraft size were significantly different. Previous chaser vehicles (Gemini, Apollo Command/Service Module and Lunar Module) were about the same size as the target spacecraft (Radar Evaluation Pod (REP), Titan second stage, Gemini VII, Agena, Augmented Target Docking Adapter, LM, Soyuz) or smaller (Saturn S-IVB, Skylab). Until the Mir and International Space Station (ISS) missions, the orbiter was much larger than its rendezvous targets.

Rather than docking at ~1 foot/second, as was done in Gemini and Apollo, satellite retrievals involved capture and berthing with a robotic arm (the Remote Manipulator System, or RMS), with nearly zero relative velocities between the two spacecraft. Robotic arm operations, capture and berthing had not been performed on previous programs. RMS design requirements were a function of orbiter stopping distance, arm joint loads and the ability of the crew to detect and control relative rates.

Shuttle docking with Mir and ISS required a contact velocity an order of magnitude lower than Gemini and Apollo, with tighter piloting tolerances on time of docking and contact velocity. Gemini and Apollo docking were axial and along the crew line-of-sight. The Gemini crew had a direct view of the docking hardware at contact, while the Apollo CSM and LM crews did not. Shuttle grappling and docking required the use of cameras to provide adequate crew visibility and cues for final control. Since target spacecraft could possibly already be in orbit during mission planning, some grapple equipment used by the Shuttle Program was designed from documentation of target spacecraft hardware, and was not mated on the ground for preflight checks as was done for Gemini and Apollo docking hardware (Figures. 7.1 and 7.2).

Development of rendezvous profiles, mission plans, and procedures is a complex systems integration problem. Rendezvous personnel do not own many of the sub-systems that are required for rendezvous since they are also used to support non-rendezvous activities. This made it more challenging for rendezvous personnel to justify and achieve sub-system design changes to accommodate new and unique aspects of shuttle rendezvous and proximity operations.

Reference

1. Pearson, D. J., "Shuttle Rendezvous and Proximity Operations," *Proceedings of the CNES International Symposium on Space Dynamics*, Centre National d'Etudes Spatiales, Paris, France, 1989, pp. 833-851.

Figure 7.1 Gemini and Agena testing before Gemini VI (Sept. 1965).

Figure 7.2 Apollo and Docking Module testing before Apollo/Soyuz (Jan. 1975).

This page intentionally left blank.

CHAPTER 8 - EARLY SHUTTLE RENDEZVOUS STUDIES

In 1969, a study of on-orbit ΔV budgeting was conducted for the Advanced Logistics System (ALS), an early name for the Space Shuttle.[1] A five-maneuver coelliptic profile (Figure 8.1) was proposed for a resupply mission to a space station in a 200 or 270 n.m. circular orbit, with an inclination of 55 degrees. Apollo and Gemini flight techniques, sensor characteristics, and flight experience was factored into the propellant budgeting estimate. The ALS terminal phase was the same as that used on most Gemini and Apollo missions (Figure 8.2).[2,3]

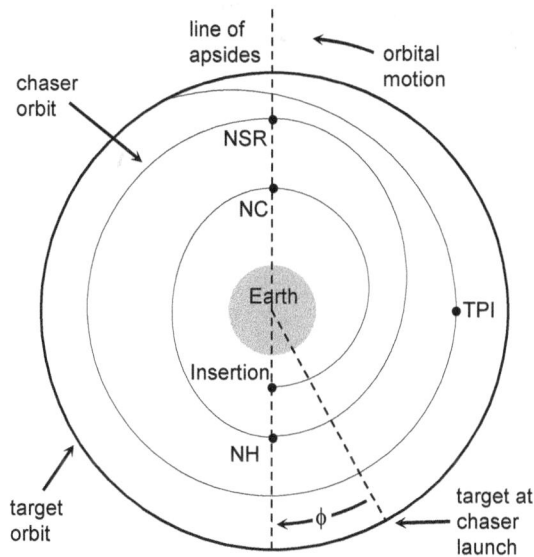

Figure 8.1   ALS profile for rendezvous (1969).

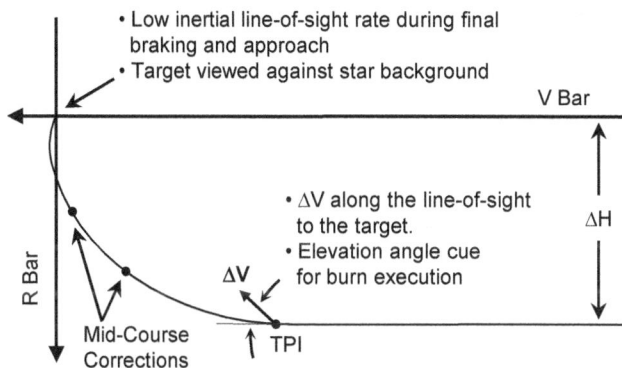

Figure 8.2   Terminal Phase for coelliptic rendezvous.

The phasing, height adjustment, and coelliptic maneuvers are normally performed at an apsis crossing, resulting in a Hohmann transfer and a theoretically optimum ΔV (180 degree transfer). The study assumed a launch directly into the plane of the station, a daily launch window, a minimum phasing perigee of 100 n.m., rendezvous within 24 hours of launch, and deorbit within 24 hours of departure from the station.

Propellant budgeting estimates were important for establishing payload capability. Additional required ΔV, beyond the theoretical minimum, had to be estimated due to operational considerations, maneuver execution errors, targeting dispersions, guidance and navigation dispersions during powered ascent and the on-orbit phases, as well as worst-case phasing scenarios for both rendezvous and landings at a specified runway.

The first phasing maneuver (NC) would be executed without the benefit of ground tracking of the ALS, or ALS tracking of the station. Subsequent error propagation could necessitate additional burns between NC and TPI to reduce dispersions. In addition to an on-orbit ΔV estimate, the study concluded ΔV for worst case phasing scenarios could be reduced by delaying rendezvous beyond 24 hours from launch, delaying launch to take advantage of phase angle change per day, or lowering the minimum phasing perigee to below 100 n.m.

The high inclination and a requirement for rendezvous in from three to seventeen orbits placed severe constraints on launch window duration and frequency. The study showed that propellant required could be significantly reduced if the requirements for every day launch, rendezvous duration and minimum perigee were relaxed.

Rendezvous with a passive target (radar using skin tracking, optical tracking via reflected sunlight) was identified as a shuttle requirement by 1969. Prior to the shuttle, target spacecraft (Titan II second stage, Gemini VII, Agena and ATDA, Apollo CSM and LM, Skylab, Soyuz) possessed active relative navigation aids (Very High Frequency or L band transponders, strobes). Using a standard Gemini/Apollo profile, skin-tracking radar would have only minutes to acquire and track a target before the TPI burn. Passive relative navigation was seen as a major challenge. Although the coelliptic profile used for most Gemini and all Apollo missions was favored for use on the shuttle, studies were conducted in 1970 to determine if other profiles might provide better geometry and tracking arcs for navigationally passive targets.

Figure 8.3a depicts a football profile that would increase the time available for radar tracking while inside a hypothetical 30 nautical mile tracking range of a navigationally passive target. A coelliptic approach with a ΔH of 5 nautical miles would provide time for acquisition and tracking before the AD1 maneuver. A smaller ΔH might have prevented rendezvous under dispersed conditions. Both the AD1 and AD2 maneuvers would place the vehicle at a TPI point with the standard geometry used on Gemini/Apollo profiles, and to provide additional tracking time prior to TPI.

Figure 8.3b illustrates another concept, a stable orbit profile. The AD2 maneuver would initiate shuttle long-range station keeping behind the target, on the –V Bar. Additional tracking would be conducted at this point.

However, the post TPI trajectory would not have the standardized terminal phase benefits provided by the TPI relative geometry of the Gemini/Apollo coelliptic profile.

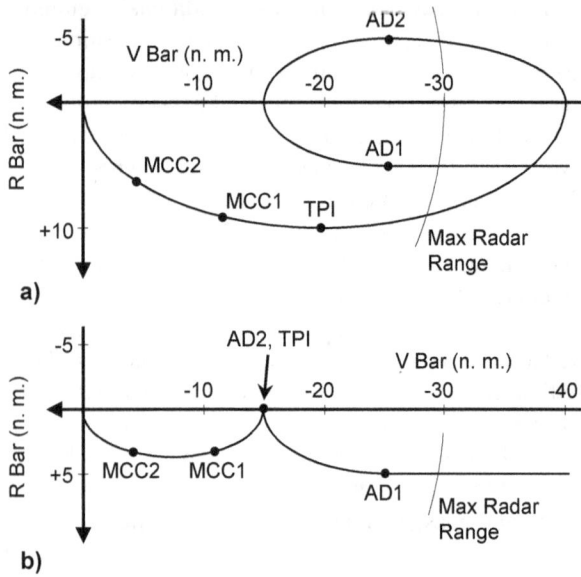

Figure 8.3 Notional profiles for a navigationally passive target (1970).

References

1. *In-Orbit Delta V Study For An Advanced Logistics System (ALS) Resupply Mission*, JSC-06020, IN-69-FM-291, NASA Manned Spacecraft Center, November 17, 1969.

2. Lunney, G. S., "Summary of Gemini Rendezvous Experience," *AIAA Flight Test, Simulation and Support Conference*, AIAA, Reston, VA, 1967.

3. Lineberry, E. C., "Gemini VI-A Rendezvous Mission Planning," *Gemini Midprogram Conference*, NASA SP-121, Manned Spacecraft Center, 1966, pp. 277-282.

CHAPTER 9 - SHUTTLE DESIGN REFERENCE MISSIONS

During the Space Shuttle Phase B studies (1970-1971), the following assumptions were made: 1) rendezvous techniques and principles were well understood, and the flight regime should not contain technical challenges; 2) the coelliptic terminal phase from Gemini and Apollo will be used; 3) a target mounted navigation transponder will allow tracking out to the maximum range achieved during the Apollo Program (~300 n.m.); 4) radar skin tracking of a passive target out to 10 n.m. was a contingency mode of operation; 5) the shuttle will be capable of autonomous rendezvous; and 6) on-board computer capacity will be significantly greater than Apollo.

By 1973, four shuttle reference missions were in use for mission planning, vehicle sizing, and subsystem requirements definition, and three of them involved rendezvous.[1] There was also a requirement (later waved) for a shuttle to rescue the crew of another shuttle stranded in orbit. Rescue was to occur no later than 96 hours after launch of the rescue vehicle. The rescue shuttle was to be able to phase from either above or below the other shuttle's orbit, depending on the initial phasing at launch.

Rendezvous For Reference Missions 1 and 2

The Mission 1 design involved a shuttle deployed space tug returning a geosynchronous satellite to an orbit coelliptic (ΔH of 10 n.m) with the shuttle, to facilitate retrieval. The shuttle would then perform a TPI maneuver and fly a terminal phase similar to Gemini and Apollo (Figure 8.2). Mission 2 was a servicing mission to an orbiting science platform.

In April of 1973, the five-maneuver profile used for Mission 2 was replaced by a Skylab based profile (Figure 9.2) that satisfied shuttle operational considerations that had been identified up to that time. Those considerations were: 1) rendezvous with a navigationally active or passive target at orbital altitudes ranging from 150 to 400 nm; 2) liftoff time selected whenever coplanar launch is possible, and will not be constrained by time-of-day; 3) minimize onboard relative navigation sensor cost, operating range, and accuracy; 4) ground tracking support requirements had not been clearly defined; 5) an optical sensor was required for inertial platform alignment; and 6) the phasing portion of the rendezvous was not to be unnecessarily large.

A change to the Skylab plan involved the insertion of a second coelliptic segment before the NCC burn (Figures 9.1, 9.2, and 9.3). This second coelliptic phase allowed the subsequent maneuver points to be chosen to maximize use of reflected sunlight for optical tracking of navigationally passive targets. The additional coelliptic segment also ensured the same relative geometry from the start of optical tracking through intercept for variations in liftoff time and target orbital altitude.

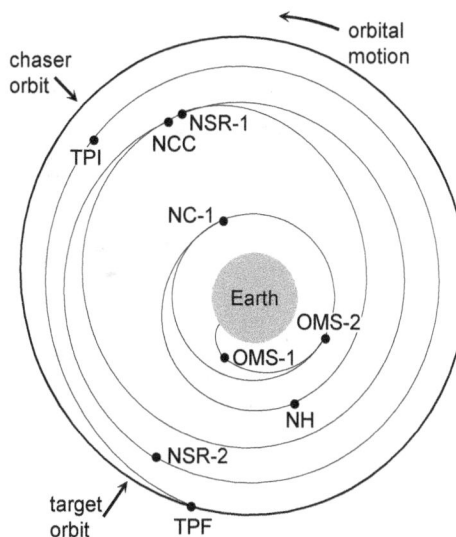

Figure 9.1 Inertial view of dual co-elliptic rendezvous (1973-1983).

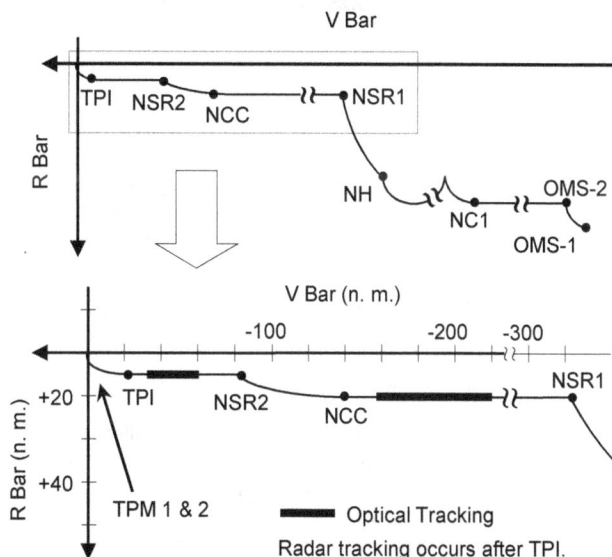

Figure 9.2 Relative view of dual co-elliptic rendezvous (1973-1983).

Relatively constant range at the first optical tracking opportunity was also important due to the lower quality of optical tracking at this point. The dual coelliptic sequence (ΔH of 20 and 10 nm) also provided enough control over lighting to minimize lighting considerations for launch window determination. A wide variation in liftoff time was permitted without resulting in an excessively long phasing period. The profile also permitted flexibility in selecting the level of ground tracking required and in the selection of on-board relative navigation sensors.

The standard terminal phase (Figure 8.2) was also used for Mission 2. One issue, however, was that the targets would probably not possess strobes, as other targets had in previous programs. Lighting requirements

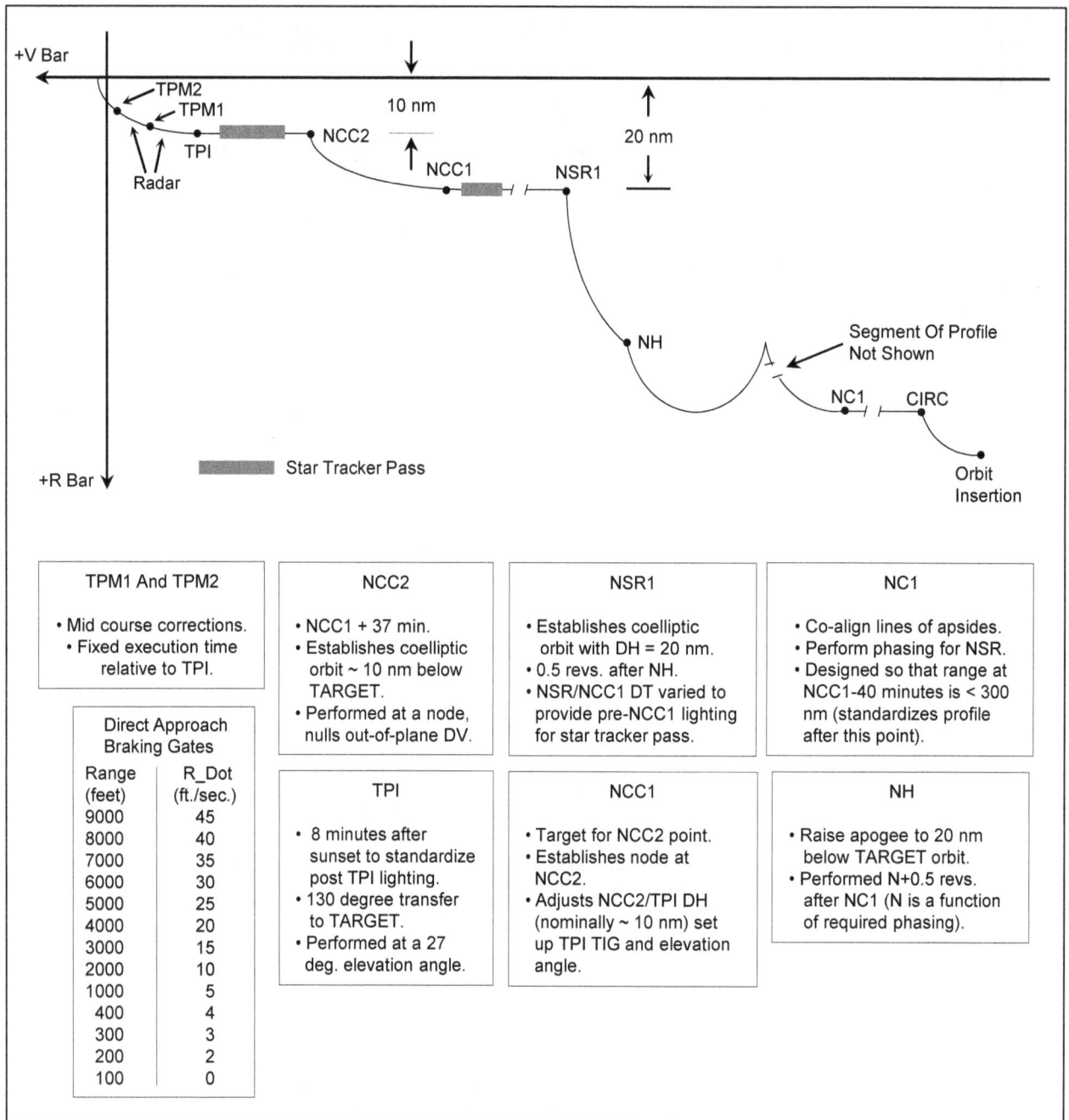

Figure 9.3 Shuttle Double Coelliptic Profile

+V Bar

TPM2
TPM1
TPI
Radar

10 nm
NCC2
NCC1
NSR1
20 nm

NH

Segment Of Profile
Not Shown

NC1   CIRC

Orbit
Insertion

+R Bar

Star Tracker Pass

TPM1 And TPM2
• Mid course corrections. • Fixed execution time relative to TPI.

NCC2
• NCC1 + 37 min. • Establishes coelliptic orbit ~ 10 nm below TARGET. • Performed at a node, nulls out-of-plane DV.

NSR1
• Establishes coelliptic orbit with DH = 20 nm. • 0.5 revs. after NH. • NSR/NCC1 DT varied to provide pre-NCC1 lighting for star tracker pass.

NC1
• Co-align lines of apsides. • Perform phasing for NSR. • Designed so that range at NCC1-40 minutes is < 300 nm (standardizes profile after this point).

TPI
• 8 minutes after sunset to standardize post TPI lighting. • 130 degree transfer to TARGET. • Performed at a 27 deg. elevation angle.

NCC1
• Target for NCC2 point. • Establishes node at NCC2. • Adjusts NCC2/TPI DH (nominally ~ 10 nm) set up TPI TIG and elevation angle.

NH
• Raise apogee to 20 nm below TARGET orbit. • Performed N+0.5 revs. after NC1 (N is a function of required phasing).

Direct Approach Braking Gates

Range (feet)	R_Dot (ft./sec.)
9000	45
8000	40
7000	35
6000	30
5000	25
4000	20
3000	15
2000	10
1000	5
400	4
300	3
200	2
100	0

for the pre-TPI optical tracking pass and the initiation of manual piloting (a few thousand feet from the target) at sunrise drove TPI to be performed after sunset. A lack of target artificial lighting meant that the backup manual procedure of pointing the vehicle thrust axis at the target to execute TPI would not be available, as it was on many trajectories flown by Gemini and Apollo vehicles. The dual coelliptic (Figure 9.3) would serve as the baseline shuttle profile for mission planning until April of 1983.

Rendezvous For Reference Mission 3B

Mission 3B was a satellite retrieval from a 100 nm. circular orbit, with launch and landing occurring at Vandenberg Air Force Base. Mission duration was about 2 hours.[1,2,3]

The insertion point (Figure 9.4) was chosen to place the shuttle at the start of a 73 degree, 18 minute transfer to the target. The relative trajectory at this point was also, by design, the same as if the shuttle had initiated a 130 degree transfer with a TPI burn from a coelliptic orbit with a delta-height of 15 nm, and had 73 degrees left to travel. This placed the shuttle on a terminal trajectory with characteristics similar to those used on terminal approaches flown on Gemini, Apollo, Skylab and Apollo-Soyuz missions (Figure 8.2).

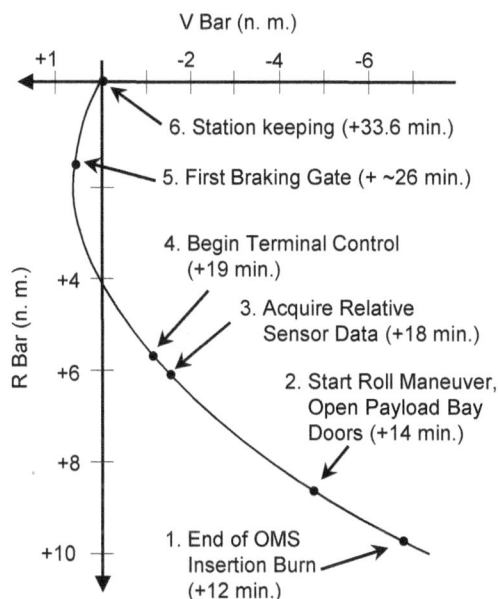

Figure 9.4 Mission 3B approach (July 1975). Times are with respect to liftoff.

Insertion point selection also took into account search capabilities of the relative sensor, maximum time allotted for rendezvous, Reaction Control System (RCS) jet duty cycles, and evaluation in off-line and pilot-in-the-loop simulations. Terminal control began with nulling of inertial angular rates. Braking gates were designed so that coast periods were long enough to accommodate

longer burn times due to dispersions or failed RCS jets, and to limit any delay in rendezvous to no longer than seven minutes.

Due to the short timeline (station-keeping at a range of 100 feet established ~21.6 minutes after orbit insertion), no ground tracking of the shuttle was to be performed, nor would the shuttle have processed relative sensor measurements in a Kalman filter. No on-board targeted maneuvers would have been performed. Radar data (range, range rate, inertial line-of-sight rates) was to have been used by the crew to fly an approach along a straight line relative to an inertial reference frame and reduce closing velocity to appropriate levels. While similar profiles had been flown on Gemini XI (Figure 3.19) and Apollo lunar missions 14 through 17 (Figure 4.14), the Mission 3B profile was much more demanding. Whether or not rendezvous, target capture with the RMS, berthing, payload bay door closure, and deorbit could have been accomplished within the timeline is questionable.

Missions 3B and 3A (a similar mission, but with a deployment rather than retrieval) were the most challenging of the reference missions, and had the most impact on shuttle systems design and performance requirements. Planning for both missions ended around October of 1975, and neither was flown.*

Reference

1. Young, K. A., "Representative Space Shuttle Missions and Their Impact on Shuttle Design," *AIAA/ASME/SAE Joint Space Mission Planning and Execution Meeting*, AIAA, Reston, VA, 1973.

2. *Space Shuttle System Baseline Reference Missions, Volume III – Mission 3A and Mission 3B, Revision 1*, JSC-07896, JSC Internal Note No. 73-FM-47, NASA JSC Mission Planning and Analysis Division, May 7, 1974.

3. *Space Shuttle System Baseline Reference Missions, Volume IV - Mission 3B, Revision 2*, JSC-07896, JSC Internal Note No. 73-FM-47, NASA JSC Mission Planning and Analysis Division, July 28, 1975.

---

* For a discussion on roughly estimating the time of Mission 3B cancelation see page 15 of Goodman, John, *Space Shuttle Proximity Operations Papers, Volume 1 of 2 (1970-1979)*, JSC-35052, in the A Note On Sources appendix.

This page intentionally left blank.

CHAPTER 10 - PLUME IMPINGEMENT

Identification of the Problem

Gemini and Apollo attitude control systems produced little cross coupling, and thrust magnitude, nozzle canting, target vehicle size, and appendages did not result in significant plume impingement issues. Lunar Module self-impingement did have to be addressed with hardware modifications (Figure 10.1). In the early 1970s, the existence of plume impingement was controversial, but analysis of Gemini XI film showing tether dynamics in response to RCS firings proved that plume impingement was real (Figure 10.2). During the first attempt on Skylab 2 to deploy a stuck solar array, the CSM was maneuvered so that a crewman standing in the hatch could reach the array with a deployment tool (Figure 10.3). Apollo CSM thrusting to null the closing velocity triggered Skylab jet firings to maintain attitude, which resulted in an opening rate between the vehicles.[1] Later film of Apollo CSM RCS effects on the Skylab thermal control parasol (Figure 10.4) triggered Russian concerns about plume impingement for the Apollo/Soyuz mission. Four of the CSM's RCS jets were inhibited within 2 seconds of contact to avoid plume loading on the Soyuz solar arrays (Figure 10.5).[2]

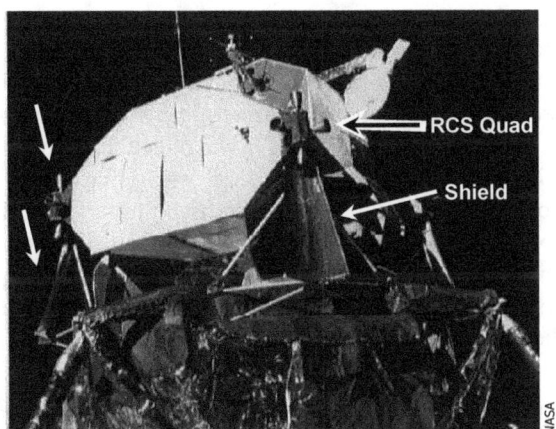

Figure 10.1 Two of the four LM RCS Quads and plume self-impingement shields (Apollo 16, April 1972).

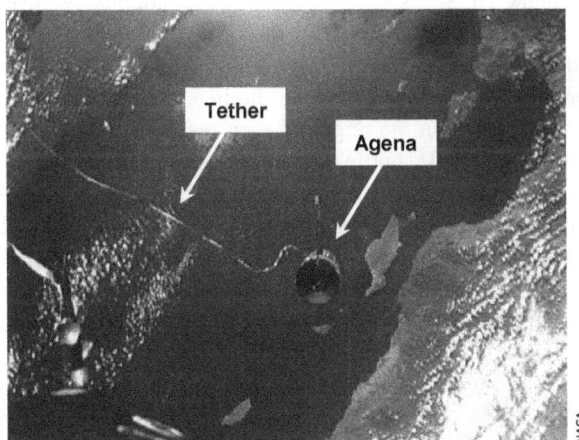

Figure 10.2 Agena tethered to the Gemini XI spacecraft (Sept. 1966).

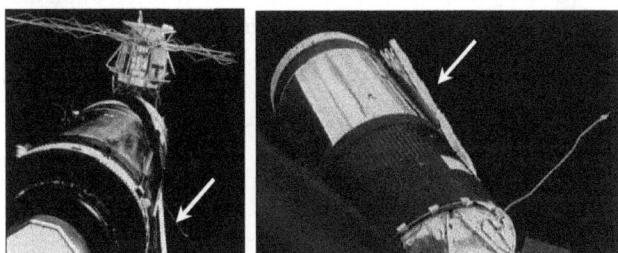

Figure 10.3 Two views of the stuck Skylab solar array taken during the Skylab 2 CSM fly-around inspection (May 1973).

Figure 10.4 Skylab with thermal parasol indicated (Feb. 1974).

Figure 10.5 Soyuz with solar arrays during the Apollo/Soyuz mission (July 1975).

From March through June of 1973, a total of 491 human-in-the-loop simulations were conducted at JSC to evaluate orbiter docking with a spacecraft (Figure 10.6). The objective was to determine if any changes to the Preliminary Requirements Review baseline orbiter design were required in the areas listed in Table 10.1. Based on the study, many recommendations were made concerning crew displays, crew station arrangement, cameras, stand-off cross targets, sensor data, target vehicle attitude control, orbiter flight control modes, RCS jet selection, propellant consumption, and cross coupling. The study concluded that more simulations were needed to take into account changing orbiter systems configuration so that docking requirements could be more precisely and realistically defined. However, the implications of plume impingement on RCS system design was not examined.

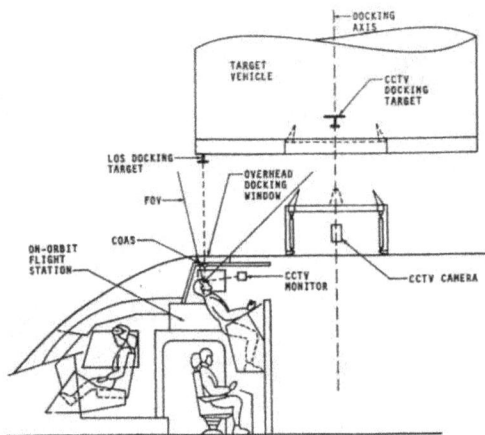

Figure 10.6 Orbiter and target configuration used in the 1973 docking simulations.

Table 10.1 1973 Docking Study Topics

- Maneuver rates
- Required minimum impulse for translation and rotation
- Flight control requirements
- Attitude hold requirements
- Effects of control modes on center of gravity variations
- Target motion
- Hand controller location and logic
- Reduced RCS thrust levels
- Target displays
- Range, range rate, and attitude display requirements
- Station-keeping
- Docking contact criteria
- RCS propellant requirements

By mid 1973, contamination of payloads by shuttle RCS jet effluents during the shuttle approach and braking phase was a concern to the payload community. Previous analysis focused on potential contamination in the payload bay at the launch site and on-orbit. An approach trajectory was proposed that minimized the expulsion of combustion by-products at the target, and therefore minimized the potential for contamination (Figure 10.7).

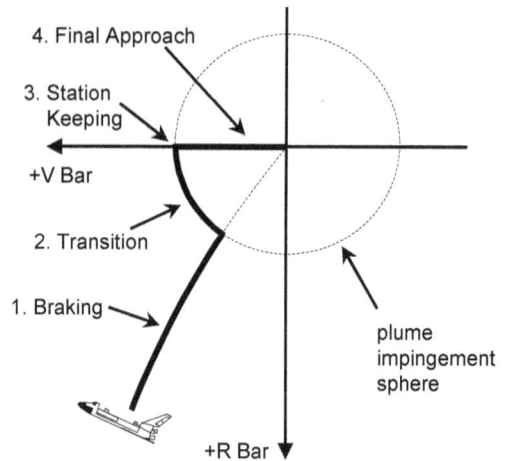

Figure 10.7 Terminal approach to minimize plume impingement on target (October 1973).

The trajectory was designed under the assumption that the target spacecraft could not be designed with features to prevent contamination (such as movable sensor covers), or that control of target attitude could not prevent contamination. A target specific minimum range at which jets could be fired in the direction of the target without a contamination concern was defined. At this point the orbiter would transition from the direct approach trajectory to a station-keeping point on the target velocity vector. After preparations for grapple with the RMS were complete, the orbiter would initiate the final approach to the target.

In 1975, work began on rendezvous procedures for the Long Duration Exposure Facility (LDEF, Figure 10.8) retrieval and Solar Maximum Mission satellite servicing (Figure 10.9), due to an anticipated deployment of LDEF on an early shuttle mission, and the approaching launch of Solar Max on a Delta booster. Issues arising out of these efforts were to have a profound impact on shuttle operational concepts. The large size of the shuttle primary RCS jets (870 pounds thrust) coupled with the small size of LDEF and Solar Max compared to the shuttle led to more concerns about RCS plume impingement effects. Plume impingement could induce attitude rates on the target or even result in separation of the target and shuttle. Targets with attitude control systems may not have been designed to maintain attitude in the presence of orbiter plumes. This was a particular concern for payloads that used gravity gradient stabilization, such as LDEF. Shuttle thruster sizing, placement, and orientation were designed to provide adequate flight control authority throughout the shuttle flight envelope, and to avoid self-impingement of aero surfaces, but impingement of target spacecraft or the RMS was not factored into the design.[3]

By June of 1976, off-line plume impingement simulations using simple math models had been conducted. Results indicated that plume impingement induced dynamics at RMS release or grapple ranges

Figure 10.8   LDEF being maneuvered with the RMS.

Figure 10.9   Attempted retrieval of the Solar Max satellite by an astronaut flying a Manned Maneuvering Unit on STS-41C (April 1984).

could make LDEF deployment and retrieval difficult and perhaps impossible. A development effort was initiated to obtain improved models of shuttle RCS jets and plume physics. New models were required to better characterize impingement effects and test trajectories, piloting techniques, new software, and identify vehicle hardware modifications needed to mitigate impingement effects.

Human in the loop simulations of approach and station-keeping were performed in the Shuttle Engineering Simulator (SES) in March, May, and June of 1976.[4] The approach techniques flown used the same type of contamination/overpressure sphere defined in October of 1973 (Figure 10.7). The simulations also evaluated three types of close-in station-keeping: 1) +V Bar using the Crew Optical Alignment Sight (COAS), 2) +V Bar using the Closed Circuit Television Camera in the payload bay, and 3) +R Bar using the COAS. SES results indicated that the standard Gemini/Apollo inertial approach in the current baseline profile did not mitigate plume concerns. The acceptable approach techniques were +V Bar and +R Bar approaches. However, the +V Bar approach required a new propulsion system to avoid

plume impingement and the +R Bar approach required accurate close-in radar.

Resolving the Plume Impingement and Forward RCS Propellant Problems

By March of 1977 the term "proximity operations" or "prox ops" was coined, and proximity operations became a distinct discipline within the Shuttle Program. Proximity operations occur close to the target (within 2,000 feet), and are characterized by nearly continuous trajectory control, whereas rendezvous control maneuvers typically occur at intervals of hours or tens of minutes.[4]

By April of 1977, after a considerable amount of lobbying by concerned technical and management personnel, potential problems with the ability of the Space Shuttle to retrieve satellites such as LDEF and Solar Max were receiving visibility at high levels within the Shuttle Program and the payloads community external to the Program.

Some proposed solutions to the plume impingement problem, such as alternate recovery techniques using new hardware (stand-off berthing using a mast or tether), a payload bay mounted cold-gas propulsion system, and hardened payloads were not acceptable due to complexity and cost. Operational work-arounds consisting of new piloting techniques and shuttle flight control system modifications were preferred. However, these options increased propellant usage and increased complexity of crew procedures and shuttle software.

Both the Gemini and Apollo vehicles carried ample propellant margins, but the shuttle was limited in terms of forward RCS propellant. The shuttle could run out of forward RCS propellant during the terminal phase (Figure 10.7) under dispersed trajectory conditions, and in the event of a radar failure.

From July to September of 1977, a study of approach and station-keeping techniques was conducted in the SES. The PDRS-III runs were the first to incorporate six degree-of-freedom RCS jet plume effects in a human-in-the-loop simulation. Inertial, +V Bar, +R Bar, and H Bar approaches and station-keeping were evaluated (Figure 10.10). Targets were LDEF and Skylab.

Results confirmed earlier studies, which indicated that an Apollo inertial approach and braking technique caused the gravity gradient stabilized LDEF to tumble. Some +R Bar approaches worked with the Apollo (inertial) approach and technique, due to the natural braking effect of orbital mechanics. Other findings were: 1) a +R Bar approach can be flown to a gravity gradient stabilized spacecraft (LDEF) using +/- X RCS jet firings for braking during –Z body axis approaches, 2) rendezvous radar is required, 3) a more accurate rendezvous radar than the one baselined for the orbiter would ease the piloting task, and 4) the payload bay Closed Circuit Television Cameras (CCTVs) were very useful for sensing small opening or closing rates.[4]

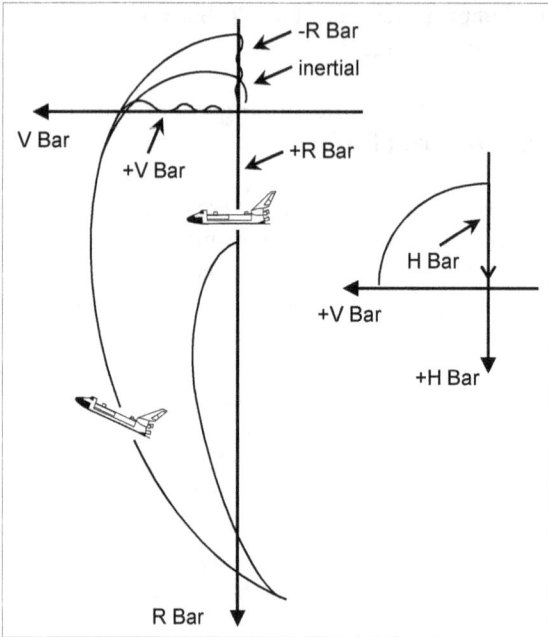

Figure 10.10   Proximity operations approaches.

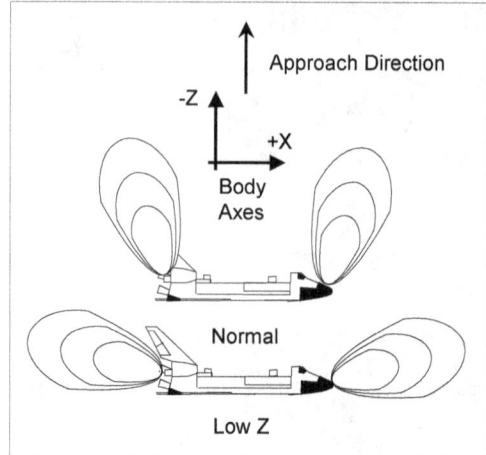

Figure 10.11   Comparison of plumes.

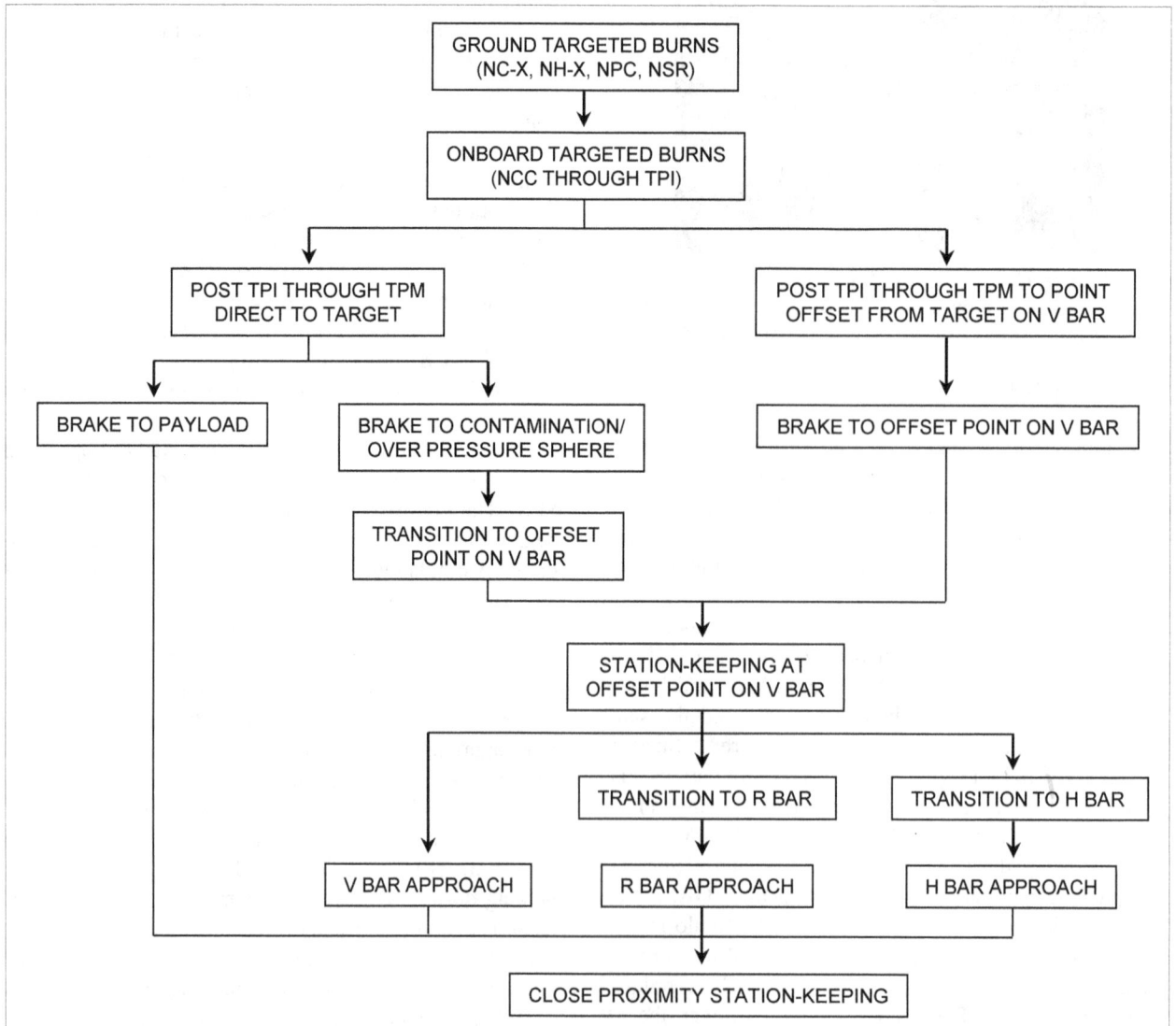

Figure 10.12 Proximity operations options as of November 1977 after the PDRS-III simulations.

The one technique that worked for approaches along all three Local Vertical Local Horizontal (LVLH) frame axes (V Bar, R Bar, H Bar) used the orbiter +/-X body axis RCS jets (Figure 10.11) for braking. These jets had a small component of thrust along the +Z body axis.

A new technique, the lateral or H Bar approach (out-of-plane) was extensively tested to bring it up to the maturity level of the R Bar and V Bar approaches (Figures 10.10 and 10.12).[4,5,6] A CCTV overlay was used as a piloting aid for H Bar approaches. Advantages of the H Bar approach were consistently good lighting conditions for piloting and Y LVLH motion that did not couple into the LVLH X and Z axes. Unlike the +R Bar approach, the H Bar approach did not have natural braking, but had natural acceleration, which necessitated frequent thrusting at the target during approach. Out-of-plane motion still occurred after relative translational rates were nulled. The H Bar approach was never baselined for operational use, due to safety, station-keeping, propellant consumption, and plume impingement concerns.

As a result the orbiter flight control system was modified to provide a "Low Z" mode. This provided some RCS braking capability while minimizing RCS plume impingement (Figure 10.11). Jets used for this mode had a thrust component that was primarily along the X body axis. The serendipitous canting of the aft X axis RCS jets was not an original design requirement for proximity operations. The braking contribution provided by the scarfed, nose mounted X axis RCS jets is negated by RCS firings to control pitch. Upward firing RCS jets were inhibited in Low Z. However, use of the Low Z mode was expensive in terms of propellant use. The ability to perform an attitude hold with respect to the LVLH frame was also added to the shuttle flight software.[4]

Figure 10.12 illustrates the proximity operations approach options that existed as of November 1977.[4] The inertial approach, brake to payload, was the legacy approach from the Gemini, Apollo, Skylab, and Apollo-Soyuz missions. The contamination sphere, V Bar, R Bar, and H Bar approaches were developed in response to shuttle plume impingement on the target spacecraft concerns. However, further proximity operations technique development would be required before the first proximity operations mission (STS-7, June 1983) and the first rendezvous mission (STS-41C, April 1984).[7,8]

The shuttle never flew Gemini or Apollo style inertial approaches or H Bar approaches. Most shuttle rendezvous missions in the 1980s to mid 1990s flew V Bar approaches. Some low energy inertial approaches were flown due to target spacecraft inertial attitude requirements. The first +R Bar approach was flown on STS-66 in November of 1994. +R Bar approaches became viable with the addition of proximity operations sensors (Trajectory Control Sensor and Hand Held Lidar) and the Rendezvous and Proximity Operations Program.

Summary of the Plume Impingement Issue and Resolution

Plume impingement was a significant technical challenge that had to be overcome before the first rendezvous and proximity operations missions could be planned and flown. The issue arose from shuttle RCS system design, grappling target spacecraft with a robotic arm, and the sizes and characteristics of target spacecraft attitude control. Although new simulation capabilities and flight control techniques had to be developed, the issue was resolved without extensive modification of the shuttle or target spacecraft. However, extensive mission specific planning and analysis was required to mitigate plume impingement risk to the wide variety of target vehicles associated with the Shuttle Program.

References

1. Conrad, C., and J. P. Kerwin, "Skylab 1 and 2 – Pilot's Report," *Proceedings of the Seventeenth Symposium of the Society of Experimental Test Pilots*, Society of Experimental Test Pilots, Lancaster, CA, 1973, pp. 211-218.

2. Ezell, E. C., and L. N. Ezell, *The Partnership: A History of the Apollo-Soyuz Test Project*, NASA SP-4209, NASA, Washington, DC, 1978.

3. Nakano, M. M., and Williams, R. L., "Space Shuttle On-Orbit Flight Control System," *Proceedings of the AIAA Guidance and Control Conference*, AIAA, Reston, VA, 1982.

4. Goodman, John, *Space Shuttle Proximity Operations Papers, Volume 1 of 2 (1970-1979)*, JSC-35052, NASA/JSC Flight Dynamics Division, July 2011.

5. Olszewski, O., *Orbiter/Payload Proximity Operations SES Postsim Report, Lateral Approach and Other Techniques*, JSC Internal Note No. IN-78-FM-16, JSC-13913, NASA/JSC Mission Planning and Analysis Division, March 28, 1978. See also Space Shuttle Proximity Operations Papers, Volume 1 of 2 (1970-1979), JSC-35052, in the A Note On Sources appendix.

6. Mosel, Duane K., *Terminal Phase Rendezvous Propellant Minimization Techniques Post Simulation Report*, NASA JSC Crew Training and Procedures Division, May 17, 1978. See reference 4 above.

7. Yglesias, J. A., *Proximity Operations Analysis - Retrieval Of The Solar Maximum Mission Observatory*, JSC Internal Note IN-80-FM-11, JSC-16411, NASA/JSC Mission Planning and Analysis Directorate, April 28, 1980.

8. Yglesias, J. A., and M. Donahoo, *Retrieval Techniques - LVLH and Inertially Stabilized Payloads*, JSC Internal Note IN-80-FM-12, JSC-16425, NASA/JSC Mission Planning and Analysis Directorate, May 28, 1980.

CHAPTER 11 - ON-BOARD SYSTEMS

Relative Navigation Sensors

The Space Shuttle presented new technical and operational challenges. The shuttle had a requirement to rendezvous with spacecraft that were not modified or equipped to support cooperative tracking. Target spacecraft in the Gemini, Apollo, Skylab, and Apollo/Soyuz Programs possessed strobes and radio-frequency transponders to support long-range cooperative relative tracking. Unlike Gemini and Apollo, shuttle rendezvous target spacecraft would not necessarily possess active navigation aids (transponders or lights), nor were many of them originally designed to support rendezvous, retrieval, and on-orbit servicing. The use of radar skin tracking and reflected sunlight for star tracking led to concerns about detection and tracking of targets during shuttle development.

The number of ground radar tracking stations would be lower than those available during Gemini and Apollo. Not all shuttle rendezvous targets were assumed to have ground radar compatible transponders. Later, ground tracking uncertainties were deemed too conservative.

Optical tracking would be provided by one of two star trackers, which were also to be used for aligning the Inertial Measurement Units.[1] The trackers had field of view restrictions based on Earth limb and bright object considerations (Sun, Moon). Availability of optical measurements, which used target reflected sunlight to facilitate acquisition and tracking, was seen as a major challenge before shuttle rendezvous missions were flown. Strobes, used on targets in previous programs for optical tracking via the human eye, were judged to be incompatible with the shuttle star trackers.

Original shuttle rendezvous navigation requirements called for a radar range of 300 nm, provided that the target was equipped with a transponder. Skin tracking (no transponder) of a target with a 1 square meter cross section out to a range of 10 nm would also be available.[2] In the mid 1970s radar development costs led to examination of deferral of radar operational capability, which would have resulted in many early rendezvous missions not having radar. The cost of Ku band radar development also motivated the study of alternative sensors. "All optical rendezvous" was studied, but simulations indicated that the probability of successful dual coelliptic rendezvous (Figure 9.2) under dispersed conditions was less than desirable due to increased RCS propellant consumption.

A proposal to equip shuttle rendezvous targets with Tactical Air Navigation (TACAN) transmitters was briefly examined in 1976. This option would take advantage of shuttle TACAN receivers and measurement processing software already under development to support shuttle landings. However, this would have required equipping target spacecraft with TACAN transmitters. Furthermore, this would not have enabled the shuttle to meet the passive relative navigation requirement.

The decision to proceed with Ku radar development in the mid 1970s was in part motivated by concerns about the proposed Skylab reboost mission that might have flown in 1979 or 1980. The Ku antenna and electronics would also be used for communications through the Tracking and Data Relay Satellite System (TDRSS).

In late 1976 consideration was given to equipping the orbiters with two Ku band radars to provide relative navigation sensor and burn targeting redundancy. However, the cost and weight penalties of two rendezvous radars did not justify the additional redundancy during rendezvous. In addition, a proposal to move the rendezvous radar from the starboard (right) side of the payload bay to the port (left) side (and moving the data link only Ku from the port side to the starboard side) was not approved due to a consideration of cost versus benefit. If approved radar tracking could have been performed in parallel with –Y star tracker tracking.

In the spring of 1977 consideration was given to modifying the Ku radar to provide proximity operations tracking at a minimum range of 35 to 50 feet, rather than the specified 30 meters. This was not pursued due to cost.

Cost overruns prevented the acquisition of target transponders and spare parts for the shuttle radar, and the passive skin tracking mode of radar operation was the normal operating mode, which in turn limited the range of the radar (10 nm specification, ~22 nm maximum range). This was a factor in the inability of the shuttle to meet rendezvous autonomy requirements. A target spacecraft that required the transponder never appeared, and the Shuttle Program requirement to support cooperative rendezvous (radar transponder) was deleted in the mid 1980s.

Crew Optical Alignment Sight (COAS) line-of-sight measurements could be processed in the Kalman filter as a backup to the radar and star trackers. Relative navigation sensor measurements from the radar, star tracker, and COAS were processed in a Kalman filter that built upon the Apollo experience.[3-5] Original filter requirements called for an optimal filter that updated both the shuttle and target state vectors, but the 1976 on-board computer requirements scrub resulted in the filtering of only one state vector, as was done on Apollo.[6] However, the shuttle rendezvous navigation sensors and software were more capable than the Apollo sensors and software. The shuttle relative navigation filter did not require modification over the life of the program. The operational envelope of shuttle rendezvous sensor data. processed by the Kalman filter is depicted in Figure 11.1.

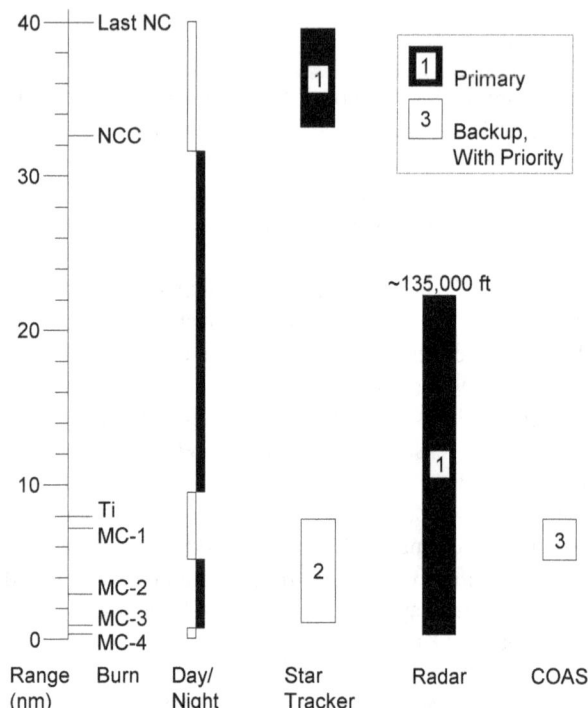

Figure 11.1 Operational use of shuttle rendezvous sensors for a typical ISS mission.

As a back-up to the radar at close range, the COAS could be used to obtain sub-tended angle measurements that would be used to obtain a rough estimate of range though the use of a chart. This method required that the target be in a known and stable attitude. The COAS would also later see extensive use during proximity operations as a piloting aid.[7] Tilt angles of the payload bay Closed Circuit Television (CCTV) cameras could also be used to obtain a rough estimate of range at close range during proximity operations.

By the mid 1970s, when the rendezvous navigation sensors and relative navigation capability was baselined, proximity operations culminated with approach and grapple of a spacecraft with a robotic arm. The baselined rendezvous radar was adequate to support this, although it lacked effective redundancy. There were concerns about the lack of a back-up range and range-rate measurement device for the Ku band rendezvous radar, particularly during proximity operations and the proposed Skylab re-boost mission (canceled in December of 1978). A number of potential off-the-shelf solutions were examined in the mid and late 1970s, such as hand-held police radar and battlefield lasers. None of the off-the-shelf options evaluated were satisfactory.

A laser rangefinder used for auto-focusing the payload bay CCTV cameras was tested during proximity operations on STS-41B and STS-41C (1984), but limitations in range and range rate accuracy limited its usefulness. A parallax rangefinder and a night vision

system were also tested on early missions in the 1980s, but performance was not adequate.

The baseline proximity operations sensors and piloting aids (radar, COAS, CCTV cameras) were successfully used to null the final approach velocity to zero to support grapple of a spacecraft by the Remote Manipulator System (RMS) on rendezvous missions. This success made it difficult to obtain sufficient priority and resources to address the concern about lack of a backup source of range and range rate measurements.

However, these sensors and piloting aids had operational limitations. Rendezvous radar tracking for even small payloads reached the end of the tracking envelope at about 80 feet. Visual ranging methods using COAS subtended angles or two CCTV camera tilt angles were imprecise, imposed a heavy crew work-load during an already busy phase of flight, and did not provide the crew with a direct range-rate measurement. On-orbit use of the CCTVs was frequently limited by extreme orbital lighting conditions that caused image blooming.

An orbit determination autonomy requirement also existed for the shuttle. Processing of TDRSS one way Doppler by the shuttle computers was considered, but not pursued due to on-board computer memory limitations. During the late 1970s, use of the Global Positioning System (GPS) was also examined, but it was not adopted due to cost and the immaturity of the technology.[8] The navigation autonomy requirement could not be met, and the shuttle would rely on ground based orbit determination using radar, and later in the program, measurements from TDRSS tracking. Single TDRS ground navigation was certified in October of 1984. Two TDRS ground navigation was certified after tests during STS-29, STS-30, and STS-32.

Burn Targeting

The ground-targeted phase of rendezvous begins after orbit insertion (Table 11.1). Rendezvous burns are computed by Mission Control using orbit determination data obtained by processing ground radar and TDRSS Doppler measurements. The length of this phase varies, and typically lasts several days. Although a ground-targeted phase burn plan is determined before launch, some adjustments may be required after launch due to shuttle ascent performance dispersions, or shuttle or target spacecraft systems problems.

The on-board targeted phase begins once shuttle sensors (the first is star tracker, Figure 11.1) are able to obtain relative measurements. Shuttle orbit adjustments are then computed on-board, while Mission Control computations are available as a back-up, in the event of an on-board system anomaly. Unlike the ground-targeted phase, activities from the beginning of on-board relative navigation to the beginning of proximity operations (at a range of ~2,000 feet) may change little from flight to flight.

FEB „72 — RENDEZVOUS TARGETING SPACE SHUTTLE GN&C EQUATION DOCUMENT NO. 7 (REV. 2)

JULY „73 — RENDEZVOUS TARGETING SPACE SHUTTLE GN&C EQUATION DOCUMENT NO. 7 (REV. 3)

JAN „75 — ORBITAL MANEUVER PROCESSOR (OMP) JSC-09339

SOFTWARE FOR ABSOLUTE AND RELATIVE NAV PHASE BURNS NC-NH-NSR1-NCC-NSR2-TPI-TPM

JULY „75 — ORBITAL MANEUVER PROCESSOR (OMP) JSC-09339, REVISION 1

SOFTWARE FOR RELATIVE NAV PHASE BURNS NCC-NSR-TPI-TPM CORE = 1.00

SCRUB

JULY „76 — RENDEZVOUS TARGETING SPECIALIST FUNCTION JSC-11446

APRIL „77 — RENDEZVOUS TARGETING SPECIALIST FUNCTION JSC-11446, REVISION 1

MARCH „78 — FUNCTIONAL SUBSYSTEM SOFTWARE REQUIREMENTS, GUDIANCE, ONORBIT I, SD 76-SH-0003A

SECTION 4.7 RENDEZVOUS TARGETING SPECIALIST FUNCTION SCRUBBED

TASKS FROM RENDEZVOUS TARGETING SPECIALIST FUNCTION
• STATE VECTOR UPDATE
• PRECISION VELOCITY REQUIRED
• ELEVATION ANGLE SEARCH

NEW TASKS
• LAMBERT
• ORTHOGONAL BRAKING

COMBINE

ORBIT TARGETING SPECIALIST FUNCTION JSC-14393 — AUG „78

NEW SECTION 4.7 ORBIT TARGETING SPECIALIST FUNCTION CHANGE REQUEST 12813A — AUG „78

COMBINE

SOFTWARE FOR RELATIVE NAV PHASE BURNS NCC-NCC-TPI-TPM PLUS PROXIMITY OPERATIONS CORE = 0.85

FUNCTIONAL SUBSYSTEM SOFTWARE REQUIREMENTS, GUDIANCE, ONORBIT I, SD 76-SH-0003A PAGE CHANGE 3 — DEC „78

JAN „77 — PROXIMITY OPERATIONS SIMULATIONS & TARGETING STUDIES

AUG „77 — PROXIMITY OPERATIONS SOFTWARE PRELIMINARY REQUIREMENTS JSC-13027

MAY „78 — PROXIMITY OPERATIONS SPECIALIST FUNCTION JSC-13993

SOFTWARE FOR PROXIMITY OPERATIONS CORE = 0.4

Based on a November 30, 1978 presentation by R. L. West of McDonnell Douglas.

JSC documents were kept in the NASA/JSC Scientific and Technical Information Center in Building 45.

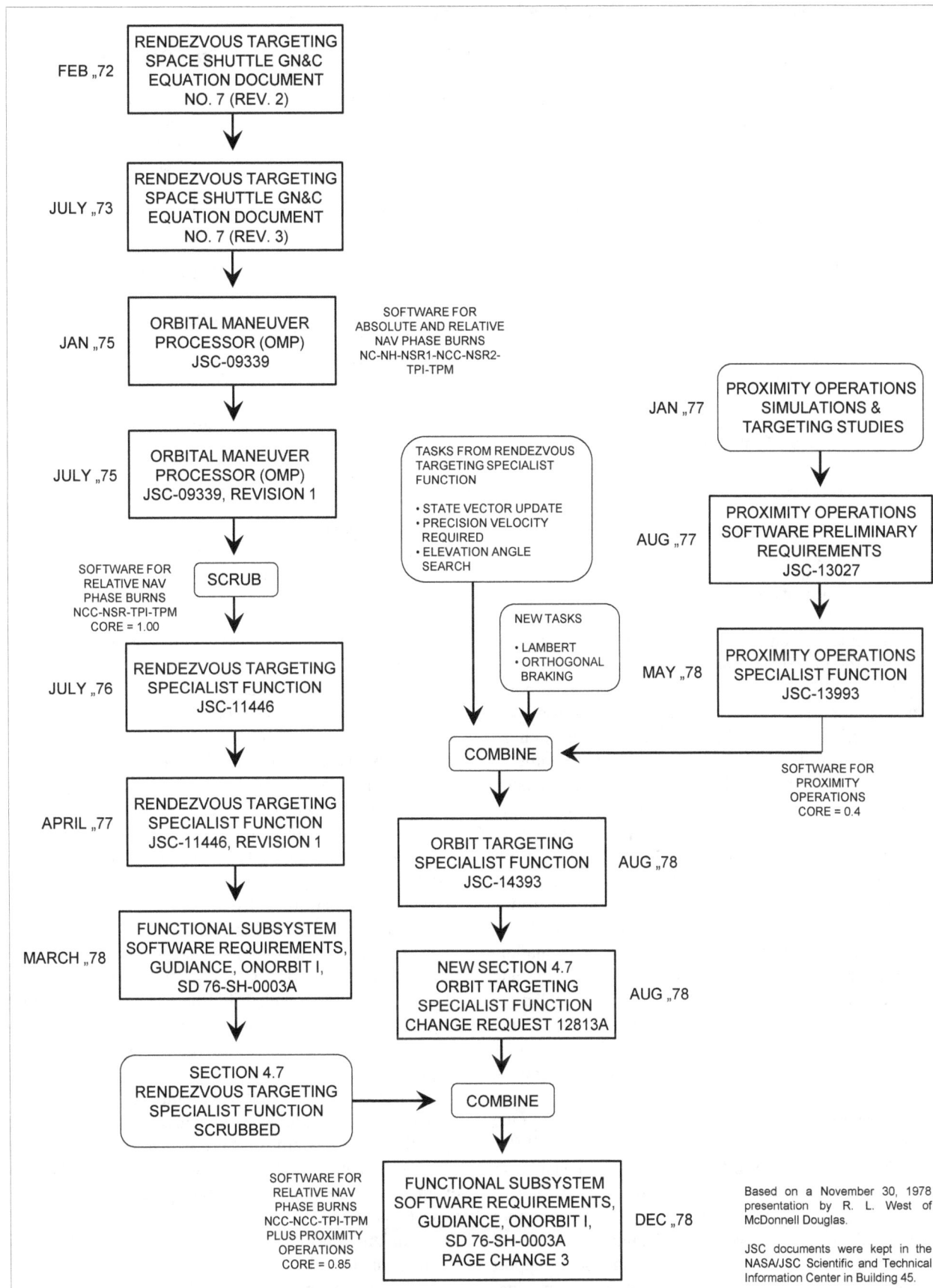

Figure 11.2 On-board targeting software evolution in terms of requirements documents. The final Orbit Targeting Specialist Function consisted of Lambert Targeting, Clohessy-Wiltshire Targeting, and Orthogonal Braking. Only Lambert targeting was used on shuttle missions. See JSC-35053 in the A Note on Sources appendix.

Table 11.1 Rendezvous and Proximity Operations Mission Regimes Circa 1978

Regime	Maneuvers or Activities	Computed By	State Determination
I	Phase Adjustment (NC1) Plane Change (NPC) Height Adjustment (NH) Coelliptic (NSR1)	Mission Control	Mission Control using S-Band, C-Band, TDRSS. On-board using one-way Doppler or GPS.
II	Phase, Height, and Plane Adjustment (NCC) Coelliptic (NSR2) Terminal Phase Initiation (TPI) Terminal Phase Mid-Course (TPM)	On-board (primary) Mission Control (backup)	Mission Control using S-Band, C-Band, TDRSS. On-board using star tracker, radar.
III	Braking Station-keeping Final Approach Terminal Station-keeping	On-board targeting or charts & crew procedures using radar data & COAS.	On-board using radar.
IV	Payload Grapple Payload Berthing in Payload Bay Payload Un-Berthing Payload Deployment	N/A	N/A
V	Terminal Station-keeping Departure Fly-around Separation	On-board targeting or charts & crew procedures using radar data & COAS.	On-board using radar.

Definition of on-board rendezvous burn targeting concepts for the Space Shuttle was underway by 1972 (Figure 11.2). The original intention for shuttle on-board targeting was that it would handle all burns from post orbit insertion through final approach under nominal conditions. In 1974, a requirement for the shuttle to conduct autonomous rendezvous (little or no support from Mission Control) existed. Mission Control burn targeting would be available as a backup for cross-checks of on-board targeting and for targeting contingency burns in the event that off-nominal performance or changes in mission plans required changes to the rendezvous profile. Astronauts were to compute a nominal series of burns and execute them without Mission Control confirmation. For off-nominal scenarios, the crew could compute and execute a rendezvous plan with inputs from checklists or Mission Control. The on-board computer would not recommend actions in response to off-nominal situations. Mission Control was still to be able to compute burns, particularly in the event of off nominal scenarios.

Algorithms to perform this on-board targeting (including the baseline dual coelliptic NC1-NH-NSR1-NCC-NSR2-TPI-TPF profile), called the Orbit Maneuver Processor (OMP), were documented in a 1975 JSC document (Figure 11.2). On-board OMP was more flexible than its predecessors (Gemini, Apollo, and Skylab Docking Initiation, or DKI) and could support different combinations of burns without reprogramming. It was also capable of targeting all orbital burns from insertion through intercept. However, limited on-board computer capacity made the targeting autonomy requirement difficult to meet. On-board OMP did not become actual software requirements due to the 1976 software scrub led by Apollo 13 astronaut Fred Haise. The 1975 on-board OMP algorithms did form the basis of a later version of OMP used in Mission Control for targeting burns during the Shuttle Program.

In April of 1977 a scrubbed version of the 1975 on-board OMP document was released called the Rendezvous Targeting Specialist Function (RTSF, see Figure 11.2). Targeting for burns performed outside of relative sensor range (phasing, or NC, and altitude, or NH) was to be performed by Mission Control (Table 11.1 and Figure 11.2). Only burns performed after on-board relative sensor measurement acquisition would be targeted on-board (NCC-NSR-TPI-TPM). These algorithms were included in the first orbit targeting Functional Subsystem Software Requirements (FSSR) document published by Rockwell in March of 1978 (Figure 11.2).

In April of 1977 concerns about Reaction Control System (RCS) jet plume impingement during proximity operations and forward RCS propellant depletion led the Mission Planning and Analysis Division (MPAD) to examine the creation of proximity operations targeting algorithms based on the Clohessy-Wiltshire equations. The proximity operations targeting function was believed to be a method for lowering forward RCS propellant consumption and permitting final approaches from any direction to minimize plume impingement (Figure 10.10). This would also provide maximum flexibility during mission planning. It was intended for targeting Mid-

Course Correction (MCC, or Terminal Phase Mid-course, or TPM) burns after the TPI burn, long range circumnavigation, and proximity operations.

However, the RTSF would not fit in the limited memory of the shuttle flight computer, nor would both RTSF and the new Clohessy-Wiltshire targeting function fit in the flight computer. RTSF was scrubbed from the software requirements and replaced with the simple addition of the Proximity Operations Specialist Function (Figure 11.2). Studies indicated that the Clohessy-Wiltshire targeting package might not be able to adequately support burns with longer transfer times, such as TPI. Later a decision was made to expand the Proximity Operations Specialist Function to include two-impulse, point-to-point Lambert targeting in order to increase the range of operations of the targeting software. To reflect the addition of the longer range Lambert targeting functionality the Proximity Operations Specialist Function was renamed the Orbit Targeting Specialist Function (Figure 11.2).

RTSF deletion also meant that the on-board targeted phase profile was changed from NCC-NSR-TPI-TPM to NCC-NCC-TPI-TPM. A NCC (corrective combination) burn was designed to function as a coelliptic (NSR) burn since the NSR targeting capability was scrubbed with the RTSF. Lambert performed point-to-point targeting and did not retarget the entire burn sequence each time targeting was performed, as was done in on-board OMP requirements (1975) and RTSF (1977). Changes to the March 1978 FSSR algorithms (RTSF deletion, addition of Lambert and Clohessy-Wiltshire targeting) were approved in August of 1978 and a new requirements documents containing the new Lambert and proximity operations targeting algorithms was published in March of 1979. At that time Clohessy-Wiltshire equations were preferred for proximity operations over Lambert targeting as it was assumed that Lambert targeting would take too long on a shuttle flight computer (30 to 60 seconds) compared to Clohessy-Wiltshire targeting requiring about 5 seconds to complete. However, later experience indicated that Lambert targeting typically took less than 15 seconds to execute.

Even with the addition of proximity operations targeting based on the Clohessy-Wiltshire equations, there was still a need for manual proximity operations piloting. Proximity operations targeting was dependent on good radar data and relative navigation using the Kalman filter in the flight software. It was expected that the quality of radar measurements with large target spacecraft could introduce variation in burn targeting solutions. Manual piloting was still needed as a backup for flying transitions to, from, and between the V Bar and R Bar axes, approaches along the V Bar and R Bar axes, and glideslope approaches. Manual piloting directly controlled range, range rate, and line-of-sight angles while Clohessy-Wiltshire targeting performed point-to-point transfers over a fixed time interval. Approach and

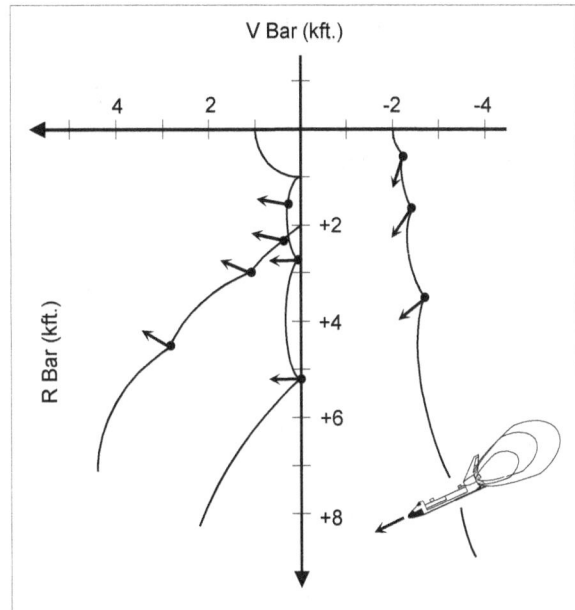

Figure 11.3 Approaches using orthogonal braking (1978). Arrows indicate DV.

separation trajectories needed to be manually flyable to preserve the backup option of manual piloting, rather than simply bouncing from point-to-point using the Clohessy-Wiltshire two point targeting. The need to have manual piloting techniques for backup and direct manual control of range, range-rate, and line-of-sight angles as opposed to point-to-point targeting over a fixed transfer time made Clohessy-Wiltshire targeting less desirable than near continuous manual piloting.

The Clohessy-Wiltshire based targeting also included an orthogonal braking algorithm. This algorithm solved for the transfer time between proximity operations braking burns so that the burn was orthogonal to the line-of-sight to the target spacecraft (Figure 11.3). The orbiter –Z body axis (out of the payload bay) was pointed at the target during final approach. Orthogonal braking provided +X body axis RCS jet burns. This technique mitigated risk of plume impingement and avoided use of forward RCS propellant. Aft RCS propellant consumption was not as critical. However, orthogonal braking was never used on shuttle missions. Analysis of orthogonal braking during +R Bar approaches indicated that performance under nominal trajectory conditions was acceptable. However, under dispersed conditions the nominally 5 minute transfer time between two burns could decrease by several minutes.

Clohessy-Wiltshire targeting, like Lambert targeting, was upgraded with a precision predictor outer loop to account for atmospheric drag and non-conic gravity. It also had a multi-revolution capability. However, the Clohessy-Wiltshire algorithm was never used in flight as it performed essentially the same function as Lambert targeting. The Lambert algorithm was used for all on-board targeted burns over the life of the Shuttle Program. The forward RCS propellant depletion problem that in

Table 11.2 Maneuver Solution Sources Available During the On-Board Targeted Phase of Rendezvous

Vehicle	CSI/NCC [A]	CDH/NSR [B]	TPI/Ti [C]	Mid-Course Corrections [D]
Gemini	• Gemini Computer [E] • Mission Control	• Gemini Computer [E] • Mission Control	• Gemini Computer • Crew Charts • Mission Control	• Gemini Computer • Crew Charts
Apollo LM	• LM PGNS • LM AGS • LM Crew Charts • CSM PGNCS • Mission Control	• LM PGNS • LM AGS • LM Crew Charts • CSM PGNCS • Mission Control	• LM PGNS • LM AGS • LM Crew Charts • CSM PGNCS • Mission Control	• LM PGNS • LM AGS • LM Crew Charts • CSM PGNCS
Apollo CSM	• LM PGNS • LM AGS • LM Crew Charts • CSM PGNCS • Mission Control	• LM PGNS • LM AGS • LM Crew Charts • CSM PGNCS • Mission Control	• LM PGNS • LM AGS • LM Crew Charts • CSM PGNCS • Mission Control	• LM PGNS • LM AGS • LM Crew Charts • CSM PGNCS
Skylab CSM	• CSM PGNCS • CSM Crew Charts • Mission Control	• CSM PGNCS • CSM Crew Charts • Mission Control	• CSM PGNCS • CSM Crew Charts • Mission Control	• CSM PGNCS • CSM Crew Charts
Apollo/Soyuz CSM	• CSM PGNCS • Mission Control	• CSM PGNCS • HP-65 Calculator • Mission Control	• CSM PGNCS • HP-65 Calculator • Mission Control	• CSM PGNCS • HP-65 Calculator
Space Shuttle	• Shuttle Computer • Mission Control	• Shuttle Computer • Mission Control	• Shuttle Computer • Mission Control	• Shuttle Computer • Mission Control

[A] Skylab and Apollo-Soyuz and the Space Shuttle performed an NCC maneuver. CSI was executed on Apollo missions 9, 10, 11, and 12. Gemini X, XI, and XII could perform on-board targeting for phasing, height, and out-of-plane corrections.
[B] NSR was executed on Gemini, Apollo 7, Skylab, and Apollo Soyuz. CDH was executed on Apollos 9, 10, 11, and 12. Gemini X and XII could perform on-board targeting of NSR.
[C] Gemini, Apollo, Skylab, and Apollo/Soyuz performed TPI maneuvers. The Space Shuttle performed Ti maneuvers.
[D] Gemini, Apollo, Skylab, and Apollo/Soyuz performed 2 mid-course corrections after the TPI maneuver. The Space Shuttle performed 4 mid-course corrections after the Ti maneuver.
[E] Gemini X executed pre-TPI targeting on-board for comparison with ground as part of a sextant navigation test. Gemini XI targeted two pre-TPI burns on-board as part of the short rendezvous. It is not clear if the Gemini X pre-TPI on-board solutions were executed or if ground targeted solutions were executed. Gemini XII targeted and executed an on-board NSR.
AGS – Abort Guidance Section, **CDH** – Constant Delta Height, **CSI** – Coelliptic Sequence Initiation, **CSM** – Command Service Module, **HP** – Hewlett Packard, **LM** – Lunar Module, **NCC** – Corrective Combination, **NSR** – Slow Rate, **PGNCS** – CSM Primary Guidance, Navigation, and Control System, **PGNS** – Lunar Module Primary Guidance and Navigation Section, **Ti** – Transition Initiation, **TPI** – Terminal Phase Initiation. See Appendix H for an explanation of burn names.

1977 was believed to be solvable with proximity operations targeting was eventually solved with the adoption of the stable orbit profile in April of 1983. The stable orbit profile lowered the closing rate during final approach by an order of magnitude.

Unlike Gemini and Apollo, the shuttle crew did not use charts or a programmable calculator to compute backup burn solutions (Table 11.2). This was due to the increased confidence in the shuttle GNC system design and the successful experience with the Apollo GNC system and relative sensors. In addition, shuttle flight computer redundancy provided an additional source of burn solutions. For shuttle rendezvous missions before Mir two flight computers executed GNC software in parallel during rendezvous and proximity operations. With the Mir flights the number was increased to three.

Crew Interface with the Flight Computer

Advances in computer technology permitted the shuttle to have a more advanced crew/flight computer interface than Gemini and Apollo. Systems displays provided the crew with item entries to execute commands and digital data. Some of the systems displays used during rendezvous and proximity operations are depicted in Figure 11.4. The keyboard is in Figure 11.5.

In the mid 1990s an effort began to upgrade the shuttle cockpit avionics. The first phase of the upgrade was to replace many of the 1970s era mechanical displays with images on flat panel displays. This was known as the glass cockpit and all orbiters in the fleet were so modified. The second phase of the upgrade, the Cockpit Avionics Upgrade (CAU) developed new crew

```
2021/033/        REL NAV          1 005/02:07:37
RNDZ NAV ENA 1*   SV UPDATE        000/00:00:00
KU ANT   ENA 2*    POS  0.02     AVG G ON     5*
MEAS     ENA 3    VEL  0.03    ┌──────GPS──────
            NAV                │      STAT  P 1σ DES
SV SEL 4 FLTR      RR GPC      │1           75 31
RNG   25.721    RNG   25.610   │2*          75 32
Ṙ   - 21.19     Ṙ   - 20.80    │3           75 33
θ     25.76     EL   -  0.4    │   SV TRANSFER
Y    + 0.08     AZ   -  0.8    │FLTR MINUS PROP
Ẏ    + 0.1     ωP     + 0.8    │ POS          4.99
NODE  2:54:00   ωṘ     + 0.1   │ VEL          5.72
            FILTER             │FLTR TO PROP  8
S TRK 12    RR 13*   COAS 14   │PROP TO FLTR  9
STAT                X - 3.4    │ORB TO TGT   10
FLTR UPDATE  15 ORB  Y +4.4    │TGT TO ORB   11
COVAR REINIT 16               └──EDIT OVRD──
         RESID  RATIO ACPT REJ   AUT  INH FOR
RNG   + 0.09    0.2  807    0   17*  18   19
Ṙ     + 0.02    0.0  807    0   20*  21   22
V/EL/Y +0.02    0.0  807    0   23*  24   25
H/AZ/X +0.01    0.0  807    0
GPS P + 0.19    0.4               42  43*  44
      V + 0.26  0.3
```

```
2021/034/      ORBIT TGT         1 001/13:01:41
                                 000/00:02:29
MNVR       TIG      ΔVX     ΔVY     ΔVZ    ΔVT
  9    1/13:27:31  -  0.1  -  0.2  +  0.4  +  0.4
                              PRED MATCH =    2
 INPUTS                           CONTROLS
1 TGT NO               9         T2 TO T1    25
2 T1 TIG       1/13:27:31        LOAD        26
6  EL              0.00          COMPUTE T1  28
7  ΔX/DNRNG    [-] 194.98        COMPUTE T2  29
8  ΔY          [+]  0.30
9  ΔZ/ΔH       [+] 34.62
10  ΔX         [+] 67.18
11  ΔY         [+]  0.25       ┌─ORBITER STATE─
12  ΔZ         [+] 14.88       218/19:39:47.578
13 T2 TIG      1/14:25:13       X  - 4756.307
17  ΔT         [+] 57.7         Y  -16423.261
18  ΔX         [-] 48.60        Z  -14042.513
19  ΔY         [ ]  0.00       VX +16.973446
20  ΔZ         [+]  1.20       VY -14.712980
21 BASE TIME   1/14:25:13      VZ +11.500532

ITEM 28 EXEC
```

```
2021/  /     ORBIT MNVR EXEC      1 001/17:27:30
OMS BOTH 1                        000/00:00:00
     L 2*          BURN ATT
     R 3           24 R 120     ΔVTOT        9.1
RCS SEL  4        25 P 232      TGO        0:11
5 TV ROLL    0    26 Y 017
TRIM LOAD         MNVR 27 AUTO    VGO X +   8.48
6  P   [-]0.1     TTG                 Y +   2.01
7  LY  [+]5.2     REI                 Z +   2.49
8  RY  [+]5.2     TTP 39:23
9 WT 244124           GMBL           HA    HP
10 TIG              L     R      TGT 195  +182
   1/17:27:30.0  P -0.3  -0.3    CUR 195  +178
TGT PEG 4        Y +5.1  -5.1
14 C1
15 C2  [] ⌐      PRI 28*  29*   35 ABORT TGT
16 HT            SEC 30   31
17 θT            OFF 32   33      FWD RCS
18 PRPLT []                        ARM   36
TGT PEG 7        GMBL CK  34       DUMP  37
19 ΔVX [+]  0.0                    OFF   38*
20 ΔVY [-]  0.3  LAMBERT          SURF DRIVE
21 ΔVZ [-]  0.0                     ON   39
LOAD 22/TIMER 23                   OFF  40*

EXEC
```

```
2011/021/        IMU ALIGN        1 006/22:39:35
IMU                               000/00:00:00
         1    2    3
STAT                            ALIGN ENA
TEMP    OK   OK   OK     IMU 1 10*  REF STAR 13*
STBY   21   22   23        2 11*    IMU       14
OPER    4*   5*   6*       3 12*  TYPE 15 TORQUE
DES     7    8    9
                                   EXEC 16*
ACC                                TERM 17
         1       2       3
X  +0.00 +0.03 -0.01           IMU BITE MASK
Y  +0.02 +0.00 +0.02            1  0000   24
Z  -0.00 -0.00 -0.01            2  0000   25
ANG                             3  0000   26
        1       2       3
X   16.47   16.65   16.72
Y  284.33  284.31  284.31
Z  186.56  186.29  186.32      NAV ΔV THRESH
ΔX + 0.04 - 0.04 + 0.03         18     3840
ΔY + 0.03 - 0.04 + 0.04
ΔZ + 0.04 - 0.04 + 0.03         MM READ  19

ITEM 16 EXEC
```

```
2011/022/      S TRK/COAS CNTL    1 005/02:07:37
                                  000/00:00:00
S TRK CNTL -Y  -Z       S TABLE    1    2    3
SELF-TEST  1   2        TRK ID     0    0    0
STAR TRK   3   3        Δ MIN      0    0    0
TGT TRK    5   6*       ANG DIF  0.0  0.0  0.0
BREAK TRK  7   8          ERR  0.00 0.00 0.00
TERM IDLE  9*  10        SEL       17   18   19
                       S TABLE CLR 20
S STRK     -Y     -Z        COAS
REQD ID 11   0  12  0    REQD ID  21  11
TRK ID      00      0    DDEG X      + 2.6
S PRES       *       *        Y     - 0.4
ΔANG      0.00    0.00   SIGHT MODE  22
THOLD  13   0  14  0    ACCEPT      23
SHUTTER    OP     OP    CAL MODE    24
    MAN OP 15     16*   DES         25*
STATUS                 POS  +X 26   -Z 27*
                       DBIAS 0.00   0.00
                       UPDATE 28    29
```

```
2011/  /        UNIV PTG          1 004/23:53:00
   CUR MNVR COMPL 23:53:20        000/00:20:54
1 START TIME __0/00:00:00       CUR FUT    CUR FUT
MNVR OPTION    START MNVR 18    RBST 25
5 R  __0.00             TRK 19*  CNCL 26
6 P   _0.00             ROT 20   DURATION 27
7 Y   _0.00             CNCL 21  _0:00:00.00
TRK/ROT OPTIONS        ATT MON
8 TGT ID __1          22 MON AXIS 1 +X
                        ERR TOT 23*
9 RA  __0.000           ERR DAP 24
10 DEC _0.000
11 LAT _0.000                ROLL  PITCH   YAW
12 LON _0.000          CUR 212.39  6.41   4.09
13 ALT ___0.0          REQD 212.44 10.53  6.70
                       ERR + 0.43 - 4.86 + 0.00
14 BODY VECT 3         RATE - 0.001 -0.169 -0.002
15 P  _90.00
16 Y   0.00
17 OM

ITEM 19 EXEC
```

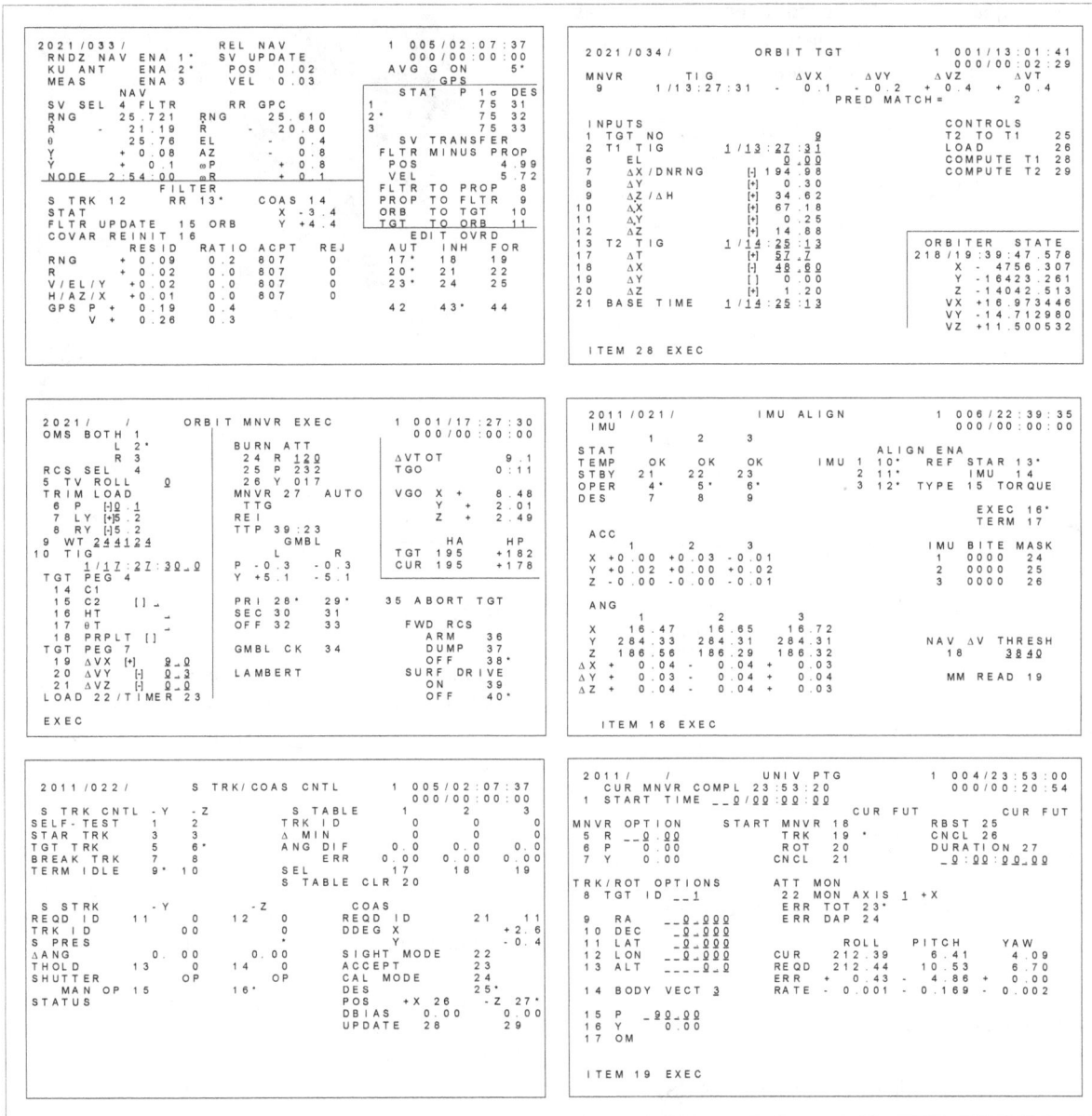

Figure 11.4 Shuttle flight computer displays used during rendezvous and proximity operations.

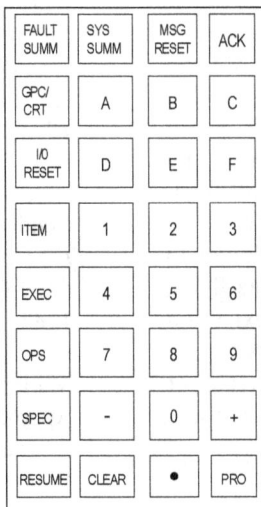

FAULT SUMM	SYS SUMM	MSG RESET	ACK
GPC/ CRT	A	B	C
I/O RESET	D	E	F
ITEM	1	2	3
EXEC	4	5	6
OPS	7	8	9
SPEC	-	0	+
RESUME	CLEAR	•	PRO

crew_interface E.cvx

Figure 11.5 Shuttle Computer Keyboard

displays to replace those developed in the 1970s, such as those in Figure 11.4. The new displays were task oriented to reduce the crew workload and supplement systems displays. The Rendezvous Task Display shown in Figure 11.6 combined data and commands from four legacy displays (Figure 11.4). The CAU was canceled in December of 2004 and the CAU task oriented displays later served as the starting point for the Orion displays.

```
 Nav Rndz                    201     MET 001/17:22:31
 ┌─── State Vector ───┐          ┌─── Star Tracker ───┐
 Select 10 Fltr    Update    0.01   20 S Trk Angles
                   Filter-Prop 0.64  21 S Trk Select    -Zm
 Range     47.068                    22 Threshold        0
 Rdot   +  7.34    Fltr to Prop 11   23 Mode            Tgt
 Y      -  0.07    Prop to Fltr 12   24 Shutter Forced OP
 Ydot   +  0.4     Orb  to Tgt  13      Tgt Present
 Node -    2:45    Tgt  to Orb  14   25 Break Track
 ┌──── Nav Filter ────┐          ┌──────── IMU ────────┐
  Radar    Raw    Resid  Inh/Aut Force        1   2   3
 Range  47.049  - 0.05  40 Aut    43   Des  31  32  33
 Rdot  +  7.35  + 0.12  41 Aut    44   Attitude  1
 V/El  +  1.50  + 0.00  42 Aut    45   ┌──────── Radar ───────┐
 H/Az  +  4.00  - 0.04
                 Covar Reinit 49       50•Status Ena
 ┌──── Targeting ────┐                    Mode  ATRK
                     ┌─── Solution ───┐  51•Radar Angles
 60 Target Set 10 TI                   52 Self Test
 61 TIG + 1/17:27:30   TIG -  00:04:59  53 Range Min/Auto
 65 El           0.00  ΔVx  +   8.9    ┌──────── COAS ───────┐
 66 ΔT       +  76.90  ΔVy  -   0.4
 67 ΔX       -   0.90  ΔVz  -   0.1    80 COAS Nav Ena/Des
 68 ΔY           0.00  ΔVtot     8.9   81 COAS Angles
 69 ΔZ       +   1.80  Miss Dist  .12  ┌─── Nav Control ───┐
 70 Compute
 71 Basetime  1/17:27:30   75 Update   90 Nav Enabled
                                       91 202 Meas Inh

 Item 70 Ent
 ┌─────┬─────┬─────┬─────┬─────┬─────┐
 │ Sys │ Nav │ IMU │Rndz │ GPS │     │
 └─────┴─────┴─────┴─────┴─────┴─────┘
```

Figure 11.6 CAU Rendezvous Task Display

Grappling Hardware

The Remote Manipulator System (RMS) is an approximately 50 foot long, six degree-of-freedom arm equipped with six joints (shoulder yaw, shoulder pitch, elbow pitch, wrist pitch, wrist yaw, and wrist roll).[12] It is located on the port side of the payload bay, and is capable of handling payloads up to 65,000 pounds. The RMS end effector on the end of the arm grapples a fixture installed on the payload. An RMS display and control panel, rotational and translational hand controllers, and associated television displays are located in the aft flight deck flight crew station. A starboard arm was also planned in the 1970s, but was never flown. In addition to deployment and retrieval of satellites and free-flying scientific payloads, the RMS is also used as an extension ladder for EVA crews (Figure 11.7), for positioning modules during ISS assembly and replenishment, and for conducting orbiter and ISS inspections using television cameras and other sensors.

Figure 11.7   Bruce McCandless being maneuvered with the RMS during STS-41B (Feb. 1984).

References

1. Smith, F. E., M. E. Campbell, T. J. Blucker, C. E. Manry, and I. Saulietis, "Shuttle Orbiter Stellar-Inertial Reference System," *AIAA Journal Of Guidance And Control*, Vol. 6, No. 6, 1983, pp. 424-431.

2. McQuillan, W. F., et al., "Rendezvous Radar For Space Shuttle Orbiter," *Proceedings of the International Telemetering Conference*, International Foundation for Telemetering, 1974, pp. 447-465.

3. Wylie, A. D., and H. G. deVezin, "Onboard Rendezvous Navigation For The Space Shuttle," *Navigation: Journal of the Institute of Navigation*, Vol. 32, No. 3, Fall 1985, pp. 197-220.

4. Kachmar, P., "Apollo and Space Shuttle On-Board Navigation Systems: Application of Kalman Filtering Techniques," *58th Annual Meeting of the Institute of Navigation and CIGTF Guidance Test Symposium*, Institute of Navigation, Fairfax, VA, 2002.

5. Little, Michael J., "Space Shuttle Orbiter Onboard Rendezvous Navigation," *Proceedings of the 1982 Position Location and Navigation Symposium (PLANS)*, IEEE, Atlantic City, NJ, December 6-9, 1982, pp. 105-109.

6. Muller, E. S., and P. M. Kachmar, "A New Approach To On-Board Navigation," *Navigation: Journal of the Institute of Navigation*, Vol. 18, No. 4, Winter 1971-1972, pp. 369-385.

Approved for public release via STI DAA 24483. See statement on title page.

7. Pearson, D. J., "Shuttle Rendezvous and Proximity Operations," *Proceedings of the CNES International Symposium on Space Dynamics*, Centre National d'Etudes Spatiales, Paris, France, 1989, pp. 833-851.

8. Goodman, J. L., "Space Shuttle Navigation in the GPS Era," *Proceedings of the Institute of Navigation National Technical Meeting*, Institute of Navigation, Fairfax, VA, 2001.

9. Clohessy, W. H., and R. S. Wiltshire, "Terminal Guidance System For Satellite Rendezvous," *Journal of the Aerospace Sciences*, Volume 27, Number 9, September 1960, pp. 653-659.

10. Battin, R. H., *An Introduction To The Mathematics And Methods Of Astrodynamics*, Revised Edition, AIAA, Reston, VA, 1999.

11. Collins, D. J., "Flight Operations For Shuttle Rendezvous Navigation and Targeting," *Proceedings of the IEEE PLANS '84 – Position Location and Navigation Symposium*, IEEE, New York, NY, 1984, pp. 106-112.

12. Logan, B. A., "Shuttle Payload Deployment and Retrieval System," *AIAA Guidance and Control Conference*, AIAA, Reston, VA, 1982.

This page intentionally left blank.

CHAPTER 12 - SKYLAB REBOOST

A final Apollo CSM RCS burn performed by the Skylab-4 crew was expected to delay a Skylab re-entry to 1981 or 1982.* However, increased solar activity caused the Skylab orbit to decay more rapidly than expected. In 1977, planning began for a Skylab re-boost mission to be flown on an early shuttle mission. A dual co-elliptic rendezvous sequence (Figure 9.2) was adopted to address the elliptical Skylab orbit, lighting, and range concerns before NCC. The rendezvous radar and software for on-board burn targeting were not expected to be available in time to support the flight, requiring Mission Control to perform all burn targeting with shuttle on-board states updated with star tracker data.

Rendezvous and proximity operations designers had long desired to use orbital mechanics to brake the orbiter during the manual phase, and reduce reliance on RCS thrusting. A +R Bar approach for the proposed Skylab reboost mission was studied. However, a +R Bar approach required a reliable range rate sensor, and the Ku Band radar was not expected to be available. Even if it was, the rendezvous radar did not have sufficient range rate accuracy or reliability to support a +R Bar approach. In addition, it was assumed that adoption of a +R Bar approach would require changes to rendezvous burn targeting.

Several proximity operations profiles were examined. In one, at a range of 500 feet, the orbiter was to transition from an inertial approach to a partial fly-around to the +V Bar. The sun was in the crew field of view on the +R Bar, and remained in the field of view during the inertial rate transition to the +V bar. While station-keeping, the Teleoperator Retrieval System (TRS) was to be deployed by the RMS (Figures 12.1 and 12.2). The orbiter was to transition to the –2000 foot point on the –V Bar, from which TRS operations (including docking) would be controlled (Figure 12.3). After the reboost, the TRS was to undock and be recovered on a later flight. Due to delays in the shuttle schedule, and the rapid orbital decay of Skylab, the re-boost mission was canceled in December 1978.

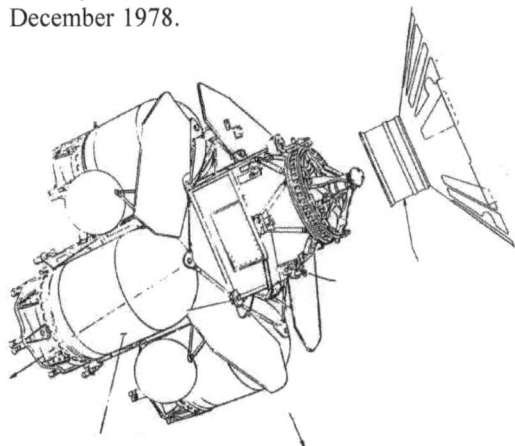

Figure 12.1 The TRS docking with the Skylab radial port.

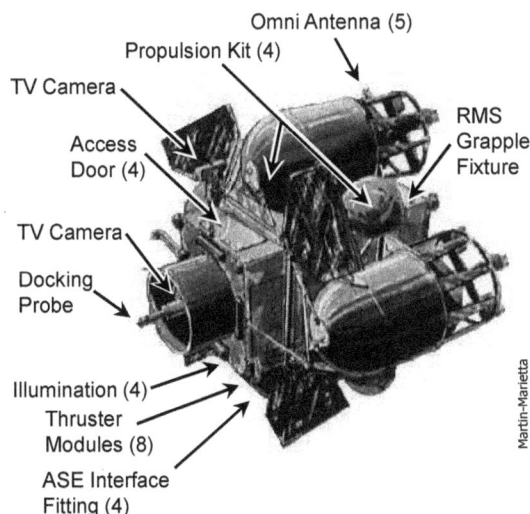

Figure 12.2 The Teleoperator Retrieval System

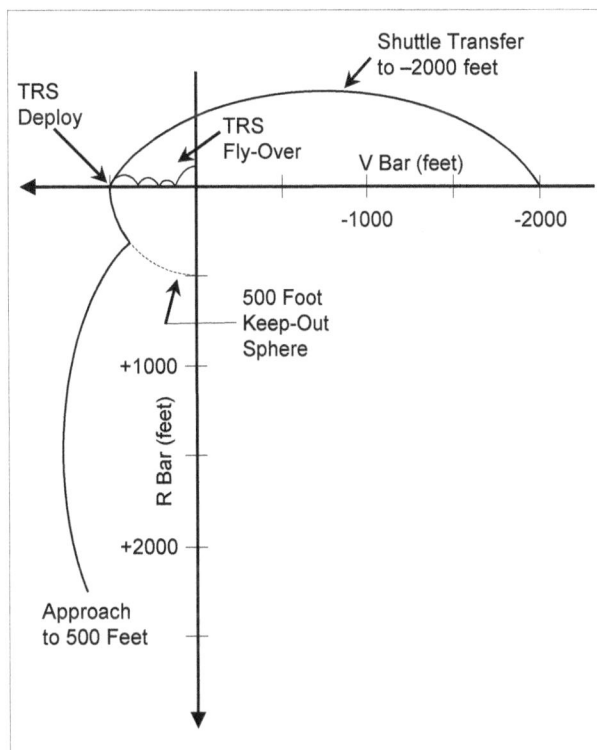

Figure 12.3 Proposed Skylab reboost proximity operations (1978).

* A mission to deorbit Skylab with an Apollo Command/Service Module (CSM) was also proposed. Skylab-5 would have been flown by the same two man crew that trained for the never flown Skylab rescue mission. The CSM Service Propulsion System (SPS) would have been used to deorbit the combined Skylab Workshop/Skylab-5 CSM stack. Skylab 5 would then undock and use the SPS to raise its orbit before it reached entry interface. The proposal was eventually discarded for safety reasons.

This page intentionally left blank.

CHAPTER 13 - COELLIPTIC VERSUS STABLE ORBIT RENDEZVOUS

Early Examination of the Stable Orbit Profile For Shuttle

In 1975 application of the stable orbit profile, first flown on Gemini XI, to the Space Shuttle was studied (Figure 13.1). Stable orbit involved the initiation of the intercept from a station-keeping point on the –V Bar, rather than from a coelliptic orbit. Stable orbit might simplify flight design and operations for missions involving deployment of a satellite, followed by retrieval of a second satellite. Contingency retrieval of a deployed payload might also be easier to perform with stable orbit. A stable orbit profile would desensitize the mission timeline from trajectory considerations. Stable orbit, long-range station-keeping (tens of miles) was preferable to close range station-keeping (tens or hundreds of feet), due to the need for continuous crew monitoring and resulting propellant expenditure. However, like dual coelliptic, the availability of sufficient tracking on a stable orbit profile for a navigationally passive target was in question. By the mid 1970s, early plans for the shuttle Orbital Flight Tests (OFTs) included rendezvous with navigationally passive targets (Figure 13.2).

Problems With the Baseline Dual Coelliptic Profile

Although the dual coelliptic (Figure 13.3) had been baselined for mission planning purposes in April of 1973, doubts about its capability to support Space Shuttle rendezvous missions persisted into the early 1980s. The ability to obtain sufficient on-board optical tracking using reflected sunlight in the presence of Earth limb and celestial bright object constraints on the field of view was questionable. By 1978, forward RCS propellant depletion due to the high relative approach velocity inherent with coelliptic was a serious concern, as was plume impingement on target spacecraft.

As a result of simulations conducted from July to September of 1977 the Low Z braking mode and LVLH attitude hold were added to the flight control software. Also in 1977 development was begun on proximity operations burn targeting software that would permit approach to a target from any direction.

In 1979 a modified dual coelliptic profile for the Solar Maximum repair mission was proposed to meet lighting requirements and reduce forward RCS propellant consumption. The delta-height of the second coelliptic segment (TPI delta-height) was decreased from 10 nm to 5 nm to reduce propellant consumption during the terminal phase. TPI targeted orbiter for a point 1 nm below target on the +R Bar. Orthogonal braking, part of the new on-board proximity operations burn targeting software, was then used to fly an approach up the +R Bar

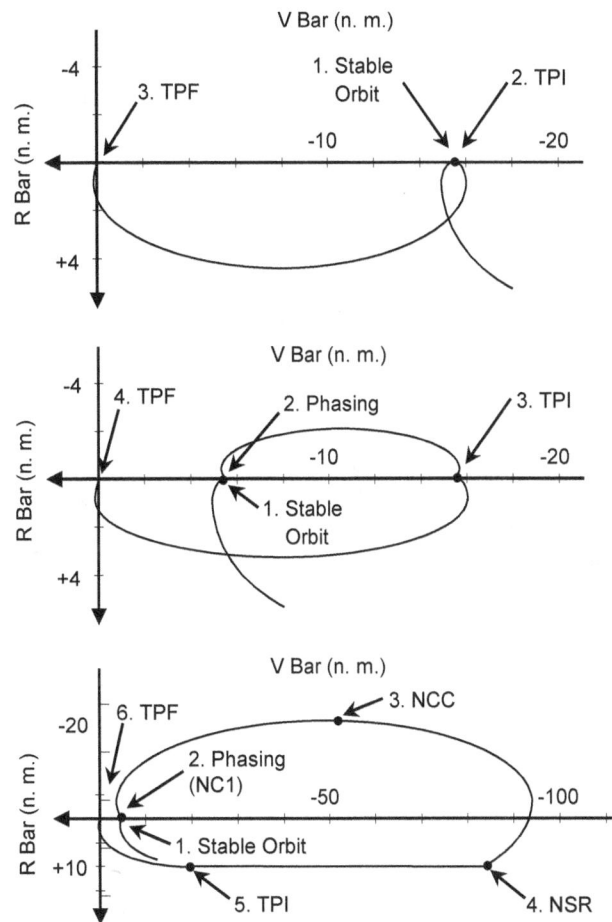

Figure 13.1 Notional stable orbit profiles studied in 1975.

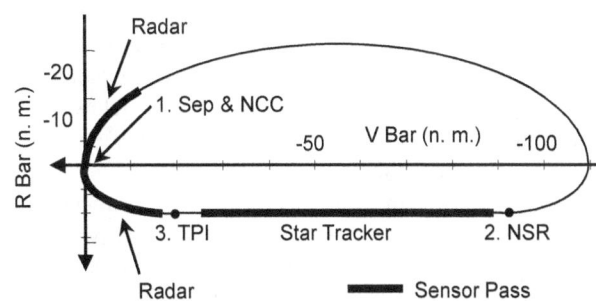

Figure 13.2 Proposed relative navigation test (1976).

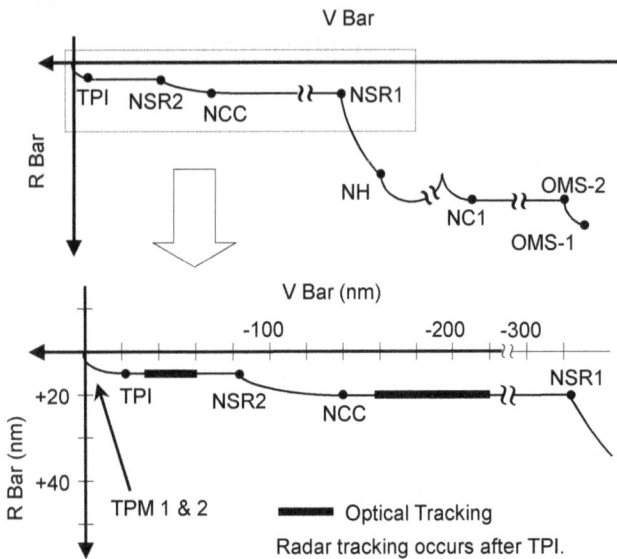

Figure 13.3 Relative view of dual co-elliptic rendezvous (1973-1983).

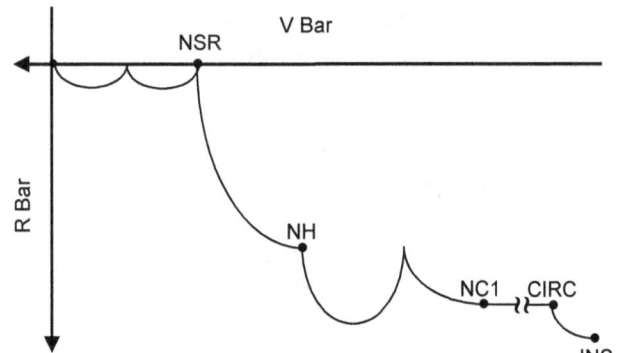

a) June 1981

using aft RCS propellant. However, by the time the transition to +V Bar range of ~200 feet was reached it was already orbital noon, resulting in +V Bar arrival near sunset with the sun in the eyes of the crew. TPI TIG slips increased from +/- 8 minutes worst case to +/- 15 minutes. These slips resulted in unacceptable lighting during proximity operations.

Introduction of proximity operations targeting, including orthogonal braking, did not resolve problems due to the conflict between terminal phase lighting requirements and forward RCS propellant consumption.

Stable Orbit Profile For Shuttle

By 1981, mission design for the LDEF deployment and Solar Max repair mission (later flown on STS-41C in 1984) was encountering difficulties. Mission planners began to adapt the stable orbit concept to overcome propellant depletion, mission timeline, and on-board tracking issues with the dual coelliptic profile (Figure 13.4).

The Solar Max repair mission also involved LDEF deployment at a higher orbital altitude than the Solar Max orbital altitude. If the dual coelliptic rendezvous profile was to be flown after LDEF deployment, the orbiter had to decrease altitude to 20 nm below Solar Max and phase 300 nm behind it. At this point the orbiter would start the first coelliptic segment and acquire line-of-sight angle measurements with the star tracker.

Mission Planning and Analysis Division personnel reasoned that rather than phasing ~300 nm behind Solar max, ground tracking should be accurate enough for Mission Control to target burns to get the orbiter to within rendezvous radar range (10 nm tracking requirement for a 1 square meter target) on the –V Bar. The point on the –V Bar would be close enough to permit radar measurement acquisition, but far enough away to

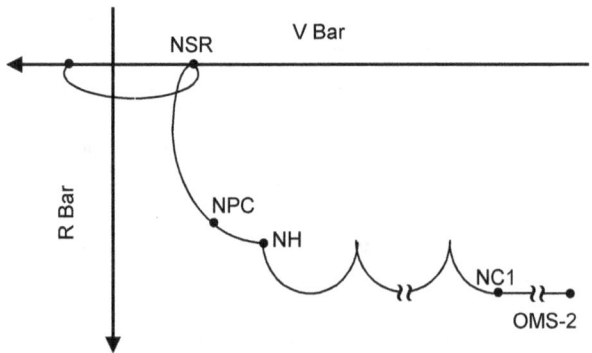

b) March 1982.

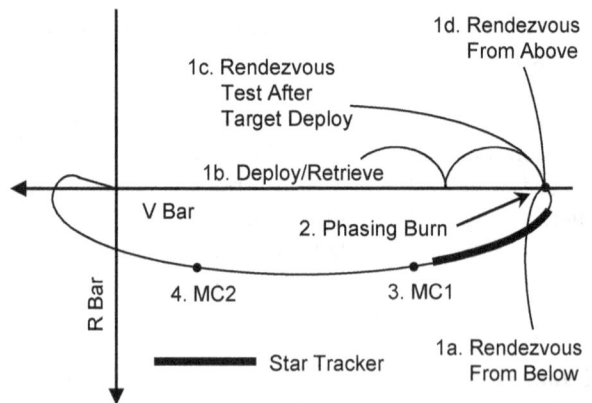

c) August 1982.

Figure 13.4 Three of many proposed stable orbit rendezvous profiles.

permit star tracker operation in the event of a radar failure. A –V Bar station-keeping point of 8 nm was selected. This was within radar range but far enough away to avoid potential target size and brightness problems with the orbiter star trackers.

Station-keeping at the 8 nm stable orbit point would be performed until orbital noon, at which point the shuttle would initiate an intercept trajectory with an on-board targeted burn. The station-keeping and the timing of the transfer would also provide control over lighting in the manual piloting phase. Station-keeping could also be extended in the event of orbiter or target systems problems. In the event of a radar failure, optical tracking with a star tracker could be performed. The transfer to intercept would require ~4 feet/second of braking, an order of magnitude reduction over the dual coelliptic, which required ~40 feet/second of braking.

The Tuned Coelliptic Profile

To address concerns with the dual coelliptic profile, coelliptic advocates designed an alternative called the tuned coelliptic (Figure 13.5). It was designed to overcome the high approach relative velocity of dual coelliptic.

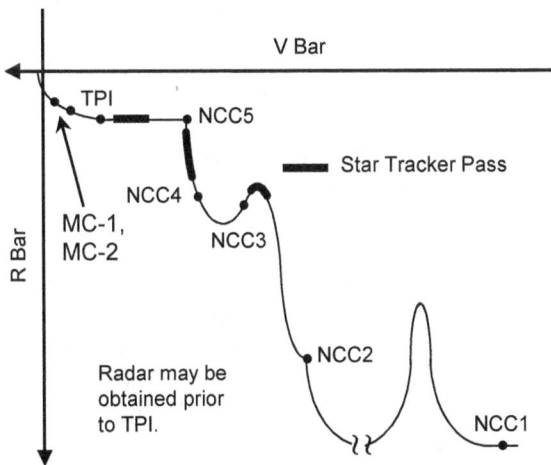

Figure 13.5 Tuned coelliptic rendezvous with a coelliptic ∆H of 2.5 nautical miles (1982).

All day-of-rendezvous burns would be on-board targeted, with a maximum star tracker tracking range of about 150 nm. The coelliptic segment delta-height was much lower than the second dual coelliptic segment delta-height (2.5 nm versus 10 nm). The lower delta-height permitted radar acquisition of the target before TPI, and provided an overlap in radar and star tracker tracking for comparison purposes.

However, the lower delta-height also increased the variability in the time (TPI TIG slip) at which the desired TPI relative geometry (elevation angle) was achieved (Figure 13.6). The profile could be tuned during the mission to control slips in TPI time and trajectory

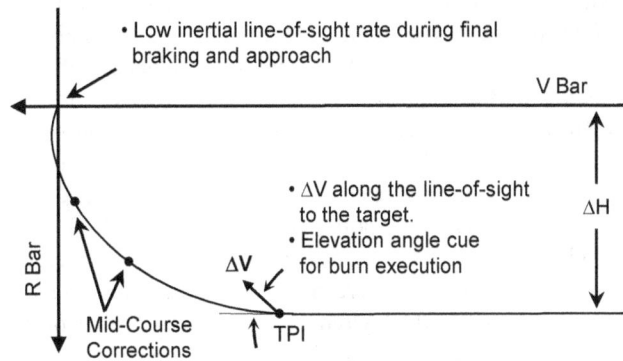

Figure 13.6 Terminal Phase for coelliptic rendezvous.

dispersions. Adjusting the placement of early phasing maneuvers increased the number of tracking periods prior to the coelliptic maneuver, and decreased TPI sensitivity to dispersions from earlier burns.

Tuned coelliptic was designed to use the on-board Lambert targeting software for height (NH) and phasing (NC) control burns on the day of rendezvous. This was motivated by a desire to meet a Shuttle Program rendezvous autonomy requirement whose meaning had been the subject of debate within the Program. The ability to target NH and NC burns on-board had been removed from the flight software when the Rendezvous Targeting Specialist Function was scrubbed from the requirements in the spring of 1978. The tuned coelliptic would enable these burns to be targeted on-board with Lambert targeting rather than by Mission Control using NC and NH targeting algorithms.

TPI placed the orbiter on the +V Bar 1,000 feet ahead of the target. Two mid-course correction burns were executed after TPI. TPI had the same elevation angle as Gemini and Apollo (~27.5 degrees) but increasing the transfer angle from 130 degrees to 160 degrees lowered the amount of braking required during final approach.

A delay (equal period football) could be designed into the tuned coelliptic in the event of a need to delay the rendezvous, but at a higher propellant cost than stable orbit.

Selection of a New Baseline Profile

A lengthy debate ensued between stable orbit proponents and tuned coelliptic supporters from the summer of 1981 to the spring of 1983.[1,2] The debate involved some of the same personnel that had been involved in the coelliptic versus tangential versus first apogee rendezvous debate during mission planning for Gemini VI in 1964.[3] Coelliptic was a proven technique, and some Mission Control personnel, as well as some astronauts, were not in favor of adopting a new profile.

Mission planners believed stable orbit provided several advantages over tuned coelliptic; 1) lower propellant consumption, 2) less complex crew and

Mission Control procedures, 3) stable station-keeping points on the -V Bar in the event of a systems anomaly or change in mission planning, and 4) elimination of the need to perform optical tracking with star trackers unless there was a rendezvous radar failure. However, pilot-in-the-loop simulations indicated that stable orbit procedures were just as complex as tuned coelliptic.

During Gemini, Apollo, Skylab, and Apollo/Soyuz rendezvous profiles had initial relative conditions (below and far behind) that were more standard than those of proposed shuttle deploy/retrieve missions. Stable orbit potentially offered more straightforward trajectory design for flights requiring rendezvous from in front or above (Figure 13.4c).

Station-keeping on the –V Bar at the 8 nm stable orbit point was eliminated in favor of performing the intercept maneuver, called Transition Initiation (Ti), when the 8 nm point on the –V Bar was reached. In the event of a systems anomaly, an equal period "football" trajectory could be initiated at Ti, called the Ti Delay burn, until it was permissible to continue the rendezvous. Like stable orbit, tuned coelliptic could be designed with a delay option, but with higher propellant consumption and increased procedural complexity.

Several variations of stable orbit terminal phase were studied. In one, Ti was targeted to place the orbiter several miles in front of the target on the +V Bar, after which the orbiter would move in along the +V Bar. In another, Ti targeted the orbiter for a point 5000 feet ahead of the target and 1500 feet above it. From there, the orbiter would fly a "glideslope approach" (Figure 13.7), which avoided RCS firings that could impinge on the target.[4] However, analysis indicated that a direct (inertial) approach could be flown with a transition to the +V Bar at a range of about 500 feet. This approach could be flown with acceptable propellant consumption and reduced risk of plume impingement. Neither a lengthy +V Bar approach nor glideslope approach was needed for stable orbit proximity operations.

The stable orbit Ti burn delta-velocity vector was not along the line-of-sight to the target spacecraft. This was a convenient "point and burn" feature of Gemini and Apollo coelliptic rendezvous in the event of a GNC systems failure. However, this was not considered to be an issue due to the increased redundancy in the orbiter GNC system.

Analysis of the stable orbit plan by advocates and critics revealed a number of weaknesses, which were corrected by changing the profile. Four Mid-course Correction (MC) burns were placed between Ti and intercept. A planar change maneuver (null out-of-plane velocity) was placed at the nodal crossing following MC-1. To reduce the size of the out-of-plane velocity null after MC-1, on-board tracking was extended before Ti to include one or two star tracker passes, starting at a range of 40 nm. This created an overlap of ground and on-board tracking for cross checking before committing to an intercept trajectory. An additional on-board targeted burn before Ti, NCC, was added to ensure that the Ti point would be in the orbital plane of the target.[5,6] The Ti maneuver point was raised above the –V Bar to make trajectory dispersions more manageable when near continuous manual piloting was initiated (~2,000 feet from the target). The MC-2 burn was targeted on elevation angle as a 130-degree transfer, providing a low inertial line-of-sight rate condition at manual takeover and facilitating the inertial line-of-sight approach technique that was used on the Gemini and Apollo terminal profiles.

Stable orbit relied more heavily on Mission Control orbit determination and burn targeting than tuned coelliptic. This did not address the concern about meeting the Shuttle Program autonomy requirement raised by tuned coelliptic advocates. However, meeting that requirement would have been difficult due to the 1976 and 1978 scrubs of the on-board targeting software.

Stable orbit was adopted as the shuttle baseline rendezvous plan at the first Rendezvous Flight Techniques Panel meeting, chaired by Flight Director Jay Greene, on April 29, 1983 (Figure 13.8), during planning for mission STS-41C. Factors influencing the decision were the inability of the Mission Control software (OMP) to support the tuned coelliptic without modification, and that the stable orbit concept was promoted by MPAD, the JSC organization responsible for trajectory design and mission planning. In the event that a second rendezvous with a target was required, stable orbit potentially incurred lower propellant expenditure than tuned coelliptic. Another factor was that if a contingency hold was required before entering proximity operations range (~2000 feet), -V Bar station-keeping was required. For tuned coelliptic, this essentially resulted in a down-mode to the stable orbit.

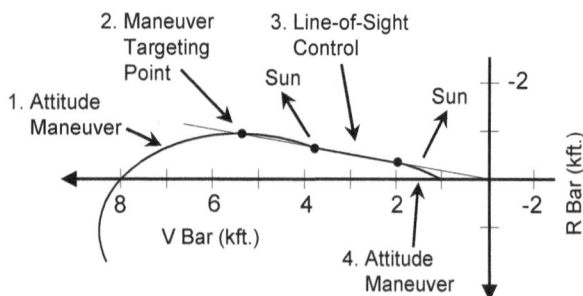

Figure 13.7 Proposed 12 degree glide-slope approach to 1000 feet (Aug. 1982).

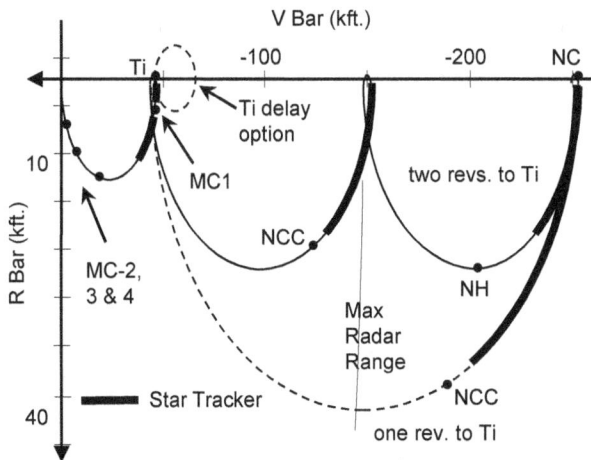

Figure 13.8 Stable orbit rendezvous (1983-1997).

Discussions of tuned coelliptic versus stable orbit continued until about the time of STS-51A (November 1985). However, coelliptic transfers were not flown during the on-board targeted phase of subsequent missions. Coelliptic transfers (NSR burns) were flown occasionally during the ground-targeted phase (range > 40 nm) to ensure that lighting conditions would be met during the later on-board targeted phase.

Starting with STS-71 (June 1995) NSR burns were no longer executed during the ground-targeted phase. Horizontal phasing burns (NCs) were performed at points off the line of apsides so that lighting requirements would be met during the on-board targeted phase. In addition, deletion of NSRs saved crew time and reduced the number of OMS engine firings.

The STS-134 mission (May 2011) did fly an Orion coelliptic profile after undocking from the International Space Station in support of the Orion relative sensor test (see Chapter 21, The STORRM DTO).

References

1. Pearson, D. J., "Baselining the Shuttle Rendezvous Technique," NASA JSC Flight Design and Dynamics Division, September 21, 1990. See JSC-35051, Volume 1, in the A Note On Sources chapter in this document.

2. Oberg, J. E., "STS Rendezvous Evolution," NASA JSC Flight Design and Dynamics Division, October 1, 1991. See JSC-35051, Volume 1, in the A Note On Sources chapter in this document.

3. Lineberry, E. C., "Gemini VI-A Rendezvous Mission Planning," *Gemini Midprogram Conference*, NASA SP-121, Manned Spacecraft Center, 1966, pp. 277-282.

4. Pearson, D. J., "The Glideslope Approach," *Orbital Mechanics and Mission Design – Advances in the Astronautical Sciences*, Vol. 69, Univelt, San Diego, CA, 1989.

5. Pearson, D. J., "Shuttle Rendezvous and Proximity Operations," *Proceedings of the CNES International Symposium on Space Dynamics*, Centre National d'Etudes Spatiales, Paris, France, 1989, pp. 833-851.

6. Collins, D. J., "Flight Operations For Shuttle Rendezvous Navigation and Targeting," *Proceedings of the IEEE PLANS '84 – Position Location and Navigation Symposium*, IEEE, New York, NY, 1984, pp. 106-112.

This page intentionally left blank.

CHAPTER 14 - FIRST PROXIMITY OPERATIONS AND RENDEZVOUS FLIGHTS

After the first flight of the Space Shuttle (STS-1) in April of 1981, and successful demonstrations of the RMS on subsequent flights, more personnel, computer resources, and simulator time became available for rendezvous and proximity operations procedure development, trajectory analysis, and issue resolution.[1]

SPAS (STS-7)

STS-7 (June 1983) performed a proximity operations demonstration using the Shuttle Pallet Satellite (SPAS-01, Figure 14.1).[2,3] Primary objectives were to demonstrate and evaluate proximity operations techniques required for deployment, separation, station-keeping, final approach and RMS capture of a free-flying payload (Figure 14.2). No computer based maneuver targeting or relative navigation data using computer processed radar measurements was available. Out-the-window cues and radar data direct from the sensor were used. Results indicated that plume impingement math models were accurate, the rendezvous radar performed better than expected, piloting using out-the-window cues and radar data was easily accomplished, and that the proximity operations tasks could be accomplished with propellant consumption falling within one sigma of predicted values. The Low Z and LVLH attitude hold flight control options were proven effective. The mission also provided the first photographs of a shuttle orbiter taken from another spacecraft (Figure 14.3).

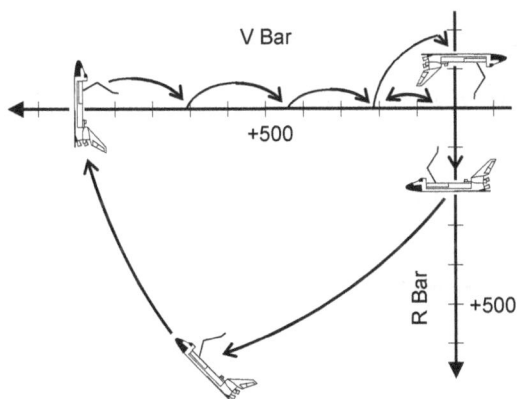

Figure 14.2 SPAS-01 as *Challenger* approaches with the RMS (left) and (right) partially obscured by a cabin window as it is about to be grappled (STS-7, June 1983).

Figure 14.1 STS-7 proximity operations with SPAS-01 (June 1983).

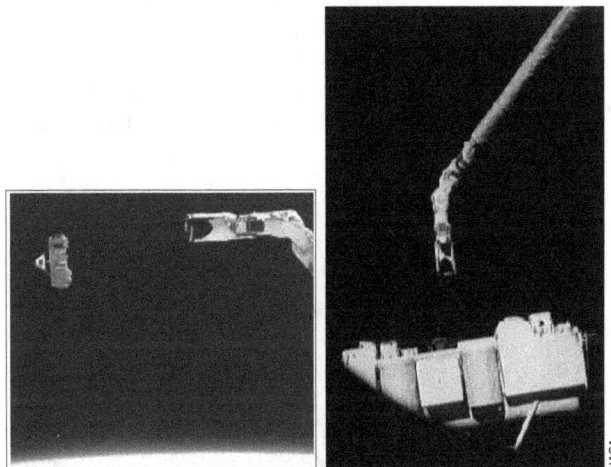

Figure 14.3 *Challenger* as photographed by SPAS-01 (STS-7, June 1983).

Integrated Rendezvous Target (STS-41B)

The first rendezvous demonstration was planned for STS-41B (February 1984), the tenth shuttle mission (Figures 14.4 through 14.8). However, the rendezvous was canceled after the Integrated Rendezvous Target (IRT) balloon burst during deployment from the shuttle payload bay (Figure 14.8). A breakout maneuver was performed to avoid any shuttle contact with the IRT debris. Rendezvous radar, star tracker and COAS data were collected on the IRT debris, and processed in the shuttle computer Kalman filter.

To support free flight of the astronaut piloted Manned Maneuvering Unit (MMU), a proximity operations astronaut/MMU rescue procedure had been developed pre-flight. The procedure was validated during retrieval of a foot restraint that floated away from the vehicle during an EVA.

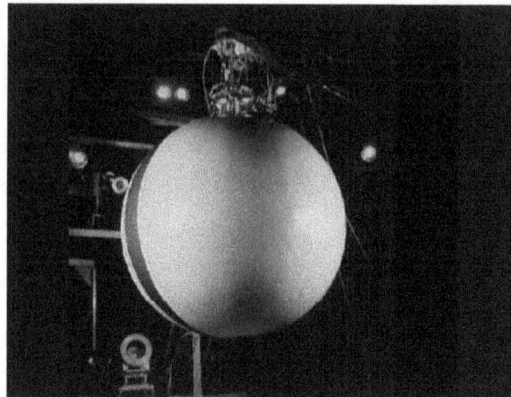

Figure 14.4 IRT backup flight article inflated in a thermal/vacuum chamber test (Dec. 1983).

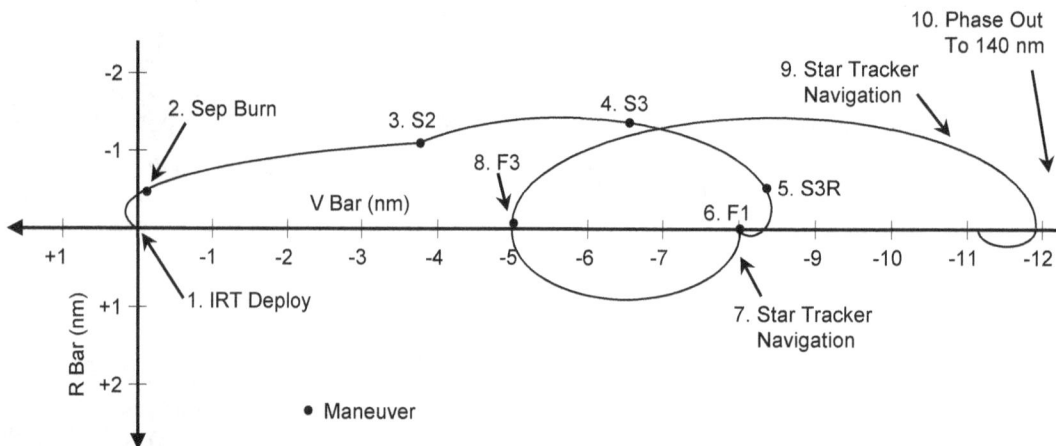

Figure 14.5   First phase of the planned STS-41B rendezvous with the IRT (Feb. 1984). Profile not executed due to IRT balloon failure.

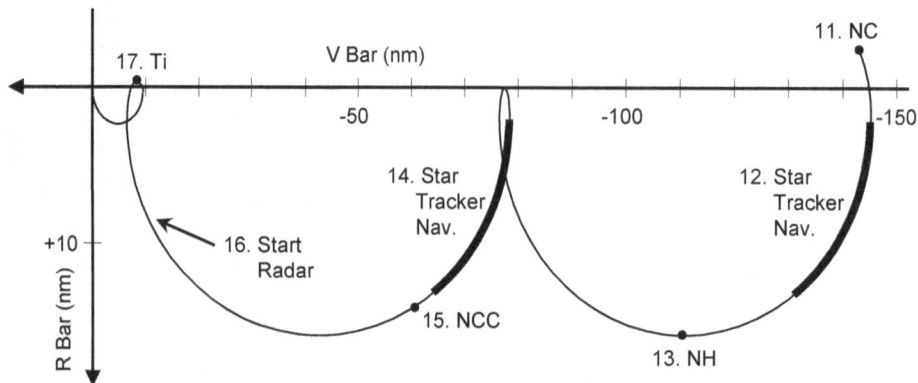

Figure 14.5 Second phase of the planned STS-41B rendezvous with the IRT (Feb. 1984). Profile not executed due to IRT balloon failure.

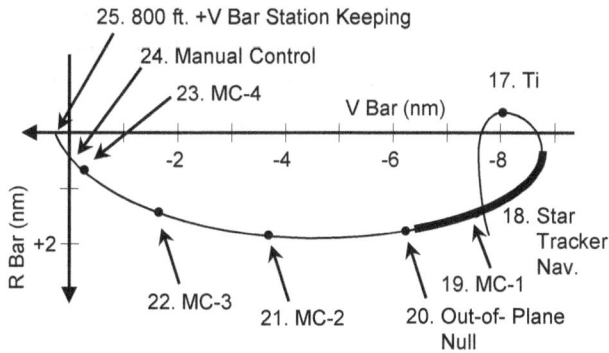

Figure 14.7 Third phase of the planned STS-41B rendezvous with the IRT (Feb. 1984). Profile not executed due to IRT balloon failure.

Figure 14.9 Solar Max under repair in the payload bay of *Challenger* (April 1984).

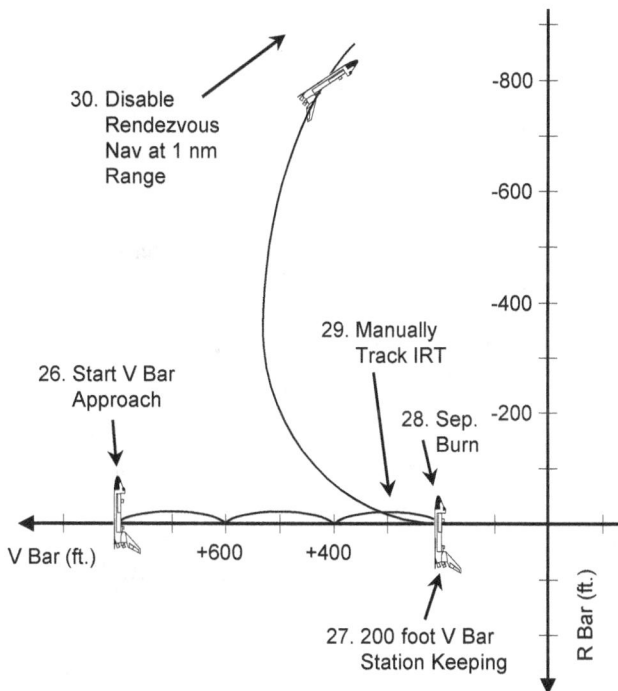

Figure 14.8 Planned proximity operations after the planned STS-41B rendezvous with the IRT (Feb. 1984). Profile not executed due to IRT balloon failure.

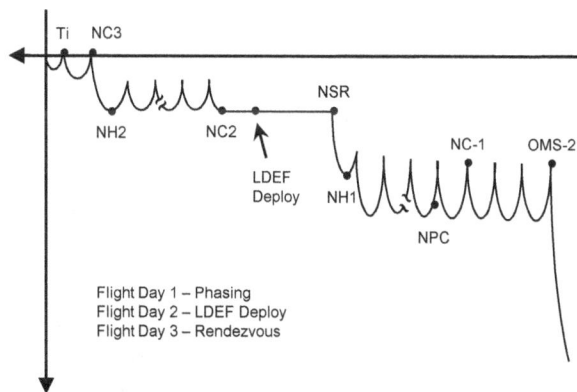

Figure 14.10 Pre-launch Solar Max rendezvous profile. LDEF deploy orbit was coelliptic with Solar Max and 10 nm below Solar Max orbital altitude.

STS-41C (Solar Max)

The Solar Max repair mission (STS-41C, April 1984, Figure 14.9) was the first all-up use of the shuttle's integrated rendezvous and proximity operations capabilities (Figures 14.10, 14.11, and 14.12). These included pre-flight trajectory design, launch window targeting, ground targeting using radar-based orbit determination, deployment of the Long Duration Exposure Facility (LDEF) during the ground-targeted phase, onboard rendezvous navigation with a navigationally passive target, onboard rendezvous targeting, and three body proximity operations involving *Challenger*, Solar Max, and an astronaut flying the Manned Maneuvering Unit.

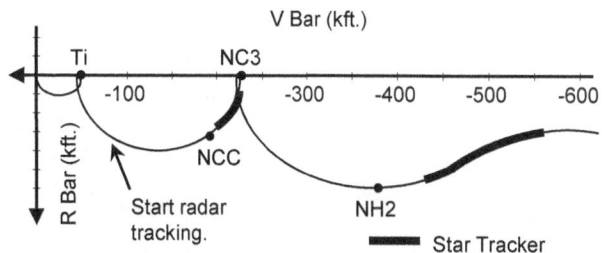

Figure 14.11 STS-41C first rendezvous, Solar Max repair (April 1984).

First Rndz -
+V Bar arrival
800 ft.

200 ft
station-
keeping.

+800   +400   -200

+V Bar (ft.)

First Rndz -
At 800 ft.
null R_DOT
& transition
to +V Bar.

+V Bar
arrival
350 ft.

+400

Second
Rndz

+800

Range	R_DOT (ft/.sec.)	
	First	Second
2000	-4.0	-4.0
1500	-2.0	-2.0
1000		
900		
800	null*	-0.8
700	-0.8	-0.8
600	-0.8	-0.8
500	-0.5	-0.5
400	-0.4	-0.4
300	-0.3	-0.3
200	null	null

+1200

+1600

+2000

+2400

+2800

+R Bar (ft.)

* Maintain target stationary
in COAS until +V Bar.

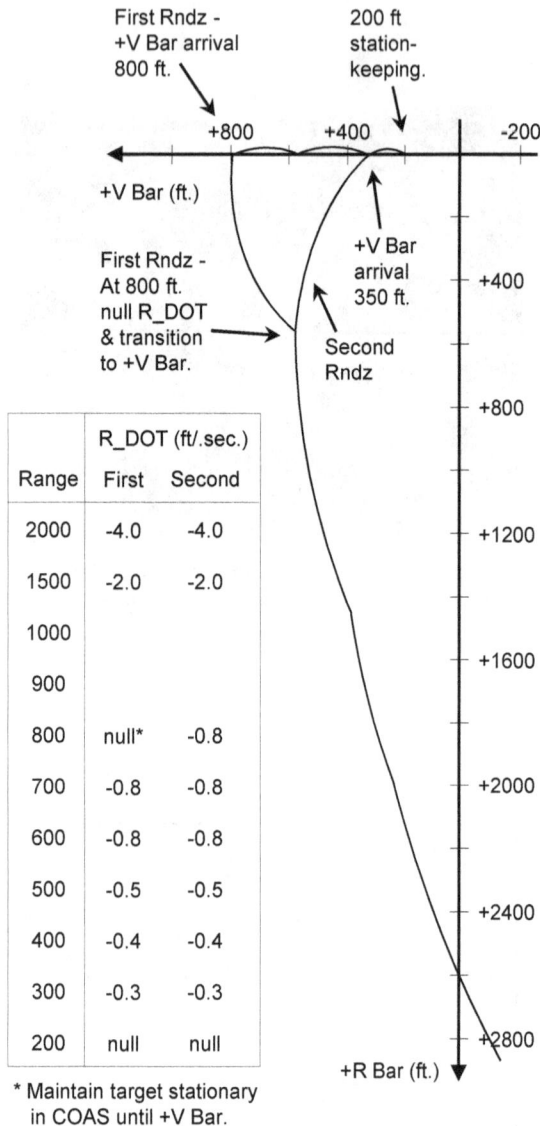

Figure 14.12 Approaches to the +V Bar for the first and second rendezvous with Solar Max. The second rendezvous arrived on the +V Bar at 350 feet to conserve propellant. Braking gate table does not represent all proximity operations procedures, such as establish +V Bar and 200 foot station-keeping.

A previously developed backup capture procedure using the RMS was used to successfully grapple Solar Max, and it was placed in the payload bay. It was later repaired and deployed.

Summary

The successful execution of proximity operations on STS-7 and STS-41C and two rendezvous profiles on STS-41C validated work performed over a decade to create piloting techniques and trajectories that overcame shuttle systems limitations, and allowed the shuttle to meet mission requirements different from those in the Gemini and Apollo programs.

References

1. Alexander, R. N., and P. B. Chiu, "Space Shuttle On-Orbit Proximity Operations," *Proceedings of the 20th Aerospace Sciences Meeting*, AIAA, Reston, VA, 1982.

2. Crippen, R. L., and F. H. Hauck, "Orbiter Operations In Close Proximity To Free-Flying Satellites Or Formation Flying In Space," *Proceedings of the 27th Symposium of the Society of Experimental Test Pilots*, Society of Experimental Test Pilots, Lancaster, CA, 1983, pp. 314-334.

3. Oberg, J. E., "Control Of RCS Plume Effects During Proximity Operations On STS Missions," *Proceedings of the AIAA Shuttle Environment And Operations Meeting*, AIAA, Reston, VA, 1983, pp. 68-71.

The Long Duration Exposure Facility was deployed on Flight Day 2. On Flight Day 3, the attempt to capture Solar Max by an astronaut flying the MMU failed. The failed docking and a "by hand" attempt to stabilize Solar Max induced attitude rates, which precluded the possibility of a capture attempt with the RMS. Enough propellant margin was available to perform a second rendezvous two days later. A break-out and long range station-keeping 40 nm behind Solar Max was performed until the second rendezvous was initiated. For the second rendezvous, the arrival on the +V Bar was reduced from 800 feet to 350 feet to lower forward RCS consumption (Figure 14.12).

CHAPTER 15 - CHALLENGES OF SUBSEQUENT RENDEZVOUS AND PROXIMITY OPERATIONS FLIGHTS

The success of STS-7 and STS-41C did not mean that later shuttle rendezvous and proximity operations missions were in any way routine. The unique characteristics of the various rendezvous targets, along with shuttle system limitations, posed technical challenges for every rendezvous mission, and necessitated mission unique analysis and procedure development. Complexity of and variation in procedures and techniques for shuttle rendezvous and proximity operations missions was far greater than during Gemini and Apollo.

The pace of rendezvous flights between STS-41C (April 1984) and the *Challenger* accident (January 1986) had not been seen since the Gemini flights in 1965 and 1966 (Table 3.2). The success of these complex missions reflected the maturity of shuttle rendezvous and proximity operations planning and execution. The loss of *Challenger* eliminated many potential commercial missions involving rendezvous and proximity operations, such as Leasecraft and the Industrial Space Facility. After the accident, rendezvous missions resumed in 1990. Missions executed included retrieval and return to Earth of orbiting satellites, deployment and retrieval of scientific payloads, and servicing.[1]

Proximity operations and ground targeted phase trajectory design varied from flight to flight, and was driven by many factors that required extensive analysis and contingency procedure (Mission Control and onboard) development, particularly if the flight involved more than one deploy/retrieve payload. Maneuver planning to provide adequate spacecraft separation for ground radar tracking, spacecraft to spacecraft communication links, and protection against collision under dispersed trajectory conditions was particularly challenging. By 1990, the availability of ground based processing of TDRSS Doppler measurements and near continuous TDRSS communications coverage enhanced orbit determination and mission activities.

Design of the onboard-targeted phase (for most flights, approximately 40 nm behind the target through manual takeover at ~2,000 feet) remained relatively stable from flight to flight. A one revolution transfer from the last ground targeted burn (a phasing burn, or NC) to Ti was used on flights to rendezvous with an already orbiting target, due to mission timeline considerations (Figure 13.8). For deploy/retrieve missions, a two revolution to Ti transfer was used to save propellant. Radar failure procedures were continually improved to maximize probability of mission success.

Propellant consumption, particularly forward RCS, was controlled through limited use of Low Z, avoiding long duration approaches, and minimizing station-keeping, fly-arounds, and attitude trim maneuvers. Keeping piloting procedures simple was a challenge.

Flying an in-plane approach, and minimizing six-degree-of-freedom maneuvers (fly-arounds, alignments), flight control system changes, and events requiring exact timing helped lower procedural complexity.

Solar lighting (sun in the eyes of the crew) and adequate artificial lighting of the target were also considerations. These were controlled by the type and direction of the approach, and performing station-keeping to wait for appropriate solar lighting. Providing stable station-keeping points ensured mission success in the event of system anomalies, and permitted re-initiation of an approach. V Bar station-keeping has been preferred over R Bar station-keeping due to procedural simplicity and lower propellant consumption. Thermal constraints on the vehicles (orbiter and target), attitude and time in attitude were taken into account during proximity operations design. Proximity operations trajectory design principles and piloting techniques were also applied to satellite deployment, emergency breakouts, and nominal separations.

Proximity operations trajectory design varied from flight to flight, and was driven by many factors. Plume impingement concerns were structural loading, target attitude stability, and contamination. Before the Mir and ISS flights, attitude stability of small targets was the main concern. Plume impingement was controlled through choice of approach direction and duration, and use of the Low Z flight control mode (Figure 10.11).

Alignment of orbiter and target grapple or capture hardware was a major driver in selecting the type of approach (Figure 10.10). Ease of performing grapple/capture operations, and target rotation relative to the orbiter were of concern. These were managed through choice of approach direction, performing close-in fly-arounds, station-keeping, or attitude trim maneuvers; and matching the target rotation rate with the orbiter. For targets that maintained a stable attitude in the LVLH frame, LVLH approaches (V Bar, R Bar) were used. Inertially stabilized targets such as the Hubble Space Telescope and EURECA (STS-57) required an inertial approach.

The LDEF retrieval mission (STS-32) used a minus R Bar approach. LDEF was gravity gradient stabilized, with the long body axis of LDEF aligned with the local vertical (R Bar). The roll angle of the RMS grapple fixture on LDEF about the R Bar was not controlled. An R Bar approach provided the easiest access to the LDEF grapple fixture, and an orbiter fly-around in only one axis. However, a long range plus R Bar approach was not possible due to the lack of a range and range rate sensor with enough precision to support a long range plus R Bar approach (the later introduction of TCS and HHL provided this capability). The LDEF proximity operations design involved a standard approach to the plus V Bar followed by a close range transition to the minus R Bar for a short range minus R Bar approach. The existing sensors and piloting techniques could support a standard approach to the plus V Bar followed by a close-range minus R Bar approach.

The SPARTAN spacecraft maintained an inertial attitude hold. However, five SPARTAN retrievals (STS-51G, STS-56, STS-63, STS-69, and STS-77) flew plus V Bar approaches. The SPARTAN grapple fixture was pointed out-of-plane towards orbital north. The plus V Bar approach provided good geometry for both nominal grapple and the contingency grapple case for a failure of the SPARTAN attitude control system. On STS-64 the orbiter flew an inertial approach to SPARTAN to keep the Sun out of the 64 degree field of view of the Lidar In-space Technology Experiment (LITE) in the payload bay. The LITE boresight was along the orbiter minus Z body axis.

Mission planning for the STS-60 Wake Shield Facility-1 (WSF-1) involved plume impingement tests by the orbiter on the plus and minus V Bar before the retrieval. After a minus V Bar approach the orbiter was to roll 90 degrees to perform the grapple from the Y LVLH axis (H Bar). However, WSF-1 was not deployed; therefore the plume impingement test and retrieval were not performed.

After plus R Bar approaches became standard practice on the missions to Mir a number of deploy/retrieve missions flew plus R Bar approaches. These were STS-72 (OAST-Flyer and SFU), STS-80 (ORFEUS-SPAS 2 and WSF-3), STS-87 (SPARTAN-201-04), and STS-95 (SPARTAN-201-05). SPARTAN was an inertially stabilized spacecraft. For STS-87 and STS-95 a procedural work-around was developed to permit SPARTAN to fly a pseudo-LVLH attitude hold with the RMS grapple fixture pointed out-of-plane north. This facilitated testing of the Video Guidance Sensor (VGS) during the plus R Bar approach by keeping the VGS target on SPARTAN pointed at the VGS on the approaching orbiter.

The shuttle's baseline rendezvous navigation hardware and software did not required modification to place the shuttle at the proximity operations initiation point for all rendezvous missions flown, in spite of the wide variety of target spacecraft.[2,3] Radar failure procedures for use during the on-board targeted phase (for most flights, approximately 40 nm behind the target through manual takeover at ~2,000 feet) were continually improved to maximize probability of mission success. This was demonstrated during the STS-92 (October 2000) and STS-131 (April 2010) missions to the ISS, due to radar failures before the day of rendezvous. During both missions the rendezvous was performed with star tracker data until TCS and HHL data became available several thousand feet from the ISS. STS-92 was the first

all optical rendezvous flown by NASA since Apollo 7 in October of 1968.

The ground-targeted phase of two flights (STS-49 in 1992 and STS-72 in 1996) used a control box rendezvous technique (Figure 15.1).[4] The target executed a series of maneuvers after the shuttle was launched to enter a control box in space at a designated time. This technique reduced shuttle propellant consumption. Once the target entered the box, it no longer maneuvered. A shuttle planar change (NPC) burn could also be performed to compensate for target planar error introduced by target phasing maneuvers.

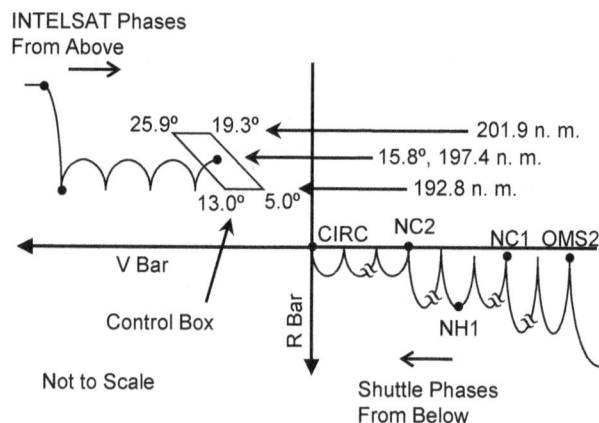

Figure 15.1   STS-49 planned relative motion until control box start time (May 1992).

References

1. Jenkins, D. R., *Space Shuttle – The History of the National Space Transportation System – The First 100 Missions*, Specialty Press Publishers, North Branch, MN, 2001.

2. Wylie, A. D., and H. G. deVezin, "Onboard Rendezvous Navigation For The Space Shuttle," *Navigation: Journal of the Institute of Navigation*, Vol. 32, No. 3, Fall 1985, pp. 197-220.

3. Kachmar, P., "Apollo and Space Shuttle On-Board Navigation Systems: Application of Kalman Filtering Techniques," *58th Annual Meeting of the Institute of Navigation and CIGTF Guidance Test Symposium*, Institute of Navigation, Fairfax, VA, 2002.

4. Gavin, R. T., "Development Of A Cooperative Operational Rendezvous Plan For Eureca And Other Maneuvering Shuttle Payloads," *Proceedings of the 38th Congress of the International Astronautical Federation*, International Astronautical Federation, Paris, France, 1987.

CHAPTER 16 - RENDEZVOUS OR PROXIMITY OPERATIONS DEMONSTRATION MISSIONS

In addition to the previously mentioned STS-7 and STS-41B, three other missions conducted demonstrations of rendezvous and proximity operations techniques (Table 16.1). The shuttle also served as a test platform for relative navigation sensors (Table 16.2).

Table 16.1 Rendezvous or Proximity Operations Demonstration Missions

Flight	Orbiter	Year	Profile	Target	Prox Ops Approach	Comments
7	*Challenger*	1983	Deploy/Retrieve	SPAS-01	+V Bar	Proximity operations only.
41B	*Challenger*	1984	Deploy/Rendezvous	IRT	+V Bar	No rendezvous due to IRT balloon failure.
51G	*Discovery*	1985	Station-Keeping	None	N/A	Station-keeping test of proximity operations autopilot.
61B	*Atlantis*	1985	Deploy/Station-Keeping	radar reflector	N/A	Station-keeping test of proximity operations autopilot.
37	*Atlantis*	1991	Deploy/Rendezvous	GRO	N/A	GRO used as target for optical navigation test.

GRO = Gamma Ray Observatory, **IRT** = Integrated Rendezvous Target , **SPAS** = Shuttle Pallet Satellite

OEX DAP (STS-51G & -61B)

The Orbital Experiments Digital Autopilot (OEX DAP) was an experimental proximity operations autopilot. On STS-51G (June 1985), the OEX DAP performed station-keeping with a phantom target. For STS-61B (Nov./Dec. 1985), a crewmember hand deployed a 15 inch radar reflector at the end of an EVA. The orbiter was moved to within about 35 feet of the target, with it centered in the field of view of the payload-bay and RMS end-effector cameras. Closed loop station-keeping was not performed since the Ku-band radar was not installed, but other tests were accomplished. The autopilot was not incorporated into the shuttle's certified avionics system.

Mid-Range Targeted Station-Keeping With GRO (STS-37)

A test of long-range station-keeping techniques for STS-39 was originally planned for STS-37. The target was to be a radar corner reflector deployed at the end of an EVA. However, a 1990 manifest change moved STS-39 ahead of STS-37, and the test was redefined. The new test involved long-range station-keeping using star tracker measurements while flying an out-of-plane profile. This technique could be used for future flights with station-keeping distances constrained by communications requirements.

The target was later changed from the radar reflector to the Compton Gamma Ray Observatory (GRO) (deployed on the same flight, Figure 16.1) for the following reasons: potential re-contact problem between GRO and the reflector, re-contact concerns if hardware was jettisoned during an EVA, possible star tracker detection difficulties with the reflector, and the possibility of adding additional orbital debris in a 28.5 degree inclination orbit.

The profile involved deployment of the GRO, followed by separation and phasing to about 125 nm behind GRO. A contingency profile was also designed to avoid re-contact if any hardware were jettisoned during an EVA later in the flight. The orbiter performed a rendezvous to a point 8 nm miles behind GRO and executed a series of out-of-plane profiles. A phasing burn was performed on each orbit to re-target the orbiter to the 8 nm VBN1 point. Star tracker relative navigation during the test was successful.

Figure 16.1 The Gamma Ray Observatory after release from the RMS (STS-37, April 1991).

Relative Sensor Demonstrations

The Space Shuttle served as a platform for tests of relative sensors (Table 16.2).

On STS-7 (June 1983) the Improved Crew Optical Sight (ICOS) was tested during proximity operations with SPAS. ICOS was proposed as a crew situational awareness aid to provide night vision during proximity operations. SPAS was only visible to the crew through the ICOS if the SPAS running lights were turned on. ICOS was not adopted as a crew proximity operations piloting aid.

STS-41B (February 1984) was to be the first test of the baseline orbiter relative navigation sensors before the Solar Max rendezvous and repair mission later flown on STS-41C (April 1984). However, the Integrated Rendezvous Target (IRT) balloon ruptured during deployment.[1] The orbiter executed a break-out burn to ensure there was no contact with the IRT debris. The re-rendezvous was canceled. However, some star tracker and rendezvous radar data was obtained during the separation. Star tracker and radar performance was better than expected. The auto-focusing laser on a payload bay camera was evaluated as a close-range proximity operations sensor. Range rate measurements were too noisy for piloting, pointing the camera at the Manned Maneuvering Unit during the EVA was tedious, and use of the camera auto-focusing laser was a full time task for one crew member.[2]

STS-41C (April 1984) was the first shuttle rendezvous and the first full demonstration of the baseline shuttle relative navigation sensors. During the Long Duration Exposure Facility (LDEF) deploy, the payload bay camera auto-focusing laser was used to track LDEF out to a range of 200 feet. During the star tracker passes on the first rendezvous with Solar Max the measurements were noisy due to attitude computations using data from three different Inertial Measurement Units (IMU). On all subsequent missions only one IMU was used as a source of attitude computation data during star tracker passes.[2]

STS-51F (July-August 1985) tested a long hand-held parallax rangefinder as a backup to the rendezvous radar. Results indicated that there was about a 10% bias in the measurements over the tested ranges from 600 to 900 feet as compared to the rendezvous radar. The rangefinder was not adopted as a radar backup.

STS-39 (April-May 1991) was the first flight of Payload Bay (PLBAY), a proximity operations piloting situational awareness program. Camera angles could be input into PLBAY by the crew. On STS-49 (May 1992) and STS-56 (April 1993) PLBAY was used to process data from two laser rangefinders under evaluation by the Shuttle Program. One unit was manufactured by Laser Technology Incorporated (LTI) and the other unit was called the Mini Eyesafe Laser Infrared Observation Set (MELIOS). The LTI unit was chosen to serve as the Hand Held Lidar (HHL) to support upcoming missions to Mir and the ISS, as well as on other rendezvous missions.

STS-51 (September 1993) was the first flight of the Trajectory Control Sensor (TCS) and the laptop computer hosted Rendezvous and Proximity Operations Program (RPOP). RPOP was based on the earlier PLBAY program and was required to process TCS data to provide the crew with relative motion cues during proximity operations. HHL, TCS, and RPOP were evaluated on several missions in 1993, 1994, and 1995 before the first operational use of them on the first docking mission with the Mir space station (STS-71, June-July 1995).

The use of relative Global Positioning System (GPS) data to support rendezvous was evaluated on four rendezvous missions from 1995 to 1997.[3] The first was flown on STS-69 (September 1995). GPS receivers were on the orbiter and the Wake Shield Facility (WSF).[4-6] The STS-80 (November-December 1996) relative GPS experiment was flown in support of relative GPS development for the European Space Agency (ESA) Automated Transfer Vehicle (ATV).[7-8] STS-84 (May 1997) and STS-86 (September-October 1996) flew the Rendezvous Sensor (RVS), Telegoniometer (TGM), and relative GPS units for further support of ATV sensor development.[9-10]

STS-87 (November-December 1997) and STS-95 (October-November 1998) flew the Video Guidance Sensor (VGS) developed by the NASA Marshall Space Flight Center.[11-14] The Advanced Video Guidance Sensor (AVGS) was developed based on the VGS flight tests on the shuttle.[15-21]

The DragonEye sensor developed by Space Exploration Technologies Corporation (SpaceX) was flown on STS-127 (July 2009) and STS-133 (February-March 2011). Dragoneye was developed for the uncrewed and crewed versions of the SpaceX Dragon spacecraft. The NASA Goddard Space Flight Center Relative navigation Sensor (RNS) flew on the last servicing mission to the Hubble Space Telescope, STS-125 (May 2009).[22-23] RNS used sensor hardware originally procured for the HST Robotic Servicing and De-orbit Mission that was canceled in February of 2005. The Canadian Space Agency TriDAR flew on STS-128 (August-September 2009), STS-131 (April 2010), and STS-135 (July 2011).

The Vision Navigation Sensor (VNS) developed for the Orion Multi-Purpose Crew Vehicle (MPCV) successfully flew on STS-134 (May 2011). The VNS Sensor Test for Orion Rel Nav Risk Mitigation Detailed Test Objective (STORRM DTO) is discussed in Chapter 21, The STORRM DTO.

Table 16.2 Relative Sensor Demonstration Missions

Flight	Orbiter	Year	Target	Sensor	Test Sponsor	Comments
41C	*Challenger*	1983	SPAS-01	ICOS	NASA/JSC	Not suitable for night vision during prox ops.
41B	*Challenger*	1984	IRT	CCTV auto-focusing laser, star tracker, rendezvous radar		Sensor test in preparation for STS-41C. No rendezvous due to IRT balloon failure. Laser range rate too noisy to support piloting.
41C	*Challenger*	1984	LDEF Solar Max	CCTV auto-focusing laser	NASA/JSC	Laser tracked LDEF out to 200 feet.
51F	*Challenger*	1985	PDP	parallax range finder	NASA/JSC	Not adopted as a radar backup.
39	*Discovery*	1991	IBSS/SPAS-II		NASA/JSC	First PLBAY flight
49	*Endeavour*	1992	INTELSAT	LTI, MELIOS	NASA/JSC	PLBAY flight.
56	*Discovery*	1993	SPARTAN-201-01	LTI, MELIOS	NASA/JSC	Last MELIOS flight. LTI later chosen to serve as shuttle HHL. PLBAY flight.
57	*Endeavour*	1993	EURECA	HHL (LTI)	NASA/JSC	Last PLBAY flight. No HHL data collected due to crew workload.
51	*Discovery*	1993	ORFEUS-SPAS 1	HHL, TCS	NASA/JSC	First flight of TCS and RPOP.
61	*Endeavour*	1993	HST	HHL	NASA/JSC	
64	*Discovery*	1994	SPARTAN-201-02	HHL, TCS	NASA/JSC	
63	*Discovery*	1995	Mir	HHL, TCS	NASA/JSC	
69	*Endeavour*	1995	WSF-2	RGPS	NASA/JSC	Relative GPS test.
80	*Columbia*	1996	ORFEUS-SPAS 2	RGPS	NASA/JSC	Relative GPS testing for ATV flights to ISS.
84	*Atlantis*	1997	Mir	RVS/TGM, RGPS	ESA	Testing for ATV flights to ISS.
86	*Atlantis*	1997	Mir	RVS/TGM, RGPS	ESA	Testing for ATV flights to ISS.
87	*Columbia*	1997	SPARTAN-201-04	VGS	NASA/Marshall	Sensor test with target on SPARTAN for AR&C Project.
95	*Discovery*	1998	SPARTAN-201-05	VGS	NASA/Marshall	Longer range proximity-operations test with AR&C Project sensor.
125	*Atlantis*	2009	HST	RNS	NASA/Goddard	Data collection during approach and deploy.
127	*Endeavour*	2009	ISS	DragonEye	SpaceX	Dragon flash LIDAR sensor development.
128	*Discovery*	2009	ISS	TriDAR	CSA	Developed by Neptec.
131	*Discovery*	2010	ISS	TriDAR	CSA	Developed by Neptec.
133	*Discovery*	2011	ISS	DragonEye	SpaceX	Dragon flash LIDAR sensor development.
134	*Endeavour*	2011	ISS	VNS	NASA/JSC	Orion sensor development (STORRM DTO).
135	*Atlantis*	2011	ISS	TriDAR	CSA	Developed by Neptec.

AR&C – Automated Rendezvous and Capture, **ATV** – Automated Transfer Vehicle, **CCTV** – Closed Circuit Television, **CSA** – Canadian Space Agency, **ESA** – European Space Agency, **HHL** – Hand Held Laser, **HST** – Hubble Space Telescope, **HTV** – H-II Transfer Vehicle, **ICOS** – Improved Crew Optical Sight, **IRT** – Integrated Rendezvous Target, **ISS** – International Space Station, **JSC** – Johnson Space Center, **LDEF** – Long Duration Exposure Facility, **LIDAR** – Light Intensification Detection and Ranging, **LTI** – Laser Technology, Incorporated, **MELIOS** – Mini Eyesafe Laser Infrared Observation Set, **PDP** – Plasma Diagnostics Package, **PLBAY** – Payload Bay, **RGPS** – Relative Global Positioning System, **RNS** – Relative Navigation Sensor, **RPOP** – Rendezvous and Proximity Operations Program, **RVS** – Rendezvous Sensor, **SPAS** – Shuttle Pallet Satellite, **STORRM DTO** – Sensor Test for Orion Rel Nav Risk Mitigation Detailed Test Objective, **TCS** – Trajectory Control Sensor, **TGM** – Telegoniometer, **VGS** – Video Guidance Sensor, **VNS** – Vision Navigation Sensor

References

1. Goodman, John L., *Lessons Learned From Seven Space Shuttle Missions*, NASA Contractor Report NASA/CR-2007-213697, NASA Johnson Space Center, January 2007.

2. Oberg, James E., *Rendezvous and Proximity Operations Handbook, Appendix: STS Rendezvous And Proximity Operations Experiences 1981-1986, Basic,* JSC-10589-A, Flight Design and Dynamics Division, Mission Operations Directorate, NASA/JSC, May 16, 1988.

3. Zyla, Lubomyr V., and Moises N. Montez, "Use of Two GPS Receivers in Order to Perform Space Vehicle Orbital Rendezvous," *Proceedings of the 6th International Technical Meeting of the Satellite Division of The Institute of Navigation (ION GPS 1993)*, Salt Lake City, UT, September 1993, pages 301-312.

4. Schroeder, C., B. E. Schutz, and P. A. M. Abusali, "STS-69 Relative Positioning GPS Experiment," *Space Flight Mechanics 1996, Volume 93 of Advances In The Astronautical Sciences*, Univelt Inc., 1996, pages 1239-1255.

5. Carpenter, J. Russell, and Robert H. Bishop, "Flight Data Results Of Estimate Fusion For Spacecraft Rendezvous Navigation From Shuttle Mission STS-69," *Space Flight Mechanics 1996, Volume 93 of Advances In The Astronautical Sciences*, Univelt Inc., 1996, pages 1257-1276.

6. Park, Y., J. Brazzel, R. Carpenter, H. Hinkel, and J. Newman, "Flight Test Results From Realtime Relative GPS Flight Experiment On STS-69," *Space Flight Mechanics 1996, Volume 93 of Advances In The Astronautical Sciences*, Univelt Inc., 1996, pages 1277-1295.

7. Hinkel, Heather, Young Park, and Wigbert Fehse, "Real-Time GPS Relative Navigation Flight Experiment," *Proceedings of the 1995 National Technical Meeting of The Institute of Navigation*, Anaheim, CA, January 1995, pages 593-601.

8. Schiesser, E., J. P. Brazzel, J. R. Carpenter, and H. D. Hinkel, "Results Of STS-80 Relative GPS Navigation Flight Experiment," *Space Flight Mechanics 1998, Volume 99 of Advances In The Astronautical Sciences*, Univelt Inc., 1998, pages 1317-1333.

9. Durand, P., T. Schirmann, L. Vaillon, and V. Pascal, "Ground and In-Orbit Testing of GPS for Space Applications," *Spacecraft Guidance, Navigation and Control Systems, Proceedings of the 4th ESA International Conference*, October 18-21, 1999 in ESTEC, Noordwijk, The Netherlands, Edited by B. Schürmann, ESA SP-425, European Space Agency, Paris, 2000, page 113.

10. Blarre, Ludovic, et al., "Description and In-Flight Performances of Rendezvous Sensors for the ATV," *Proceedings of the 32nd Annual AAS Guidance and Control Conference*, January 30 - February 4, 2009, Breckenridge, CO, in *Guidance and Control 2009, Advances in the Astronautical Sciences*, Volume 133, Univelt, Inc., 2009, pages 369-384.

11. Jackson, J. L., R. T. Howard, H. J. Cole, R. A. Belz, "Automatic Docking System Sensor Analysis & Mission Performance", *Proceedings of the SPIE Laser Radar Technology and Applications III*, edited by G. Kamerman, SPIE Vol. 3380, April 1998.

12. Howard, R., T. Bryan, and M. Book, "Video Guidance Sensor Flight Experiment Results", *Proceedings of the SPIE Laser Radar Technology and Applications III*, edited by G. Kamerman, SPIE Vol. 3380, April 1998.

13. Howard, R. T., T. C. Bryan, and M. L. Book, "The Video Guidance Sensor – A Flight Proven Technology," *Proceedings of the Annual AAS Guidance and Control Conference*, February 3-7, 1999, Breckenridge, CO, in *Guidance and Control 1999, Advances in the Astronautical Sciences*, Volume 101, American Astronautical Society, Univelt, Inc., 1999, pages 281-298.

14. Howard, R. T., T. C. Bryan, and M. L. Book, "On-Orbit Testing of the Video Guidance Sensor," in *Proceedings of SPIE Laser Radar Technology and Applications IV*, edited by G. Kamerman, SPIE Proceedings Vol. 3707, Bellingham, WA, 1999.

15. Howard, R. T., A. S. Johnston, T. C. Bryan, and M. L. Book, "Advanced Video Guidance Sensor (AVGS) Development Testing," in *Spaceborne Sensors*, edited by R. D. Habbit, Jr. and P. Tchoryk, Jr., SPIE Conference Proceedings Vol. 5418, Bellingham, WA, 2004, pages 50-60.

16. Howard, R. T., and T. C. Bryan, "DART AVGS Flight Results," in *Sensors and Systems for Space Applications*, edited by R. T. Howard and R. D. Richards, SPIE Conference Proceedings Vol. 6555, Bellingham, WA, 2007, pages 1-10.

17. LeCroy, J. E., D. S. Hallmark, and R. T. Howard, "Effects of Optical Artifacts in a Laser-Based Spacecraft Navigation Sensor," in *Sensors and Systems for Space Applications*, edited by R. T. Howard and R. D. Richards, SPIE Conference Proceedings Vol. 6555, Bellingham, WA, 2007, pages 1-11.

18. Kennedy, III, Fred G., "Orbital Express: Accomplishments and Lessons Learned," *Proceedings of the 31st Annual AAS Guidance and Control Conference*, February 1-6, 2008, Breckenridge, CO, in *Guidance and Control 2008, Advances in the Astronautical Sciences*, Volume 131, Univelt, Inc., 2008, pages 575-586.

19. Pinson, Robin, Richard Howard, and Andrew Heaton, "Orbital Express Advanced Video Guidance Sensor: Ground Testing, Flight Results and Comparisons," *AIAA Guidance, Navigation and Control Conference and Exhibit*, Honolulu, Hawaii, August 18-21, 2008.

20. Howard, Richard T., Thomas C. Bryan, Jimmy Lee, and Bryan Robertson, "Next Generation Advanced Video Guidance Sensor Development and Test," *Proceedings of the 32nd Annual AAS Guidance and Control Conference*, January 30 - February 4, 2009, Breckenridge, CO, in *Guidance and Control 2009, Advances in the Astronautical Sciences*, Volume 133, Univelt, Inc., 2009, pages 409-422.

21. Howard, Richard T., and Thomas C. Bryan, "The Next Generation Advanced Video Guidance Sensor: Flight Heritage and Current Development," *Space, Propulsion & Energy Sciences International Forum*: SPESIF-2009, February 24-26, 2009, Huntsville, AL.

22. Naasz, Bo J., et al., "The HST SM4 Relative Navigation Sensor System: Overview and Preliminary Testing Results from the Flight Robotics Lab," *F. Landis Markley Astronautics Symposium, Advances in the Astronautical Sciences*, Univelt, San Diego, CA, 2009.

23. Naasz, Bo J., et al., "Flight Results from the HST SM4 Relative Navigation Sensor System," *Proceedings of the 33rd Annual AAS Guidance And Control Conference*, February 5-10, 2010, Breckenridge, CO, in *Guidance and Control 2010, Advances in the Astronautical Sciences*, Volume 137, Univelt, Inc., 2010, pages 723-744.

This page intentionally left blank.

CHAPTER 17 - SATELLITE SERVICING MISSIONS

Satellite servicing missions flown by the shuttle (Table 17.1) required close coordination and planning between rendezvous personnel, proximity operations personnel, Extra Vehicular Activity (EVA) specialists, satellite manufacturers, and satellite operators. EVA preparation and execution occurred simultaneously with rendezvous and proximity operations tasks. The previously mentioned Solar Max repair (STS-41C) was the first servicing mission.

Table 17.1 Satellite Servicing Missions

Flight	Orbiter	Year	Profile	Target	Prox Ops Approach	Comments
41C	*Challenger*	1984	Ground-Up	Solar Max	+V Bar	Retrieved with the RMS during the second rendezvous. Solar Max repaired and deployed.
51D	*Discovery*	1985	Deploy/ Rendezvous	SYNCOM IV-3	+V Bar	Contingency rendezvous after SYNCOM activation failure.
51I	*Discovery*	1985	Ground-Up	SYNCOM IV-3	+V Bar	Rendezvous & EVA planned in four months. SYNCOM in elliptical orbit.
49	*Endeavour*	1992	Ground-Up	INTELSAT VI (F-3)	+V Bar	Hybrid Control Box. Three rendezvous. Lambert targeting problem. Original capture technique failed. Captured by 3 EVA crew.
61	*Endeavour*	1993	Ground-Up	Hubble	Inertial	Servicing Mission 1
82	*Discovery*	1997	Ground-Up	Hubble	+R Bar/Inertial	Servicing Mission 2
103	*Discovery*	1999	Ground-Up	Hubble	+R Bar/Inertial	Servicing Mission 3A
109	*Columbia*	2002	Ground-Up	Hubble	+R Bar/Inertial	Servicing Mission 3B
125	*Atlantis*	2009	Ground-Up	Hubble	+R Bar/Inertial	Servicing Mission 4

EVA = Extra Vehicular Activity, **INTELSAT** = International Telecommunications Satellite, **RMS** = Remote Manipulator System, **SYNCOM** = Synchronous Communication

SYNCOM IV-3 (STS-51D)

After deployment of the SYNCOM IV-3 satellite by *Discovery* on STS-51D (April 1985), the SYNCOM on-board sequencer did not initiate antenna deployment, the spin-up maneuver or the perigee kick motor firing. A contingency rendezvous and EVA was planned, and the flight was extended by two days. The crew had received some rendezvous training nine months before the flight, and nominal rendezvous procedures and cue cards were sent to crew via the teleprinter. Based on ground instruction, a flyswatter was constructed by the crew, which was placed on the RMS during an EVA (Figure 17.1). The SYNCOM separation switch was successfully snared three times with the flyswatter, but the SYNCOM sequencer did not activate.

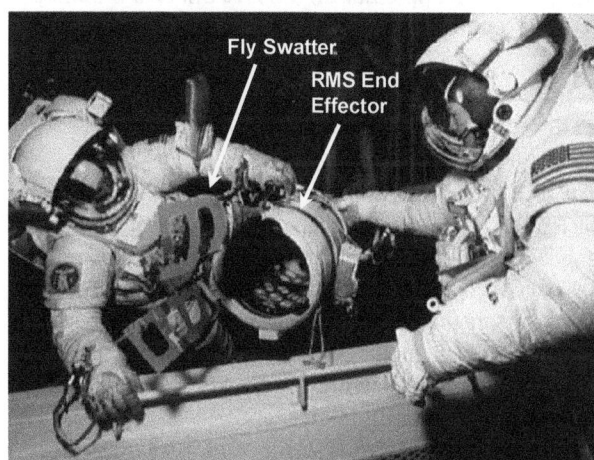

Figure 17.1 Ground rehearsal of flyswatter attachment to the RMS (STS-51D, April 1985).

SYNCOM IV-3 (STS-51I)

The mission of *Discovery* on STS-51I (August-September 1985) was modified to include rendezvous with and repair of the SYNCOM IV-3 satellite that malfunctioned after deployment on STS-51D. The time available to plan the SYNCOM rendezvous, EVA, and repair was four months. *Discovery* deployed two communications satellites on the first day of the mission, and a third on the second day, after which ground targeted rendezvous maneuvers began. The circular orbits required for satellite deployment and the elliptical (160 x 235 nm) orbit of SYNCOM complicated rendezvous profile design. Concerns about the length of the crew workday and EVA time drove a decision to have one revolution between the final NC and Ti burn. A close-in fly-around of SYNCOM was performed upon arrival. Inadvertent pluming of the SYNCOM, which induced attitude rates, complicated the retrieval. The capture, repair and redeployment were successful (Figure 17.2).

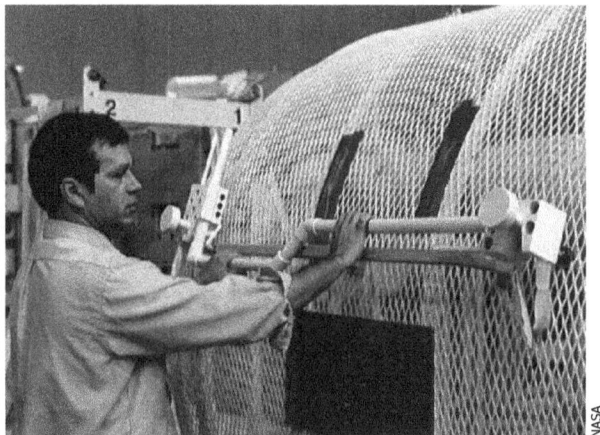

Figure 17.2 Grapple bar on a SYNCOM IV-3 mock-up in preparation for the STS-51I mission (1985).

INTELSAT VI F-3 (STS-49)

On March 14, 1990, after launch on a Titan III, the INTELSAT-VI (603) communications satellite was stranded in low Earth orbit after it failed to separate from the Titan second stage. Ground controllers separated the satellite from the second stage at the interface between the perigee kick motor and INTELSAT, and used the limited propulsion capability of INTELSAT to boost it into a slightly higher orbit.

The mission of the first flight of *Endeavour* (May 1992) was to rendezvous with the stranded INTELSAT, so that a new perigee kick motor could be installed, and the satellite boosted to a suitable orbit. INTELSAT was not designed to support retrieval by the shuttle. A capture bar was designed to enable an astronaut on the end of the RMS to capture the satellite and stop it's

rotation rate. The INTELSAT would then be grappled and maneuvered into the payload bay.

This mission also used a hybrid control box rendezvous, with the INTELSAT maneuvered after *Endeavour* launch from a 299 nm by 309 nm orbit into a control box in a 200 nm by 210 nm, 28.35 degree inclination orbit (Figure 17.3, Figure 17.4). Before the rendezvous, the INTELSAT spin rate was reduced from about 10.5 to about 0.65 revolutions per minute.

The two EVA crewman entered the payload bay about 1.5 hours prior to the first capture attempt. Close coordination of complex rendezvous and EVA procedures was required throughout the mission. The Flight Day 4 capture attempt failed and *Endeavour* performed a breakout and phased away (Figure 17.4). The Mission Management Team approved another rendezvous and capture attempt on Flight Day 5, with appropriate propellant management to protect for a third rendezvous and capture attempt, if it was needed.

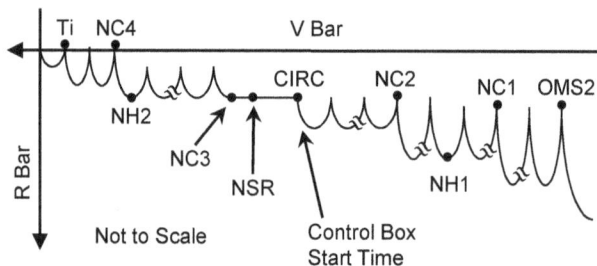

Figure 17.3 STS-49 shuttle relative motion profile (May 1992).

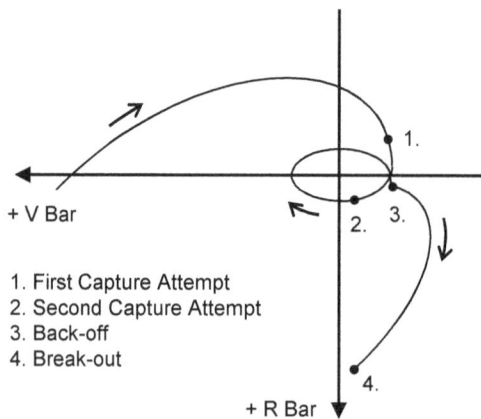

1. First Capture Attempt
2. Second Capture Attempt
3. Back-off
4. Break-out

Figure 17.4 Sketch of proximity operations for the first STS-49 rendezvous on Flight Day 4.

INTELSAT performed a maneuver to ensure that it stayed ahead of *Endeavour*. The Flight Day 5 capture attempt also failed, after several tries during two orbital daylight periods (Figure 17.5 and 17.6). During the two orbits of proximity operations, several fly-arounds were flown to optimize the relative geometry for the capture attempts. However, INTELSAT rotational dynamics continued to degrade, with the satellite in a flat spin at the last capture attempt.

Figure 17.5 EVA crewman on the RMS attempts to capture INTELSAT (right). The COAS is on the left (STS-49, May 1992).

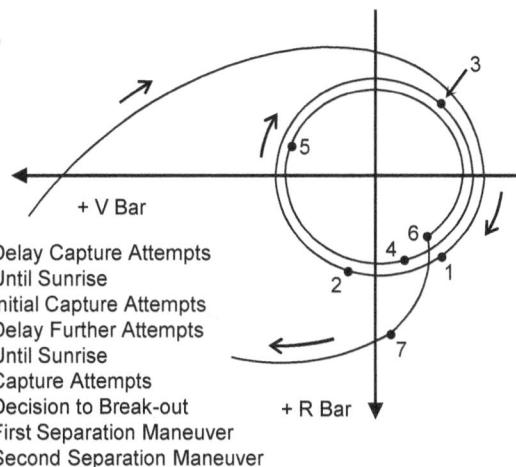

1. Delay Capture Attempts Until Sunrise
2. Initial Capture Attempts
3. Delay Further Attempts Until Sunrise
4. Capture Attempts
5. Decision to Break-out
6. First Separation Maneuver
7. Second Separation Maneuver

Figure 17.6 Sketch of proximity operations for the second STS-49 rendezvous on Flight Day 5 (May 1992).

For the next rendezvous and capture attempt on Flight Day 7, a new capture procedure was developed, using three EVA crew, rather than two, to capture the satellite by hand, without using the capture bar. Appropriate grab areas on the INTELSAT were located and safety issues were identified and addressed. Simulations were conducted to verify the new procedure in the Weightless Environment Training Facility at NASA Johnson, and analysis of shuttle plume impingement on INTELSAT dynamics was also studied.

After the NCC burn on Flight Day 7, several on-board Ti burn targeting attempts failed to converge on a solution.[1] An attempt to clear up the problem by reloading the computer software failed, and a Ti delay maneuver was performed (the first and so far only Ti delay to be conducted). Propellant use eliminated the possibility of a fourth rendezvous, if the Flight Day 7 capture attempt failed. Targeting failures continued to occur. Navigation data from the shuttle flight computers was used by Mission Control to compute the Ti and Mid-Course maneuvers, and the burn solutions were voiced to the crew. Before the capture attempt, *Endeavour* was maneuvered so that the INTELSAT spin axis was into the payload bay. The capture was successful (Figure 17.7 and 17.8), and the perigee kick motor was attached. After INTELSAT deploy, due to limited propellant remaining, *Endeavour* performed a retrograde separation maneuver to place the perigee in the correct hemisphere and minimize propellant required for deorbit. The INTELSAT perigee kick motor was fired when *Endeavour* was below and about 300 n.m. in front of it, and INTELSAT eventually reached an operational orbit.

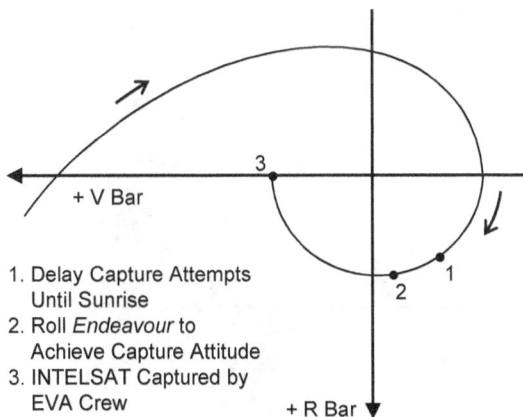

1. Delay Capture Attempts Until Sunrise
2. Roll *Endeavour* to Achieve Capture Attitude
3. INTELSAT Captured by EVA Crew

Figure 17.7 Sketch of proximity operations for the third STS-49 rendezvous on Flight Day 7 (May 1992).

Figure 17.8 Three person EVA finally captures INTELSAT VI (STS-49, May 1992).

Situational awareness and propellant preservation during proximity operations was enhanced through the use of a new Hand Held Laser (HHL) rangefinder and a new laptop computer program called Payload Bay. The new capability was especially useful during orbital night, as the INTELSAT was intrinsically dark, and depth perception degraded at night. STS-49 set a new shuttle record for the number of rendezvous profiles flown (three) and the total amount to proximity operations time (~8 hours).

Hubble Space Telescope (STS-61, -82, -103, -109, & -125)

Between 1993 and 2009 five missions were flown to successfully service the Hubble Space Telescope (HST). These complex servicing missions enhanced and ensured the ability of HST to provide significant scientific data and breathtaking photography (Figures 17.9 to Figure 17.14).[2] The first HST servicing mission flew an inertial approach (Figure 10.10). Later HST flights used a +R Bar LVLH approach, and the orbiter went into an inertial attitude hold shortly before grapple to achieve alignment at the appropriate time. Chapter 24 provides more detail on the Hubble servicing missions.

Figure 17.10 HST after RMS grapple, before berthing in *Columbia's* payload bay (STS-109, March 2002).

Figure 17.11 Hoffman with the original Wide Field/Planetary Camera that was replaced on STS-61 (Dec. 1993).

Figure 17.9 Story Musgrave about to be elevated to the top of the HST (STS-61, Dec. 1993).

Figure 17.12 HST after deployment from *Endeavour* (STS-61, Dec. 1993).

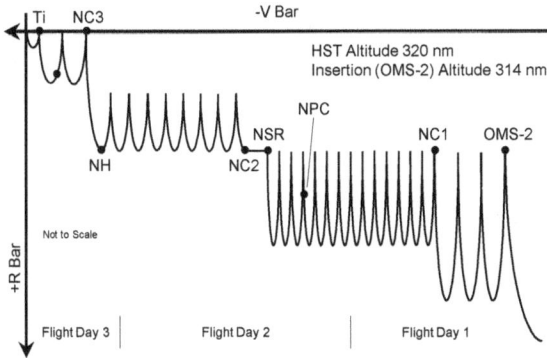

Figure 17.13 STS-82 Ground Up Rendezvous Profile

References

1. John L. Goodman, *Lessons Learned From Seven Space Shuttle Missions*, NASA Contractor Report NASA/CR-2007-213697, NASA Johnson Space Center, January 2007. See the NASA Technical Reports server at http://ntrs.nasa.gov/, or the Johnson Technical Reports server at http://ston.jsc.nasa.gov/collections/TRS/.

2. Lee, S., et al., "Hubble Space Telescope Servicing Mission 3A Rendezvous Operations," *Guidance and Control 2001, Advances in the Astronautical Sciences*, Vol. 107, Univelt, San Diego, CA, 2001, pp. 615-633.

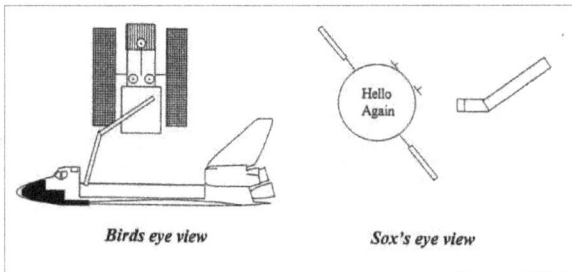

Figure 17.14 Illustration created by STS-82 rendezvous instructor Alan Fox. Ken "Sox" Bowersox was pilot of STS-61 and commander of STS-82. Source: STS-82 Flight Specific Briefing, August 26, 1996.

This page intentionally left blank.

CHAPTER 18 - DEPLOYMENT AND RETRIEVAL OF SCIENTIFIC PAYLOADS

Sixteen missions were flown involving the deployment and retrieval of from one to two science packages (Table 18.1). The eight types of deploy/retrieve payloads flown concerned astronomy, space physics, atmospheric physics, and missile defense research. Parallel execution of deploy/retrieve profiles, satellite deployments, EVAs, and multiple research tasks coordinated with multiple ground facilities made these the most complex of the shuttle missions to plan and execute. Dual shift, 24-hour crew operations on some missions further complicated planning and real-time operations.

Table 18.1 Deployment and Retrieval of Scientific Payloads

Flight	Orbiter	Year	Target	Prox Ops Approach	Comments
51G	*Discovery*	1985	SPARTAN-101	+V Bar	Incorrect SPARTAN attitude at retrieval.
51F	*Challenger*	1985	PDP	+V Bar	On-board Lambert targeted proximity operations.
39	*Discovery*	1991	IBSS-SPAS II	+V Bar	Most complex deploy/retrieve profile flown.
56	*Discovery*	1993	SPARTAN-201-01	+V Bar	Laser range and range rate sensor test.
51	*Discovery*	1993	ORFEUS-SPAS 1	+V Bar	Long range, in-front and behind station-keeping.
60	*Discovery*	1994	WSF-1	-V Bar	WSF-1 problems prevented deployment.
64	*Discovery*	1994	SPARTAN-201-02	Inertial	First successful test of Trajectory Control Sensor (TCS) laser.
66	*Atlantis*	1994	CRISTA-SPAS 1	+R Bar	Football for data collection. +R Bar Mir approach corridor test.
63	*Discovery*	1995	SPARTAN-204	+V Bar	Deploy the day after Mir rndz. Trajectory designed to avoid Mir.
69	*Endeavour*	1995	1. SPARTAN-201-03 2. WSF-2	1. +V Bar 2. +V Bar	1. Incorrect SPARTAN attitude at retrieval. 2. Long range, in-front station-keeping.
72	*Endeavour*	1996	OAST-Flyer	+R Bar	Gas venting by an experiment complicated ground tracking.
77	*Endeavour*	1996	1. SPARTAN-207-IAE 2. PAMS-STU*	1. +V Bar 2. –V Bar SK	1. Inflatable Antenna Experiment (IAE) 2. Three rendezvous and station-keeping (650 meters) periods.
80	*Columbia*	1996	1. ORFEUS-SPAS 2 2. WSF-3	1. +R Bar 2. +R Bar	1. Relative GPS test for ISS ESA Automated Transfer Vehicle. 2. Long range, in-front station-keeping.
85	*Discovery*	1997	CRISTA-SPAS 2	TORVA/+V Bar	Tested ISS TORVA and +V Bar corridor approach using payload bay keel camera.
87	*Columbia*	1997	SPARTAN-201-04	+R Bar	SPARTAN activation failure, EVA retrieval. VGS test.
95	*Discovery*	1998	SPARTAN-201-05	+R Bar	VGS test.

* Deploy/rendezvous or "proxy-vous." No retrieval.
CRISTA = Cryogenic Infrared Spectrometers and Telescopes for the Atmospheric, **ESA** = European Space Agency, **EVA** = Extra-Vehicular Activity, **GPS** = Global Positioning System, **IBSS** = Infrared Background Signature Survey, **ISS** = International Space Station, **OAST** = Office of Aeronautics and Space Technology , **ORFEUS** = Orbiting and Retrievable Far and Extreme Ultraviolet Spectrometer, **PAMS-STU** = Passive Aerodynamically Stabilized Magnetically Damped Satellite-Satellite Test Unit, **PDP** = Plasma Diagnostics Package, **SK** = Station-Keeping, **SPARTAN** = Shuttle Pointed Autonomous Tool For Astronomy, **SPAS** = Shuttle Pallet Satellite, **TORVA** = Twice Orbital Rate V Bar Approach, **VGS** = Video Guidance Sensor, **WSF** = Wake Shield Facility

Plasma Diagnostics Package (STS-51F)

During STS-51F (July/Aug. 1985) the Plasma Diagnostics Package (PDP) experiment (Figure 18.1, 18.2) explored the plasma environment around *Challenger*.[1] The mission required the development of complex nominal and contingency (such as radar fail and delayed deploy) procedures, and close coordination with scientific investigators. Precise proximity operations burn targeting was performed using the shuttle computer's Lambert targeting algorithm. An abort-to-orbit due to the shutdown of a main engine during ascent resulted in a lower orbital altitude, forcing a redesign of on-board Lambert targeting data by Mission Control. The challenging trajectory was successfully flown (Figure 18.3), but the third orbit of *Challenger* about the PDP was canceled due to increased propellant consumption during ascent.

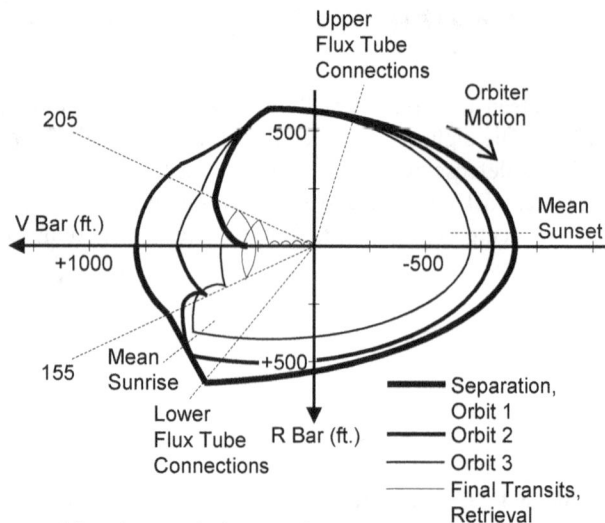

a) In-plane relative motion.

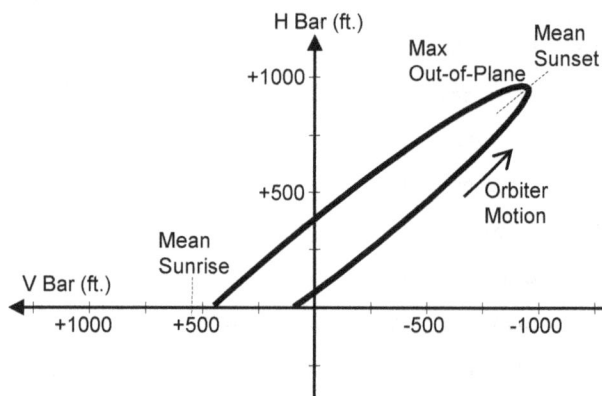

b) Out-of-plane relative motion for Orbit 1.

Figure 18.3 STS-51F station keeping with PDP (July/Aug. 1985).

Figure 18.1 PDP on the end of the RMS (STS-51F, July/Aug. 1985)

Figure 18.2 PDP after release from the RMS (STS-51F, July/Aug. 1985).

SPARTAN (STS-51G, -56, -64, -63, -69, -72, -77, -87, & -95)

The Shuttle Pointed Autonomous Tool For Astronomy (SPARTAN) was a free-flying astronomical observatory that flew on nine shuttle missions.[2]

After deployment on it's first mission, STS-51G, (June 1985) radar and star tracker data were obtained by *Discovery*. When *Discovery* approached SPARTAN-101 two days later, the crew noted that the RMS grapple fixture was not pointed in the expected direction (Figure 18.4). Rather than yaw *Discovery* to facilitate retrieval, the RMS procedures were modified to adjust for the unexpected SPARTAN-101 attitude.

After deployment from *Discovery* on STS-56 (April 1993), SPARTAN-201-01 conducted two days of data collection concerning solar wind and the sun's corona. The trajectory design was not constrained by a SPARTAN-201-01 communication constraint during approximately two days of detached operations.

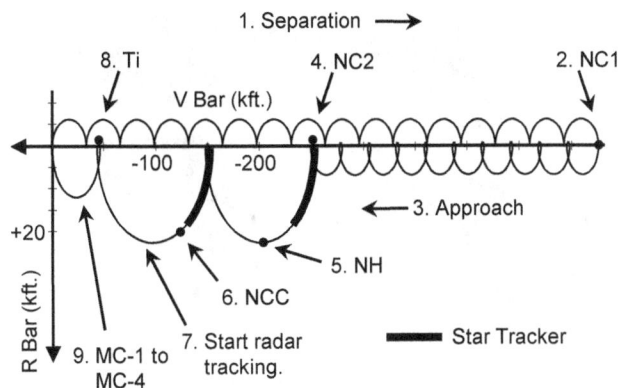

Figure 18.4 STS-51G SPARTAN deploy and retrieval profile (June 1985).

The shuttle borne Lidar In Space Technology Experiment (LITE) on STS-64 (*Discovery*, September 1994) required a low eccentricity orbit during data takes, which constrained design of *Discovery's* trajectory profile during detached operations of SPARTAN-201-02 (Figure 18.5). Data takes were scheduled during two crew sleep periods and a crew day between deploy and retrieval. Difficulty establishing radar tracking after the deploy (50 minutes of failed attempts, acquisition finally obtained at a range of 2760 feet) resulted in moving the last ground targeting phasing maneuver (NC) of the rendezvous from 40 n.m. trailing to 34 n.m. trailing to provide more time for evaluation of radar data. On the day of rendezvous initial acquisition was about 30 minutes before the second star tracker pass and the radar performed well. The Trajectory Control Sensor (TCS) lidar was also successfully tested during the approach to SPARTAN-201-02, with radar and TCS data showing excellent agreement. An inertial approach to SPARTAN-201-02 was used to protect the LITE sensor from sunlight.

Discovery (STS-63, February 1995) deployed SPARTAN-204 the day after the rendezvous and +V Bar approach to Mir. An orbit adjust maneuver was conducted following the separation from Mir to ensure that *Discovery* and SPARTAN-204 maintained a safe distance from Mir during SPARTAN-204 operations. SPARTAN-204 was retrieved after two days of free flight.

SPARTAN-201-03 (STS-69, September 1995) conducted two days of detached operations after deployment from *Endeavour* (Figure 18.6). During the rendezvous, the NCC burn used more propellant than expected due to a flight software algorithm performance issue. As a result, *Endeavour* missed the desired Ti point by 0.96 nm.* SPARTAN-201-03 was not in the nominal retrieval attitude, and a fly-around was required to position *Endeavour* and the RMS for capture.

Figure 18.6 SPARTAN before capture (STS-69, Sept. 1995).

The day after the STS-72 retrieval of the Space Flyer Unit in January of 1996, *Endeavour* deployed the Office of Aeronautics and Space Technology Flyer (OAST-Flyer, SPARTAN-206) for two days of detached operations (Figure 18.7). Significant trajectory dispersions were induced by an OAST-Flyer experiment that performed a vent previously advertised as non-propulsive. The retrieval was successful.

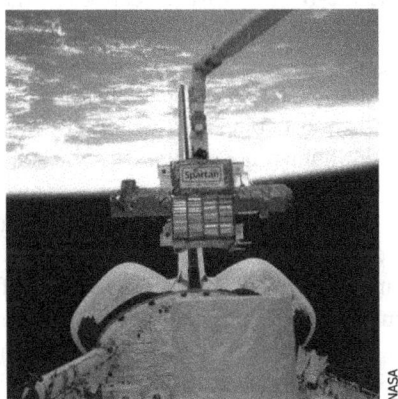

Figure 18.5 SPARTAN before deployment (STS-64, Sept. 1994).

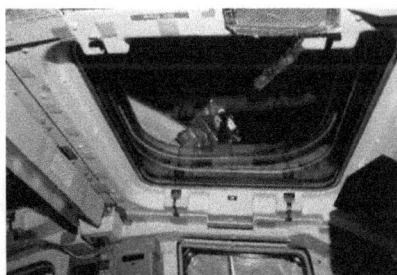

Figure 18.7 SPARTAN-206 (OAST-Flyer) seen from an aft flight deck window (Jan. 1996).

---

* See John L. Goodman, *Lessons Learned From Seven Space Shuttle Missions*, NASA Contractor Report NASA/CR-2007-213697, NASA Johnson Space Center, January 2007.

STS-77 (*Endeavour*, May 1996) set a Shuttle Program record for most rendezvous profiles flown (four) and longest total proximity operations time (21 hours). After the SPARTAN-207 deploy on flight day two, the orbiter performed minus R bar station-keeping to observe the Inflatable Antenna Experiment (IAE) (Figure 18.8). The IAE was then jettisoned from the SPARTAN. SPARTAN-207 was retrieved on flight day three. Three additional rendezvous profiles were flown for the PAMS-STU experiment.

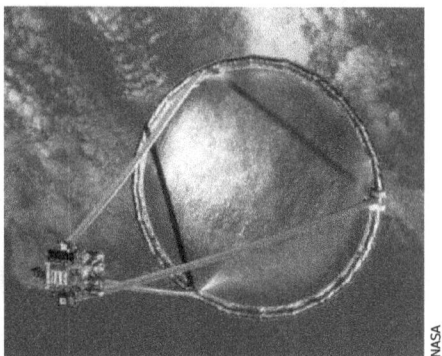

Figure 18.8 Inflatable Antenna Experiment and SPARTAN-207 (STS-77, May 1996).

After the SPARTAN-201 deploy by *Columbia* (Figure 18.9) on STS-87 (Nov./Dec. 1997), the SPARTAN attitude control system failed to activate. Attitude rates prevented capture by the RMS soon after the deploy. The activation failure prevented the accomplishment of science objectives. Later in the mission, a rendezvous was performed and SPARTAN was manually captured by two astronauts during an EVA previously planned to demonstrate ISS assembly techniques (Figure 18.10). Video Guidance Sensor (VGS, an experimental proximity operations sensor) data was obtained while the SPARTAN was attached to the RMS, but not during the retrieval due to the SPARTAN activation failure.[3]

Figure 18.9 SPARTAN before release (STS-87, Nov./Dec. 1997)

Figure 18.10 Two EVA crew about to catch SPARTAN (STS-87, Nov./Dec. 1997).

STS-95 was a re-flight of the SPARTAN payload from STS-87, and VGS data during SPARTAN retrieval was obtained (Figure 18.11).[3] An improved version of VGS, called the Advanced Video Guidance Sensor, was later developed for the Demonstration of Autonomous Rendezvous Technology (DART) and Orbital Express programs.

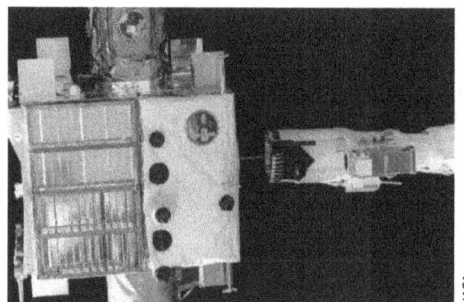

Figure 18.11 SPARTAN and the RMS (STS-95, Oct./Nov. 1998)

IBSS (STS-39)

Discovery flew a dedicated Department of Defense mission in late April and early May of 1991. The Shuttle Pallet Satellite-II (SPAS-II), carrying the Infrared Background Signature Survey (IBSS) experiment, was deployed, flew for almost two days and was then retrieved (Figure 18.12). The 38 hour deploy/retrieve profile was the most complex flown by the Space Shuttle, and involved numerous on-board and ground-targeted maneuvers (Figure 18.13). Mission planning, dual shift crew operations and observations by ground stations were coordinated.

The mission plan for *Discovery*, after IBSS deployment, involved phasing out to the far-field (10 kilometers, 5.4 n.m.) point on the –V Bar and conduct three OMS burns and one translational RCS burn to be observed by the IBSS experiment. Ideally, to meet an IBSS observation requirement for minimal relative line-of-sight rates, the orbiter would perform station-keeping on the –V Bar before each plume observation maneuver. However, since *Discovery* could not be placed on the

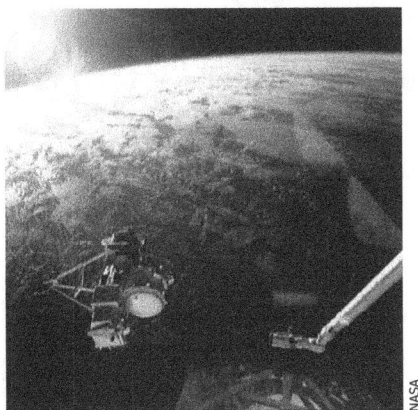

Figure 18.12 SPAS-II/IBSS after release from the RMS (STS-39, April/May 1991).

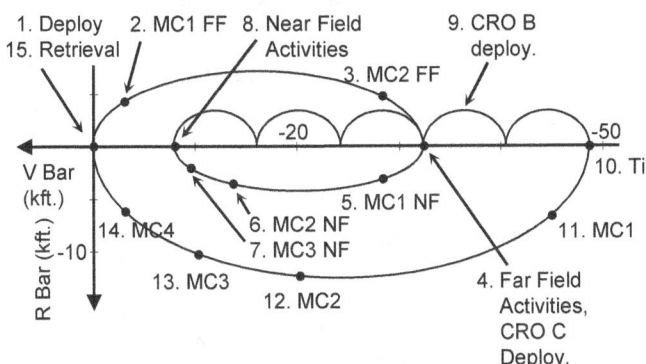

Figure 18.13 STS-39 planned profile for IBSS detached activities with maneuvers indicated (April/May 1991).

–V Bar with perfect accuracy, a maneuver was executed to place it in a trajectory coelliptic with the SPAS-II, which controlled the relative line-of-sight rates. The plume observation maneuvers were out-of-plane, and were followed by a "fast flip" attitude maneuver and another burn targeted to place *Discovery* back on the -V Bar. Another coelliptic maneuver was planned to set-up for the next plume observation maneuver.

Discovery then transitioned to the near-field plume observation point, 2.25 kilometers (1.2 n.m.) behind the SPAS-II, where two plume observation maneuvers were conducted in the same manner as the far-field maneuvers. After the near field activities were completed, *Discovery* phased out to the normal 8 nm Ti range and conducted a rendezvous with IBSS for the retrieval.

Two Chemical Release Observation (CRO B and C) sub-satellites were deployed during the IBSS detached operations, and a third (CRO A) was deployed after SPAS-II was retrieved. The sub-satellites were not tracked by the shuttle's relative navigation sensors, but were tracked by ground radar until there was no longer a possibility of re-contact with *Discovery*. After IBSS retrieval, CRO-B could not be located by ground radar tracking, and *Discovery* executed two additional maneuvers to protect against re-contact. These

maneuvers had been planned prior to the mission for just such a situation.

Insufficient knowledge of IBSS and orbiter vents, and SPAS-II attitude and drag characteristics complicated activities during the SPAS-II detached operations phase. On-board targeting worked well, but predicting long-term relative motion to support ground-targeted maneuvers was challenging (Figure 18.14 and Figure 18.15). While all objectives during the SPAS-II detached operations phase were met, overall relative motion differed from the pre-mission plan.

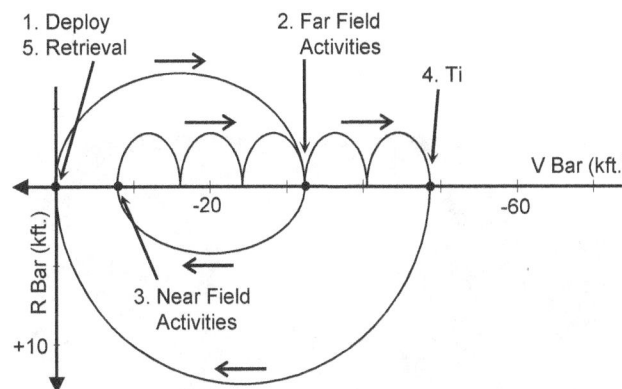

Figure 18.14 Planned STS-39 IBSS profile (April/May 1991).

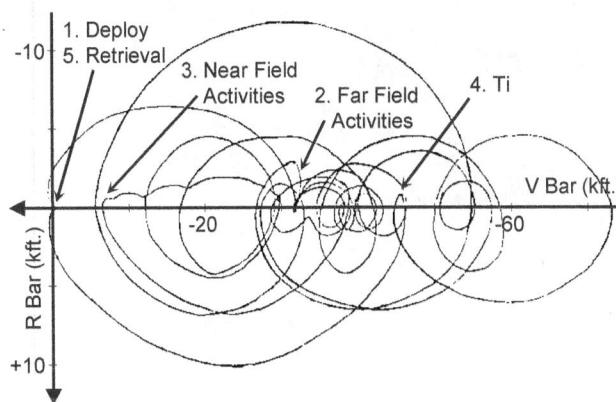

Figure 18.15 Flown STS-39 profile (April/May 1991).

ORFEUS-SPAS (STS-51 & -80)

ORFEUS-SPAS (Orbiting and Retrievable Far and Extreme UV Spectrometer) was another free-flying astronomical payload.[4]

For STS-51 (*Discovery*, September 1993), SPAS was deployed on flight day 2. *Discovery* transitioned from long-range, in front to behind station-keeping on flight day 6. Maximum station-keeping range was constrained to meet communications constraints. On flight day 8, ORFEUS-SPAS drag uncertainties resulted in the start of

123

the rendezvous occurring 40 nm behind SPAS, rather than the pre-mission planned range of 30 nm. Before the grapple with the RMS (Figure 18.16), attitude rates induced by shuttle RCS jet firings (plume impingement) were detected by the SPAS attitude control system, but it was able to maintain SPAS attitude.

On STS-80 (November-December 1996), SPAS was deployed 8.25 hours after the launch of *Columbia*. Wake Shield Facility 03 (WSF-03) was deployed about 3 days and 5.7 hours after launch, and retrieved 3 days later. SPAS was retrieved about 14 days and 12.5 hours after launch. Extended three body operations during the flight presented a challenge to mission planning, but was successful. SPAS trajectory dispersions were most likely due to translational accelerations from the SPAS attitude control thrusters.

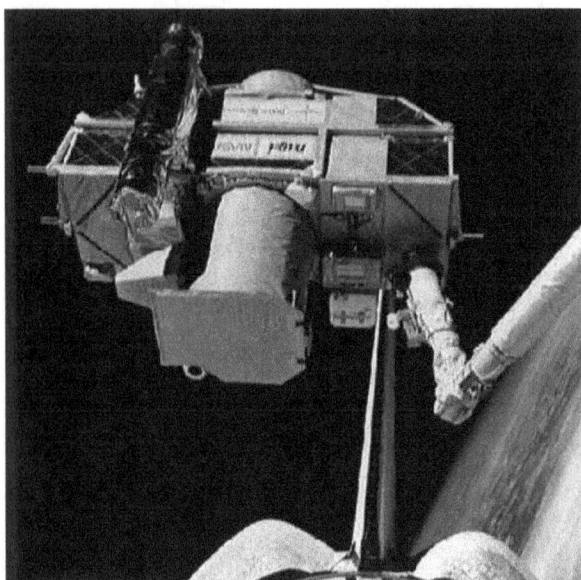

Figure 18.16 ORFEUS/SPAS after capture by the RMS (STS-51, Sept. 1993).

CRISTA-SPAS (STS-66 & STS-85)

The Cryogenic Infrared Spectrometers and Telescopes for the Atmospheric Shuttle Pallet Satellite (CRISTA-SPAS) flew on STS-66 (*Atlantis*, November 1994) and STS-85 (*Discovery*, August 1997) (Figure 18.17). The STS-66 separation profile was designed to provide 5 hours of continuous communications between the SPAS and the shuttle. To avoid contamination of the CRISTA-SPAS during science periods, the orbiter had to maintain at least an 11 n.m. separation. CRISTA-SPAS instrument field of view constraints mandated in-front station-keeping. *Atlantis* phased out to 67 n.m. in front of CRISTA-SPAS, but the range had to be reduced to improve space-to-space communications. Maneuver planning was complicated by variable CRISTA-SPAS drag, which may have been due to cryostat vent self impingement. During the rendezvous, a one-orbit

football trajectory was initiated at the Ti point to allow the Middle Atmosphere High Resolution Spectrograph Investigation (MAHRSI) instrument on CRISTA-SPAS to observe the orbiter. The +R bar approach designed for the subsequent (November 1995) STS-74 mission to Mir was flown during the approach to CRISTA-SPAS.

Figure 18.17 CRISTA-SPAS after being grappled by the RMS (STS-66, Nov. 1994).

For STS-85 (August 1997), profile design constraints were similar, but a football was not flown during the rendezvous. Orbiter attitude maneuvers were limited to reduce trajectory dispersions. During the approach to CRISTA-SPAS (Figure 18.18), a Twice Orbital Rate +V Bar Approach (TORVA) and a plus V Bar approach corridor were flown to test proximity operations procedures for missions to the ISS.

Figure 18.18 CRISTA-SPAS before retrieval with the RMS (STS-85, Aug. 1997).

Wake Shield Facility (STS-60, -69, & -80)

An example of mission-specific trajectory design were the Wake Shield Facility (WSF) flights (Figures 18.19 and Figure 18.20). The WSF structure created an enhanced vacuum on the downwind side of the vehicle to support thin film epitaxial growth and materials purification. Long-range station-keeping was performed ahead of the WSF, rather than behind, to avoid WSF contamination by shuttle RCS firings and water dumps (Figure 18.20). There was also a requirement for the payload bay to be visible to the WSF for communications purposes. Extended station-keeping with the orbiter windows and radiators pointed opposite the velocity vector (toward the WSF) was also desirable to minimize orbital debris impacts on those surfaces.

PAMS-STU (STS-77)

On flight day four, the Passive Aerodynamic-Magnetically Stabilized Satellite Test Unit (PAMS-STU) was deployed (Figure 18.21). Station-keeping and rendezvous profiles were flown for data collection on flight days four, seven and eight. The PAMS-STU rendezvous and station-keeping profiles (also called "proxy-vous") were specifically designed and flown to collect data for the experiment. New on-board targeting procedures were developed to ensure that the orbiter intercepted the –V Bar station-keeping box (Figure 18.22). New piloting procedures were also developed to lower propellant consumption (particularly forward RCS) during the extended station-keeping periods. PAMS-STU was not retrieved.

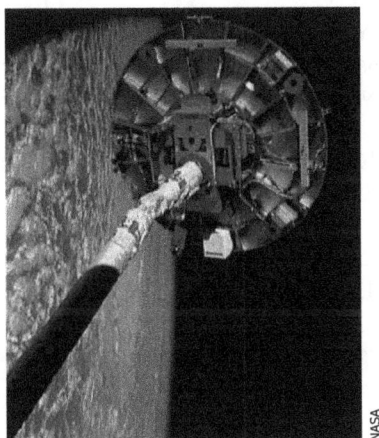

Figure 18.19 Wake Shield on the RMS (STS-60, Feb. 1994).

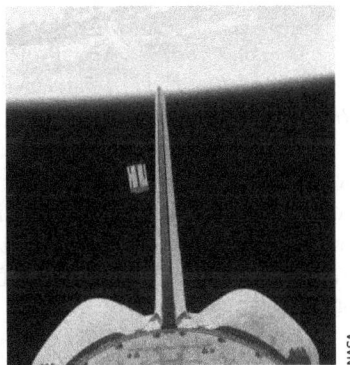

Figure 18.21 PAMS-STU after deployment from *Endeavour* (May 1996).

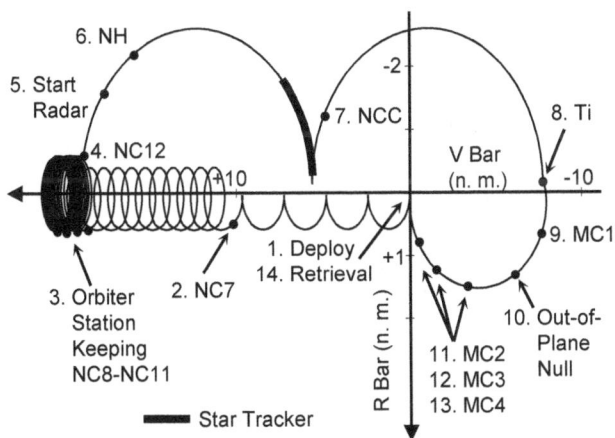

Figure 18.20 STS-80 deploy/retrieve profile for the Wake Shield Facility (Nov./Dec. 1996).

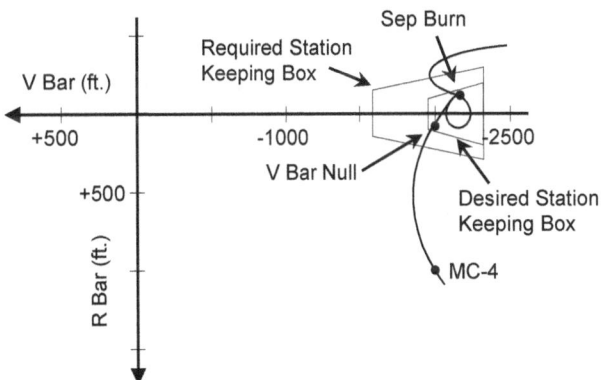

Figure 18.22 STS-77 station-keeping (proxy-vous) with PAMS-STU (May 1996).

Relative GPS Demonstrations

Several deploy/retrieve missions were used to evaluate relative GPS technology for application to future rendezvous vehicles. During STS-69 (Sept. 1995), *Endeavour* carried a Collins 3M receiver and the Wake Shield Facility a Osbourne/Jet Propulsion Laboratory TurboRogue receiver.[5-7] On STS-80 (1996), *Columbia* carried a TANS Quadrex receiver and the ORFEUS-SPAS II a Laben Tensor receiver in support of the European Space Agency (ESA) Automated Transfer Vehicle (ATV) program.[8]

References

1. Kurth, W. S., and L. A. Frank, "Spacelab 2 Plasma Diagnostics Package," *Journal of Spacecraft and Rockets*, Vol.27 No.1, 1990, pp. 70-75.

2. Carson, D. E., "SPARTAN Project Overview," *Proceedings of the First Annual NRO-OSL/CSFC-ATS Rideshare Conference*, April 15-16, 1999.

3. Howard, Richard T., Thomas C. Bryan, and Michael L. Book, "On-Orbit Testing of the Video Guidance Sensor," *Laser Radar Technology and Applications IV, Proceedings of SPIE*, Volume 3707, May 1999, pp. 290-300.

4. Wattenbach, R., and K. Moritz, "Astronomical Shuttle Pallet Satellite (ASTRO-SPAS)," *Acta Astronautica*, Volume 40, Number 10, May 10, 1997, pp. 723-732.

5. Carpenter, J. Russell, and Robert H. Bishop, "Flight Data Results Of Estimate Fusion For Spacecraft Rendezvous Navigation From Shuttle Mission STS-69," *Space Flight Mechanics 1996, Volume 93 of Advances In The Astronautical Sciences*, Univelt Inc., 1996, pp. 1257-1276.

6. Schroeder, C., B. E. Schutz and P. A. M. Abusali, "STS-69 Relative Positioning GPS Experiment," *Space Flight Mechanics 1996, Volume 93 of Advances In The Astronautical Sciences*, Univelt Inc., 1996, pp. 1239-1255.

7. Park, Y., J. Brazzel, R. Carpenter, H. Hinkel, and J. Newman, "Flight Test Results From Realtime Relative GPS Flight Experiment On STS-69," *Space Flight Mechanics 1996, Volume 93 of Advances In The Astronautical Sciences*, Univelt Inc., 1996, pp. 1277-1295.

8. Schiesser, E., J. P. Brazzel, J. R. Carpenter and H. D. Hinkel, "Results Of STS-80 Relative GPS Navigation Flight Experiment," *Space Flight Mechanics 1998, Volume 99 of Advances In The Astronautical Sciences*, Univelt Inc., 1998, pp. 1317-1333.

CHAPTER 19 - RETRIEVAL AND RETURN TO EARTH OF A SATELLITE

Five satellites were retrieved by the Space Shuttle and returned to Earth. Two of them, Palapa-B2 and Westar-VI, were not originally designed for retrieval by the shuttle. The others were designed to support shuttle retrieval.

Table 19.1 Retrieval and Return to Earth of a Satellite*

Flight	Orbiter	Year	Target	Prox Ops Approach	Comments
51A	*Discovery*	1984	1. Palapa-B2 2. Westar-VI	1. +V Bar 2. +V Bar	Both maneuvered to meet downrange and planar constraints and were retrieved by an astronaut flying the MMU.
32	*Columbia*	1990	LDEF	-R Bar	Hot final approach due to radar procedure issue.
57	*Endeavour*	1993	EURECA (ESA)	Inertial	Solar array latch failure, corrected during EVA.
72	*Endeavour*	1996	SFU (Japan)	+R Bar	Hybrid control box. Solar array retraction failure & jettison.

* All were ground-up rendezvous.
LDEF = Long Duration Exposure Facility, **EURECA** = European Retrievable Carrier, **EVA** = Extra Vehicular Activity, **MMU** = Manned Maneuvering Unit, **SFU** = Space Flyer Unit

Palapa-B2 and Westar-VI (STS-51A)

The Palapa-B2 and Westar-VI communications satellites were deployed by *Challenger* on STS-41B (February 1984) but both were stranded in low orbit due to failures of the Payload Assist Modules (PAM) on both satellites. *Discovery* on STS-51A (November 1984) deployed the Anik D-2 and SYNCOM (LEASAT) IV-1 satellites on Flight Days 2 and 3, and then performed a rendezvous with Palapa on Flight Day 5. An astronaut flying an MMU inserted a stinger device into the apogee motor nozzle and captured Palapa. The same procedure was performed on Flight Day 7 for Westar-VI (Figure 19.1).[1]

Figure 19.1   Dale Gardner about to dock with Westar-VI (STS-51A, Nov. 1984).

Plume impingement on Westar and Palapa did not cause problems, but motion of thermal protection material on the satellites could be observed when orbiter RCS jets were fired. Prior to the mission, studies were conducted to determine the appropriate inertial attitude for the satellites, which was a trade-off between visibility to the star tracker, the rendezvous radar, and the human eye. Both satellites were visible to the crew, using the COAS as a cue, at a range of over 100 n.m. A total of 44 orbital change maneuvers were executed, and the heavy workload of rendezvous and EVA activities extended the crew days well into the scheduled sleep periods.

The retrieval and return to Earth of the Palapa-B2 and Westar-VI satellites demonstrated the ability of the Shuttle Program to rapidly respond to new requirements involving vehicles not designed to support shuttle activities. Planning for the dual rendezvous mission was further complicated by the deployment of two other communications satellites prior to the rendezvous and servicing phase, and the combination of proximity operations with free-flying (MMU) EVA crew capturing and maneuvering the satellites for grapple using the RMS. Detailed mission preparation and real-time re-planning enabled the rendezvous with, retrieval and return to Earth of the satellites within a tight propellant budget. Both Palapa-B2 and Westar-VI maneuvered to meet downrange and planar offset conditions before the launch of *Discovery*.

Long Duration Exposure Facility (STS-32)

LDEF (Figure 19.2), deployed on STS-41C (April 1984), was supposed to have been retrieved in early 1985 (STS-51D), then in September of 1986 (STS-61I, canceled due to the loss of *Challenger*), and finally the summer of 1989 (STS-32, *Columbia*). *Columbia* was finally launched in January of 1990. LDEF orbital decay due to the solar maximum, *Columbia* launch delays and the SYNCOM IV-5 deploy two days before LDEF rendezvous complicated mission planning. Orbit prediction of the LDEF had a high degree of uncertainty, and experience with Skylab in 1978 and 1979 heightened concerns that LDEF could reenter the atmosphere before retrieval.

Figure 19.2 LDEF after capture by the RMS (STS-32, Jan. 1990).

During the rendezvous, poor quality radar data at long range resulted in a dispersed trajectory, and a faster final approach that required additional braking.[2,3] The gravity gradient stabilized LDEF was in the expected attitude, and the retrieval and return to Earth was successful, with only a few weeks of LDEF orbital lifetime left. After the mission, radar data incorporation procedures were changed based on the rediscovery of a radar hardware limitation.

EURECA (STS-57)

The European Retrievable Carrier (EURECA) was deployed on STS-46 (*Atlantis*, July-August 1992), and retrieved on STS-57 (*Endeavour*, June-July 1993, Figure 19.3). Before and during the flight, close coordination on mission planning and procedures was required between Mission Control in Houston and the European Space Operations Center (ESOC) in Darmstadt, Germany. EURECA completed an orbit adjustment program in preparation for the rendezvous seven days prior to the launch of *Endeavour*. However, a control box rendezvous was not performed. A phase repeating orbit was used to establish periodic launch windows and ease mission planning. In the event of an off-nominal shuttle orbit insertion; plans were developed for EURECA to lower its orbital altitude to facilitate a rendezvous and retrieval.[4]

Endeavour established communication with EURECA at a range of 44 nm. Between the shuttle MC-1 and MC-2 burns, the EURECA batteries were at full charge and the solar arrays were retracted and latched. Antenna retraction was performed via ground command when the *Endeavour* reached 200 feet, but they failed to latch. EURECA was grappled with the RMS, but not berthed until it was confirmed that berthing could be safely done with the antennas unlatched. They were later manually stowed during a previously scheduled EVA.

Figure 19.3 EURECA above KSC after deployment from *Atlantis* (STS-46, July/Aug. 1992).

Space Flyer Unit (STS-72)

STS-72 (January 1996) retrieved the Japanese Space Flyer Unit (SFU, Figure 19.4), which had been launched from the Tanegashima Space Center by an H-2 booster on March 18, 1995. A hybrid control box rendezvous was performed. A 4 ft./sec. posigrade maneuver was executed ~22 hours after launch to ensure a safe separation distance from an orbiting object. The burn was designed to have minimal impact on the rendezvous. The two SFU solar arrays were jettisoned before retrieval when sensors indicated improper latching after array retraction.

Figure 19.4   SFU about to be berthed in the payload bay of *Endeavour*. OAST-Flyer is below SFU (STS-72, Jan. 1996)

References

1. Hauck, F. H., and D. A. Gardner, "Space Salvage – A Report On Shuttle Mission STS 51-A," *Proceedings of the 29th Symposium of the Society of Experimental Test Pilots*, Society of Experimental Test Pilots, Lancaster, CA, 1985, pp. 236-261.

2. John L. Goodman, *Lessons Learned From Seven Space Shuttle Missions,* NASA Contractor Report NASA/CR-2007-213697, NASA Johnson Space Center, January 2007. See the NASA Technical Reports server at http://ntrs.nasa.gov/, or the Johnson Technical Reports server at http://ston.jsc.nasa.gov/collections/TRS/.

3. Oberg, J. E., "Rendezvous in Space," *Air & Space Smithsonian*, Vol. 8, No. 3, Aug./Sept. 1993, pp. 44–52.

4. Dreger, F., et al., "Eureca: The Flight Dynamics of the Retrieval," *ESA Bulletin*, No. 76, Nov. 1993, pp. 92-99.

This page intentionally left blank.

CHAPTER 20 - MIR AND THE INTERNATIONAL SPACE STATION

Docking of the Space Shuttle with notional space stations was studied in the early 1970s, as well as docking in support of space rescue motivated by the Apollo/Soyuz Test Project. Much of the work done to prepare the shuttle to support Space Station Freedom was applied to the Mir and ISS missions (Tables 20.1 and 20.2).

Table 20.1 Space Shuttle Flights to Mir*

Flight	Orbiter	Year	Prox Ops Approach	Comments
63	*Discovery*	1995	+V Bar	+V Bar approach to 37 feet. No docking planned. Leaking RCS jet problem.
71	*Atlantis*	1995	+R Bar	Docked to Buran port on Kristall Module. Crew exchange.
74	*Atlantis*	1995	+R Bar	Installed Shuttle Docking Module on Kristall.
76	*Atlantis*	1996	+R Bar	Resupply & U.S. crew delivery.
79	*Atlantis*	1996	+R Bar	Resupply & U.S. crew exchange.
81	*Atlantis*	1997	+R Bar	Resupply & U.S. crew exchange.
84	*Atlantis*	1997	+R Bar	Resupply & U.S. crew exchange. GPS & laser test for ESA ATV.
86	*Atlantis*	1997	+R Bar	Resupply & U.S. crew exchange. GPS test for ESA ATV. First ORBT flight.
89	*Endeavour*	1998	+R Bar	Resupply & U.S. crew exchange.
91	*Discovery*	1998	+R Bar	Resupply & U.S. crew return.

* All were ground-up rendezvous. **ATV** = Automated Transfer Vehicle, **ESA** = European Space Agency, **GPS** = Global Positioning System, **ORBT** = Optimized R-Bar Targeted Rendezvous

Docking Hardware

The Androgynous Peripheral Docking Assembly (APAS) unit (Figure 20.1) is a descendent of the APAS-75 unit jointly developed by the Soviet Union and the U.S. for the Apollo/Soyuz Test Project. The later generation APAS was originally intended for use on a Soyuz class vehicle and the Buran shuttle. Soyuz TM-16 (January-February 1993) docked with one of the two Kristall Mir module ports equipped with the APAS. For the U.S. shuttle, the APAS is mounted on the Orbiter Docking System (ODS) in the payload bay. APAS was used for dockings to both Mir and ISS. A centerline camera mounted in the ODS with a bore sight through the ODS hatch window provides the shuttle crew with a view of a docking target mounted on the Mir and ISS hatches.[1]

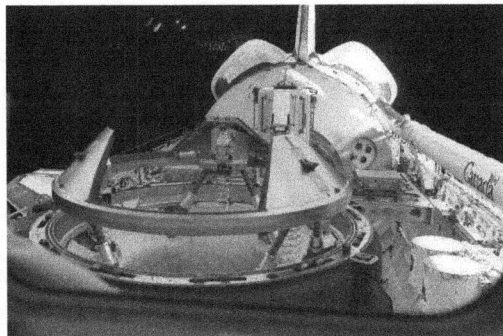

Figure 20.1 APAS on the Orbiter Docking System in the payload bay. The RMS is on the right.

For the first shuttle docking with Mir (*Atlantis*, STS-71, June 1995), the Kristall module was moved from a longitudinal port to an axial port to provide enough clearance between *Atlantis* and Mir solar arrays. After the departure of *Atlantis*, Kristall was moved back to the original position to support Soyuz and Progress dockings. A Russian built docking module was attached to the Kristall axial docking port during STS-74 (November 1995) to allow Kristall to remain attached to the Mir axial port while providing enough clearance for the shuttle.[2] The ISS is equipped with two APAS units, mounted on two Pressurized Mating Adapters (PMA).

New Sensor Development and New Challenges

In the mid 1970s, when the rendezvous navigation sensors and relative navigation capability was baselined, proximity operations culminated with approach and grapple of a spacecraft with a robotic arm. The baselined rendezvous radar was adequate to support this, although it lacked effective redundancy.

The lack of a backup range and range rate sensor providing better data and ease of operation than COAS subtended angles and Closed Circuit Television (CCTV) tilt angles had been a source of concern since the late 1970s. Attempts to obtain a new source of range and rate data were limited to off-the-shelf options (stadimetric binoculars, a parallax rangefinder, laser range finders in

Table 20.2 ISS Assembly and Replenishment Missions*

Flight	Orbiter	Year	Prox Ops Approach	Comments
88 (2A)	*Endeavour*	1998	TORRA/-R Bar	Captured Zarya with RMS, attached Unity Node with PMA 1 & 2.
96 (2A.1)	*Discovery*	1999	TORRA/-R Bar	First docking with ISS. ISS resupply and outfitting.
101 (2A.2a)	*Atlantis*	2000	TORRA/-R Bar	ISS resupply and outfitting.
106 (2A.2b)	*Atlantis*	2000	TORRA/-R Bar	ISS resupply and outfitting.
92 (3A)	*Discovery*	2000	TORRA/-R Bar	Radar failure. Z1 Truss, PMA 3, Ku comm, & CMGs installed.
97 (4A)	*Endeavour*	2000	+R Bar	Delivered P6 truss (with solar arrays & radiators).
98 (5A)	*Atlantis*	2001	+R Bar	Delivered Destiny lab.
102 (5A.1)	*Discovery*	2001	TORVA/+V Bar	MPLM resupply. Crew exchange. Tail forward approach.
100 (6A)	*Endeavour*	2001	TORVA/+V Bar	Installed robotic arm. MPLM resupply. Tail forward approach.
104 (7A)	*Atlantis*	2001	TORVA/+V Bar	Delivered Quest Airlock (installed with ISS robotic arm).
105 (7A.1)	*Discovery*	2001	TORVA/+V Bar	MPLM resupply. Crew exchange.
108 (UF-1)	*Endeavour*	2001	TORVA/+V Bar	MPLM resupply. Crew exchange.
110 (8A)	*Atlantis*	2002	TORVA/+V Bar	Delivered S0 truss and Mobile Transporter.
111 (UF-2)	*Endeavour*	2002	TORVA/+V Bar	MPLM resupply. Mobile base installation. Crew exchange.
112 (9A)	*Atlantis*	2002	TORVA/+V Bar	Delivered S1 truss, radiators & CETA cart A.
113 (11A)	*Endeavour*	2002	TORVA/+V Bar	Delivered P1 truss, radiators & CETA cart B. Crew exchange.
114 (LF-1)	*Discovery*	2005	TORVA/+V Bar	MPLM Resupply. CMG replacement. First RPM.
121 (ULF-1.1)	*Discovery*	2006	TORVA/+V Bar	MPLM Resupply. ISS repairs via EVA. Add third ISS crewmember.
115 (12A)	*Atlantis*	2006	TORVA/+V Bar	Install P3/P4 truss.
116 (12A.1)	*Discovery*	2006	TORVA/+V Bar	P5 Truss, SPACEHAB
117 (13A)	*Atlantis*	2007	TORVA/+V Bar	S3/S4 Truss
118 (13A.1)	*Endeavour*	2007	TORVA/+V Bar	S5 Truss
120 (10A)	*Discovery*	2007	TORVA/+V Bar	U.S. Node 2, first flight of Lambert guidance upgrade.
122 (1E)	*Atlantis*	2008	TORVA/+V Bar	Columbus Laboratory
123 (1J/A)	*Endeavour*	2008	TORVA/+V Bar	Kibo Logistics Module, Dextre Robotics System
124 (1J)	*Discovery*	2008	TORVA/+V Bar	Kibo Pressurized Module, Japanese Remote Manipulator System
126 (ULF2)	*Endeavour*	2008	TORVA/+V Bar	MPLM
119 (15A)	*Discovery*	2009	TORVA/+V Bar	S6 truss segment
127 92J/A)	*Endeavour*	2009	TORVA/+V Bar	Kibo JEM EF, Kibo Japanese ELM-ES
128 (17A)	*Discovery*	2009	TORVA/+V Bar	Leonardo MPLM, LMPESSC, Vernier RCS failure.
129 (ULF3)	*Atlantis*	2009	TORVA/+V Bar	ELC1, ELC2
STS-130 (20A)	*Endeavour*	2010	TORVA/+V Bar	Tranquility Node 3, Cupola. TCS failure during approach.
STS-131 (19A)	*Discovery*	2010	TORVA/+V Bar	Leonardo MPLM, radar fail.
STS-132 (ULF4)	*Atlantis*	2010	TORVA/+V Bar	ICC, MRM1, COAS bulb replacement.
STS-133 (ULF5)	*Discovery*	2011	TORVA/+V Bar	ELC4, PMM
STS-134 (ULF6)	*Endeavour*	2011	TORVA/+V Bar	ELC3, AMS-2, STORRM DTO
STS-135 (ULF7)	*Atlantis*	2011	TORVA/+V Bar	Raffaello MPLM, LMC, return to Earth of failed ammonia pump. ISS yaw maneuver after orbiter undocking to facilitate engineering photos during orbiter half-lap fly-around.

* All were ground-up rendezvous. **A** = Assembly, **AMS** = Alpha Magnetic Spectrometer, **ATV** = Automated Transfer Vehicle, **CETA** = Crew and Equipment Translation Aid, **CMG** = Control Moment Gyro, **DTO** = Detailed Test Objective, **ELC** = EXPRESS Logistics Carrier, **ELM-ES** = Experiment Logistics Module - Exposed Section, **EVA** = Extra Vehicular Activity, **ICC** = Integrated Cargo Carrier, **JEM EF** = Japanese Experiment Module Exposed Facility, **LF** = Logistics Flight, **LMC** = Lightweight Multi-purpose Carrier, **LMPESSC** = Lightweight Multi-Purpose Experiment Support Structure Carrier, **MPLM** = Multi-Purpose Logistics Module, **MRM** = Mini Research Module, **ORBT** = Optimized R-Bar Targeted Rendezvous, **PMA** = Pressurized Mating Adapter, **PMM** = Permanent Multi-Purpose Module, **Rndz** = Rendezvous, **RPM** = R Bar Pitch Maneuver, **STORRM** = Sensor Test for Orion RelNav Risk Mitigation, **TORRA** = Twice Orbital Rate R Bar Approach, **TORVA** = Twice Orbital Rate V Bar Approach, **UF** = Utilization Flight, **ULF** = Utilization & Logistics Flight

use by the military, or auto focusing lasers on the payload bay camera system) due to budget restrictions. None of the off-the-shelf options evaluated were satisfactory. Once shuttle rendezvous and proximity operations missions began, their success made it difficult to obtain sufficient priority and resources to address the concern.

The baseline sensors (radar, COAS, CCTV) were successfully used to null the final approach velocity to zero to support grapple of a spacecraft by the RMS on rendezvous missions. However, these sensors and piloting aids had operational limitations. Rendezvous radar tracking for even small payloads reached the end of the tracking envelope at about 80 feet. Visual ranging methods using COAS subtended angles or two CCTV camera tilt angles were imprecise, imposed a heavy crew work-load during an already busy phase of flight, and did not provide the crew with a direct range-rate measurement. On-orbit use of the CCTVs was frequently limited by extreme orbital lighting conditions that caused image blooming. There was no orbiter sensor capability that would permit the crew to achieve a specific non-zero approach velocity with precision. Such a capability would be required for docking the shuttle to a space station.

In 1987, studies of shuttle docking with Space Station Freedom indicated that a better proximity operations sensor than the Ku Band radar was needed. This provided the justification needed to begin development of custom built sensors and procurement of off-the-shelf sensors that were more advanced than sensors evaluated in the late 1970s and early 1980s (Figures 20.2 and 20.3).[3]

Hand Held Lidar (HHL) first flew on STS-49 (May 1992). The Trajectory Control Sensor (TCS) lidar flew as a Detailed Test Objective (DTO) on STS-51, STS-63 and STS-64. The first official flight as a payload was on STS-71 (June-July 1995) for rendezvous and docking with the Mir Space Station. TCS and HHL provided the precise range and range rate measurements needed to meet Mir and ISS docking conditions.

Though raw data was adequate to meet docking requirements, HHL, TCS, and legacy sensor data (radar, centerline camera, CCTV cameras) were processed in a laptop computer using a software package known as the Rendezvous and Proximity Operations Program (RPOP). RPOP provided a relative motion display and proximity operations piloting cues not available in the legacy shuttle avionics system (Figures 20.2 through 20.4).[3-5] TCS, HHL, and RPOP became known as "rendezvous tools." HHL, TCS, and RPOP data were not provided to the GNC flight computer. Even though HHL data could be acquired before MC-3 and TCS data before MC-4, these burns were Lambert targeted independently of RPOP data (Figure 20.2).

In addition to TCS and HHL measurements, RPOP also could processes centerline camera vertical angles and CCTV tilt angles. The one foot CCTV range ruler marks could be processed if the crew pressed a key each time a

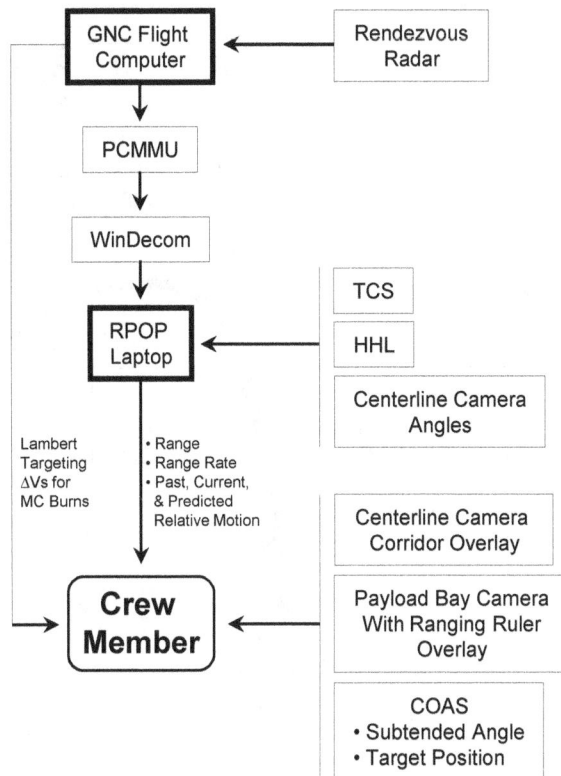

Figure 20.2 Simplified GNC and rendezvous tools architecture. RPOP, TCS, HHL, and cameras provide piloting cues after the last Lambert targeted burn, MC-4.

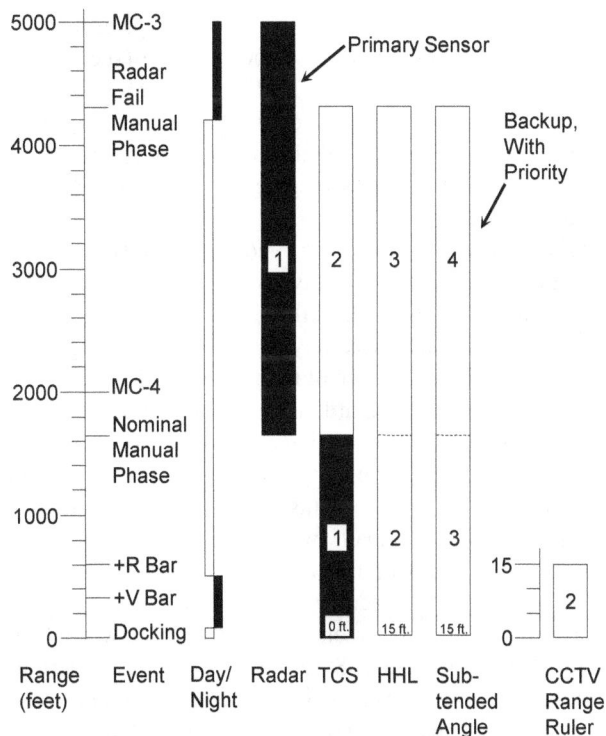

Figure 20.3 Operational use of shuttle proximity operations sensors for a typical ISS mission.

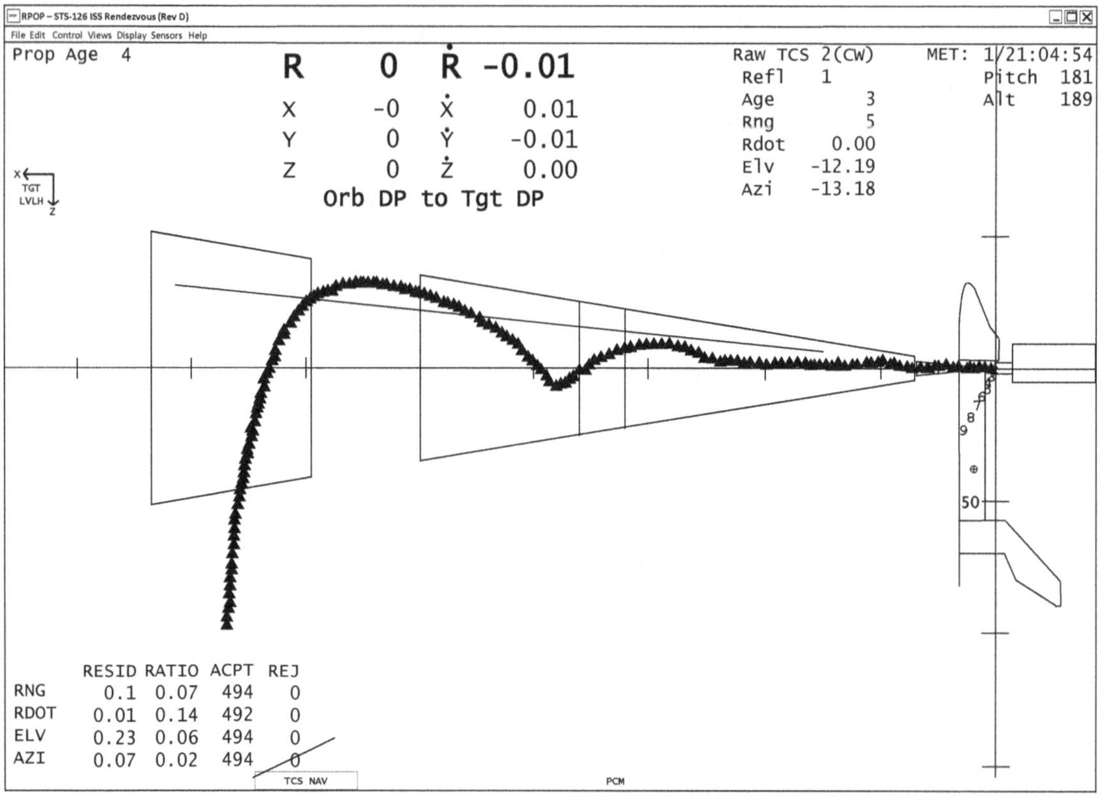

Figure 20.4 Recreation of Mission Control RPOP display showing the STS-126 +V Bar arrival and final approach to docking (Nov. 17, 2008).

landmark on the ISS crossed a one foot mark on the ruler. However, normally only the centerline camera vertical angles were processed in conjunction with HHL range as a backup to TCS.

The operational envelope of proximity operations sensors is illustrated in Figure 20.3 for a typical mission to the ISS. In the event of a radar failure (such as on STS-92 and STS-131), TCS, HHL, and COAS subtended angle (Figure 20.5) are used earlier in the profile than on a nominal mission. A ranging ruler overlay on an aft cockpit monitor provides ranging during the last 15 feet (Figure 20.6).

To save money, the new rendezvous tools were not certified to the same criticality level as the baseline 1970s era rendezvous GNC capability. Crews were expected and trained to fly proximity operations, docking, and undocking without the use of rendezvous tools.

While the Space Shuttle effectively flew missions to the Mir and ISS, the staggered integration of rendezvous (mid 1970s) and proximity operations (early 1990s) relative navigation algorithms and sensors resulted in a high crew work load with most crew members participating in proximity operations (Figure 20.7).

The rendezvous radar was usable with small targets down to ranges of between 80 to 100 feet. However, the size of Mir and the ISS resulted in beam wandering, which degraded measurement quality out to ranges of a few thousand feet depending on the size of the target. For ISS missions rendezvous radar measurements typically were not processed by the Kalman filter after

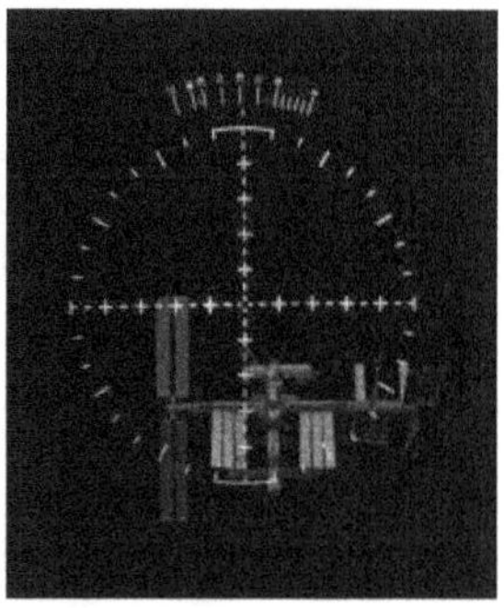

Figure 20.5 ISS viewed through the COAS on STS-126.

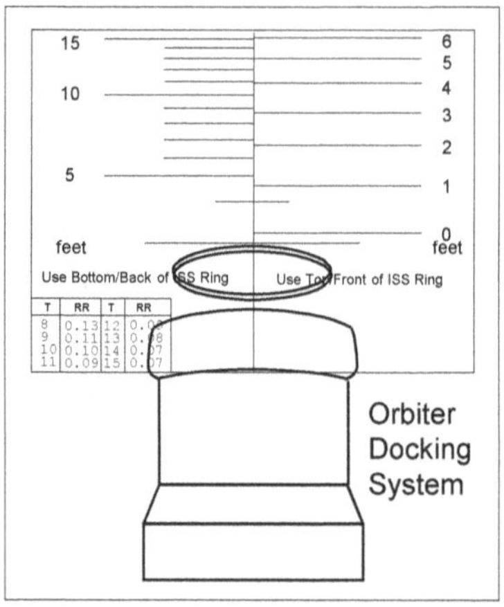

Figure 20.6 Space Shuttle camera ranging ruler overlay for use during docking with the ISS.

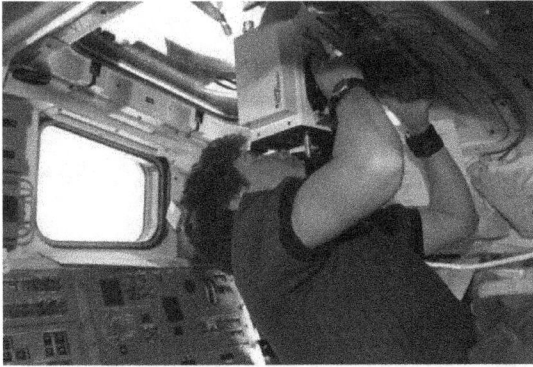

Figure 20.7 Wendy Lawrence using an HHL during STS-114 (July/Aug. 2005).

Figure 20.8 View from *Discovery* of Valeriy V. Polyakov looking out a Mir window during +V Bar approach (STS-63, Feb. 6, 1995).

the +R Bar Pitch Maneuver (RPM). However, the radar data was often still available to the crew up to the time the orbiter arrived on the +V Bar as the radar range rate proved to be accurate at close ranges and was monitored by the crew. At some point during proximity operations the Ku band antenna was transitioned from radar mode to communications mode to transmit video to Mission Control through TDRSS. TCS and HHL exhibited better performance during proximity operations than the Ku radar. The availability of TCS and HHL measurements was essential to ensure safe and successful approaches to Mir and the ISS (Figures 20.5 and 20.8).

It was also recognized that Mir and ISS brightness and size issues could complicate or prevent use of daytime star tracker measurements for relative navigation after the Ti maneuver, in the event of a radar failure (Figure 11.1). Night star tracker data was obtained between the MC-1 and MC-3 burns during the STS-64 rendezvous with SPARTAN. Analysis techniques verified with the collected flight data were applied to data collected during the STS-63, STS-71, and STS-74 missions to Mir.

Analysis of these missions indicated that the 18 lights of varying intensity and character (flashing and non-flashing) distributed across Mir provided a suitable target for the shuttle star tracker. Post Ti contingency night star tracker navigation procedures were first flown on STS-79 (September 1996). A tracking light was added to the ISS Zvezda ("Star") Service Module to enable contingency star tracking during orbital night for ISS missions. Night star tracker navigation was performed during STS-92 (October 2000) and STS-131 (April 2010) due to the radar failures.

Although shuttle orbiters were equipped with GPS receivers for use on-orbit and during entry, and the ISS was equipped with GPS as well, GPS was not used for shuttle rendezvous or proximity operations with the ISS until the STORRM DTO flown on STS-134 (see Chapter 21).[6]

Flight Control and Plume Challenges

All missions to Mir and ISS required extensive flight control and plume impingement analysis of the various configurations during approach, mated flight, assembly, and separation.[7-12] For example STS-88, the first ISS assembly flight, involved the attachment of the U.S. built Unity node to the previously launched, Russian manufactured Zarya (FGB) module (Figure 20.9). Unity was docked to the ODS using the RMS before the rendezvous with Zarya. Shuttle flight control analysis was required to ensure that execution of rendezvous maneuvers would not violate structural loading constraints on Unity and the ODS. Zarya was later grappled with the RMS, and docked to Unity (Figure 20.10). At 42,000 pounds, Zarya was the largest object ever manipulated with the RMS. Analysis was also performed to ensure that ISS orbit raising with shuttle RCS jets could be successfully performed.[10]

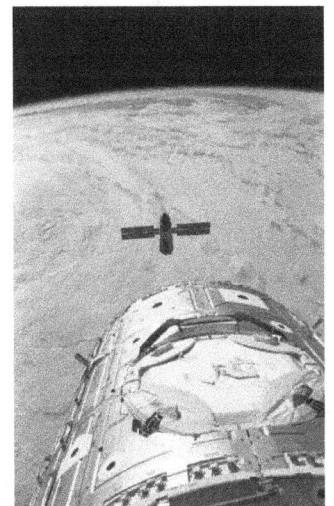

Figure 20.9 Zarya in the distance as *Endeavour* approaches with Unity in the foreground (Dec. 1998).

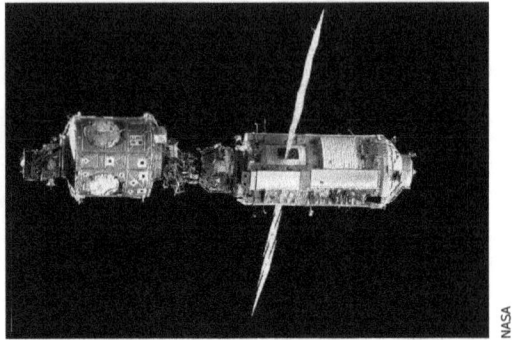

Figure 20.10 Zarya and Unity after deployment from *Endeavour* (Dec. 1998).

Figure 20.11 *Atlantis* docked to Mir during STS-71, as seen from Soyuz TM-21 (June 1995).

New Profile Development

The stable orbit rendezvous profile was designed for mainly inertial and +V Bar approaches (a transition to the –R Bar could be performed upon arrival at the +V Bar). A difficulty with the stable orbit approach was the increased amount of propellant required for braking in Low Z mode (Figure 10.11) and greater sensitivity to plume impingement loads of Mir and ISS. Reducing plume concerns (static, dynamic, thermal, contamination) was critical, particularly for solar arrays.

Planning for Mir and ISS rendezvous missions prompted renewed study of the direct +R Bar approach in 1993 (Figure 10.10). Use of orbital mechanics to reduce the needed braking, rather than using RCS jet firings, would lower plume impingement and provide propellant savings. An additional benefit was that a +R Bar separation could also take advantage of orbital mechanics, requiring fewer jet firings. Studies indicated that the new approach could be performed without changing on-board computer targeting constants for the stable orbit profile. The availability of laser sensors (TCS, HHL) provided range and range rate measurement redundancy which was not available when the direct +R Bar approach was considered for the Skylab reboost mission in the late 1970s.

After extensive analysis, procedure development, and efforts to overcome programmatic resistance, the direct +R Bar approach was approved by the Shuttle Program in April of 1994. It was first flown in November of that year during the STS-66 retrieval of CRISTA-SPAS. Direct +R Bar approaches were flown on all docking missions to Mir.[4,13,14] The Mir missions (Figure 20.11) validated shuttle proximity operations and docking analysis originally performed for Space Station Freedom.

Further analysis led rendezvous designers to investigate changes to the rendezvous profile itself, before the proximity operations phase, to further reduce propellant consumption and increase shuttle payload capability. The stable orbit profile, like its' predecessor the coelliptic profile, was a "high energy" profile

designed to support a terminal phase inertial approach and direct intercept. Additional propellant and procedures were required for R Bar or V Bar activities. A new profile was designed which was optimized for the +R Bar approach.

Optimized R-Bar Targeted Rendezvous (ORBT) differed from stable orbit in several ways (Figure 20.12). ORBT was designed to optimally set up initial conditions for a low energy coast up the +R Bar (Figures 20.13, 20.14, and 20.15). By targeting the Ti, and first three mid-course maneuvers for the manual takeover point at 2,000 feet, rather than for intercept, manual phase trajectory dispersions were reduced and propellant consumption was lowered. The Ti point for ORBT was below the V Bar so that the subsequent MC-4 ΔV vector would be primarily in the +X body axis direction (Figure 10.11), saving propellant. The MC-4 maneuver targeted the orbiter for a point 600 feet below the target, on the +R Bar. ORBT did not require as many +R Bar stabilization

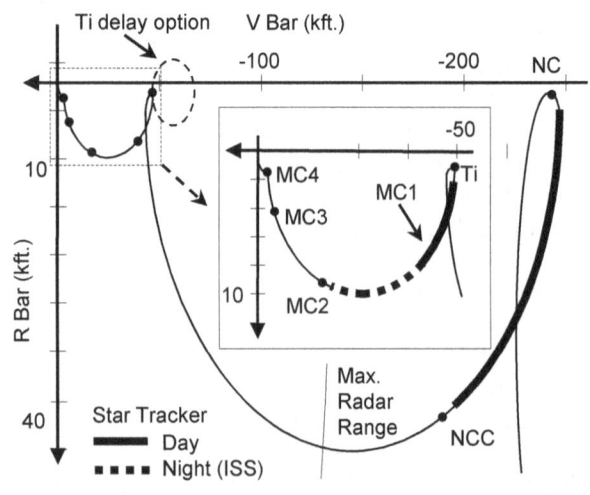

Figure 20.12 Optimized R-Bar Targeted Rendezvous (1997 to end of program).

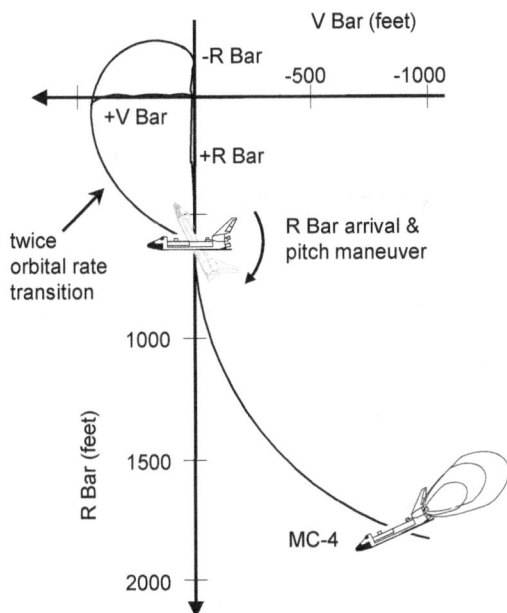

Figure 20.13 Approaches to ISS.

Figure 20.14 ISS viewed from *Endeavour* on the +R Bar during STS-113 (Nov./Dec. 2002).

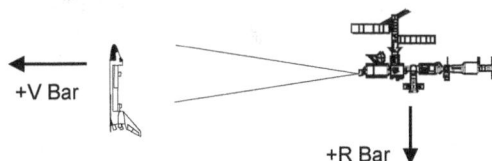

Figure 20.15 Entering ISS approach corridor at ~400 feet.

burns or as many braking burns as were needed with the stable orbit profile. The first ORBT flight was STS-86 to Mir (September-October 1997).

Proximity Operations and Docking

Final approach to the Mir (+R Bar) and ISS (+V Bar, +R Bar, or -R Bar, depending on the ISS configuration, Figure 20.13) involved flying a precise range and range rate profile.[14] An 8-degree, followed by a 5-degree, approach corridor centered on the Mir or ISS docking hatch target was flown (Figure 20.15). An angular fly-out could be performed at a range of 30 feet to achieve the required alignment for docking. Station-keeping points existed during the approach to allow delays to ensure proper lighting, gain time to work systems issues, or obtain visibility to ground communication stations.

Rather than performing orbital rate inertial fly-arounds to transition between the R Bar and V Bar axes, twice orbital rate fly-arounds were conducted. This technique was originally developed for the WSF fly-around on STS-60, but was not flown due to the unsuccessful WSF deployment. Twice orbital rate fly-arounds permitted faster transfers with lower propellant consumption and plume impingement, and was also used for post-undocking fly-arounds of Mir and ISS. The faster rate compared with inertial transitions prevented the sun from continuously staying in the field of view of the crew. Post-undocking fly-arounds were used to obtain photography of the Mir and ISS, if sufficient propellant was available.

The ability of the shuttle to perform station-keeping during proximity operations with Mir and early ISS missions (through STS-92) was required so that the

docking would occur within the visibility of a Russian ground communications station. Mir attitude control was transitioned to free drift at docking by the Mir crew. Ground commanding by Russian flight controllers served as a backup to the crew. The shuttle crew flew timed proximity operations approaches to ensure that final approach and docking occurred during a Russian communications window.

The attitude of the shuttle relative to the Mir or ISS while docked was defined by the clocking angle. Determination of the clocking angle depended on several considerations. These were: 1) Ability of the station solar arrays to generate solar power in a given attitude, 2) Thermal control capabilities of the shuttle and station, 3) Station and shuttle communications while docked (i.e. antenna visibility), 4) Adequate clearance of station and shuttle hardware to avoid undesirable contact, 5) Shuttle RCS jet plume impingement on the station, and 6) How well the mated stack could be controlled by the station or shuttle flight control systems. The Mir or ISS attitude during the approach was determined through a trade study comparing station solar power generation and thermal control capabilities for a given attitude with the shuttle flight control system margin (controllability).

At a range of 30 feet on all Mir and ISS missions station-keeping could be performed (if required) to perform an angular attitude alignment correction to ensure that the shuttle and Mir (or ISS) docking hardware was properly aligned. Station-keeping at 30 feet could also include waiting for appropriate orbital lighting for the shuttle crew to read the docking target on Mir (or ISS) to determine the relative misalignment. In addition, for flights that required docking with communications through a Russian ground station, the 30 foot station-

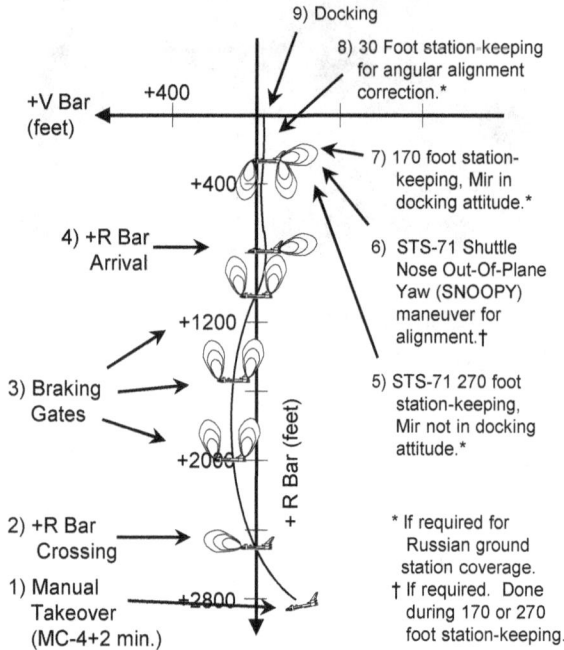

Figure 20.16 Plus R Bar approach to Mir from Stable Orbit profile, STS-71 (June 1995) & STS-74 (Nov. 1995).

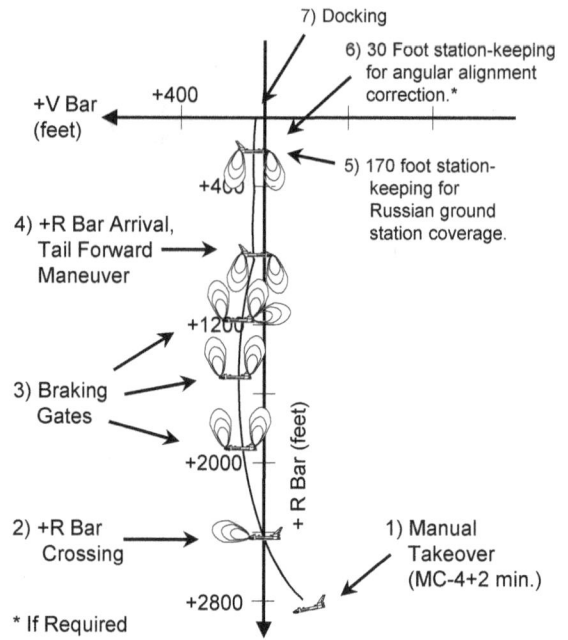

Figure 20.17 Plus R Bar approach to Mir from Stable Orbit profile, STS-76 (March 1996), STS-79 (Sept. 1996), STS-81 (Jan. 1997), & STS-84 (May 1997).

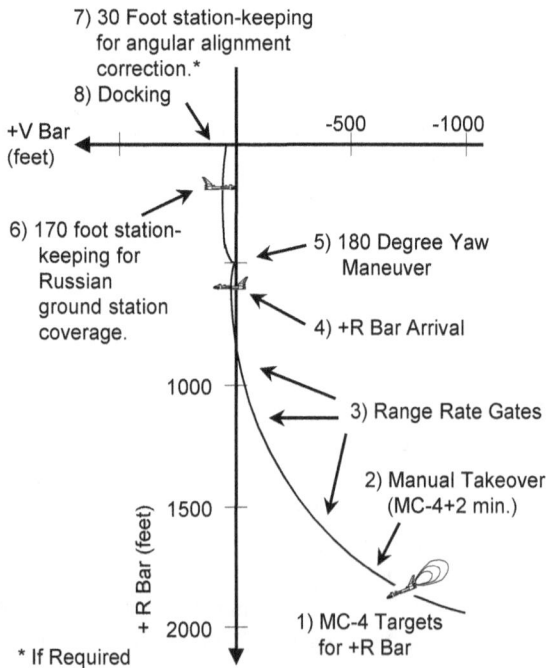

Figure 20.18 Plus R Bar approach to Mir from ORBT profile, tail forward docking, STS-86 (Sept./Oct. 1997) & STS-91 (June 1998).

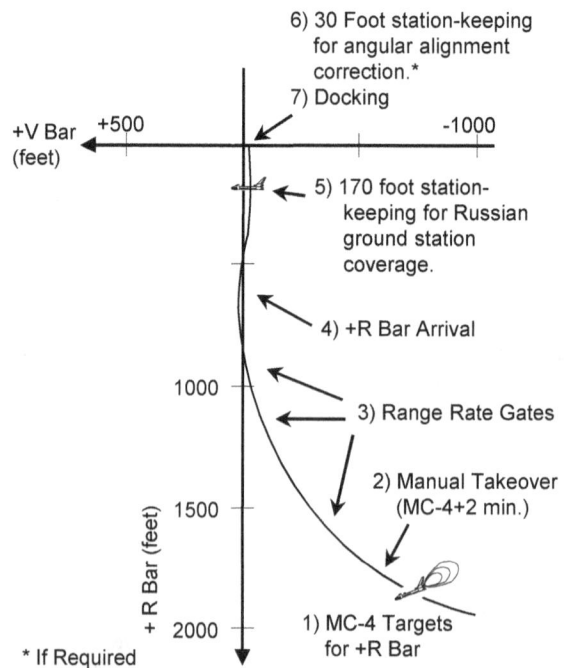

Figure 20.19 Plus R Bar approach to Mir from ORBT profile, nose forward docking, STS-89 (Jan. 1998).

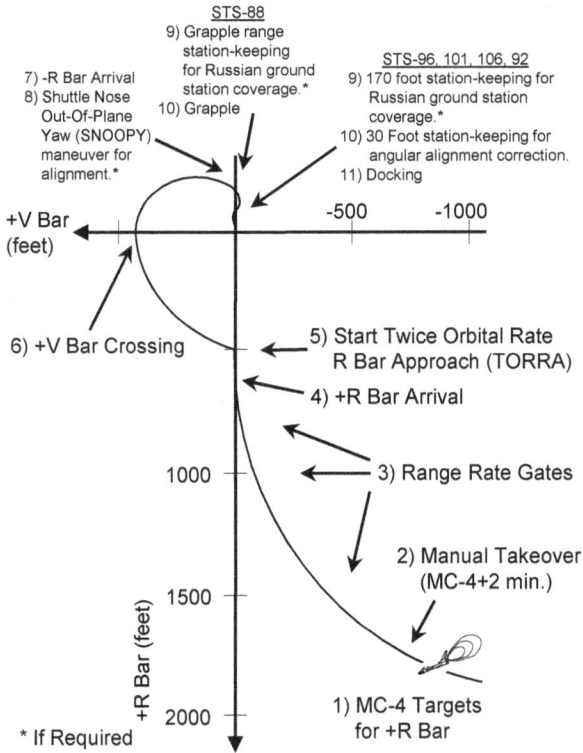

Figure 20.20 Minus R Bar approaches to ISS, STS-88 (Dec. 1998) to STS-92 (Oct. 2000).

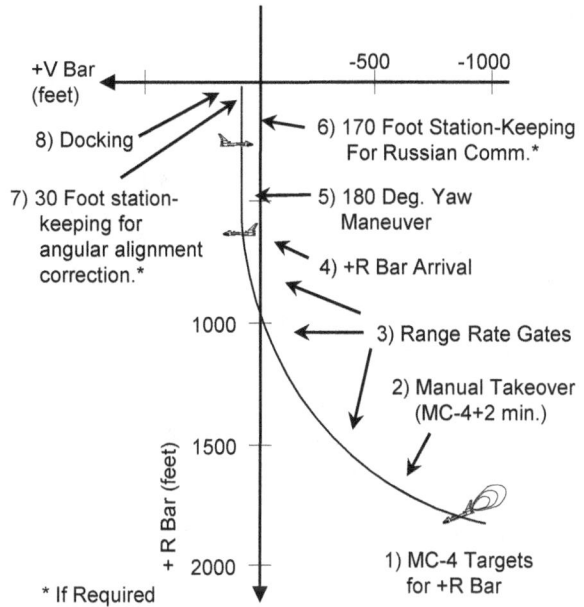

Figure 20.21 Plus R Bar approaches to ISS, STS-97 (Nov./Dec. 2000) & STS-98 (Feb. 2001).

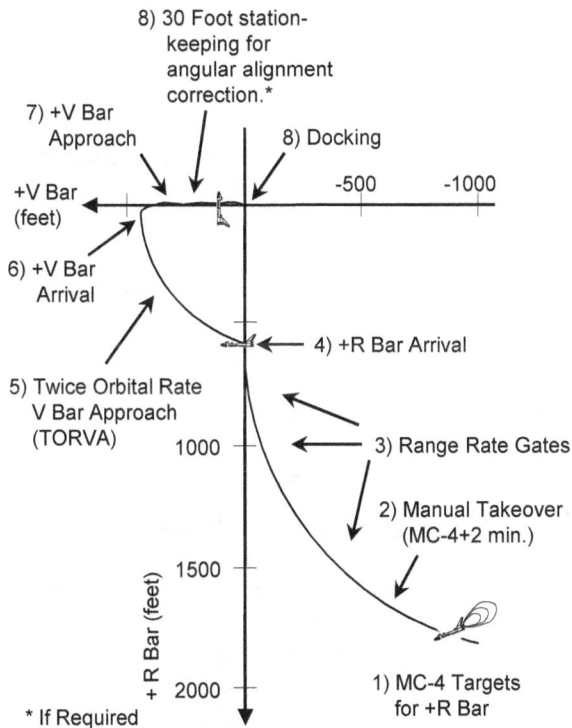

Figure 20.22 Plus V Bar approaches to ISS, STS-102 (March 2001) to STS-113 (Nov./Dec. 2002).

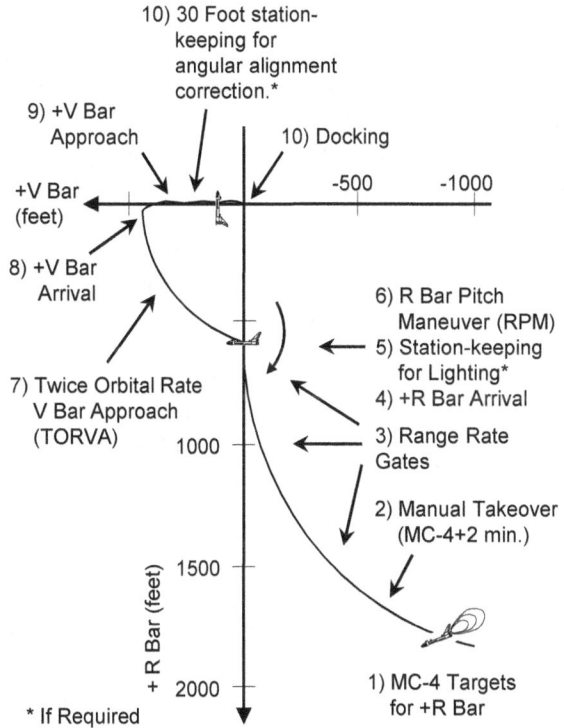

Figure 20.23 RPM and plus V Bar approaches to ISS, STS-114 (July/Aug. 2005) to end of program (2011).

139

keeping could be used to absorb time of arrival dispersions.

For the first docking mission to Mir (STS-71, June 1995) station-keeping was performed at a range of 270 feet on the plus R Bar (Figure 20.16) if Mir was not in the docking attitude. If Mir was in the docking attitude station-keeping was performed at a range of 170 feet. For beta angles higher than 30 degrees docking would be performed in a nose out-of-plane attitude. This was driven by orienting the Mir solar arrays out-of-plane. The actual beta angle was driven by the launch date. A delay in the STS-71 launch resulted in a lower beta angle, and the docking was performed with the shuttle nose in plane (nose forward). For STS-74 (Nov. 1995) docking was performed nose forward (Figure 20.16) and station-keeping was performed at a range of 170 feet.

Missions STS-76, STS-79, STS-81, and STS-84 (Figure 20.17) flew tail forward dockings to improve Mir communications with the ground. STS-86 and STS-91 also flew a tail forward dockings (Figure 20.18). STS-86 (Sept./Oct. 1997) was the first flight of the Optimized R Bar Targeted (ORBT) rendezvous. STS-89 (Jan. 1998) flew a nose forward docking (Figure 20.19) to test the new single minus X RCS jet firing procedure. This procedure was required for future ISS missions to reduce shuttle RCS plume impingement on the ISS. The procedure had not yet been analyzed to determine if it could provide adequate flight control for the tail forward approach. Tail forward dockings were highly desirable but not mandatory.

STS-88 (Dec. 1998), STS-96 (May/June 1999), STS-101 (May 2000), STS-106 (Sept. 2000), and STS-92 (Oct. 2000) all flew minus R Bar approaches (Figure 20.20) to avoid orbiter obscuration of Russian communications antennas on ISS. These missions also required nose out-of-plane grapple (STS-88) or dockings for beta angles greater than 45 degrees due to pointing ISS solar arrays out-of-plane. However, these missions were launched when the beta angle was low and the shuttle nose out-of-plane attitude was not required. All dockings (96, 101, 106, and 92) and the STS-88 grapple using the RMS were performed with the nose in-plane.

The STS-88 mission to the ISS required Russian ground station coverage during grapple for Russian commanding of the ISS attitude control system to free drift. There was no backup method of transitioning to free drift. For STS-96, STS-101, STS-106, and STS-92 the primary means of moding to free drift was automatic, with Russian ground commanding as the backup.

STS-97 (Nov./Dec. 2000) and STS-98 (Feb. 2001) flew plus R Bar approaches (Figure 20.21) with tail forward dockings. Tail forward was the only attitude that allowed a PMA-3 docking for these flights. PMA-3 dockings were required for installation of the P6 truss and the Destiny laboratory. For these flights the Russian ground station communications requirement was changed to highly desirable as long as some form of air-to-air

communication existed between the Mir and shuttle crews. The primary means of moding to free drift was automatic, with ISS crew commanding as the second method. The third method was Russian ground commanding. Station-keeping could still be performed if Russian ground communications became a requirement. In this event the shuttle crew would have flown a timed approach.

STS-102 (March 2001) was the first flight to fly the plus V Bar approach (Figure 20.22) that would be flown on all missions to the ISS through the end of the Shuttle Program. Also beginning with STS-102 Russian ground station communications coverage was neither required nor highly desirable. Commanding of the ISS attitude control system to free drift at docking was performed automatically, with ISS crew commanding and Mission Control Houston commanding as the backup methods. Timed final approach procedures were no longer carried on the shuttle. The ORBT rendezvous profile placed the orbiter on the plus R Bar at a range of 600 feet. A twice orbital rate fly-around from the plus R Bar to the plus V Bar was performed, followed by the plus V Bar approach and docking.

In the wake of the *Columbia* tragedy, options for inspecting the thermal protection system on the bottom of the orbiters were examined. For flights to the ISS, one option entailed visual and photographic inspection from the ISS as the orbiter approached for docking, or during the post-undocking fly-around. The post-undocking survey was eventually discarded due to propellant costs, procedural complexity for both nominal and off nominal (i.e. systems anomalies) scenarios, limited consumables remaining to support any required post-redocking EVAs for repairs, and the limited time available to assess the photographic survey before a decision to re-dock was required. The ISS crew photographed the orbiter through windows in the Zvezda Service Module.

A 360-degree pitch maneuver after arrival at the 600 foot point (before docking) on the +R Bar was selected (Figures 20.13, 20.23, 20.24).[5,15,16] After maneuver initiation, the flight control system is placed in free drift (no RCS firings) to avoid ISS window contamination and plume impingement. A minimal closing rate at maneuver initiation, coupled with +R Bar braking effects protects

Figure 20.24 *Discovery* **photographed from the ISS during the RPM on STS-114 (July/Aug. 2005).**

against an undesirable closing trajectory while the ISS is out-of-view of the shuttle crew and relative navigation sensors. A rotation rate of 0.75 degree/second was chosen to allow sufficient time for photography, minimize propellant usage, and limit trajectory dispersions while in free drift. After maneuver completion, a normal twice orbital rate transition from the +R Bar to the +V Bar is performed, followed by the standard +V Bar approach and docking.

The +R Bar tile inspection maneuver was attractive since it could be flown with existing hardware, software and crew capabilities, and could be certified for operational use to support the earliest return to flight. Development and certification of the procedure also involved consideration of RCS plume loads on the ISS, propellant consumption, stability of the orbiter pitch rotation in the presence of dispersed attitude, rates, and mass properties (pitch is unstable because it is about an intermediate axis of inertia), use of ISS windows, ability of cameras to provide imagery of sufficient quality, orbital lighting, breakout procedures, and procedures for handling system anomalies (such as sensor or RCS jet failures).

A new requirement to perform shuttle thermal protection repair at the ISS also drove extensive proximity operations analysis and procedure development. The shuttle RMS grapples a fixture on the ISS and the shuttle is rotated to an appropriate position relative to the ISS for repair. An ISS attitude was defined that would facilitate a safe separation (no undesirable contact with or pluming of ISS and Soyuz structure) and re-docking in the event a RMS or other failure resulted in a contingency separation from the ISS.[17]

In the event an orbiter docked to the ISS had suffered thermal protection system damage that could not be repaired, an unmanned undocking procedure was developed so that the docking port could be cleared for another shuttle to pickup the damaged shuttle's crew and bring them home. The damaged shuttle would be deorbited for a South Pacific destructive entry.[18]

Launch Windows and Mission Planning

All ground-up rendezvous missions flown by the shuttle before the Mir flights were to an orbital inclination of ~28.5 degrees. This provided long launch windows since the Earth fixed launch site was near the desired orbital plane for several hours. For example, launch windows for HST servicing missions were approximately one hour long. However, the 51.6 degree orbital inclination of the Mir station resulted in launch windows of 10 to 12 minutes duration. For steep inclinations the launch site approached, passed through, and departed the desired orbital plane in a few minutes. Launching inside such short windows was considered a challenge. For the initial Mir missions launch was targeted to occur at the beginning of the launch window to maximize the probability that the shuttle could launch

on a given day, even though this required propellant for yaw steering during ascent.

Shuttle missions to Mir were successfully launched on time and confidence in the ability of the shuttles and associated ground support systems to support short launch windows grew. Starting with STS-86 (September 1997) the practice of targeting launch for the beginning of the launch window was changed to targeting launch when the targeted orbital plane crossed the launch site, even though this decreased the time available in the launch window. Launch at this time, called the in-plane time, required little or no yaw steering and increased ascent performance margin through propellant savings. Most Mir and ISS launches were successfully performed at the in-plane launch time. However, there were exceptions to this practice due to other mission planning considerations and an unplanned launch hold for STS-110 (April 2002).

Mission planning for ISS missions was a complex process, with many factors such as ISS logistics, ISS hardware maintenance, ISS orbit maintenance, shuttle ascent abort, rendezvous and proximity operations considerations, and visits of other vehicles (Soyuz, Progress, ATV, HTV) to the ISS that must be considered.[19-20] Ascent propulsion problems (such as an early main engine shutdown) could limit the ability of the shuttle to fly the planned rendezvous profile (Figure 20.25).[21-22] In coordination with the Russians, contingency plans existed for the ISS to lower its orbit.

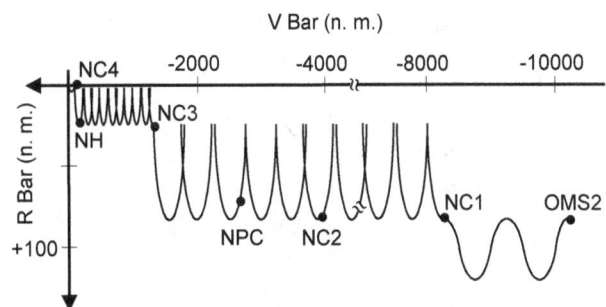

Figure 20.25 Typical ISS Flight Day 3 rendezvous ground targeted phase (1998 to end of program).

The *Columbia* Accident Investigation Board (CAIB) recommended that the Shuttle Program obtain photography of the orbiter TPS and ET and downlink the imagery during the mission to facilitate detection of foam shedding and determination of associated risk. In addition to on-orbit inspection with the OBSS and during the RPM, this resulted in photography of the shuttle during ascent (using ground based cameras and cameras mounted on NASA WB-57F aircraft flying at ~60,000 feet) and External Tank (ET) photography after separation. This in turn led to daylight launch and acceptable ET lighting requirements that placed constraints on launch windows.

Before the *Columbia* accident lighting constraints were not placed on ground-up rendezvous launch windows since such constraints would too severely

Approved for public release via STI DAA 24483. See statement on title page.

141

constrain the launch windows. Crews were trained to fly ascent abort and nominal end of mission landings at night, if they were required.

Motion picture and still cameras had been carried in the ET umbilical wells for ET photography well before the return to flight, STS-114 (July-August 2005). For STS-114 and subsequent flights the right umbilical well film camera was replaced by a digital camera with image downlink capability (Figure 20.26). ET photography was also performed after ET separation by a 16 mm motion picture camera mounted in the left ET umbilical well of

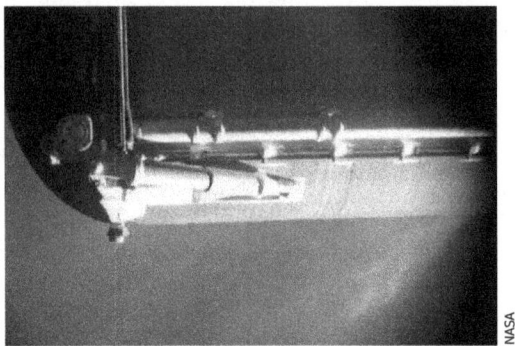

Figure 20.26 STS-131 External Tank photo taken from the right External Tank umbilical well camera (April 2010).

the orbiter. The crew performed ET photography through the aft cockpit overhead windows. The orbiter performed +X RCS translation to initiate relative motion that ensured a good ET hand-held photo opportunity.

SRB cameras that had been carried on some previous shuttle missions were reinstated on the SRBs. A small video camera was added to the ET. It provided supplemental imagery to that obtained by the OBSS. In addition, a camera mounted on the ET liquid oxygen feedline faring provided real-time video of the underside of the orbiter, wing leading edges, and part of the ET. This camera system had first flown as a technology demonstration on STS-112 (October 2002).

The availability of time periods (launch seasons) with acceptable launch and post-ET separation lighting conditions for photography, as well as on-orbit solar beta angle was summarized in a chart (Figure 20.27).* For most missions the absolute value of the solar beta angle was restricted to less than 60 degrees for thermal control considerations. For STS-115 and STS-116 the limit was 50 degrees due to the unique configuration of the ISS

---

* This management friendly launch date summary was first created by Cindy Oliver at the request of Charles K. Knarr, United Space Alliance Vice President of Flight Operations.

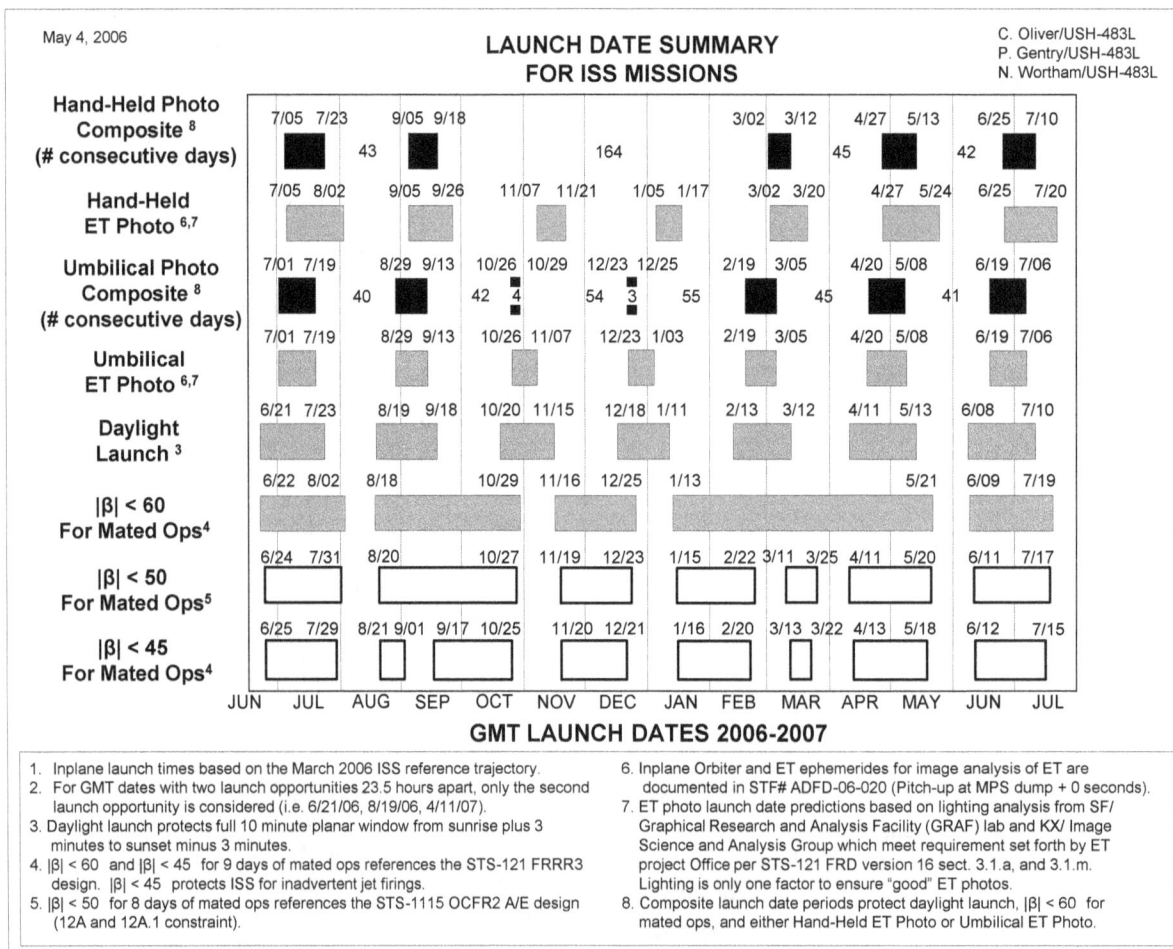

Figure 20.27 Lighting chart for launch dates from June 2006 through July 2007.

during those missions. For some missions the limit was raised to 65 degrees to gain extra days of launch capability. Raising the limit required special thermal analysis for each flight. This chart was generated every three months and provided a summary of launch conditions over a 12 month period. Only the ISS planar launch windows which met these lighting conditions were acceptable. This severely restricted launch dates available for ISS missions, creating launch seasons.[21]

Missions STS-114 (July-August 2005), STS-121 (July 2006), and STS-115 (September 2006) required lighted launch, ascent, and ET separation that supported photography. Starting with STS-116 (December 2006) ET photography lighting no longer drove launch date selection but umbilical well and crew hand held photography was performed if post ET separation lighting conditions permitted. STS-116 was also the last flight that was photographed by the NASA WB-57F during ascent. Starting with STS-123 (March 2008) flash units were mounted in the ET left umbilical well of each orbiter to ensure that ET photographs could be obtained in darkness using the umbilical well cameras. These changes provided the Shuttle Program with more flexibility in choosing launch dates and permitted night launches. The first night launch after the STS-114 return to flight was STS-116 (8:47 pm EST, December 9, 2006). If data such as in Figure 20.27 indicated that lighting would not support hand-held photography the crew would not perform the procedure.

Rendezvous Systems Failures

While all shuttle missions with rendezvous and proximity operations objectives have successfully accomplished these objectives, four missions to the ISS encountered failures in systems used during rendezvous and proximity operations. Backup procedures and well-trained crews and Mission Control personnel enabled mission success in spite of the failures.

STS-92 Radar Fail

The STS-92 (October 2000) mission to the ISS successfully installed the Zenith Z1 Truss and Pressurized Mating Adapter (PMA-3). PMA-3 was used as the docking port for subsequent shuttle missions. On flight day 2 the Ku communications system failed. This also resulted in a failure to acquire and track the ISS with the Ku radar during the rendezvous on flight day 3.

The pre-Ti burn day light star tracker pass was successfully executed. Post-Ti day and night star tracker passes were then performed as a part of the radar fail procedure. The night star tracker pass was ended shortly before MC-2 and no star tracker data was taken after MC-2.

The MC-2 burn time-of-ignition slip was large enough that MC-2 was burned at the 7 minute late time rather than on elevation angle. This indicated a short trajectory and such a large MC-2 time-of-ignition slip was not unusual.

At MC-3 plus two minutes the crew checked the position of the ISS in the COAS as part of the radar fail trajectory correction burn procedure. Since the shuttle orbiter and ISS navigation state vectors are used to point the –Z body axis at the ISS the position of the ISS in the COAS is an indication of navigation error. The radar fail correction burn was then executed based on ISS position in the COAS. ISS was 6 degrees high in the COAS which indicated a short trajectory. The crew performed 12 +X Translational Hand Controller pulses that provided a delta-velocity of 1.2 feet/second.

TCS tracking began near 1700 ft, a lower range than normal due to the short trajectory and the geometry of the TCS retro-reflectors on the ISS at that time. TCS data indicated that the orbiter was closer to the ISS and closing at a higher rate than the crew and MCC previously realized. TCS also confirmed a shorter trajectory than expected, requiring extra firings to reach the R-bar, including extra Low Z braking while setting up for the Twice Orbital Rate R-Bar Approach (TORRA). The minus R Bar approach was nominal.

While the rendezvous and docking was successful propellant consumption was higher than that expected for a radar fail case. Post-flight analysis indicated that it would have been preferable to keep the ISS higher in the COAS. Procedures for controlling ISS motion in the COAS were revised and training for misleading sensor data cases was improved. RPOP HHL data incorporation was updated to eliminate dependency on radar or TCS angles.

STS-128 Vernier RCS Jet Failure

The mission of STS-128 (August-September 2009) was to deliver the Leonardo Multi-Purpose Logistics Module and the Lightweight Multi-Purpose Experiment Support Structure Carrier to the ISS. Early on flight day one vernier (VERN) RCS jet F5R failed due to a propellant leak. The manifold was isolated and VERN jets were considered failed for the rest of the mission. This did not impact mission objectives and resulted in the first VERN fail rendezvous and docking in the Shuttle Program.

Propellant margins supported the per-procedure use of the Alternate Digital Auto-Pilot (ALT DAP) primary RCS jet configuration during the rendezvous until just after the MC-4 burn when the orbiter arrived on the +R Bar. The primary DAP RCS jet configuration was used for proximity operations and approach, which was successful. A relative attitude fly-out at a 30 foot range before docking was not required. The rendezvous and docking were successful with a slight increase in propellant consumption over a nominal rendezvous and docking using VERN jets.

Normally mated stack attitude control was performed with the VERN jets. However, after docking ISS Russian segment thrusters were used to maneuver the mated stack to the Torque Equilibrium Attitude but with high propellant consumption. During the mission extensive analysis was performed to determine if the ALT DAP could be used to maneuver the mated stack to the undocking and collision avoidance burn attitudes. A new procedure was developed to maneuver to a retrograde collision avoidance burn attitude, if it were required.

The ALT DAP was used to maneuver the mated stack to the undocking attitude. ALT DAP was also used for undocking and the separation along the +V Bar. The crew transitioned to the primary RCS jet configuration once the fly-around range was reached. The rest of the fly-around and separation was flown using this configuration.

STS-130 Partial TCS Failure

The February 2010 mission of STS-130 delivered the Tranquility node and its cupola to the ISS. After the MC-4 burn the crew noted that TCS range rate measurements were noisy. Normally TCS transitioned from pulse laser to continuous wave (CW) laser tracking before arriving on the +R Bar. However, TCS remained mostly in pulse mode with only occasional periods of CW tracking. Noisy measurements were edited by the TCS filter and the resulting range and range rate values were as expected.

During the TORVA, between the RPM and +V Bar arrival, some continuous CW tracking occurred. It was clear that the CW laser was not functioning properly and there were jumps in range and range rate. At a range of 250 feet on the +V Bar the crew forced the TCS to track in CW rather than pulse mode. Further troubleshooting attempts were called off so that the crew and Mission Control personnel could focus on completing the approach and docking.

The HHL was used as the primary source of range data for RPOP during the final approach along with centerline camera vertical angle measurements until a range of 12 feet. Occasional instances of TCS data agreed with HHL data. It was obvious to the crew and ground personnel when CW measurements were bad. An auto-angular fly-out to correct relative attitude miss-alignment was not required. At 12 feet the crew transitioned to the ranging ruler overlay on a closed circuit television screen. The docking was successful.

After docking an investigation determined that an electronic tone board in the CW laser was emitting a bad bit that resulted in range measurement spikes. Undocking and separation procedures were modified to reflect the CW laser failure. The pulse laser was to be used even though it was less accurate and noisier than the CW laser. However, HHL had always been the primary sensor during fly-arounds since the ISS retro-reflectors do not

provide complete coverage of a fly-around. The undocking, fly-around, and separation were successful. STS-130 was the first partial TCS fail rendezvous, docking, and separation in the Shuttle Program.

STS-131 Radar Failure

The mission of STS-131 (April 2010) was to deliver science equipment and cargo to the ISS using the Leonardo Multi-Purpose Logistics Module. On flight day one the Ku system failed in the communications mode (TDRSS). This resulted in the loss of all downlinked video, periods of ratty communications over S-Band, and loss of shuttle Orbital Communications Adapter capability until ISS communications assets were available after docking. Both the crew and Mission Control personnel reviewed the radar fail procedures during flight day two.

On flight day three the Ku radar self-test failed and the radar failed to acquire the ISS after the NCC burn. Per the radar fail procedure the post-Ti burn day and night star tracker passes were performed. At MC-3 plus two minutes the crew checked the position of the ISS in the COAS as part of the correction burn procedure. The shuttle orbiter and ISS state vectors maintained by shuttle on-board relative navigation are used to point the –Z body axis at the ISS. ISS position in the COAS is an indication of navigation error. The crew reported that the ISS was dead center in the COAS and exhibited a variation of +/-0.5 degrees, presumably due to shuttle attitude dead-banding. This indicated that the star tracker relative navigation was of high quality, therefore the correction burn was not required. The docking was nominal and no relative attitude fly-out correction maneuver was required at 30 feet. After docking the crew hand transferred to the ISS a hard disk containing the flight day two thermal protection system inspection imagery. It was then transmitted to Mission Control using the ISS communications system. STS-131 was the second radar fail rendezvous in the Shuttle Program.

STS-132 Procedural Work-Arounds

On STS-132 (May 2010) *Atlantis* delivered an Integrated Cargo Carrier and the Russian-built Mini Research Module to the ISS. At the beginning of Flight Day 3 the crew noted that the Crew Optical Alignment Sight (COAS) bulb was burned out. It was replaced.

During the docked phase of the mission the crew noted that the primary laptop (RPOP1) used to run the Rendezvous and Proximity Operations Program (RPOP) had a cracked screen but did not produce debris. The damage was limited to the screen. The backup laptop (RPOP2) was used as the primary RPOP laptop for undocking and fly-around.

During the docking the Orbiter Docking System (ODS) vestibule lights worked properly. However, before undocking the crew noted that the ODS vestibule lights did not turn on, preventing the crew from verifying the configuration of the centerline camera by observing the docking target on the ISS docking hatch. The vestibule lights could not have been accessed for in-flight maintenance unless the docking tunnel was re-pressurized and the shuttle docking hatch opened. This option was not desirable from a crew timeline perspective. The crew mounted a LED headlamp on the centerline camera so that the camera configuration could be verified. The headlamp was normally used by the crew in the cockpit. Later the headlamp was removed part way through the fly-around since it partially obscured the camera field of view.

STS-135 GPC3 Fail to Synch

Early on Flight Day 3 the crew was to increase the number of flight computers running Guidance, Navigation, and Control (GNC) software from one to three. Rendezvous and docking with Mir and ISS was normally flown with three GNC flight computers in a redundant set.

The mode of the GPC was controlled by a three position switch whose positions were HALT, STBY (stand-by), and RUN. In the HALT position, the GPC is not running software. In RUN, the GPC is running software. The STBY position act as a go- between and either initializes the GPC, or shuts down the GPC depending on what the previous mode was (HALT or RUN).

Before the set expansion General Purpose Computer 1 (GPC1) was running on-orbit GNC software and GPC4 was running Systems Management software (communications, life support, etc.). GPC1 and GPC4 made up what was called the common set. GPC2 and GPC3 were loaded with on-orbit GNC software but were in HALT (application software not executing) mode. GPC5 contained Backup Flight System (BFS) software in HALT mode. BFS supported ascent, orbit insertion, deorbit, entry, and landing.

The set expansion involved moding GPC2 and GPC3 to RUN and joining them to the common set of GPC1 and GPC4. As the crew took GPC3 from HALT to STBY there was a moment between the switch positions where none of the three discrete from the switch were set. When the hardware saw no discrete set it set one of the modes. In this case GPC3 was set to RUN and it began to join the common set. Then the switch reached the STBY position and GPC3 saw a transition from RUN to STBY and shut down. GPCs 1, 2, and 4 saw GPC3 momentarily join the common set, then saw it leave. The failure of GPC3 to synchronize with the common set resulted in a GPC3 fault message. Since neither the crew nor Mission Control had insight to the switch discretes, all they saw was the GPC3 fault message, the fact that GPC3 was not in the common set, and the GPC3 mode talkback was

barberpole (indicating it was not running). Mission Control opted to safe GPC3 and not attempt to recover it.

The Mission Control Data Processing System (DPS) officer wanted to dump the contents of GPC1 to confirm the cause of the GPC3 problem. However, DPS did not want to wait until after docking. The 2.5 minute dump was performed after the NH burn and before the NC-4 burn.

Since a re-Initial Program Load (re-IPL) of a GPC could take 20 to 25 minutes, the Flight Director decided to execute the rendezvous with two GPCs running GNC software, rather than the normal three. The amount of time required to re-IPL GPC3 would negatively impact the crew timeline for rendezvous. If there was a problem with one of the two GNC GPCs during the rendezvous GPC3 could be re-IPLed. Due to post-docking robotics activity the Flight Activities officer (FAO) preferred that the GPC be recovered on the morning of Flight Day 4. The GPC3 memory was dumped and re-IPLed on the morning of Flight Day 4 and then freeze dried with primary deorbit and entry software per normal procedures. Analysis of the dumps and reconstruction of the scenario showed that the cause of the issue was the switch timing from HALT to STBY which was a known and documented condition.

The rendezvous and docking with ISS was successfully accomplished with only two GPCs executing on-orbit GNC software.

STS-135 TRIDAR Relative Sensor Test

STS-135 was the third flight of the TRIDAR relative navigation sensor. On Flight Day 3 after the NCC burn the crew was not able to activate the TRIDAR due to a laptop computer communications problem. The problem was isolated to a laptop computer port and Ethernet cable. The crew taped the cable and nominal TRIDAR activation occurred. TRIDAR performance during the Flight Day 3 rendezvous and docking and the Flight Day 12 undocking, fly-around, and separation was nominal.

STS-135 Fly-Around and HHL Sensor Transfer to the ISS

The shuttle normally performed a fly-around of the ISS after undocking to obtain photography of the ISS for engineering purposes. On STS-135, to obtain photographs of parts of the ISS that had not been photographed before, the ISS was to perform a 27 minute, 90 degree yaw maneuver. The orbiter was to perform station-keeping at a range on the +V Bar between 600 and 700 feet during the start of the ISS attitude maneuver. After completion of the yaw maneuver, on Mission Control command, the orbiter was to begin a 23 minute twice orbital rate half lap fly-around, up through the -R Bar and down to the -V Bar. Nominal undocking was to be performed 32 minutes before sunrise to ensure good

Approved for public release via STI DAA 24483. See statement on title page.

145

lighting during station-keeping and for photography during the fly-around.

The fly-around would be ended by the SEP-1 burn of 1.5 feet/second radial down executed on the -V Bar. At SEP-1 the ISS was to maneuver back to the standard Torque Equilibrium Attitude (TEA). A 10 foot/second SEP-2 burn was to be executed one hour and 50 minutes after undocking at a range greater than 6,000 feet, when the orbiter was below and in front of the ISS. STS-135 was the first time the shuttle performed a fly-around of the ISS during an ISS attitude maneuver. Furthermore, the shuttle had never performed station-keeping while the ISS was executing an attitude maneuver.

Two months before the flight of STS-135 NASA/JSC personnel supporting commercial cargo vehicles for the ISS suggested that the backup Hand Held Lidar (HHL), Night Vision Scope (NVS), and a HHL to Payload General Support Computer (PGSC) data cable carried on STS-135 be left on the ISS. This would provide future ISS crew members with a sensor for monitoring automated or crewed commercial vehicles approaching or departing the ISS. The data cable would enable the HHL to send data to situational awareness software used by the crew to monitor Visiting Vehicle relative motion. Previous HHL transfers had been conducted on STS-127 (the unit was returned to Earth on STS-131) and STS-133.

Analysis was conducted after STS-135 docking by Mission Operations personnel to determine the impacts of flying the undocking and half fly-around with only the primary HHL. Normally HHL was the primary sensor during a fly-around due to gaps in TCS reflector coverage. However, TCS coverage was expected to be good after the first 30 degrees of the fly-around. There could be occasional TCS data dropouts of no more that 2 minutes due to blockage by ISS solar arrays and radiators that could cause temporary loss of data or reflector swaps, particularly at the beginning and the end of the half-lap fly-around.

A flight rule governing shuttle sensor requirements during proximity operations with the ISS stated that a functioning HHL was required to begin a fly-around. If the last functioning HHL failed during a fly-around the fly-around was to be aborted and a break-out burn performed to take the orbiter safely away from the ISS. However, for STS-135 the HHL part of the flight rule was waved. If the sole remaining HHL on-board failed during station-keeping or the fly-around, the crew could use TCS as the primary sensor. The TCS in conjunction with the rendezvous radar, COAS, centerline camera, and payload bay cameras could also be used to maintain a range of greater than 600 feet to mitigate shuttle RCS plume impact on the ISS.

Both the primary and backup HHL units and the NVS were checked out during the Flight Day 2 rendezvous tools checkout. Primary HHL performance during the Flight Day 3 rendezvous and docking was nominal.

Primary HHL performance during the undocking and separation on Flight Day 12 was nominal. TCS performance during the fly-around was better than expected.

References

1. Sampaio, C. E., "A Lighting and Visibility Evaluation of the Shuttle/Mir Docking Target," *Proceedings of the 39th Human Factors and Ergonomics Society Annual Meeting*, Human Factors and Ergonomics Society, Santa Monica, CA, 1995, pp. 45-49.

2. Portree, D. S. F., *Mir Hardware Heritage*, NASA Reference Publication 1357, NASA Johnson Space Center, March 1995.

3. Clark, F., P. Spehar, H. Hinkel, and J. Brazzel, "Laser-Based Relative Navigation and Guidance for Space Shuttle Proximity Operations," *Guidance and Control 2003, Advances in the Astronautical Sciences*, Vol. 113, Univelt, San Diego, CA, 2003.

4. Zimpfer, D., and P. Spehar, "STS-71 Shuttle/Mir GNC Mission Overview," *Spaceflight Mechanics 1996, Advances In The Astronautical Sciences*, Vol. 93, Part I, Univelt, San Diego, CA, 1996, pp. 441-460.

5. Brazzel, J., F. Clark, and P. Spehar, "RPOP Enhancements to Support the Space Shuttle R-Bar Pitch Maneuver for Tile Inspection," *AIAA Guidance, Navigation, and Control Conference*, AIAA, Reston, VA, 2005.

6. Goodman, J. L., *GPS Lessons Learned From The ISS, Space Shuttle and X-38*, NASA Contractor Report NASA/CR-2005-213693, NASA Johnson Space Center, November 2005. See the NASA JSC Technical Reports server at http://ston.jsc.nasa.gov/collections/TRS/

7. Dagen, J. D., "Structural Integration of the Space Shuttle and the Mir Space Station," *Spaceflight Mechanics 1996, Advances in the Astronautical Sciences*, Vol. 93, Part I, Univelt, San Diego, CA, 1996, pp. 461-471.

8. Jackson, M., D. Zimpfer, and J. Lepanto, "Identification of Shuttle/Mir Structural Dynamics For Notch Filter Tuning," *Spaceflight Mechanics 1996, Advances in the Astronautical Sciences*, Vol. 93, Part I, Univelt, San Diego, CA, 1996, pp. 493-509.

9. Zimpfer, D., et al., "Shuttle Stability and Control for the STS-71 Shuttle/Mir Mated Configuration," *Spaceflight Mechanics 1996, Advances in the Astronautical Sciences*, Vol. 93, Part I, Univelt, San Diego, CA, 1996, pp. 473-492.

10. Hall, R., et al., "Flight Control Overview Of STS-88, The First Space Station Assembly Flight," *Astrodynamics 1999, Advances in the Astronautical Sciences*, Vol. 103, Part II, Univelt, San Diego, CA, 1999, pp. 1063-1081.

11. Rochelle, W. C., E. A. Reid, T. L. Carl, R. N. Smith, and F. E. Lumpkin III, "Thermal Analysis For Orbiter And ISS Plume Impingement On International Space Station," *35th AIAA Thermophysics Conference*, AIAA, Reston, VA, 2001.

12. Ghofranian, S., M. S. Schmidt, J. McManamen, J. Schliesing, and T. Briscoe, "Space Shuttle Docking to Mir Mission-1," *Proceedings of the AIAA/ASME/ASCE/AHS/ASC Structures, Structural Dynamics and Materials Conference*, New Orleans, LA, pp. 333-339.

13. Gibson, R. L., and C. J. Precourt, "The First Space Shuttle to Mir Docking Mission," *Proceedings of the 39th Symposium of the Society of Experimental Test Pilots*, Society of Experimental Test Pilots, Lancaster, CA, 1995, pp. 386-419.

14. Gibson, Robert L. "Hoot", "Pilot Report - How to Fly a Space Shuttle Rendezvous and Docking," *Aviation Week and Space Technology*, Volume 172, No. 44, December 6, 2010, pages 66-67.

15. Walker, S., J. LoPresti, M. Schrock, and R. Hall, "Space Shuttle Rbar Pitch Maneuver for Thermal Protection System Inspection Overview," *AIAA Guidance, Navigation, and Control Conference*, AIAA, Reston, VA, 2005.

16. Jenkins, D. R., and J. R. Frank, *Return to Flight: Space Shuttle Discovery Photo Scrapbook*, Specialty Press, North Branch, MN, 2006.

17. Machula, M., "Orbiter Repair Maneuver Contingency Separation Methods and Analysis," *AIAA Guidance, Navigation, and Control Conference*, AIAA, Reston, VA, 2005.

18. Bigonesse, R., and W. R. Summa, "Unmanned Orbiter Undocking: Method for Disposal of a Damaged Space Shuttle Orbiter," *Spaceflight Mechanics 2006, Advances In The Astronautical Sciences*, Univelt, Inc., San Diego, CA.

19. Hale, N. W., and B. A. Conte, "Considerations in Rendezvous Launch Window Management," *53rd International Astronautical Congress*, International Astronautical Federation, Paris, France, 2002.

20. Jones, R., "Architecture In Mission Integration, Choreographing Constraints," *30th International Conference on Environmental Systems*, Society of Automotive Engineers, Warrendale, PA, 2000.

21. Adamo, D. R., "ISS Rendezvous Phasing Considerations Pertaining to Optimal STS-114/LF1 Launch Opportunities," *AIAA Guidance, Navigation, and Control Conference*, AIAA, Reston, VA, 2005.

22. Adamo, D. R., "Contingency ISS Rendezvous Recovery Planning By Houston And Moscow Control Centers," *Astrodynamics 1999, Advances in the Astronautical Sciences*, Vol. 103, Part III, Univelt, San Diego, CA, 1999.

CHAPTER 21 – THE STORRM DTO

Introduction

The Orion (also called the Multi-Purpose Crew Vehicle, or MPCV) spacecraft required a different approach to relative navigation than the shuttle.[1-12] Cost, available power, and available spacecraft volume prevented the use of a rendezvous radar like the shuttle. An automated rendezvous and proximity operations requirement, coupled with a single crew member piloting requirement, called for a more integrated approach to relative navigation and proximity operations than the shuttle. Automated docking required measurement of relative attitude, a function not performed by the shuttle TCS or HHL sensors. An HHL must be held so that its bore-sight is orthogonal to the window pane. This could be done in the shuttle aft cockpit, but the placement and design of the Orion crew windows did not permit the use of a shuttle type of HHL lidar.

After the August 31, 2006 award of the Orion contract to the Lockheed Martin team the contractor team members and NASA personnel began development of the Vision Navigation Sensor, or VNS. The VNS flash lidar was designed to provide range and line-of-sight angle measurements to the Orion relative navigation software. The nominal range for first measurement incorporation was 5 km (16,500 feet). The VNS also was to provide relative attitude measurements from 15 meters (50 feet) through docking. VNS measurements were to support both automated and manually piloted proximity operations and docking.

The STORRM DTO

NASA Johnson Space Center personnel studied the lessons learned and experiences of several flight programs with relative navigation sensors. Experience had shown that it was difficult for a ground test facility to duplicate all aspects of the space environment that impact relative sensor performance. On-orbit testing may be required in addition to ground testing, to subject new hardware and software to a wider range of flight conditions (particularly on-orbit lighting) than can be created in a ground laboratory.

To mitigate risk during Orion development NASA flew a test of the Orion VNS flash lidar and docking camera on the STS-134 Space Shuttle mission (May 2011) to the ISS. This test was called the Sensor Test for Orion RelNav Risk Mitigation Detailed Test Objective, or STORRM DTO. The objective was to collect data from the sensors in the vicinity of the ISS during shuttle approach, separation from the ISS, and part of an Orion proximity operations approach profile. Shuttle relative navigation data was also collected to serve as a source of truth data.[13,14]

DTO Hardware

The prototype VNS and docking camera were mounted on the Orbiter Docking System truss next to the TCS in the shuttle payload bay. Commanding and data handling was performed using a Payload General Support Computer (PGSC) in the shuttle crew cabin that was dedicated to the DTO. The PGSC was networked so that it received orbiter flight computer and TCS data. The PGSC display was downlinked to Mission Control via still sequential video. STORRM DTO data was also recorded on-board in an avionics package mounted on the payload bay side wall and later downlinked to Mission Control. The data was analyzed post-flight to assess the performance of the VNS and docking camera.

DTO Hardware Testing and installation

During STS-130 proximity operations (February 2010) the STORRM DTO team performed a test of the DTO laptop computer software in Mission Control. Shuttle TCS and orbiter GNC data was provide to DTO software on a laptop using the RPOP Windecom data cables. The test was successful.

The DTO required that five VNS compatible retro-reflectors be placed on the ISS docking target used by the shuttle. The VNS retro-reflectors were opaque to the shuttle TCS lidar so that the VNS flash lidar would not interfere with shuttle relative navigation. The retro-reflectors would be visible to the VNS during the +V Bar approach and docking on Flight Day 3, and during the undocking and +V Bar back-away later in the mission. The retro-reflectors would not be visible to the VNS during the re-rendezvous after undocking.

The STORRM DTO reflective elements kit was taken into orbit aboard the shuttle *Discovery* on mission STS-131 in April of 2010. The DTO retro-reflectors were installed by Expedition 23 Flight Engineer Soichi Noguchi approximately 3.5 hours after docking.

DTO Activities and Rendezvous Profile

This section provides an overview of the nominal STORRM DTO activities and re-rendezvous relative motion profile as defined during mission planning.

The VNS, docking camera, and associated DTO PGSC hardware and software was to be checked out during the rendezvous tools checkout on Flight Day 2. The rendezvous radar, TCS, HHL, and APAS docking ring were also to be checked out at this time. During the shuttle ORBT rendezvous and proximity operations approach to the ISS on Flight Day 3 data would be collected from the VNS and docking camera. The crew would monitor the DTO PGSC and perform procedures to enable the VNS to acquire the ISS by a range of 5 km.

Approved for public release via STI DAA
24483. See statement on title page.

149

There were callouts for STORRM DTO procedures in the rendezvous Flight Data File (crew procedures). The STORRM crew procedures were approximately 65 pages long. However, the DTO was to be conducted on a non-interference basis and rendezvous activities would not be changed to accommodate resolution of DTO issues. STORRM DTO procedure calls from Mission Control to the crew were to be handled through the Assembly and Checkout (ACO) officer in Mission Control. The Rendezvous Guidance and Procures Officer (RGPO) and associated backroom support would not handle STORRM DTO procedures, in accordance with the non-interference policy. The VNS and docking camera were powered off about 10 minutes after docking.

Before removal of the stand-off cross and hatch opening the crew would obtain photogrammetry of the docking target by photographing it. DTO data recorded on-board during the rendezvous would be transmitted to Mission Control for analysis to determine if any DTO software parameters needed to be changed to support the DTO after orbiter separation from the ISS and the subsequent execution of the re-rendezvous profile.

On the day before undocking the crew would perform STORRM DTO tools checkout at the same time as the rendezvous tools checkout. The crew would again perform photogrammetry of the docking target before and after re-installation of the stand-off cross.

The VNS, docking camera, and PGSC software were powered up by the crew about 30 minutes before undocking. Undocking time was determined by fly-around and STORRM DTO lighting requirements. Undocking was to occur at orbital midnight.

Data was to be collected during the fly-around. However, the fly-around was only to be conducted if enough propellant was available. SEP1 at the end of the fly-around on the +V Bar was the same 1.5 foot/second radial up burn executed on ISS missions. At a range of 1,000 feet the orbiter would maneuver to a minus Z body axis target track attitude to facilitate sensor measurement acquisition.

Radar data was to be taken from the minus V Bar crossing during the fly-around through the SEP3 burn plus 20 minutes. In the event of a rendezvous radar failure state vector uplinks were to be performed to ensure sufficient relative navigation accuracy so that the orbiter minus Z body axis could be accurately pointed at the ISS ensuring that the ISS would be in the VNS field of view.

The post-separation DTO profile was designed to match the Orion rendezvous and proximity operations profile for ISS missions within 20,000 feet of the ISS (Figure 21.1).[13] The approach profile flown within 20,000 feet of the ISS was the current Orion baseline profile as of 2010. The DTO was independent of the execution of a post-undocking ISS fly-around by the orbiter.

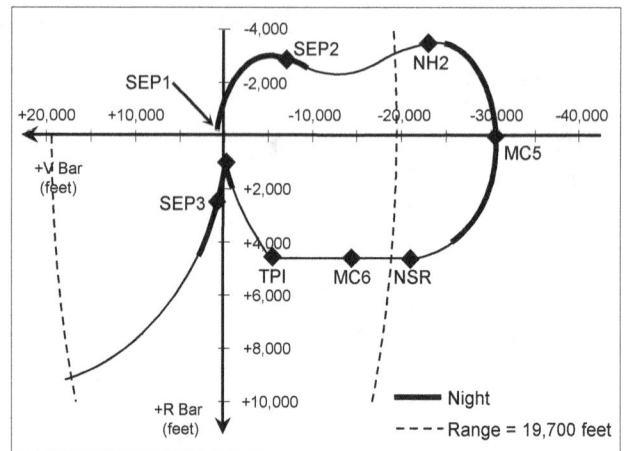

Figure 21.1 STORRM DTO relative motion profile. The pre-SEP1 ISS fly-around is not shown.

The remaining STORRM DTO profile consisted of six multi-axis RCS burns. The SEP2 burn was a ground targeted NC (phasing) burn targeted for the NSR down-range value. NH2 was a ground targeted burn that targeted the orbiter for the relative altitude at the NSR point. MC5 was on-board Lambert targeted to correct for any trajectory dispersions and ensure that the orbiter arrived at the desired relative position for NSR. The MC5 delta-velocity was nominally zero and above the minus V Bar for a non-dispersed trajectory. In a radar fail scenario star tracker data would be taken between NH2 and NSR.

The on-board Lambert targeted burn, NSR, set up a coelliptic trajectory and targeted the vehicle for the TPI point. Since the on-board Lambert software did not have a true NSR targeting capability the relative position offset targets and transfer time were computed to provide a coelliptic transfer. However, in the presence of trajectory dispersions, the NSR to TPI transfer trajectory could be non-coelliptic. Consideration was given to using the Mission Control Orbital Maneuver Processor (OMP) to target the burn. OMP had a true coelliptic targeting capability. However, since the previous MC5 and subsequent MC6 and TPI burns were on-board targeted, it was decided to compute NSR on-board to maintain consistency in the crew and ground procedures.

After NSR, at a range of 6 km, VNS sensor acquisition would be initiated. The Sun was not to be in the VNS field of view from 6 km through TPI. An on-board Lambert targeted MC6 burn ensured that the orbiter arrived at the desired TPI relative position. TPI was on-board targeted using the Lambert elevation angle option for the TDA point 330 feet behind and 1,000 feet below the ISS. The TPI elevation angle was 42 degrees with no TIG slip limits. TPI was nominally 3.3 feet/second, mostly posigrade and with a small radial up component.

During the DTO the ISS was not to maneuver to the docking attitude, nor was it to feather solar arrays. No burn was executed at the TDA point since the orbiter was not to proceed inside of 600 feet. Range rate gates were included in the crew procedures to ensure that the orbiter did not approach inside of 600 feet. At 15 minutes after the TDA point a SEP3 burn would be executed to ensure safe relative motion away from the vicinity of the ISS. The DTO hardware was to be powered off once a range of 6 km was reached during the final departure from the ISS.

Contingency Procedures

Several contingency procedures were developed to ensure that the DTO could be conducted in the event of a late undocking or a rendezvous radar failure. Contingency procedures were verified in mini-simulations using rendezvous personnel, an RPOP laptop, and a shuttle proximity operations simulator on a work-station. These mini-sims were instrumental in verifying crew and Mission Control procedures before STORRM DTO integrated simulations were held in Mission Control.

Late Undocking

For beta angles from -15 to +15 degrees the DTO profile was designed to keep direct sunlight out of the 11 degree field-of-view of the VNS. However, the possibility of a late undocking could result in violating the Sun out of the VNS field-of-view requirement later in the DTO profile. Contingency procedures were devised to permit DTO execution with the required lighting condition within a range of 6 km of the ISS in the event of a delayed undocking. There was to be no adjustment of the on-board Lambert targeted burn (MC5, NSR, MC6, TPI) data to accommodate an undocking time slip.

If the undocking time slipped by 10 minutes or less the delta-velocity of the SEP2 burn would be adjusted to keep the TPI TIG at the nominal time and meet the lighting requirement. The SEP2 DV adjustment was a continuous curve of increasing SEP2 DV to reduce the transfer time between SEP2 and NH2. The 15 minutes gained by adjusting the SEP2 DV was added to the 14 minute TPI lighting window. The 10 minute point was chosen since the SEP2 DV adjustment after that point rapidly increased.

After a 10 minute slip the TPI TIG could be slipped up to 14 minutes and still preserve appropriate lighting for VNS. A 14 minute TPI burn execution window existed that provided appropriate lighting. The 14 minute TPI lighting window allowed DTO objectives to be met within a range of 6 km of the ISS. This applied to the approach to the ISS as well as after the final SEP3 burn.

If the undocking time slipped beyond 24 minutes there was a 5 minute window available in which the SEP2 burn delta-velocity could be adjusted to meet the lighting requirement.

For a slip beyond 29 minutes TPI would be delayed one revolution by delaying the time of NH2 burn execution. The post-undocking fly-around could also be canceled in the event of a late undocking in order to meet the lighting requirement.

Radar Fail

Procedures were developed so that the DTO could be flown in the event of a shuttle rendezvous radar failure. In the event of a radar failure TCS and HHL could not support shuttle relative navigation required to support on-board Lambert targeted burns. Neither TCS nor HHL data were processed by the shuttle flight computer relative navigation function that provided state vectors to Lambert burn targeting. Furthermore, the short ranges of TCS and HHL could not support relative navigation over most of the DTO re-rendezvous profile (Figure 21.1).

The SEP1 burn delta-velocity was determined pre-mission and was part of the crew procedures. State vectors improved by filtering TCS measurements would be used by Mission Control to target the SEP2 burn. However, the short range of TCS meant that it could not support targeting of the MC5, MSR, MC6, and TPI burns. Mission Control based orbit determination using ground C-band radar and TDRSS data could not be used since it took the orbit determination process too long to converge on a solution.

Outside of TCS range an alternate method was needed to provide the shuttle flight computer with state vectors accurate enough to support Mission Control burn targeting and on-board Lambert burn targeting. Shuttle personnel developed a new navigation technique that used shuttle and ISS GPS receiver data to provide orbiter and ISS state vectors for shuttle relative navigation initialization. The shuttle was not equipped to perform relative GPS navigation with the ISS.[15,16] Neither the ISS or shuttle orbiter GPS receivers were equipped with the necessary filtering algorithms to perform precision orbit determination. However, the Mission Control based Spacecraft Position Optimal Tracking (SPOT) filter was to be used to improve the shuttle and ISS orbital estimates by filtering the on-board shuttle and ISS GPS receiver position vectors. SPOT was certified for both shuttle and ISS use in 2008. SPOT state vectors computed in Mission Control would be used to target the NH2 burn, or ground targeting of any other burns in a contingency, such as a NH Delay or TPI. The TCS and SPOT methods were successfully tested in Mission Control during shuttle separations from the ISS on STS-131 (April 2010), STS-132 (May 2010), and STS-133 (February-March 2011).

Use of star tracker in the event of a radar failure was complicated by the close range to the ISS, ISS size and brightness that limited star tracker to night passes, and the presence of the Earth behind the ISS during potential day star tracker passes. A night star tracker pass could

Approved for public release via STI DAA 24483. See statement on title page.

151

be executed between the NH2 and NSR burns in the event of a radar failure (Figure 21.1). A light on an ISS truss would be turned on by the ISS crew to support the night star tracker pass. The Service Module tracking light normally available during rendezvous would not be visible to the shuttle orbiter during the re-rendezvous. However, the crew could be told not to perform the star tracker pass if Mission Control had confidence in the SPOT data. If Mission Control did not have confidence in either the SPOT vectors or the on-board relative navigation state vectors, the TPI burn would not be executed. The coelliptic trajectory would ensure that the shuttle would pass below the ISS at a safe distance without risk of collision.

Delay Burns

If undocking were delayed or the fly-around were deleted an NH2 delay burn could be computed by Mission Control. An NH2 delay results in the orbiter continuing to phase away from the ISS while above the minus V Bar. However, a NH2 Delay could result in orbital lighting impacts for the VNS.

If shuttle, STORRM DTO, or ISS systems performance problems prevented execution of the DTO profile, an NSR delay burn would be computed by Mission Control and executed by the crew. An NSR delay would result in a relative motion football similar to a Ti Delay football.

NH2 and NSR delay burns would provide safe relative motion from which the DTO profile could be resumed when appropriate.

Vernier RCS Jet Failure

In the event of a Vernier RCS jet failure the crew would use the ALT DAP mode of the primary RCS jets.

Orbiter Boom Sensor System

Shuttle missions after the loss of *Columbia* performed a Flight Day 2 inspection of the Thermal Protection System (TPS) to check for TPS damage that occurred during ascent. On ISS missions after undocking and separation were complete a late inspection was performed to check for TPS damage due to micro-meteoroid impacts. Both inspections used the Orbiter Boom Sensor System (OBSS) that was grappled and maneuvered using the Remote Manipulator System (RMS). Since STS-134 was the next to the last shuttle mission the nominal plan was to leave the OBSS on the ISS rather than return it to Earth. There was a possibility that the OBSS could not be left on the ISS and could not be put in the stowed (Earth return) position in the payload bay before undocking. In this case the orbiter would undock with the OBSS in the standard undock

position, on the RMS and lying across the width of the payload bay.

However, analysis indicated that this position could subject the OBSS to undesirable structural loads while the crew executed a burn component in the Z body axis direction. An alternate OBSS and RMS position was defined that would protect the OBSS from structural loads during the re-rendezvous burns. This alternate plan involved positioning the OBSS vertically above the payload bay.

Flight Results

STS-134 launched on time on Monday, May 16, 2011, at 7:56 am Central Daylight Time. This section provides highlights of the mission.

Rendezvous and Docking

On Flight Day 1 the Ku Band antenna was successfully deployed and placed in operation for TDRSS communications. The Flight Day 2 centerline camera installation and alignment, docking ring extension, and rendezvous tools checkout (TCS and HHL) were successful. A successful checkout of the STORRM DTO PGSC, VNS, and docking camera was also performed. Docking camera images and raw VNS data were obtained.

During the on-board targeted phase of rendezvous relative navigation (star tracker, radar) and Lambert burn targeting performance were nominal. TCS acquisition occurred at a range of 5903 feet. At a range of 3920 feet the crew reported that the TCS and HHL range measurements matched. The TCS transition from pulse to Continuous Wave (CW) tracking mode nominally occurs at a range of about 950 feet, before the RPM. However, the transition did not occur. Pulse mode loss of tracking events occurred at ranges of 1020 feet and 895 feet. The crew forced a transition to CW tracking at a range of 648.29 feet, after the RPM. No loss of tracking events occurred between this point and docking.

On past missions loss of tracking events were associated with direct or indirect sunlight on the TCS receiver. However, TCS performance just before loss of tracking did not support that scenario. The loss of tracking incidents and difficulty transitioning from pulse to CW mode could not be explained. STS-134 was the last flight of the particular unit flown (it last flew on STS-126). The anomalies were judged not to be a constraint on use of the TCS as the primary proximity operations sensor during the subsequent undocking and plus V Bar back-out on STS-134. HHL was the primary sensor during fly-around. Nor were the STS-134 TCS tracking problems judged to impact use of TCS on the future STS-135 mission.

STORRM DTO performance was outstanding. The VNS automatically detected the ISS at a range of 5.4 km.

The docking camera obtained images of the ISS starting at a range of 60 km through docking. A total of 340 Giga-Bytes of data were collected and later downlinked for analysis.

Docked Phase

On Flight Day 5 the ISS American GPS (SIGI) unit 2 experienced an unrecoverable hardware failure. GPS 1 continued to provide GPS state vector data. This reduced the level of ISS GPS redundancy to support the shuttle radar fail procedure during the STORRM DTO on the day of undocking. During the re-rendezvous the Mission Control shuttle ground navigator would run SPOT to filter shuttle GPS position vectors, while the ISS Trajectory Operations Officer (TOPO) would filter ISS GPS position data with SPOT.

During the docked phase of the mission a STORRM checkout procedure revealed a card failure in Data Recording Unit 3 (DRU3). This DRU was used to record docking camera data and process commands sent to the docking camera. A high DRU3 temperature and higher than expected read/write activity were noted during the later part of the Flight Day 3 rendezvous. This could have been an initial indication of the problem but the docking camera performed well during the entire rendezvous. The VNS and associated DRU1 performed well during the rendezvous and the STORRM checkouts during the docked phase. New procedures were developed and sent up to the crew to accommodate the DRU3 failure. The plan was for the crew to power on DRU3 about 30 minutes before undocking on Flight Day 15. If DRU3 did not recover it and the STORRM docking camera would be powered off for the remainder of the undock and re-rendezvous.

On Flight Day 14 the shuttle rendezvous tools checkout was successfully performed. A STORRM tools checkout was not performed out of concern that cycling the hardware could damage it. The hatches between the ISS and the orbiter were closed.

Undocking and Re-rendezvous

On Flight Day 15 the change of shift briefing (Orbit 3 to Orbit 1) was held for the on-coming flight controllers that would oversee the undocking and re-rendezvous. The Flight Dynamics Officer (FDO) stated that the STORRM DTO burns (SEP-2, NH-2, NSR, TPI, SEP-3) had been cleared and there were no potential conjunctions with orbiting debris or other spacecraft. The Rendezvous Guidance and Procedures Officer (RGPO) reported that the undock messages and event summaries were sent to the crew on the previous flight day by the Orbit 2 shift and the crew had no questions after reviewing the transmitted material. In response to a question from the Flight Director the RGPO stated that if the rendezvous radar failed and the backup SPOT

process did not work as planned, he would ask the ISS Communication and Tracking Officer (CATO) to turn on the ISS lights so a night star tracker pass could be performed between NH-2 and NSR. The Assembly Checkout Officer (ACO) reported that a STORRM procedures update was uplinked to crew. STORRM photogrammetric photos were taken by the crew and downlinked during the hatch closing yesterday. From an ACO perspective all was go. The off-going Orbit 3 Flight Director stated that the Orbit 3 shift was the calm before the STORRM.

Before undocking the Guidance, Navigation, and Control hardware (GNC) officer reported that a good IMU alignment had been performed. The Propellant (PROP) officer reported that propellant margins were sufficient to support the fly-around. TCS activation was completed, along with undocking mechanism power-up and undocking preparation. When the crew performed the STORRM hardware checkout DRU3 failed again. Both DRU3 and the STORRM docking camera were powered off. The crew was given a go for undocking.

After physical separation flight controllers reported that TCS was tracking the ISS and the shuttle was in attitude control. STORRM DTO personnel stated that VNS was functioning as expected. During the backout VNS successfully transitioned through the range bins as expected. At a range of 115 feet the crew reported that the HHL range and the TCS range measurements matched. Docking mechanism power-down was completed on schedule before the fly-around started.

As expected the crew performed the plus V Bar separation at a slightly higher opening rate than the reference profile. The fly-around was started 2.5 minutes ahead of the reference time, but well within the +/-5 minute limits set to bound trajectory and lighting dispersions. After the minus V Bar crossing the Ku-band antenna system was taken from communications mode to radar mode and locked on immediately. The crew commanded the processing of rendezvous radar data by the orbiter relative navigation Kalman filter. After the plus R Bar crossing a Kalman filter covariance re-initialization was performed since elevation angle measurements were being rejected. This is expected behavior that occurs due to radar wandering over the surface of the ISS. STORRM DTO personnel reported that the VNS data recording looked good.

The SEP-1 burn was completed. The covariance was re-initialized again due to rejected elevation angle measurements. The FDO reported that the final SEP-2 burn solution that targeted for the down-track position at the NSR burn time was +1.5, 0.0, and 0.0 feet/second LVLH. SEP-2 was successfully performed. During the separation VNS continued to successfully transition through the range bins, as expected.

One of the periodic fuel cell self-tests failed. However, the Electrical Generation and Illumination (EGIL) officer reported that all other fuel cell parameters

Approved for public release via STI DAA 24483. See statement on title page.

153

were nominal and a subsequent self-test passed. EGIL reported that personnel were continuing to watch the fuel cell. Radar was performing well and it was reported that the MC-5 burn would occur at a range of about 29,000 feet. The FDO reported that the final NH-2 burn solution was -1.8, 0.0, and 0.0 feet/second LVLH. NH-2 was successfully performed.

The Mission Evaluation Room (MER) reported that the fuel cell issue was similar to an STS-130 issue with the same set of fuel cell hardware. Paper work had been written on the STS-130 anomaly and there was no need to bring in people to examine the issue this evening. A total of four fuel cell self-test failures had been seen today. A data take was scheduled for the next day per a fuel cell flight rule.

The final on-board Lambert targeted MC-5 burn solution was 0.1, 0.0, and 0.5 feet/second LVLH. The crew was given a go to burn the final on-board solution. MC-5 was successfully performed.

The TARGET position reported that the preliminary on-board Lambert targeted NSR burn solution looked good at -2.5, 0.1, and -0.3 feet/second LVLH. It was also reported that if a NSR delay burn was needed it was a no-burn. In other words, the post MC-5 relative trajectory provided the required relative motion for a NSR delay. The crew had entered the STORRM re-rendezvous phase re-initialization procedure. VNS had transitioned to the furthest range bin and was in stand-by mode (at a range of ~30,000 feet). The ISS had maneuvered to the STORRM DTO LVLH attitude of -3.0, 0.0, and 0.0 degrees pitch/yaw/roll.

The final on-board Lambert targeted NSR burn solution -2.5, 0.1, 0.0 feet/second LVLH. The crew was given a go to burn the on-board solution. The NSR burn was successfully performed. The final MC-6 burn solution was 0.0, 0.0, and 0.0 feet/second LVLH. It was declared a no burn, indicating excellent trajectory performance.

STORRM personnel reported that data was being recorded at the nominal 15 MB/second rate. VNS was believed to have acquired the ISS at a range of ~16,500 feet.

Mission Control gave the crew a go to perform the TPI burn. The crew was advised that during the burn TDRSS communications quality between the orbiter and Mission Control could be substandard, but TDRSS communications quality between ISS and Mission Control that also carried orbiter to Mission Control communications (called "The Big Loop") should be acceptable. The final on-board Lambert targeted TPI burn solution (targeted on elevation angle) was 0.3, 0.1, and -3.4 feet/second LVLH. The preliminary, intermediate, and final TPI burn solutions all had TIG slips of 33 seconds early. TPI was successfully performed.

TCS acquired at a range of 5,300 feet. After an initial period of ratty data TCS soon locked on and exhibited

solid performance. The Mission Control version of the Rendezvous and Proximity Operations Program (RPOP) that attempts to mimic the onboard version using Orbiter and TCS telemetry did not initially accept TCS data. The problem cleared after a few minutes and worked fine after that. The onboard RPOP did not experience the same problem.

Trajectory performance was about as close to the nominal mission plan as possible. The crew did not have to execute any braking gates to ensure that the orbiter did not approach closer than 600 feet to the ISS. The crew left the flight controller power off. Based on Mission Control RPOP data the closest approach after TPI was 955 feet.

The GNC officer reported that the TCS was running hot and may have to be shut down earlier than planned during the separation. Later the GNC officer reported that the TCS temperature had dropped a degree and the crew needed to take no action. The temperature trend appeared to be leveling off. STORRM DTO personnel reported that the VNS was successfully transitioning through the range bins.

The ISS maneuvered back to the nominal attitude. The SEP-3 burn of -1.0, 0.0, 0.0 feet/second LVLH was successfully performed. This resulted in the orbiter phasing away from the ISS at a rate of 9 nm per revolution of the Earth. After SEP-3 the GNC officer had the crew deactivate TCS to prevent a possible auto-shutdown.

Summary

Separation and re-rendezvous trajectory performance was very good with the actual trajectory virtually on top of the reference trajectory. The on-board targeting and burns, including the on-board targeted coelliptic burn (NSR) and the TPI burn (elevation constraint), all worked as planned. Propellant consumption was at the expected values. The VNS, DRU1, and STORRM software performed flawlessly throughout undock, re-rendezvous and final separation. The VNS successfully transitioned through all range bins.

The STORRM DTO team was very happy with VNS performance throughout the rendezvous, docking, undocking, fly-around, separation, and re-rendezvous. A total of 260 giga-bytes of VNS data were collected during the undock and re-rendezvous. On Flight Day 3 a total of 108 giga-bytes of VNS and 230 giga-bytes of docking camera data were recorded. Preliminary analysis indicated that the 5 km acquisition range requirement for the VNS was achieved on Flight Days 3 and 15. In addition, VNS tracked to within 6 feet of the ISS. STS-134 crew member Drew Feustel had complimentary words for the STORRM DTO team.

Endeavour landed on the first opportunity at the Kennedy Space Center on Wednesday, June 1, 2011.

References

1. Goodman, John L., J. P. Brazzel, and D. A. Chart, Ph.D., "Challenges of Orion Rendezvous Development," *AIAA Guidance, Navigation, and Control Conference*, AIAA, Reston, VA, 2007.

2. Goodman, J. L., Goodman and J. P. Brazzel, "Rendezvous Integration Complexities of NASA Human Flight Vehicles," *32nd Annual AAS Guidance And Control Conference, Guidance and Control 2009, Advances in the Astronautical Sciences*, Univelt, San Diego, CA, 2009.

3. Brazzel, Jack P., Chris D'Souza, Peter T. Spehar, Fred D. Clark, Chad Hanak and Tim Crain, "NASA CEV Rendezvous Proximity Operations and Docking GN&C Analysis," *Guidance and Control 2007, Advances in the Astronautical Sciences*, Univelt, San Diego, CA, 2007, pp. 491-510.

4. D' Souza, Chris, Chad Hannak, Pete Spehar, TX; Fred Clark, and Mark Jackson, "Orion Rendezvous, Proximity Operations and Docking Design and Analysis," *AIAA Guidance, Navigation and Control Conference and Exhibit*, Hilton Head, SC, August 20-23, 2007, AIAA, Reston, VA, 2007, AIAA, Reston, VA, 2007.

5. Crain, Timothy, Michael Begley, Mark Jackson and Joey Broome, "Guidance, Navigation, and Control System Design in a Mass Reduction Exercise," *Guidance and Control 2008, Advances in the Astronautical Sciences*, Univelt, San Diego, CA, 2008, pp. 407-424.

6. Goodman, John L., H. R. Mamich, and D. W. Saley, "Orion On-Board Navigation Architecture and Operations Concepts," *Guidance and Control 2008, Advances in the Astronautical Sciences*, Univelt, San Diego, CA, 2008, pp. 425-444.

7. Sostaric, Ronald R., and Robert S. Merriam, "Lunar Ascent and Rendezvous Trajectory Design," *Guidance and Control 2008, Advances in the Astronautical Sciences*, Univelt, San Diego, CA, 2008, pp. 607-630.

8. Mamich, Harvey, and Chris D'Souza, "Orion Preliminary Navigation System Design," *AIAA Guidance, Navigation and Control Conference and Exhibit*, Honolulu, Hawaii, Aug. 18-21, 2008, AIAA, Reston, VA, 2008.

9. Chambers, Robert, and Tim Straube, "Seven Vehicles in One: Orion GN&C," *AIAA SPACE 2008 Conference and Exposition*, San Diego, CA, September 9-11, 2008, AIAA, Reston, VA, 2008.

10. Jordan, Steven, "Healthy Tension: The Role of Operations in the Orion Spacecraft Design," *AIAA SPACE 2009 Conference and Exposition*, Pasadena, CA, September 14-17, 2009, AIAA, Reston, VA, 2009.

11. D'Souza, Christopher, and Michael Weeks, "Design and Implementation of the Automated Rendezvous Targeting Algorithms for Orion," *AIAA Guidance, Navigation, and Control Conference*, Toronto, Ontario, August 2-5, 2010, AIAA, Reston, VA, 2010.

12. Ruiz, Jose, and Jeremy Hart, "A Comparison Between Orion Automated and Space Shuttle Rendezvous Techniques," *AIAA Guidance, Navigation, and Control Conference*, Toronto, Ontario, August 2-5, 2010, AIAA, Reston, VA, 2010.

13. Stuit, Timothy D., "Designing the STS-134 Re-Rendezvous: A Preparation for Future Crewed Rendezvous Missions," *AIAA Space 2011 Conference and Exposition*, Long Beach, CA, September 26-29, 2011.

14. Christian, John A., Heather Hinkel, and Sean Maguire, "The Sensor Test for Orion RelNav Risk Mitigation Development Test Objective," *AIAA Guidance Navigation and Control Conference*, Portland, OR, August 8-11, 2011.

15. Goodman, John L., "Space Shuttle Navigation in the GPS Era," *Proceedings of the National Technical Meeting 2001*, Institute Of Navigation, Long Beach, CA, January 22-24, 2001, pages 709-724.

16. Goodman, John L. and Carolyn A. Propst, "Operational Use of GPS Navigation for Space Shuttle Entry," *IEEE/ION PLANS 2008 Conference*, Monterey, CA, May 5-8, 2008.

This page intentionally left blank.

CHAPTER 22 - SUMMARY

Shuttle rendezvous and proximity operations technique development was able to respond to new program requirements, but the development process was not always straight forward. The success of the Space Shuttle in fulfilling new, challenging, and unforeseen requirements was due to extensive analysis conducted by integrated, interdisciplinary teams; and continuous development of new nominal and contingency procedures for a vehicle and ground support system that possessed a high degree of flexibility.

However, the success of shuttle rendezvous and proximity operations came at the expense of some of the original objectives and goals of the Shuttle Program. These included simplified and standardized mission planning and training, lower number of mission support personnel, high flight rates, elimination of extensive flight-to-flight analysis, no computation of flight specific trajectory data, and no generation of customized onboard charts for each mission. Successful adaptation of proven rendezvous principles to meet new and emerging operational and programmatic constraints was in part due to the carry over of experienced personnel from the shorter duration Gemini and Apollo programs. Later generations of engineers that joined the Shuttle Program successfully applied knowledge learned from these experienced personnel and their own shuttle experience to solving the complex systems integration challenges associated with satellite repair missions, Mir docking missions, and ISS construction and re-supply missions. These personnel possessed extensive experience in the development and analysis of vehicle and subsystem performance specifications, requirements, and operations concepts.

This page intentionally left blank.

PART III – SPECIAL FOCUS CHAPTERS

This page intentionally left blank.

CHAPTER 23 - STS-39, THE MOST COMPLEX DEPLOY/RETRIEVE MISSION

Introduction

The Space Shuttle flew 16 missions with deployment and retrieval mission objectives. The April/May 1991 flight of STS-39 was the most complex deploy/retrieve mission flown with approximately 38 hours of relative motion, rendezvous, and proximity operations. One payload, the SPAS-II satellite (Figure 23.1), was deployed and retrieved while three smaller payloads were deployed. Rendezvous, proximity operations, deployed satellite commanding by the crew, and numerous experiments using the primary and secondary payloads required dual shift, 24 hour crew scheduling. This chapter reviews the STS-39 mission objectives, relative motion design, mission events, and lessons learned.

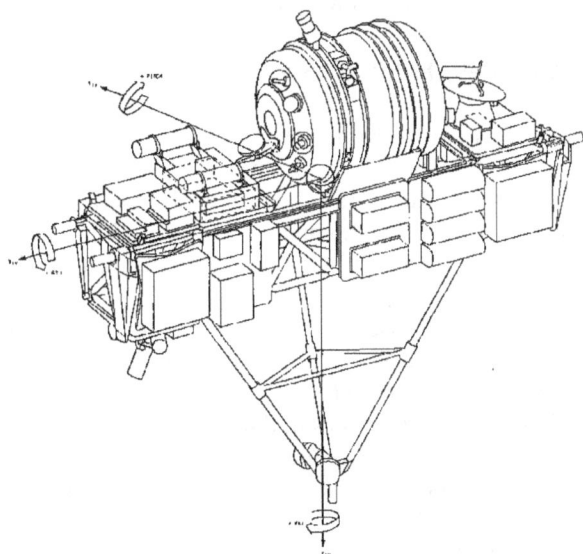

Figure 23.1 SPAS-II with the IBSS payload.

Another April 1991 mission, STS-37, tested an out-of-plane star tracker pass technique that was part of the STS-39 radar fail procedure. The evolution of the STS-37 Mid-Range Station-Keeping Detailed Test Objective mission plan is covered along with mission performance.

The Mission

The April/May 1991 flight of STS-39 was the eighth dedicated Department of Defense (DOD) mission and the first shuttle DOD mission to be flown unclassified. The orbiter *Discovery* was launched from launch pad 39A into a 57 degree inclination, 140 nm circular orbit. A direct insertion was performed (no Orbital Maneuvering System 1 (OMS-1) burn, just OMS-2). The primary mission objective was to collect visible light, infrared, X-Ray, and ultraviolet data on orbiter OMS and primary Reaction Control System (RCS) plumes. In addition, observations

of aurora, Earth limb, airglow, chemical and gas releases, the orbiter environment, and celestial objects (such as galaxies, nebula, and stars) were performed. Data from these observations were used for development of sensors and other systems for the Strategic Defense Initiative Organization (SDIO). The two primary payloads were the Infrared Background Signature Survey (IBSS) and Air Force Program (AFP) 675. Attached cargo operations were performed with the AFP-675 in the payload bay. AFP-675 activities included aurora viewing.[1]

Two of the three elements of IBSS were deployable, the SPAS-II satellite and the three Chemical Release Observation (CRO) sub-satellites. The third, the Critical Ionization Velocity (CIV) experiment, was mounted in the payload bay. Some CRO observation hardware was also mounted in the payload bay of *Discovery*.[1]

The SPAS-II (Figure 23.1) was deployed and later retrieved by *Discovery*. It was an improved version of the SPAS-I flown on STS-7 (*Challenger*, June 1983) and STS-41B (*Challenger*, February 1984). SPAS-II could be commanded by either the crew or ground. It carried two visible light television cameras, an ultraviolet multispectral sensor, and a cryogenically cooled infrared sensor. A large dewar on SPAS-II contained liquid Helium used to cool the IBSS infrared sensor. Precision attitude control by SPAS-II was required to support experiment observations. In addition, SPAS-II/IBSS activities were conducted with the spacecraft attached to and maneuvered by the shuttle Remote Manipulator System (RMS) robotic arm. SPAS-II commanding was performed by both the crew and ground personnel.[1]

The CRO experiments were developed to collect infrared, visible light, and ultraviolet data of chemicals that could be released by spacecraft for obscuration purposes. In addition, the observations were useful for characterizing signatures of propellants escaping from damaged boosters. Three CRO sub-satellites were deployed from the orbiter. The CRO sub-satellites were deployed one at a time to accommodate Vandenberg Air Force Base commanded chemical release and viewing opportunities. Two chemical releases were viewed by the SPAS-II IBSS payload while it was deployed and the third while SPAS-II was on the RMS. In addition to the SPAS-II sensors, observations of the chemical releases were also made from Vandenberg Air Force Base and from aircraft.[1]

Each CRO sub-satellite was equipped with a chemical tank, antennas, an optical beacon, solar cells, and a radar reflector mounted on a 5 foot boom. The CRO sub-satellites were not equipped with attitude control systems. CRO-C was loaded with 15 pounds of nitrogen tetroxide (N2O4), CRO-B contained 52 pounds of unsymmetrical dimethyl hydrazine (UDMH), and CRO-A had 60 pounds of monomethyl hydrazine (MMH).[1]

Space Test Payload-1 (STP-1) and the classified Multi-Purpose Experiment Canister (MPEC) were secondary payloads. MPEC was deployed from *Discovery* late in the mission. It was housed in a Get-Away Special (GAS) canister in the payload bay.

STS-39 was the debut mission of the new IBM AP-101S General Purpose Computers (GPCs). These computers replaced the original AP-101B GPCs. The AP-101S possessed 2.5 times as much memory and provided an up to three times improvement in processing speed. STS-39 was the second flight of primary flight software Operational Increment 8F (OI-8F).[1]

The Crew and Mission Control

Six of the seven STS-39 crew members were assigned to a red team or a blue team. Each team member worked 12 hour shifts with 12 hours off duty. The commander was not assigned to a team and was free to adjust his work hours as required. Crew assignments are listed in Table 23.1. Table 23.2 lists the planned major activities for each team by flight day. Figure 23.2 is the Mission Control shift schedule for the nominal April 23, 1991 launch date. The shift schedule for the actual April 28 launch was similar. Figures 23.3, 23.4, and 23.5 are the overview timelines illustrating red and blue team activities by flight day during the mission. Note that both teams were to be awake for SPAS-II deployment, the plume burns, and SPAS-II retrieval.

Flight Plan and Relative Motion Profile

This section provides an overview of the pre-flight mission plan for STS-39, with emphasis on deployed payloads (SPAS-II, CRO sub-satellites, and MPEC). Topics covered include the launch window, SPAS-II detached phase relative motion, the rendezvous radar fail

profile and procedures, CRO sub-satellite deployments and observations, MPEC deploy, and nominal end of mission. Attached payload operations are not covered.

Launch Window

The launch window open time was driven by a daylight Kennedy Space Center (KSC) launch requirement (sunrise + 15 minutes). Launch window close was driven by daylight CRO-B observation on orbit 56 (Flight Day 4) over Vandenberg AFB. All aborts (Return to Launch Site (RTLS), Trans-oceanic Abort Landing (TAL), and Abort Once Around (AOA)) were in daylight. The launch window also protected for all daylight end of mission descending orbit opportunities for the nominal end of mission day and two additional days. The launch window provided for a minimum of three aurora viewing opportunities on Flight Day 1. It also provided for a minimum of 30 minutes of umbra per orbit from orbit 15 through orbit 120.[2]

SPAS-II Detached Phase Mission Plan

SPAS-II was to be deployed using the RMS on orbit 31 (flight day 3) and retrieved with the RMS on orbit 56 (flight day 4). The overall planned relative motion profile is shown in Figures 23.6 and 23.7. Table 23.3 contains acronym definitions of burns and other planned mission events.[3]

The SPAS-II IBSS payload was to observe orbiter OMS out-of-plane and RCS plume burns at ranges of 10 km (5.4 nm, the far field) and 2.25 km (1.2 nm, the near field) while the orbiter was trailing SPAS-II on the minus V Bar. The customer requested one plume observation several hours before the rest of the plume observations. The SPAS-II detached phase was planned to last approximately 36 hours.[3]

Table 23.1 STS-39 Crew

Crew Member	Role	Team	Responsibilities
Michael L. Coats	Commander	N/A	Mission decisions, deploy, separation, rendezvous, orbiter systems, and detailed test objectives.
Blaine Hammond	Pilot	Red	Piloting tasks and orbiter systems.
Gregory J. Harbaugh	MS-1, EV-1*	Blue	IBSS, RMS
Donald R. McMonagle	MS-2, EV-2*	Blue	Orbiter systems, piloting tasks, and RME-III.
Guion S. Bluford	MS-3, IVA*	Blue	AFP-675, STP-1, and MPEC.
Charles L. Veach	MS-4	Red	AFP-675 and CLOUDS-1A.
Richard J. Hieb	MS-5	Red	IBSS and RMS.

* If required.

AFP-675 – Air Force Program 675, **CLOUDS-1A** – Cloud Logic to Optimize Use of Defense Systems 1A, **EV** – Extra-Vehicular Activity, **IBSS** – Infrared Background Signature Survey, **IVA** – Intra-Vehicular Activity, **MPEC** – Multi-Purpose Experiment Canister, **MS** – Mission Specialist, **N/A** – Not Applicable, **RME** – Radiation Monitoring Equipment, **RMS** – Remote Manipulator System, **STP-1** – Space Test Payload 1

Table 23.2 Planned STS-39 Activities By Flight Day

Flight Day	Red Team	Blue Team
Flight Day 1	• Ascent • Orbit insertion • Ku-Band antennae deploy • Group B powerdown • RMS powerup and checkout • Aft controller checkout • STP-1 activation • AFP-675 init and checkout • RME-III activation • IBSS checkout	• Ascent • Orbit insertion • AFP-675 operations
Flight Day 2	• AFP-675 operations	• AFP-675 operations • SPAS-II/IBSS predeploy checkout • SPAS-II/IBSS grapple, unberth
Flight Day 3	• SPAS-II/IBSS release, attitude control checkout • Separation to far field (10 km) • OMS plume sequence 1 • Far field station-keeping • IBSS operations (Earth scan, Earth limb, CO2 Earth sweep)	• Far field OMS plume sequences 2 and 3 • Far field RCS plume sequence • IBSS experiment operations • CRO-C deploy • Transition to near field (2 km) • Near field OMS plume sequences 4 and 5
Flight Day 4	• Orbiter systems redundant component checkout • IBSS experiment operations • CIV operations • Phase out to rendezvous initiation range • CRO-C observation • CRO-B deploy • SPAS-II/IBSS rendezvous • CRO-B observation • SPAS-II/IBSS capture	• SPAS-II/IBSS berth • Orbit adjust for CRO-C avoidance • STP-1 operations • AFP-675 operations
Flight Day 5	• AFP-675	• AFP-675 • CRO-A deploy
Flight Day 6	• AFP-675 • SPAS-II/IBSS checkout • SPAS-II/IBSS attached operations • CRO-A observation • STP-1 operations	• SPAS-II/IBSS attached operations • CIV operations • Orbiter environment • SPAS-II/IBSS berth
Flight Day 7	• SPAS-II/IBSS in-bay operations • AFP-675	• AFP-675
Flight Day 8	• Flight control system checkout • AFP-675 operations • STP-1 dedicated operations	• STP-1 dedicated operations • AFP-675 deactivation • MPEC deploy • Cabin stow
Flight Day 9	• Payload deactivation • Deorbit preparation • Entry and landing	• Payload deactivation • Deorbit preparation • Entry and landing

AFP – Air Force Program, **CIV** – Critical Ionization Velocity, **CRO** – Chemical Release Observation, **CO2** – Carbon Dioxide, **IBSS** – Infrared Background Signature Survey, **km** – kilometer, **MPEC** – Multi-Purpose Experiment Canister, **OMS** – Orbital Maneuvering System, **RCS** – Reaction Control System, **RME** – Radiation Monitoring Equipment, **SPAS** – Shuttle Pallet Satellite, **STP** – Space Test Payload

STS-39 FLIGHT CONTROL TEAM SCHEDULE
PRELAUNCH SUPPORT IS AS FOLLOWS (BASED ON APRIL 23 LAUNCH):

L-2 DAYS (SUNDAY, 4/21/91)

1900 CDT	~L-35 hrs	COMM ACTIVATION	ORBIT 3 TEAM (FD/INCO/GC)

L-1 DAY (MONDAY, 4/22/91)

0430 CDT		PRE-BRIEFING TEAM TAG UP	ASCENT TEAM
0500 CDT		L-1 DAY SYSTEMS BRIEFING	ASCENT TEAM
0545 (ASAP After SYS Brfg)		L-1 DAY WEATHER BRIEFING	ASCENT TEAM
1700 CDT	~L-13 hrs	MCC MANNING	ORBIT 3 TEAM

L-0 DAY (TUESDAY, 4/23/91)

0100 CDT	~L-5 hrs	COUNT/LAUNCH	ASCENT TEAM

POST LIFTOFF SUPPORT IS AS FOLLOWS (CALL 483-1077 FOR UPDATES):

FLIGHT DAY	TEAM	HOURS	MET ON	MET OFF	DAY	CDT ON	CDT OFF
1	ASCENT	8.0	L--5	0/03:00	TU/23	23/0100	23/0900
	ORBIT 2	8.0	0/02:00	0/10:00		23/0800	23/1600
	ORBIT 3	9.0	0/09:00	0/18:00		23/1500	24/0000
2	ORBIT 1	9.0	0/17:00	1/02:00	WE/24	23/2300	24/0800
	ORBIT 2	9.0	1/01:00	1/10:00		24/0700	24/1600
	ORBIT 3	9.0	1/09:00	1/18:00		24/1500	25/0000
3	ORBIT 1	10.0	1/17:00	2/03:00	TH/25	24/2300	25/0900
	ORBIT 2	11.0	2/02:00	2/13:00		25/0800	25/1900
	ORBIT 3	10.5	2/12:00	2/22:30		25/1800	26/0430
4	ORBIT 1	8.5	2/21:30	3/06:00	FR/26	26/0330	26/1200
	ORBIT 2	10.0	3/05:00	3/15:00		26/1100	26/2100
	ORBIT 3	9.0	3/14:00	3/23:00		26/2000	27/0500
5	ORBIT 1	9.0	3/22:00	4/07:00	SA/27	27/0400	27/1300
	ORBIT 2	9.0	4/06:00	4/15:00		27/1200	27/2100
	ORBIT 3	9.0	4/14:00	4/23:00		27/2000	28/0500
6	ORBIT 1	9.0	4/22:00	5/07:00	SU/28	28/0400	28/1300
	ORBIT 2	9.0	5/06:00	5/15:00		28/1200	28/2100
	ORBIT 3	9.0	5/14:00	5/23:00		28/2000	29/0500
7	ORBIT 1	9.0	5/22:00	6/07:00	MO/29	29/0400	29/1300
	ORBIT 2	9.0	6/06:00	6/15:00		29/1200	29/2100
	ORBIT 3	9.0	6/14:00	6/23:00		29/2000	30/0500
8	ORBIT 1*	9.5	6/22:00	7/07:30	TU/30	30/0400	30/1330
	ORBIT 2	8.5	7/06:30	7/15:00		30/1230	30/2100
	ORBIT 3	10.0	7/14:00	8/00:00		30/2000	01/0600
9	ENTRY		7/23:00		WE/1	01/0500	

*ENTRY TEAM SUPPORT WILL BE IDENTIFIED BY THE ENTRY FLIGHT DIRECTOR

Figure 23.2 Mission Control shift schedule for the April 23, 1991 launch date. Launch actually occurred on April 28.

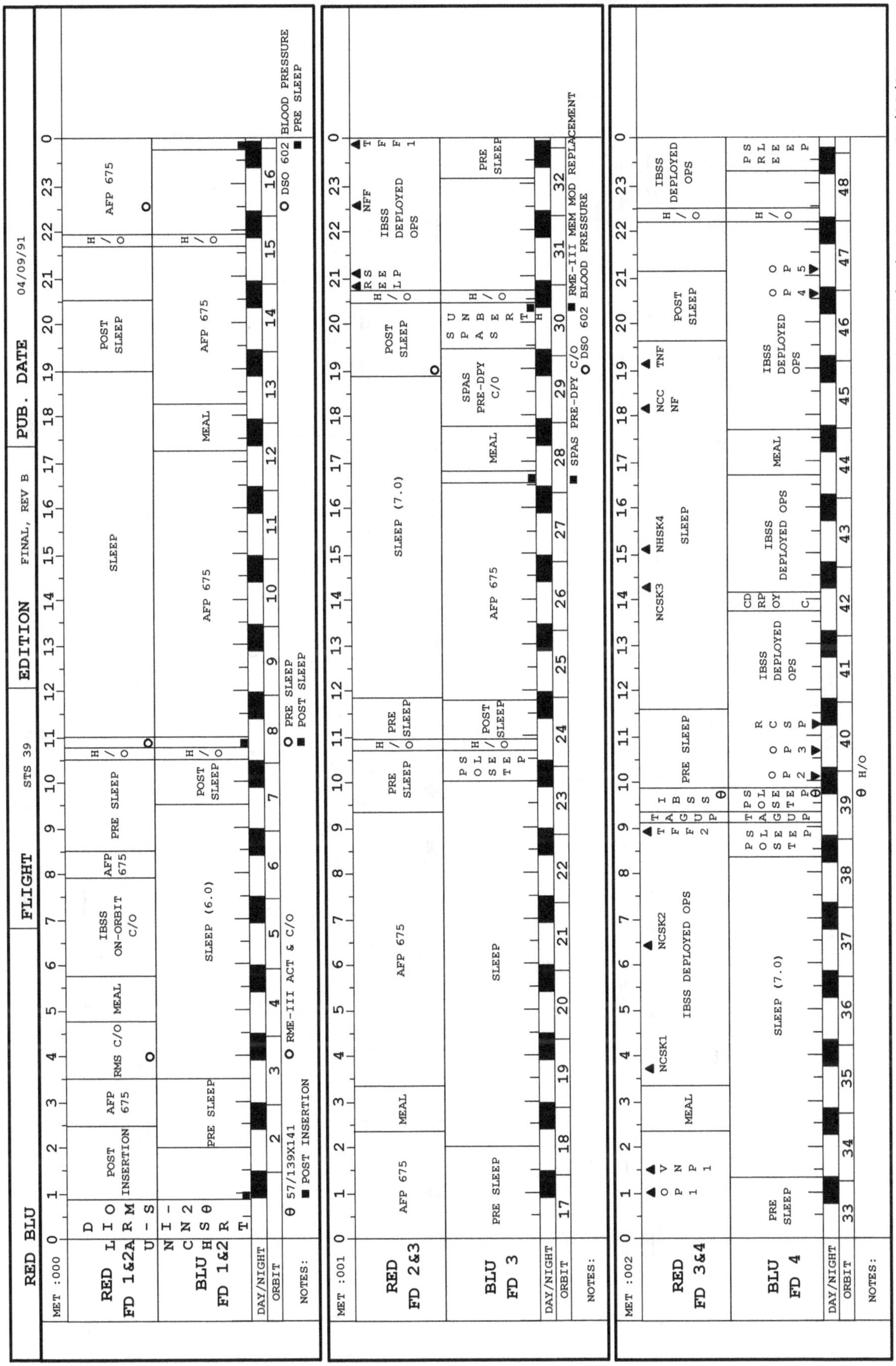

Figure 23.3 Original pre-mission overview timeline for flight days 1 through 4. A flight day began when the crew woke up.

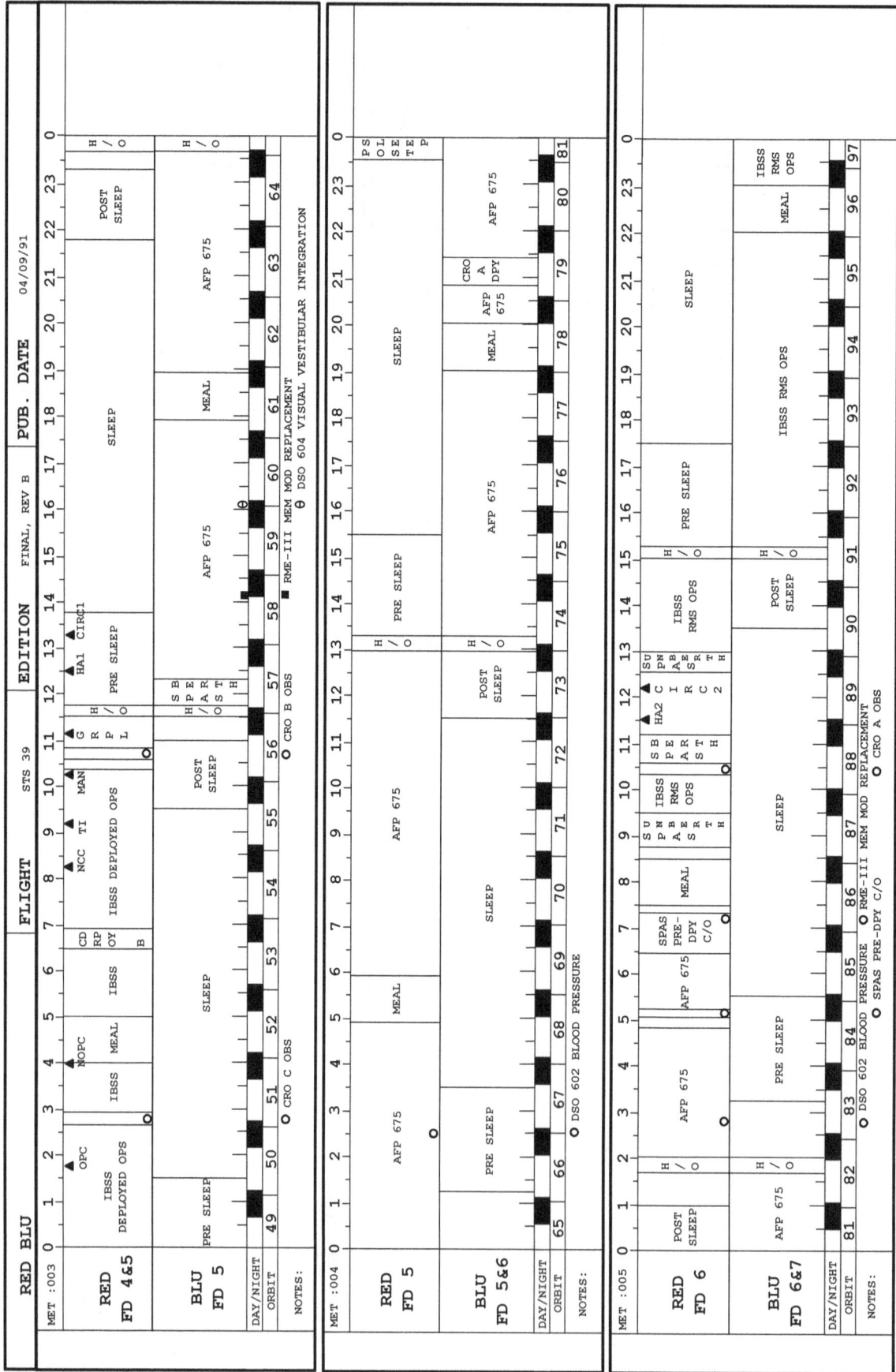

Figure 23.4 Original pre-mission overview timeline for flight days 4 through 7.

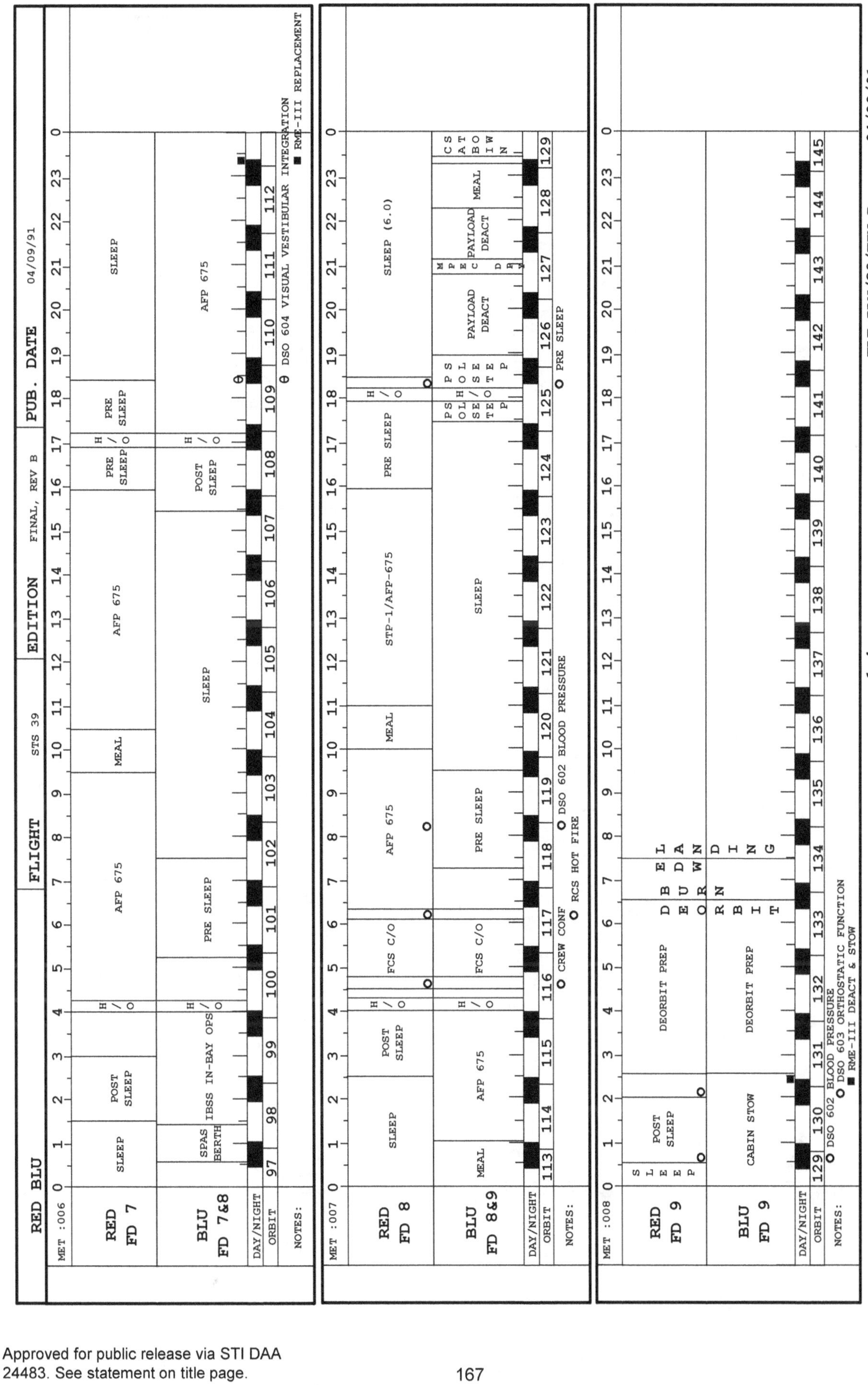

Figure 23.5 Original pre-mission overview timeline for flight days 7 through 9.

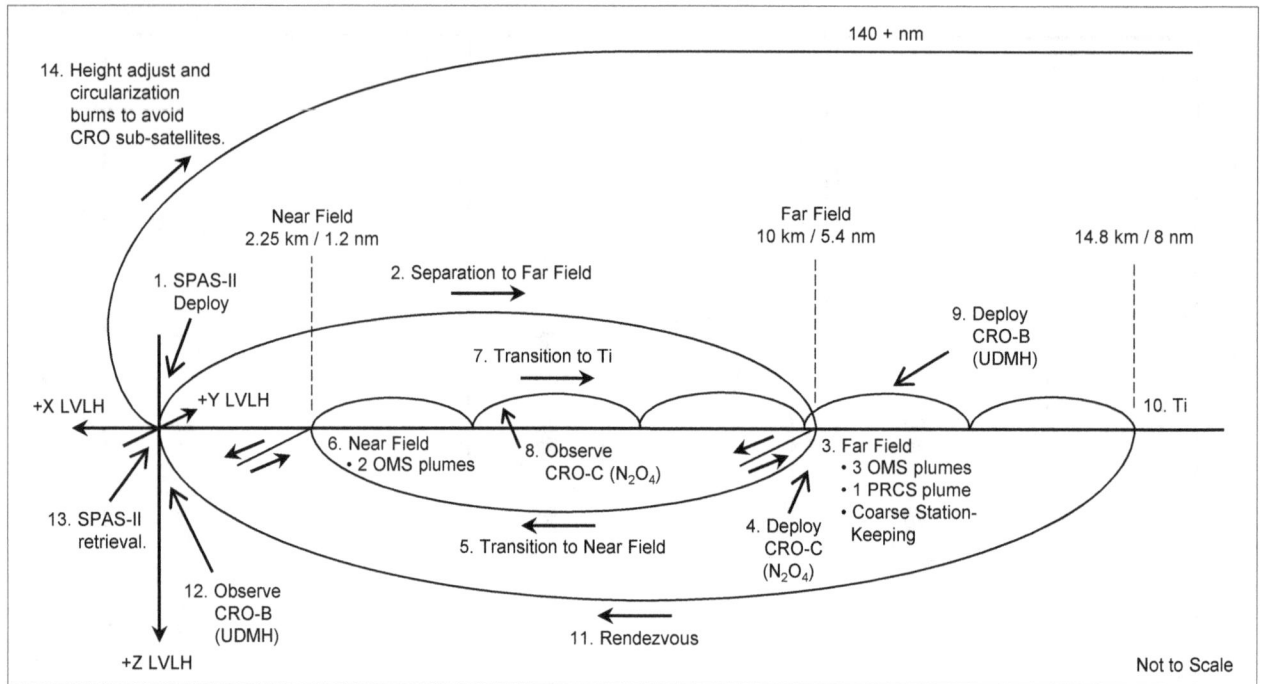

Figure 23.6 Relative motion during IBSS detached operations and after IBSS retrieval.

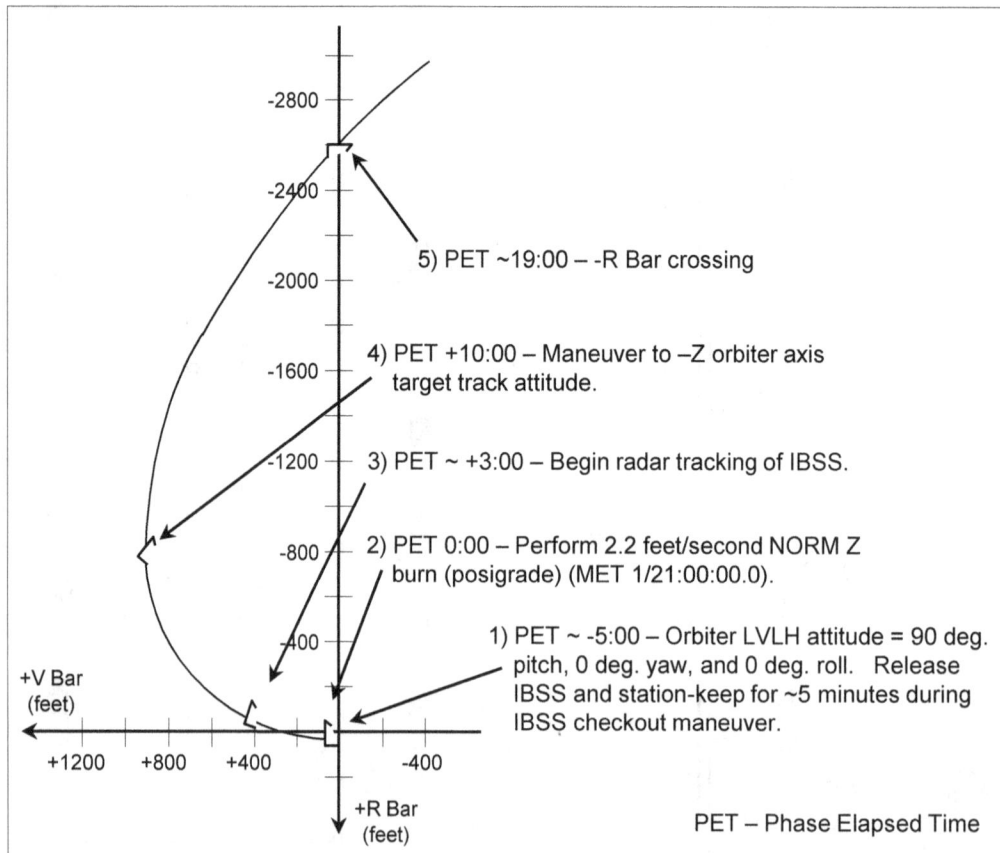

Figure 23.7 IBSS deploy and separation proximity operations.

Table 23.3 STS-39 Event Descriptions

Event	Description	Event	Description
Launch		MC1NF	Mid-Course Correction 1 to Near Field
OMS-2	Orbit Insertion	MC2NF	Mid-Course Correction 2 to Near Field
SEP	Separation from SPAS	MC3NF	Mid-Course Correction 3 to Near Field
MC1FF	Mid-Course Correction 1 to Far Field	NNF	Null Near Field
MC2FF	Mid-Course Correction 2 to Far Field	OP4	OMS Plume Observation 4
NFF0	Null at Far Field 0	NOP4	Out-of-Plane Null From Plume 4
TFF1	Transition to Far Field 1	MCP4	Out-of-Plane Mid-Course From Plume 4
MCCF1	Mid-Course Correction to Far Field 1	VNP4	V Bar Null From Plume 4
NFF1	Null at Far Field 1	OP5	OMS Plume Observation 5
NCSK1	Phasing/Station-Keeping 1	NOP5	Out-of-Plane Null From Plume 5
NCSK2	Phasing/Station-Keeping 2	MCP5	Out-of-Plane Mid-Course From Plume 5
TFF2	Transition to Far Field 2	NSR5	Coelliptic (Slow Rate) 5
MCFF2	Mid-Course Correction to Far Field 2	OPC/NC	Out-of-Plane and Phasing Correction
NFF2	Null at Far Field 2	NCSK5	Phasing/Station-Keeping 5
OP1	OMS Plume Observation 1	NOPC	Null Out-of-Plane
NOP1	Out-of-Plane Null From Plume 1	CRO-B	Deploy CRO-B
MCP1	Out-of-Plane Mid-Course From Plume 1	NCC	Corrective Combination
VNP1	V Bar Null From Plume 1	Ti	Transition Initiation
OP3	OMS Plume Observation 3	MC1	Mid-course Correction 1
NOP3	Out-of-Plane Null From Plume 3	MC2	Mid-course Correction 2
MCP3	Out-of-Plane Mid-Course From Plume 3	MC3	Mid-course Correction 3
VNP3	V Bar Null From Plume 3	MC4	Mid-course Correction 4
OP2	OMS Plume Observation 2	Grapple	Grapple SPAS
NOP2	Out-of-Plane Null From Plume 2	HA1	Height Adjust 1
MCP2	Out-of-Plane Mid-Course From Plume 2	CIRC1	Circularization 1
VNP2	V Bar Null From Plume 2	CRO-A	CRO-A Deploy
RCSP	RCS Plume Observation	HA2	Height Adjust 2
NRCS	Null RCS Plume Observation	CIRC2	Circularization 2
CRO-C	Deploy CRO-C	AIS	Out-of-Plane Burn
NCSK3	Phasing/Station-Keeping 3	UVPI	Orbit Raising
NHSK4	Height/Station-Keeping 4	MPEC	MPEC Depoy Burn
NCCNF	Corrective Combination for Near Field	SEP	Separation from MPEC
TNF	Transition to Near Field	Deorbit	Deorbit burn for KSC Landing
		Landing	KSC Landing

CRO – Extra-Vehicular Activity, **KSC** – Intra-Vehicular Activity, **MPEC** – Multi-Purpose Experiment Canister, **OMS** – Orbital Maneuvering System, **RCS** – Reaction Control System, **SPAS** – Shuttle Pallet Satellite

Rendezvous radar tracking was to be maintained throughout the SPAS-II/IBSS detached operations phase except during IBSS boresight and radiometric calibrations. The relative motion profile was designed to ensure a 20 km (10.8 nm) Payload Interrogator (PI) communications range link margin.

Figure 23.7 shows orbiter relative motion after SPAS-II deploy and during separation. One separation burn was planned, a 2.2 foot/second posigrade burn. On-board relative navigation and rendezvous radar tracking was to be commenced soon after the separation burn. Figure 23.8 is a pre-mission plot of planned relative motion from separation through arrival at the far field. Two on-board Lambert targeted Mid-course Correction burns (MC1FF and MC2FF) were to ensure arrival at the far field with low trajectory dispersions. Once at the far field the

on-board targeted NFF0 burn was planned to set-up a relative motion football. Ground targeted NCSK1 and NCSK2 burns were planned to adjust phasing (or down-track, Figure 23.9). The subsequent TFF1 and MCFF1 burns were to transfer the orbiter from the relative motion football back to the far field point for the first OMS plume sequence. NFF1 was a coelliptic (NSR) burn to lower orbiter line-of-sight rates as viewed from SPAS for the first plume sequence.

At the far field three OMS single engine out-of-plane plume burns and one RCS plume burn were to be conducted for IBSS observation. The first plume sequence (Figure 23.10) consisted of the ground targeted OMS plume burn (OP1), followed by the on-board targeted out-of-plane null burn (NOP1) designed to return the orbiter to the minus V Bar.

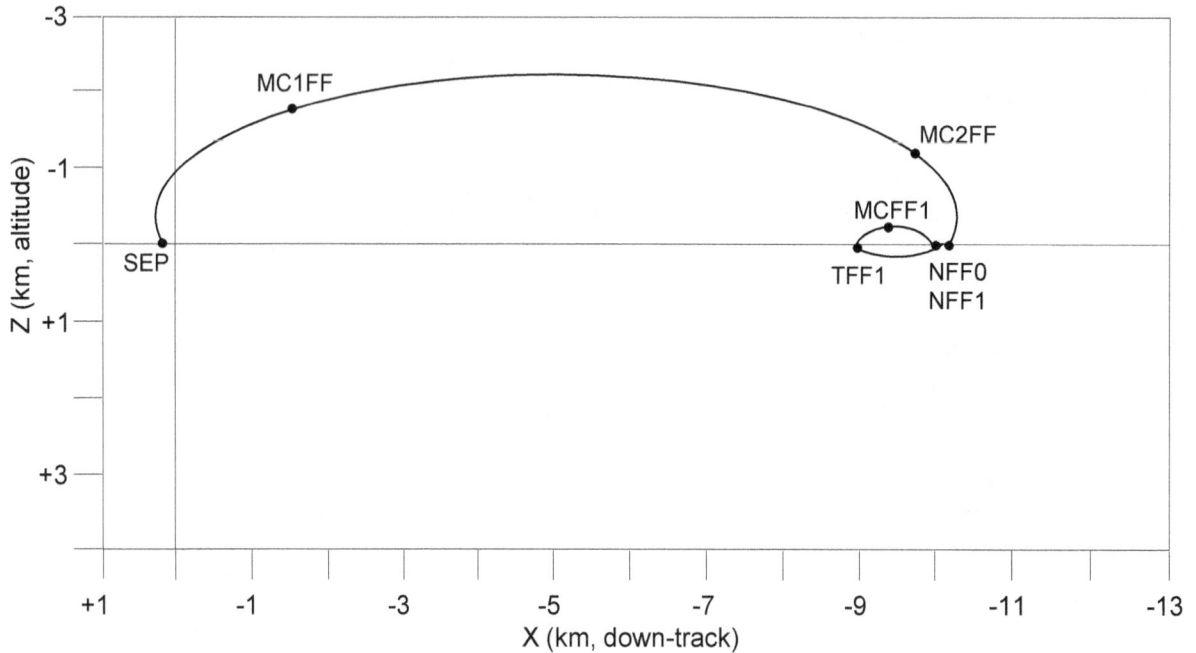

Figure 23.8 Separation to far field, Local Vertical Curvilinear (LVC) coordinate frame.

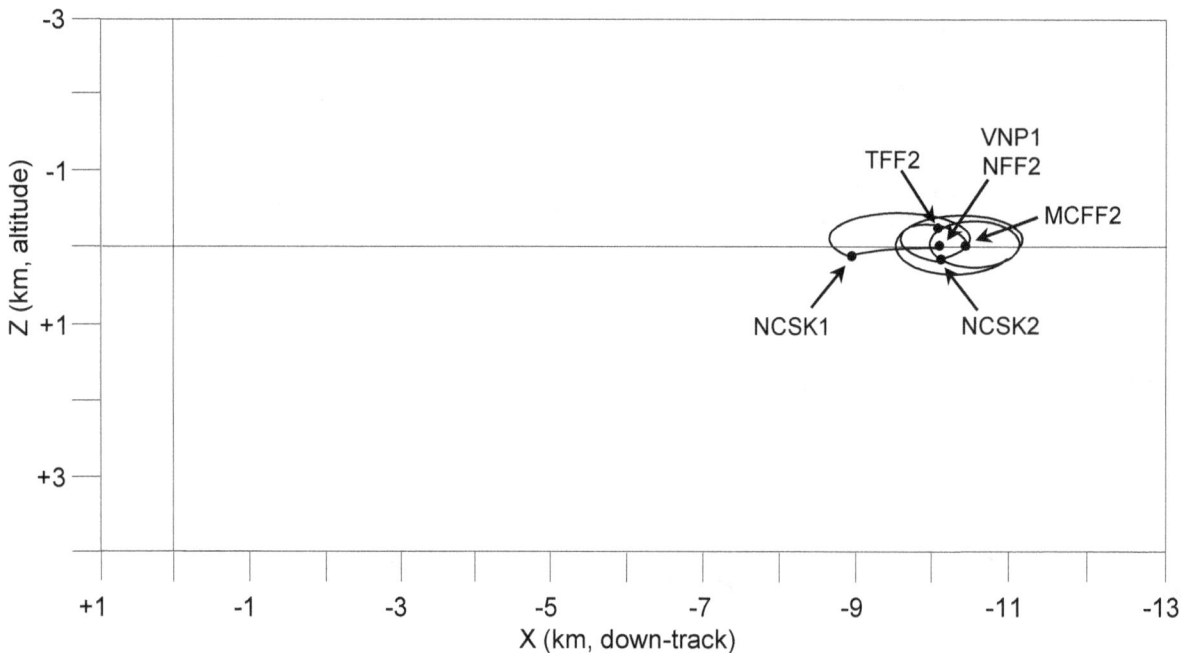

Figure 23.9 Far field coarse station-keeping, LVC coordinate frame.

Between the OP1 and NOP1 burns the orbiter performed a fast flip attitude maneuver of approximately 180 degrees to achieve the NOP2 burn attitude. The combination of out-of-plane motion and the approximately 180 degree attitude maneuver was known as the Malarkey Milkshake (see also Figure 23.11). It was named for the developer, STS-39 lead Rendezvous Guidance and Procedures Officer John Malarkey.

An on-board targeted mid-course correction burn (MCP1) was planned to tweak the arrival point on the minus V Bar. VNP1 was an on-board targeted football burn executed upon arrival at the minus V Bar.

A TFF2-MCFF2-NFF2 burn sequence was then planned to set up for the second plume sequence. These three burns performed the same function as the earlier TFF1-MCFF1-NFF1 burn sequence. Execution of the second (OP2-NOP2-MCP2-VNP2) and third (OP3-NOP3-MCP3-VNP3) OMS plume burn sequences followed. An RCS plume burn was planned after the third OMS plume burn. Figure 23.11 illustrates far field planned relative motion for the second and third OMS plume burns and the RCS plume burn.[3]

Coarse station-keeping was then to be performed at the far field with the ground targeted NCSK3 (phasing) and NHSK4 (altitude) burns. Sub-satellite CRO-C (N2O4) was to be deployed as well. Figure 23.12 depicts

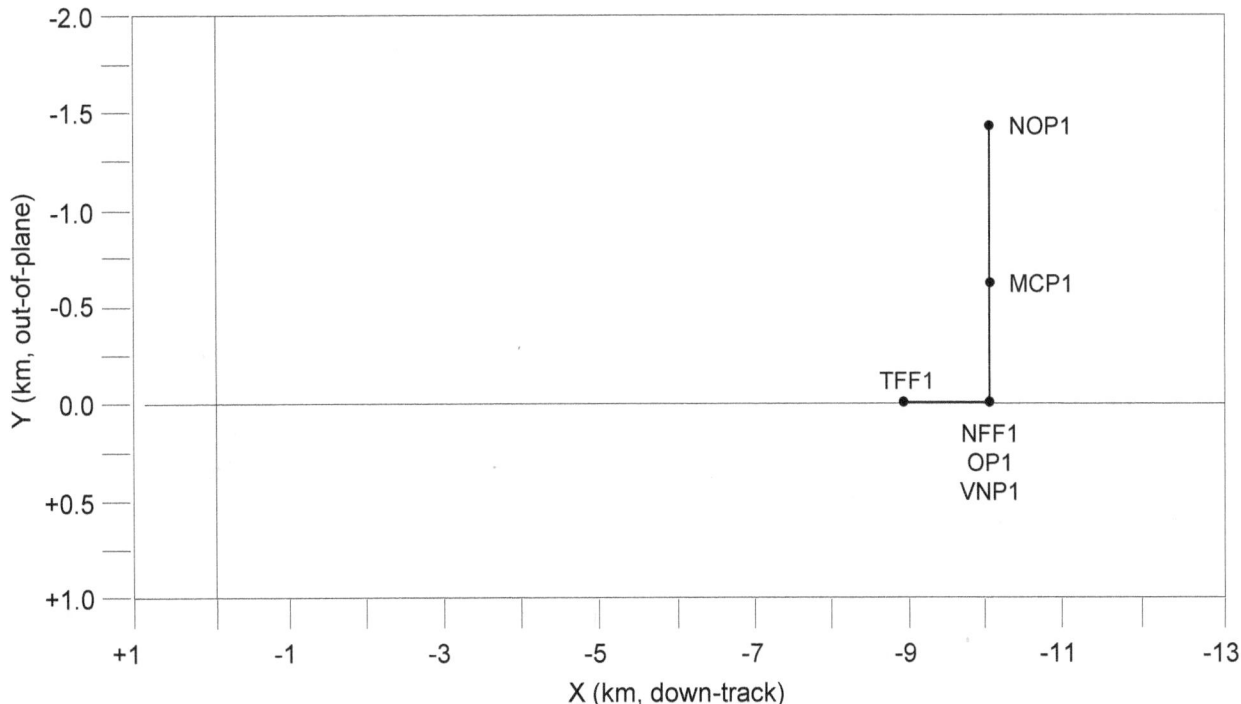

Figure 23.10 First OMS burn at far field, LVC coordinate frame.

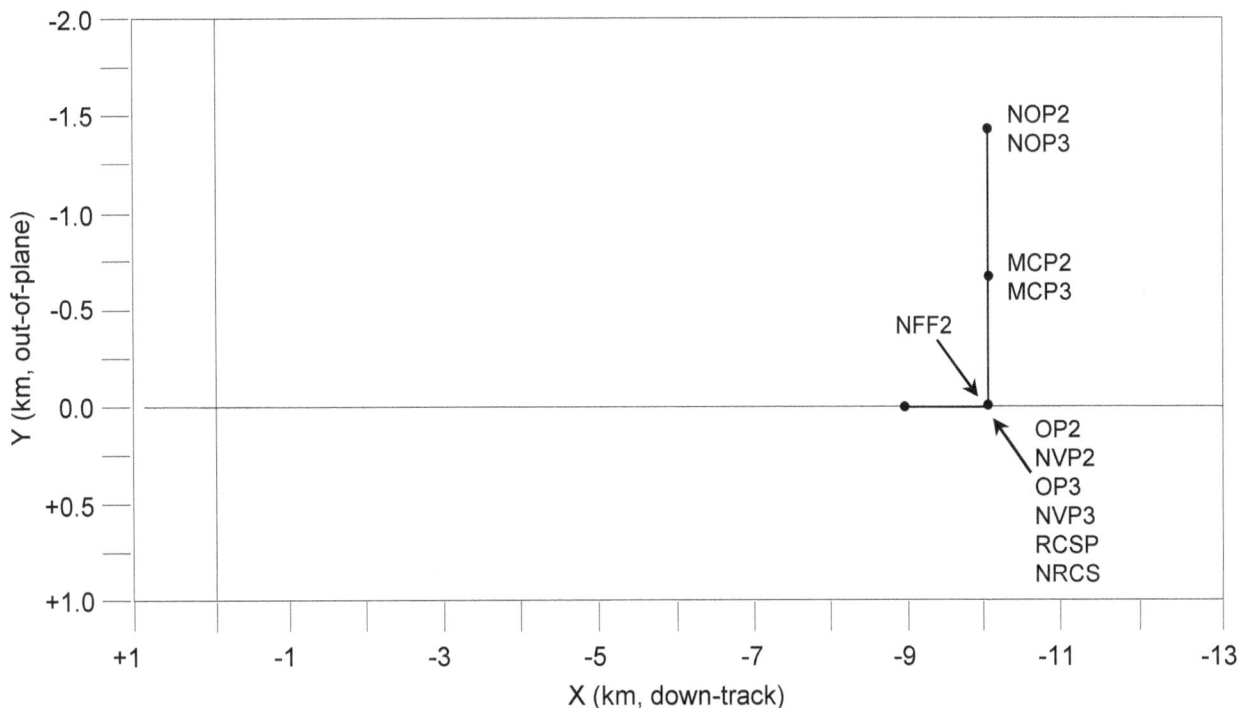

Figure 23.11 Second and third OMS burns and RCS plume at Far Field, LVC coordinate frame.

far field station-keeping after the far field plume observations.

The orbiter was then to transition to the 2.25 km (1.2 nm) near field point for more IBSS observation activities. Figure 23.12 depicts the transfer from the far field to the near field. The Transition to Near Field (TNF) burn was followed by three mid-course correction burns (MC1NF, MC2NF, and MC3NF) to ensure arrival at the near field point with small trajectory dispersions. Upon arrival at

the near field the on-board targeted NNF burn was planned to establish a coelliptic orbit with SPAS-II to lower line-of-sight rates for the fourth OMS plume sequence (OP4-NOP4-MCP4-VNP4). A fifth OMS plume sequence was also planned (OP5-NOP5-MCP5-VNP5). Figure 23.13 depicts planned near field relative motion for the fourth and fifth plume sequences. The fifth plume sequence ended with an on-board targeted NSR5 to establish a coelliptic trajectory (Figure 23.14).

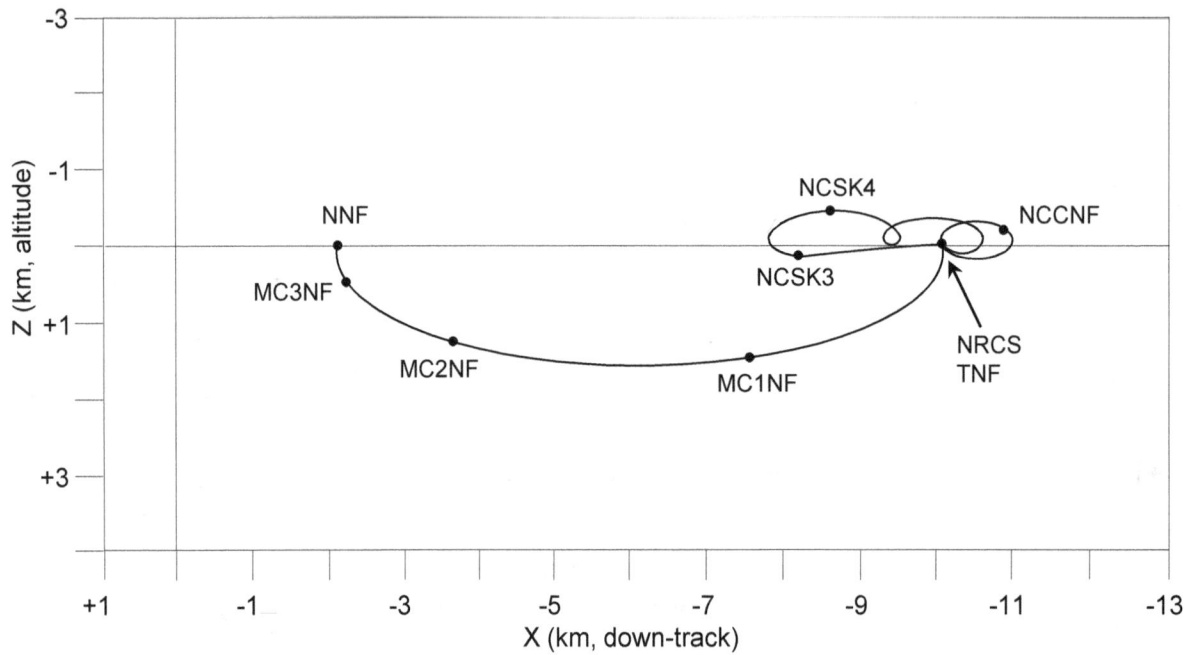

Figure 23.12 Far field station-keeping and transition to near field, LVC coordinate frame.

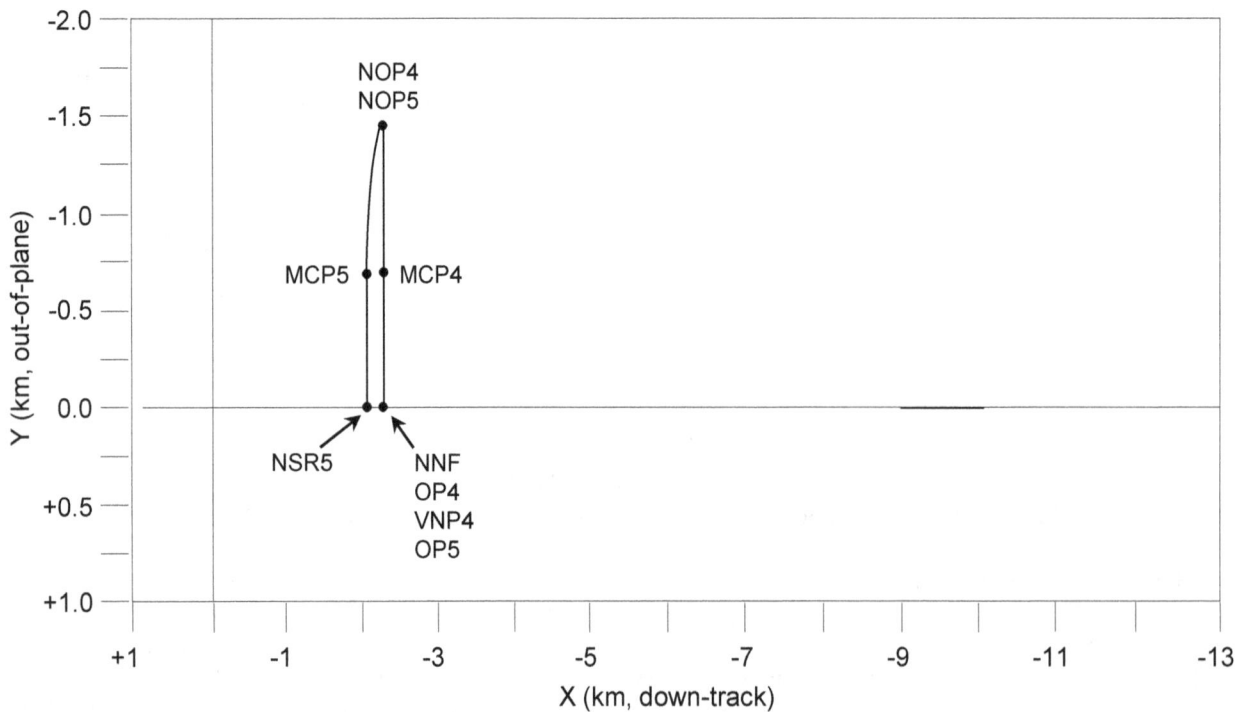

Figure 23.13 Near field OMS burns, LVC coordinate frame.

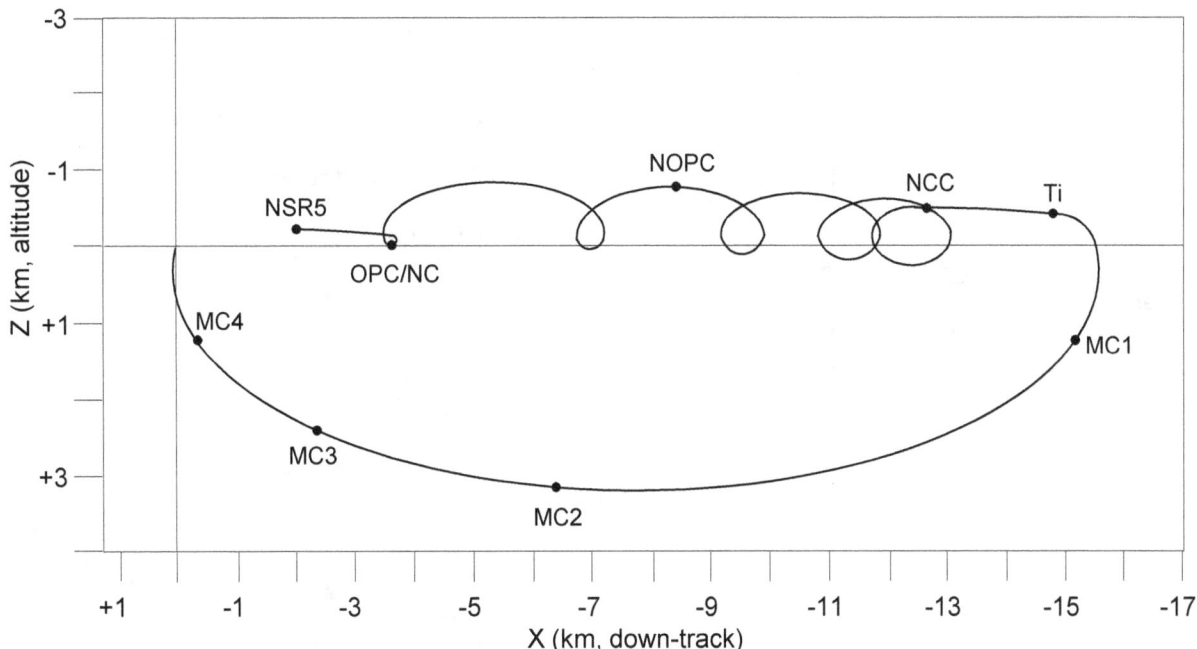

Figure 23.14 Transition from near field to 14.8 km (8 nm) Ti point followed by rendezvous with IBSS, LVC coordinate frame.

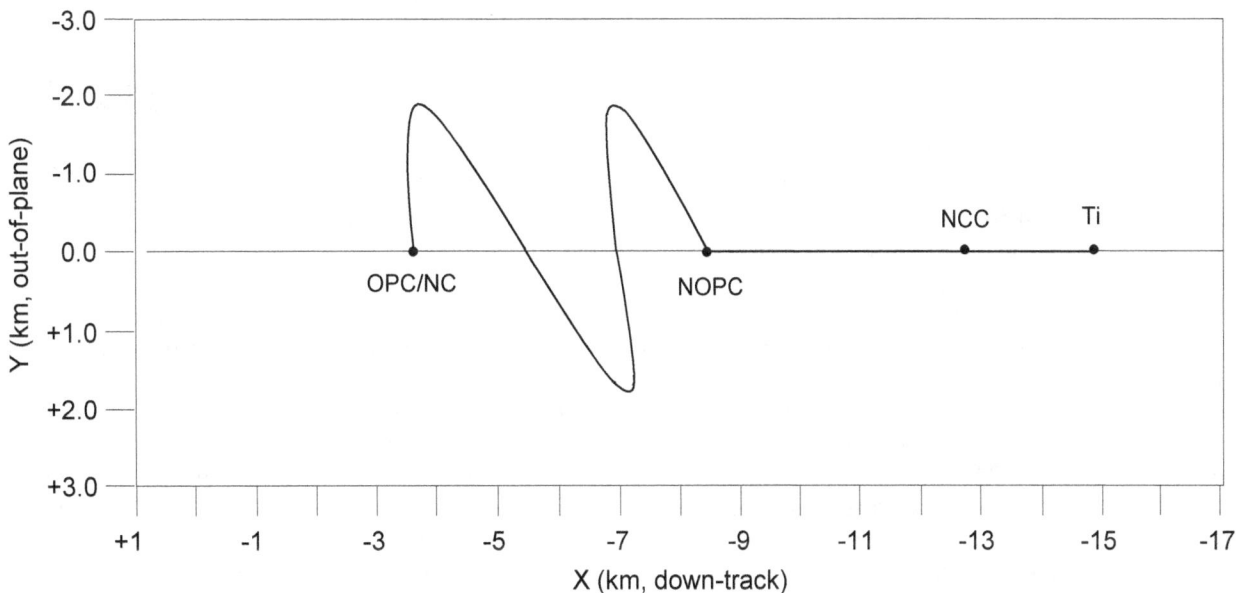

Figure 23.15 Out-of-plane motion during phasing from near field to Ti point, LVC coordinate frame.

The orbiter was then to phase away from the near field to the standard shuttle stable orbit rendezvous profile Transition Initiation (Ti) burn point at 8 nm (14.8 km) on the minus V Bar. The OPC/NC, NCSK5, and NOPC burns were planned to ensure correct phasing and out-of-plane relative motion to support the CRO-C observation and arrive at the Ti burn point (Figure 23.14). Figure 23.15 illustrates planned out-of-plane motion during the transfer from the near field to the 8 nm Ti point. The out-of-plane burn component was designed to place the orbiter 12 degrees out-of-plane in 1.25 revolutions to support the CRO-C observation. The CRO-C (N2O4) chemical release was to be observed on orbit 51. Before arriving at the Ti point (8 nm behind SPAS-II) the CRO-B (UDMH) sub-satellite was to be deployed.

The rendezvous with SPAS-II was initiated from the Ti burn point. Nominally Ti was executed 8 nm behind the target spacecraft and 1,200 feet above the minus V Bar.

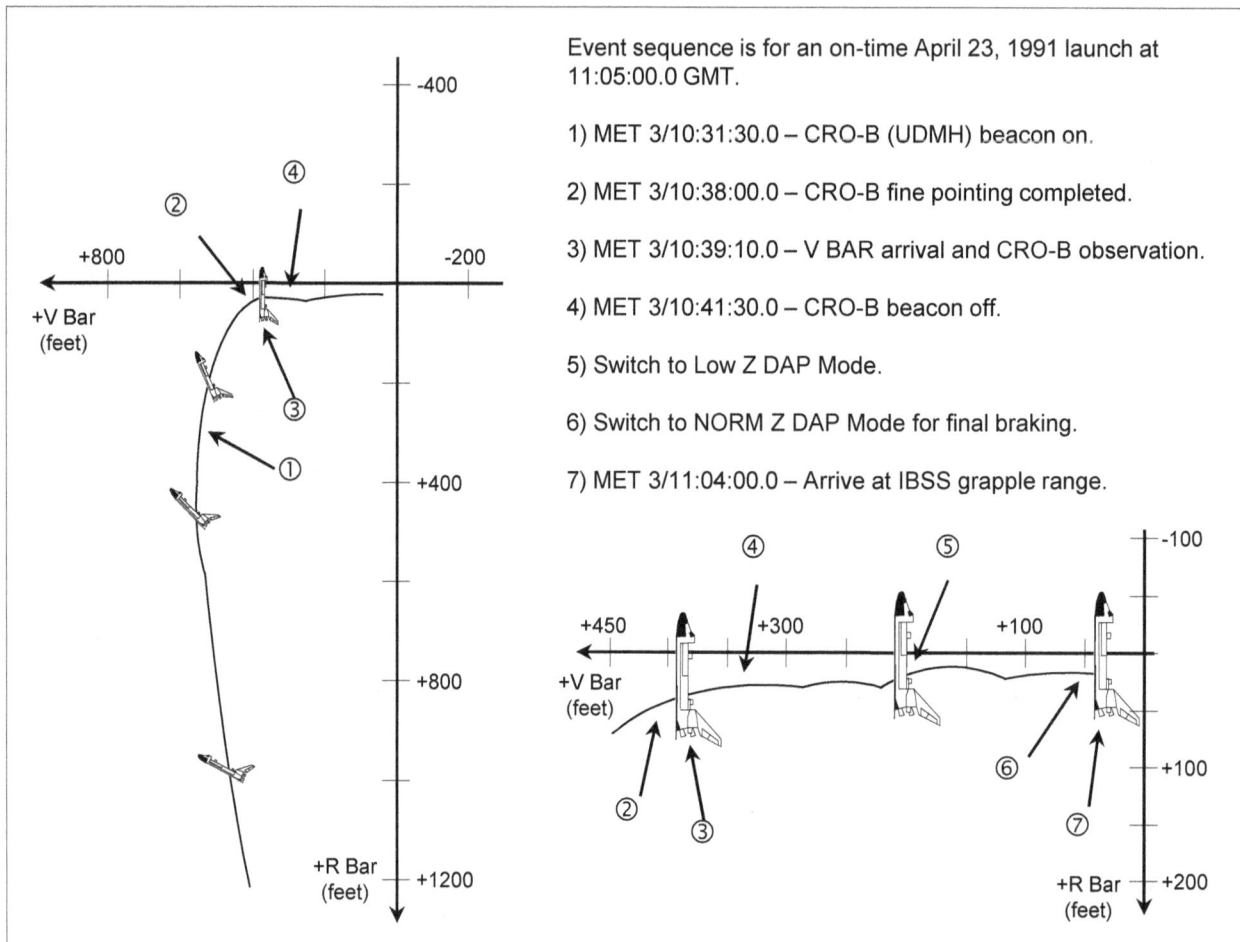

Event sequence is for an on-time April 23, 1991 launch at 11:05:00.0 GMT.

1) MET 3/10:31:30.0 – CRO-B (UDMH) beacon on.

2) MET 3/10:38:00.0 – CRO-B fine pointing completed.

3) MET 3/10:39:10.0 – V BAR arrival and CRO-B observation.

4) MET 3/10:41:30.0 – CRO-B beacon off.

5) Switch to Low Z DAP Mode.

6) Switch to NORM Z DAP Mode for final braking.

7) MET 3/11:04:00.0 – Arrive at IBSS grapple range.

Figure 23.16 IBSS retrieval proximity operations from MC4 + 2 minutes to arrival at grapple range.

A standard set of four Mid-course Correction (MC1, MC2, MC3, and MC4) burns followed Ti (Figure 23.14). The proximity operations phase for SPAS-II retrieval used standard manual piloting techniques and a plus V Bar approach (Figure 23.16) culminating in a grapple of SPAS-II by the RMS on orbit 56. The CRO-B (UDMH) sub-satellite chemical release was also to be observed on this orbit while the orbiter conducted proximity operations leading to SPAS-II retrieval.

For CIV activities no dedicated station-keeping burns were planned but a slow relative motion opening rate was established to support the experiment.

Rendezvous Radar Fail Profile

A mission plan and crew and ground procedures were developed for the rendezvous radar fail case. It was determined that shuttle star tracker angle measurements could support relative navigation during the SPAS-II deployed phase. The out-of-plane star tracker technique was proven on the earlier STS-37 mission of Atlantis (April 1991) using the Gamma Ray Observatory (GRO) as a target spacecraft.[3, 4, 5]

The entire radar fail profile could be supported by ground tracking. Expected trajectory dispersions were within acceptable limits and the radar fail profile was within the SPAS-II PI communications link margin of 20 km (10.8 nm).[3]

Radar Fail Before SPAS-II Deploy – If the rendezvous radar failed before SPAS-II deploy the crew would proceed with the SPAS-II deploy and orbiter separation to the far field (Figure 23.17). A 15 foot/second out-of-plane component would be added to the NFF0 burn. After NFF0 execution a star tracker pass would be performed that took advantage of the out-of-plane relative motion set up by NFF0. Ten minutes after the end of the star tracker pass and before the V Bar crossing a Mission Control targeted trajectory correction burn would be performed. This would be followed by the on-board targeted TFF1 and MCFF1 burns that were designed to return the orbiter to the V Bar.[5]

The NFF1 burn at V Bar arrival would null the out-of-plane motion. A plume burn (OP1) would then be performed. However, the NOP1 burn and fast flip attitude maneuver (the Malarkey Milkshake) would not be performed. NOP1 was to be followed by a star tracker pass. Half a revolution after OP1 a Mission Control targeted phasing burn would be executed to place the orbiter at the 8 nm Ti burn point in one revolution. Mission Control coarse station-keeping burns would be executed as required.[3,5]

If on-board relative navigation performance with only

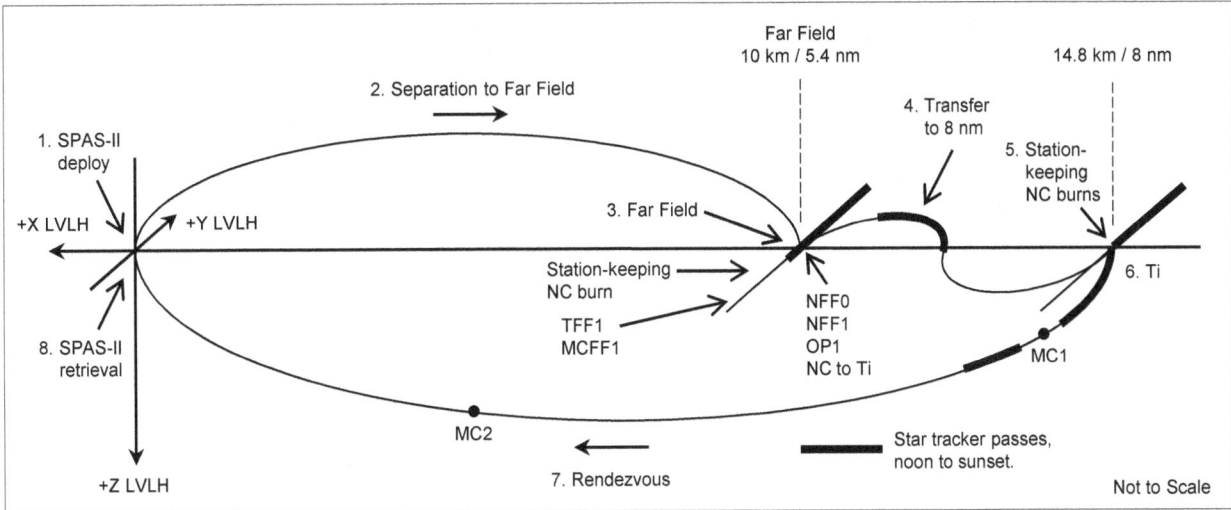

Figure 23.17 Relative motion during IBSS detached operations for the rendezvous radar fail before SPAS-II deploy case.

star tracker data was good enough the orbiter could phase to the 5.4 nm (10 km) far field point and perform the OP2 and OP3 plume observation burns. Additional far field plume burns could be performed in place of the near field plume burns. At the appropriate time the orbiter would phase to the 8 nm Ti burn point to initiate the rendezvous with SPAS-II.[5]

Radar Fail During the Far or Near Field – If the rendezvous radar failed at the near or far field points a 15 foot/second out-of-plane burn would be performed. Mission Control targeted phasing burns would be executed to place the orbiter at the 8 nm Ti burn point. Out-of-plane star tracker passes would be performed.[5]

Radar Fail After Initiation of the Transfer to the Near Field – If radar fail occurred after the MC1NF and MC2NF burns had been successfully targeted and executed the orbiter would proceed to the V Bar at the near field. The OP4 plume observation burn would be performed. Ten minutes after OP4 a 2 foot/second posigrade phasing burn would be performed to phase the orbiter to the 8 nm Ti burn point. No near field null burns or fast flip attitude maneuvers would be performed. If radar fail occurred before MC1NF or a good MC2NF burn could not be targeted the crew would perform a 5 foot/second out-of-plane burn no later than ten minutes after the scheduled MC2NF burn time. Five minutes later a 5 foot/second posigrade burn would be performed. The orbiter would phase to the 8 nm Ti burn point.[5]

Radar Fail Before or During the Rendezvous – Radar fail procedures for the rendezvous were the same as for a standard shuttle radar fail rendezvous. Star tracker measurements would be taken from after Ti until orbital sunset between the MC1 and MC2 burns. The MC3 and MC4 burns would not be performed. The crew would perform trajectory control using the target position in the Crew Optical Alignment Sight (COAS) and a chart to determine range rate corrections.[4]

CRO Deployments and Observations

The CRO sub-satellites were to be deployed one at a time to meet chemical release viewing and commanding requirements by assets located at Vandenberg Air Force Base, California. The CRO-C and CRO-B observations were to be performed during the SPAS-II detached phase. The CRO-A observation was to be performed after SPAS-II retrieval.

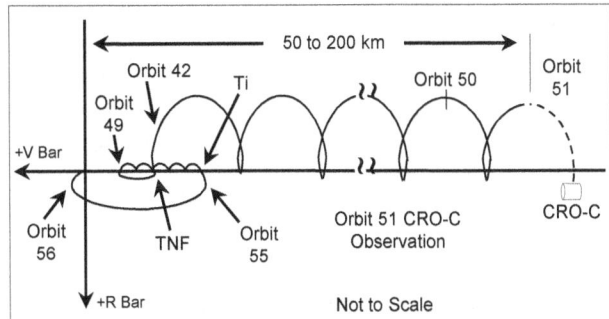

Figure 23.18 CRO-C relative motion during IBSS detached phase.

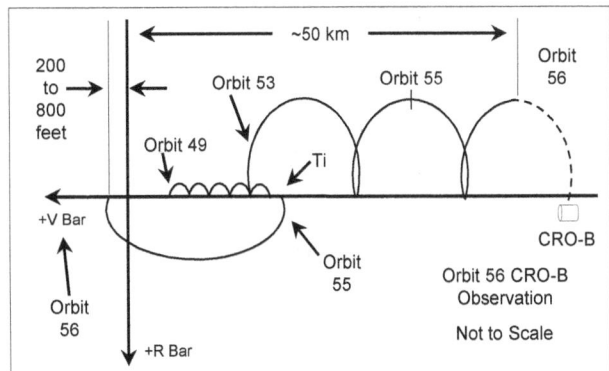

Figure 23.19 CRO-B relative motion during IBSS detached phase.

Nominal CRO sub-satellite deploys were scheduled on revolution 42 for CRO-C, revolution 53 for CRO-B, and revolution 79 for CRO-A. Nominal observations were planned for revolution 51 for CRO-C, revolution 56 for CRO-B, and revolution 88 for CRO-A. Figure 23.18 depicts planned CRO-C relative motion and Figure 23.19 depicts planned CRO-B relative motion during the SPAS-II deployed phase.

A SPAS-II pointing plan was developed for each chemical release observation. For CRO-B (UDMH) a television camera on SPAS-II was used to perform fine pointing. The television signal link margin was weak and uncertain when *Discovery* was near 180 degrees from the SPAS-II/IBSS sensor pointing direction. The exact CRO-B observation time during the final approach of *Discovery* to SPAS-II (Figure 23.16) was a function of orbital lighting during retrieval and the MC-2 Time of Ignition (TIG) slip. The nominal retrieval time could be shifted by 10 minutes to accommodate the CRO-B observation. An earlier manual phase shifted the CRO-B observation time closer to the +V Bar. A later manual phase shifted the CRO-B observation time in the direction of MC-4 (Figure 23.16). SPAS-II could be rotated by 180 degrees if *Discovery* was to be above the +V Bar during fine pointing. The observation was to be attempted even if fine pointing could not be performed. Adequate television link margin was assumed to exist between MC-4 and the +V Bar.

CRO sub-satellites required ground tracking until it was determined that they no longer presented a re-contact hazard for *Discovery*. Two burns were planned to avoid contact with the empty CRO sub-satellites and to meet AFP-675 requirements. These were a height adjustment burn (HA-1) on orbit 57 and a 140 nm circularization (CIRC-1) burn on orbit 58.

MPEC Deploy

MPEC was to be deployed retrograde on orbit 127, near the end of the mission. Deployment was scheduled for flight day 8 but could have been deployed earlier in the flight due to a contingency. MPEC was to be ejected

from the GAS canister with an estimated velocity of 2.7 feet/second. Planned MPEC deploy relative motion is illustrated in Figure 23.20. Figure 23.21 depicts MPEC relative motion with respect to *Discovery* over 24 hours.

Nominal End of Mission

Nominal end of mission was for a descending daylight landing at Edwards Air Force Base, California, on orbit 134. Mission duration was planned for 8 days, 7 hours, and 24 minutes.

Creative Use of Lambert Targeting

The SPAS-II detached phase involved more on-board Lambert targeted burns than any other shuttle rendezvous mission. Table 23.4 lists the Lambert targeting inputs for the far field, while Table 23.5 contains the Lambert targeting inputs for the near field burns and the rendezvous. A normal rendezvous had only six Lambert targeted burns, the six at the bottom of Table 23.5. For STS-39 there were a total of 37 Lambert targeted burns. Not all burns performed during the SPAS-II detached phase are listed in Tables 23.4 and 23.5 since some were targeted by Mission Control.[5]

The first column lists the acronym of the burn name. BASETIME was a reference time entered by the crew on the ORBIT TGT display (Figure 23.22). The reference time was used by the targeting software with the times in the T1 column to compute the Time of Ignition (TIG) for each burn. This provided the ability to easily shift the rendezvous sequence in time if mission activities had to be re-scheduled. The crew would simply enter a new BASETIME provided by Mission Control into the ORBIT TGT display.

LVLH X, Y, Z TARGET SET OFFSETS were the desired relative position components at the end of the transfer expressed in a target spacecraft centered LVLH curvilinear (or Local Vertical Curvilinear, or LVC) coordinate frame. This relative aimpoint was converted into inertial coordinates by the targeting software since the Lambert algorithm computed the required velocity in an inertial frame. TRANSFER TIME was the time in

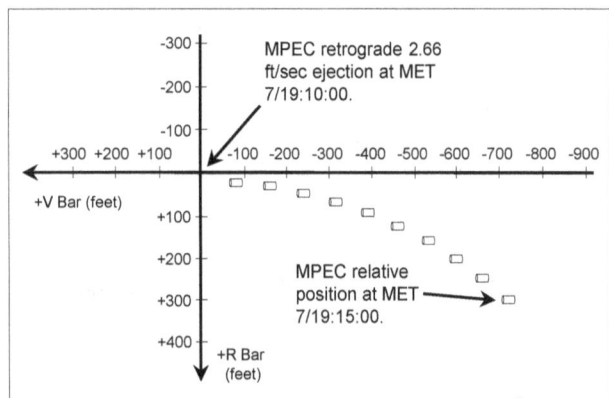

Figure 23.20 Planned MPEC deploy relative motion in an orbiter centered LVLH frame over 5 minutes.

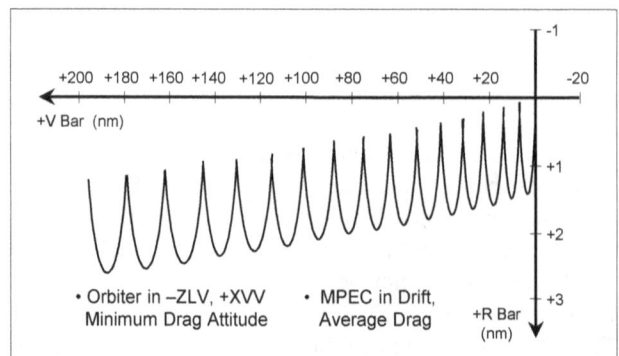

Figure 23.21 Planned MPEC relative motion over 24 hours, orbiter centered LVLH frame.

Table 23.4 Targeting Data for Far Field On-Board Targeted Lambert Burns

BURN	TARGET SET	LVLH X, Y, Z TARGET OFFSETS (FT)	T1 (MINUTES)	TRANSFER TIME MINUTES	TIG (MET)	ΔV TOTAL (FT/SEC)
BASETIME = 1/22:30:00						
SEP					1/21:00:00	2.2
MC1FF	20	-32808, 0, 0	-60.0	60.0	1/21:30:00	0.4
MC2FF	21	-32808, 0, 0	-20.0	20.0	1/22:10:00	0.0
BASETIME = 1/22:03:00						
NFF0 (FOOTBALL)	27	-32808, 0, 0	+27.0	22.4	1/22:30:00	2.0
BASETIME = 2/00:55:00						
TFF1	20	-32808, 0, 0	-60.0	60.0	1/23:55:00	3.0
MCFF1	21	-32808, 0, 0	-20.0	20.0	2/00:35:00	0.0
NFF1 (NSR)	22	-32808, 0, 0	0.0	19.0	2/00:55:00	2.6
BASETIME = 2/01:02:00						
OP1					2/01:02:00	16.8
NOP1	23	-32808, 0, 0	+5.0	22.4	2/01:07:00	15.8
MCP1	24	-32808, 0, 0	+17.0	10.4	2/01:19:00	0.0
VNP1 (FOOTBALL)	27	-32808, 0, 0	+27.0	22.4	2/01:29:00	5.8
BASETIME = 2/09:59:00						
TFF2	20	-32808, 0, 0	-60.0	60.0	2/08:59:00	0.6
MCFF2	21	-32808, 0, 0	-20.0	20.0	2/09:39:00	0.0
NFF2 (NSR)	22	-32808, 0, 0	0.0	19.0	2/09:59:00	1.2
BASETIME = 2/10:06:00						
OP2					2/10:06:00	16.8
NOP2	23	-32808, 0, 0	+5.0	22.4	2/10:11:00	15.8
MCP2	24	-32808, 0, 0	+17.0	10.4	2/10:23:00	0.0
VNP2 (NSR)	25	-32808, 0, 0	+27.0	19.0	2/10:33:00	5.7
BASETIME = 2/10:40:00						
OP3					2/10:40:00	16.8
NOP3	23	-32808, 0, 0	+5.0	22.4	2/10:45:00	15.6
MCP3	24	-32808, 0, 0	+17.0	10.4	2/10:57:00	0.0
VNP3 (NSR)	25	-32808, 0, 0	+27.0	19.0	2/11:07:00	5.8
RCSP					2/11:14:00	3.0
NRCS (FOOTBALL)	26	-32808, 0, 0	+34.0	22.4	2/11:14:30	3.1

minutes from TIG to the arrival at the desired relative position. Finally, each set of Lambert targeting inputs for a burn was identified by a TARGET SET number. This number was input by the crew into the ORBIT TGT display to designate which burn was to be targeted.[6] The TIGs and delta-velocity magnitudes listed were based on the nominal pre-flight mission design.

Observation of orbiter plume burns by the IBSS payload on SPAS-II required minimizing the orbiter relative motion with respect to SPAS-II. Ideally the orbiter would execute a burn to completely null relative motion on the minus V Bar, then perform a plume burn. However, due to navigation errors and cross coupling of

rotational RCS jet firings into translation the orbiter could not arrive exactly on the minus V Bar (Figure 23.23). To minimize relative motion in the presence of small altitude dispersions rendezvous designers chose to execute either a coelliptic (NSR, or Slow Rate) or a football burn before each plume burn (Figure 23.24). In Tables 23.4 and 23.5 the NSR and football burns are indicated in the BURN name column. Table 23.6 lists the burns designed to minimize line-of-sight rates and the corresponding plume burns. This list is based on the actual sequence of the burns during the flight.

The on-board coelliptic (NSR) burn targeting capability was deleted from the shuttle on-board software

Table 23.5 Targeting Data for Near Field On-Board Targeted Lambert Burns

BURN	TARGET SET	LVLH X, Y, Z TARGET OFFSETS (FT)	T1 (MINUTES)	TRANSFER TIME MINUTES	TIG (MET)	ΔV TOTAL (FT/SEC)
BASETIME = 2/19:04:00						
NCCNF	29	-32808, 0, 0	56.0	56.0	2/18:08:00	0.2
TNF	30	-7382, 0, 0	80.0	80.0	2/19:04:00	3.2
MC1NF	31	-7382, 0, 0	55.0	55.0	2/19:29:00	0.0
MC2NF	32	-7382, 0, 0	30.0	30.0	2/19:54:00	0.0
MC3NF	33	-7382, 0, 0	15.0	15.0	2/20:09:00	0.0
NNF (NSR)	34	-7382, 0, 0	19.0	19.0	2/20:24:00	1.7
BASETIME = 2/20:31:00						
OP4					2/20:31:00	16.8
NOP4	35	-7382, 0, 0	22.4	22.4	2/20:36:00	15.8
MCP4	36	-7382, 0, 0	10.4	10.4	2/20:48:00	0.0
VNP4 (NSR)	37	-7382, 0, 0	19.0	19.0	2/20:58:00	5.7
BASETIME = 2/21:05:00						
OP5					2/21:05:00	16.8
NOP5	38	-6562, 0, -400	22.4	22.4	2/21:10:00	15.7
MCP5	39	-6562, 0, -400	10.4	10.4	2/21:22:00	0.0
NSR5 (NSR)	40	-6562, 0, -400	19.0	19.0	2/21:32:00	6.1
BASETIME = 3/09:07:04						
NCC	9	-48600, 0, -1200	-55.9	55.9	3/08:11:10	0.9
Ti	10	0, 0, 0	0.0	79.6	3/09:07:04	3.8
MC1	11	0, 0, 0	19.3	60.3	3/09:26:22	0.0
MC2	12	0, 0, 0	47.1	32.5	3/09:54:10	0.1
MC3	13	0, 0, 0	10.0	22.5	3/10:04:10	0.0
MC4	14	0, 0, 0	20.0	12.5	3/10:14:10	0.0

requirements in the spring of 1978.[7] The on-board targeted phase stable orbit profile adopted in April of 1983 did not use NSR burns, but NSRs were computed by Mission Control and executed during the ground targeted phase. For STS-39 there was a desire to compute NSRs on-board to provide the crew with more autonomy, less reliance on communications with Mission Control, and permit timely burn targeting using on-board relative navigation state vectors during a phase of flight with a complex crew timeline.

To avoid modifying the on-board software to target a NSR burn a procedure was developed to use the Clohessy-Wiltshire equations to modify the trajectory constraints input to the Lambert targeting algorithm. These modified constraints resulted in a delta-velocity that provided a coelliptic transfer. However, this did not involve use of the onboard Clohessy-Wiltshire burn targeting algorithm, which was never used during a shuttle mission.[8]

The initial and final conditions for the coelliptic transfer in the LVLH reference frame are given below (see also Figure 23.24). To simplify the crew procedure for modifying the Lambert targeting constraints the final

down-track position x_t was set to the initial down-track position x_0 plus twice the initial LVLH altitude z_0.[8]

$$x_t = x_0 + 2z_0 \qquad z_t = z_0$$
$$y_t = 0 \qquad \dot{z}_t = 0$$
$$\dot{y}_t = 0$$

The Clohessy-Wiltshire equation for altitude rate (z dot) was used to solve for the desired down-track velocity (x_0 dot) at the start of the transfer.[9]

$$\dot{z}_t = [3\omega \sin \omega t]z_0 - [2\sin \omega t]\dot{x}_0$$
$$+ [\cos \omega t]\dot{z}_0$$
$$\dot{x}_t = \frac{3}{2}\omega z_0$$

An additional expression for the final down-track position x_f was needed to solve for the transfer time. This was obtained using the Clohessy-Wiltshire equation for down-track.[9]

```
2011 / 034 /          ORBIT  TGT          1  002 / 10 : 09 : 40
                                             000 / 00 : 06 : 20
   MNVR      TIG          ΔVX      ΔVY      ΔVZ        ΔVT
   22      2 / 10 : 16 : 00    +  1.7    + 0.2    - 0.3    +  1.8
                                    PRED  MATCH =           0

     INPUTS                                   CONTROLS
     1  TGT  NO                  22          T2  TO  T1     25
     2  T1  TIG        2 / 10 : 16 : 00       LOAD           26
     6     EL                  0.00          COMPUTE  T1  27
     7     ΔX / DNRNG    [-]  33.61          COMPUTE  T2  28
     8     ΔY           [-]   0.10
     9     ΔZ / ΔH      [-]   0.69
    10     ΔẊ           [-]   2.34
    11     ΔẎ           [-]   0.15
    12     ΔŻ           [+]   1.94
    13  T2  TIG        2 / 10 : 35 : 00      ┌──────────────────────┐
    17     ΔT           [+]  19.0            │ ORBITER    STATE     │
    18     ΔX           [-]  32.81           │ 307 / 18 : 24 : 32.514│
    19     ΔY           [ ]   0.00           │    X  + 13 811.111   │
    20     ΔZ           [ ]   0.00           │    Y  + 16 780.766   │
    21  BASE  TIME     2 / 10 : 16 : 00      │    Z  -  1 127.418   │
                                             │   VX  -  9.936051    │
                                             │   VY  +  9.619350    │
                                             │   VZ  + 21.345692    │
     ITEM  27  EXEC                          └──────────────────────┘
```

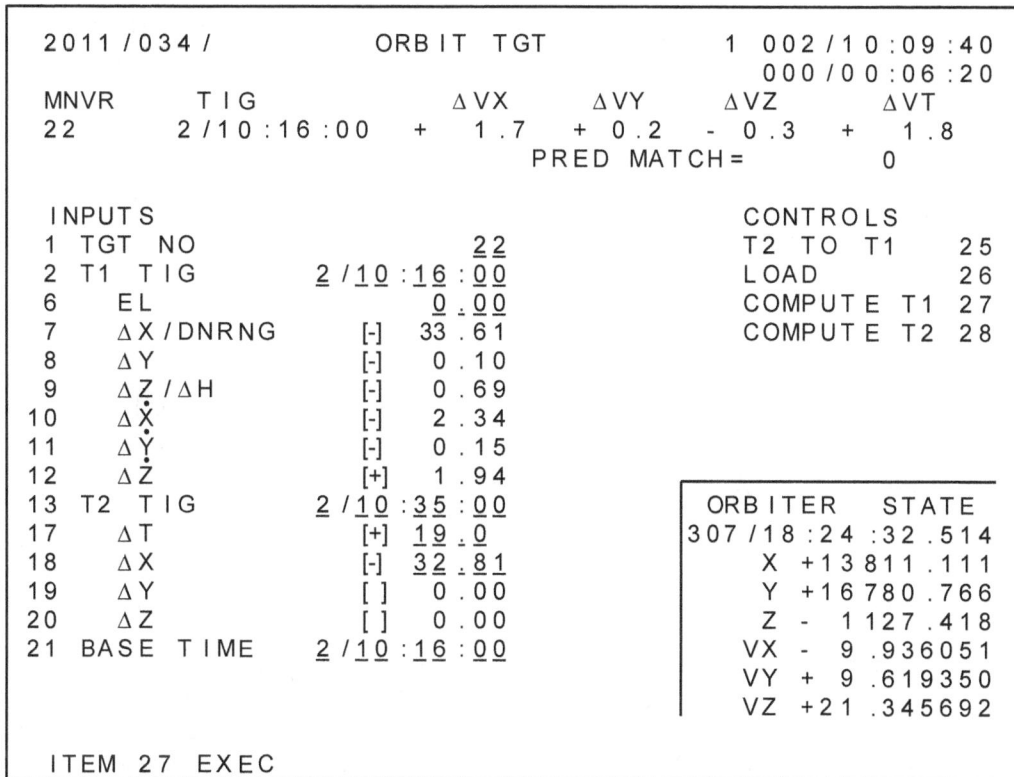

Figure 23.22 Results of preliminary NFF2 burn Lambert targeting.

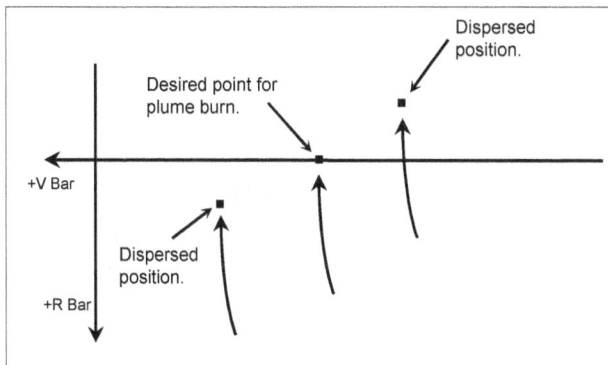

Figure 23.23 The desired position for performing plume burns was on the minus V Bar where relative motion could be completely nulled and long-term station-keeping established. However, slight trajectory dispersions would prevent nulling relative motion for any length of time.

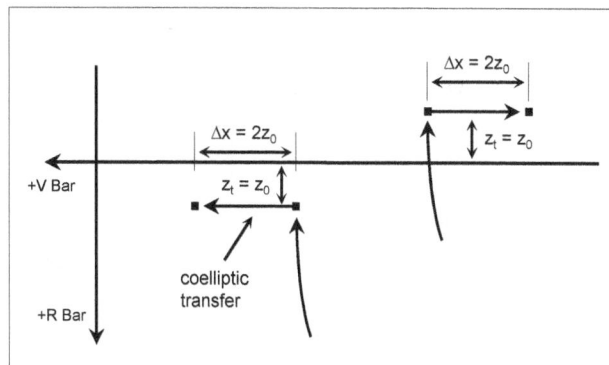

Figure 23.24 Since nulling relative motion on the minus V Bar before plume burns was not practical, coelliptic transfers were targeted on-board to minimize relative motion. The predicted miss distance z_0 was used to modify Lambert targeting constraints and obtain a coelliptic burn delta-velocity.

Table 23.6 Null Burns and Corresponding Plume Burns

Null Burn	Burn Type	Plume Burn
NFF2	NSR	OP1
VNP1	football	OP3
VNP3	NSR	OP2
VNP2	NSR	RCSP
NNF	NSR	OP4
VNP4	NSR	OP5

$$x_t = x_0 + 6[\omega t - \sin \omega t]z_0$$
$$+[\frac{4}{\omega}\sin \omega t - 3t]\dot{x}_0$$
$$+[\frac{2}{\omega}(1-\cos \omega t)]\dot{z}_0$$
$$x_t = x_0 + \frac{3}{2}\omega t z_0$$

The resulting expression and the pre-defined equation for final down-track position were then equated to solve for the transfer time.[9]

$$x_0 + 2z_0 = x_0 + \frac{3}{2}\omega t z_0$$
$$t = \frac{4}{3}\omega$$

The value for the transfer time t was ILOADed into the Lambert burn target sets for the V Bar Null (VBNx) burns. The crew performed an initial burn computation to obtain the predicted values of relative down-track position x and altitude z at the burn time (items 7 and 9 in Figure 23.22). These values were then used by the crew to modify the LVLH down-track and altitude targets input to Lambert targeting (items 18 and 20 in Figure 23.25). The resulting delta-velocity from Lambert targeting resulted in a coelliptic transfer. The equations used by

the crew in terms of the Orbit Targeting (ORBIT TGT) display parameters are shown below. The slashes do not represent division but are part of the parameter names on the display (Figures 23.22 and 23.25).[4]

$$\Delta X = \Delta X / DNRNG + 2\Delta Z / \Delta H$$
$$\Delta Z = \Delta Z / \Delta H$$

The Lambert targeting software also did not have the ability to target burns that established football (matched period) relative motion. A technique similar in concept to the NSR modification of the Lambert position targets was used by the crew. An initial targeting was performed to obtain the predicted LVLH altitude z (item 9) of the burn. The down-track (V Bar) position target x (item 18) was then modified by twice the value of the predicted altitude at TIG (Figure 23.26). Note that the x values at the far field and near field were negative, while the altitude values could be positive or negative. The transfer time was one quarter of a revolution. The slashes in the below equation used by the crew do not represent division but are part of the parameter names on the display (Figures 23.22 and 23.25). The altitude target x was not adjusted.[4]

$$\Delta X = \Delta X / DNRNG + 2\Delta Z / \Delta H$$

```
2011/034/           ORBIT  TGT           1 002/10:14:25
                                            000/00:01:35
MNVR      TIG          ΔVX     ΔVY    ΔVZ       ΔVT
22      2/10:16:00   +  1.1  + 0.2  - 2.2   +   2.4
                         PRED MATCH =          0

INPUTS                                CONTROLS
1 TGT NO              22              T2 TO T1      25
2 T1 TIG      2/10:16:00             LOAD          26
6   EL              0.00             COMPUTE T1    27
7   ΔX/DNRNG   [-]  33.54            COMPUTE T2    28
8   ΔY         [-]   0.10
9   ΔZ/ΔH      [-]   0.62
10  ΔX         [-]   2.20
11  ΔY         [-]   0.14
12  ΔZ         [+]   1.97
13 T2 TIG     2/10:35:00          ORBITER   STATE
17   ΔT        [+]  19.0          307/18:29:16.674
18   ΔX        [-]  34.99            X +10283.681
19   ΔY        []    0.00            Y +18546.065
20   ΔZ        [-]   0.69            Z + 4888.706
21 BASE TIME  2/10:16:00            VX - 14.661531
                                    VY +  2.691486
                                    VZ +20.605848

  ITEM 27 EXEC
```

Figure 23.25 Results of final NFF2 burn Lambert targeting with adjusted aim point targets (items 18 and 20).

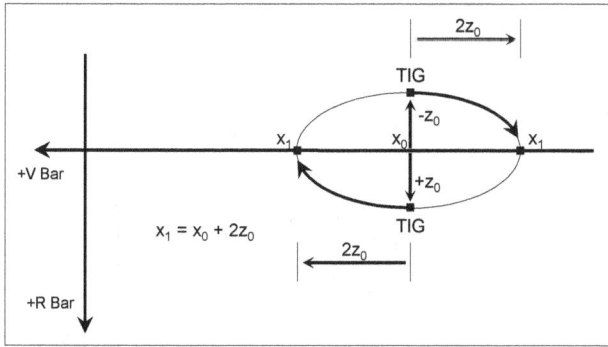

Figure 23.26 The crew modified the LVLH x_1 (V Bar) component of the Lambert aimpoint using the predicted altitude z_0 and downtrack x_0 position of the burn (TIG) from an initial burn targeting. This modified aimpoint along with a one quarter revolution transfer time resulted in a football relative trajectory.

Plume Sequence and On-Board Burn Targeting Example

Figures 23.22, 23.25, and 23.27 through 23.31 are crew displays depicting key events during a simulation of the far field OMS Plume 2 (OP2) burn sequence. The displays were generated during Rockwell Space Operations Company Level 8 Flight Software testing for STS-39.[10] Figure 23.22 contains the results of the initial computation of the Null Far Field 2 (NFF2) burn. The trajectory was targeted to place the orbiter at the far field point on the minus V Bar. However, results of the NFF2 computation indicated an altitude miss of -690 feet at the NFF2 burn time (see item 9 on the display). Figure 23.25 presents results of the final NFF2 computation using X (item 18) and Z (item 20) Lambert targeting constraints modified based on the predicted altitude miss displayed in Figure 23.22. This burn was then executed to establish a coelliptic trajectory and minimize relative motion.

Figure 23.27 is the Maneuver Execute (MNVR EXEC) display for the next burn, OMS Plume Observation 2 (OP2). This burn was not Lambert targeted but used a pre-defined delta-velocity of minus 16.8 feet/second out of plane (item 20 in the figure). MNVR EXEC was used by the crew for all OMS and translational primary RCS burns but only the MNVR EXEC display for the OP2 burn is shown.

Figure 23.28 is the final Lambert targeting computation for the next burn, Null OMS Plume Observation 2 (NOP2). This burn targeted the orbiter to return to the far field point on the minus V Bar. Figure 23.29 presents Lambert burn targeting results for the Out-of-Plane Mid-Course From Plume 2 (MCP2) burn. This burn served as a mid-course correction to fine tune the orbiter arrival at the desired far field minus V Bar point.

Figures 23.30 and 23.31 present preliminary and final Lambert targeting results for the V Bar Null 2 (VBN2) burn. VBN2 served the same purpose as NFF2, set-up a coelliptic trajectory to minimize relative motion. As with NFF2 the predicted altitude at the VBN2 point (item 9 in

```
2021/   /        ORBIT MNVR EXEC          1 002/10:22:56
OMS BOTH 1                                  000/00:00:04
        L 2        BURN ATT
        R 3*        24 R 199             ΔVTOT       16.8
RCS SEL  4          25 P 309             TGO         0:20
5 TV ROLL   90      26 Y 313
TRIM LOAD           MNVR 27   AUTO       VGO X +    15.92
 6 P  [+]1.1                                 Y -     3.32
 7 LY [+]4.9        REI                      Z +     4.24
 8 RY [-]4.9        TTC     :
9 WT 229929              GMBL                HA       HP
10 TIG                  L      R         TGT 137    +133
  2/10:23:00.0      P +1.0   +1.0        CUR 137    +133
TGT PEG 4           Y +4.8   -4.8
 14 C1
 15 C2   [] .       PRI 28*   29*      35 ABORT TGT
 16 HT       .      SEC 30    31
 17 θT       .      OFF 32    33          FWD RCS
 18 PRPLT []                                 ARM   36
TGT PEG 7           GMBL CK   34             DUMP  37
 19 ΔVX []   0.0                             OFF   38*
 20 ΔVY [-] 16.8    EXT ΔV                SURF DRIVE
 21 ΔVZ []   0.0                             ON    39
LOAD 22/TIMER 23                             OFF   40*

EXEC
```

Figure 23.27 Maneuver Execute display 4 seconds before the OP2 burn. Note 16.8 foot/second out-of-plane delta-velocity.

```
2011 / 034 /          ORBIT  TGT           1  002 / 10 :24 :22
                                              000 / 00 :03 :38
MNVR       TIG            ΔVX      ΔVY     ΔVZ        ΔVT
23        2 /10 :28 :00    +   0.6   + 15.8   + 3.8    + 16.3
                              PRED MATCH =           0


  INPUTS                                    CONTROLS
  1 TGT NO                    23            T2 TO T1      25
  2 T1  TIG        2 /10 :28 :00            LOAD          26
  6    EL                    0 .00          COMPUTE T1  27
  7    ΔX / DNRNG   [-]    34 .49           COMPUTE T2  28
  8    ΔY          [-]     4 .83
  9    ΔZ / ΔH     [-]     1 .20
  10   ΔẊ          [-]     1 .92
  11   ΔẎ          [-]    15 .84
  12   ΔŻ          [-]     0 .94
  13 T2  TIG       2 /10 :50 :23          ┌─────────────────────
  17    ΔT         [+]    22 .4           │ ORBITER    STATE
  18    ΔX         [-]    32 .81          │ 307 /18 :39 :15 .714
  19    ΔY         [ ]     0 .00          │   X  -    218 .208
  20    ΔZ         [ ]     0 .00          │   Y + 15 666 .589
  21 BASE TIME     2 /10 :23 :00          │   Z + 15 093 .464
                                          │   VX - 18 .946661
                                          │   VY - 11 .912036
                                          │   VZ + 12 .066660

  ITEM 27 EXEC
```

Figure 23.28 Results of final NOP2 burn Lambert targeting.

```
2011 / 034 /          ORBIT  TGT           1  002 / 10 :38 :51
                                              000 / 00 :01 :32
MNVR       TIG            ΔVX      ΔVY     ΔVZ        ΔVT
24        2 /10 :40 :23    -   0.5   -  0.2   + 0.3    +  0.7
                              PRED MATCH =           0


  INPUTS                                    CONTROLS
  1 TGT NO                    24            T2 TO T1      25
  2 T1  TIG        2 /10 :40 :23            LOAD          26
  6    EL                    0 .00          COMPUTE T1  27
  7    ΔX / DNRNG   [-]    33 .76           COMPUTE T2  28
  8    ΔY          [-]     3 .09
  9    ΔZ / ΔH     [+]     0 .18
  10   ΔẊ          [+]     2 .14
  11   ΔẎ          [+]     4 .39
  12   ΔŻ          [+]     0 .32
  13 T2  TIG       2 /10 :50 :23          ┌─────────────────────
  17    ΔT         [+]    10 .0           │ ORBITER    STATE
  18    ΔX         [-]    32 .81          │ 307 /18 :53 :43 .554
  19    ΔY         [ ]     0 .00          │   X - 13 890 .724
  20    ΔZ         [ ]     0 .00          │   Y -    366 .678
  21 BASE TIME     2 /10 :23 :00          │   Z + 16 722 .023
                                          │   VX -  9 .810504
                                          │   VY - 21 .822746
                                          │   VZ -  8 .639466

  ITEM 27 EXEC
```

Figure 23.29 Results of final MCP2 burn Lambert targeting.

```
2011/034/            ORBIT TGT            1  002/10:43:40
                                             000/00:06:20
MNVR      TIG          ΔVX      ΔVY      ΔVZ        ΔVT
25        2/10:50:00   - 1.4   - 5.5    + 0.8    +  5.8
                                PRED MATCH =         0


INPUTS                                       CONTROLS
1 TGT NO                    25               T2 TO T1      25
2 T1 TIG        2/10:50:00                   LOAD          26
6    EL                   0.00               COMPUTE T1    27
7    ΔX/DNRNG       [-] 32.64                COMPUTE T2    28
8    ΔY             [-]  0.14
9    ΔZ/ΔH          [+]  0.13
10   ΔẊ             [+]  1.49
11   ΔẎ             [+]  5.57
12   ΔŻ             [-]  1.17
13 T2 TIG       2/11:09:00
17   ΔT             [+] 19.0            ┌─────────────────────────
18   ΔX             [-] 32.81           │ ORBITER   STATE
19   ΔY             [ ]  0.00           │307/18:58:35.394
20   ΔZ             [ ]  0.00           │   X - 15897.940
21 BASE TIME    2/10:23:00              │   Y -  6591.408
                                        │   Z +13282.854
                                        │   VX -  3.810001
                                        │   VY - 20.420525
                                        │   VZ - 14.700988

   ITEM 27 EXEC
```

Figure 23.30 Results of preliminary VNP2 burn Lambert targeting.

```
2011/034/            ORBIT TGT            1  002/10:45:24
                                             000/00:04:36
MNVR      TIG          ΔVX      ΔVY      ΔVZ        ΔVT
25        2/10:50:00   - 1.2   - 5.6    + 1.4    +  5.9
                                PRED MATCH =         0


INPUTS                                       CONTROLS
1 TGT NO                    25               T2 TO T1      25
2 T1 TIG        2/10:50:00                   LOAD          26
6    EL                   0.00               COMPUTE T1    27
7    ΔX/DNRNG       [-] 32.70                COMPUTE T2    28
8    ΔY             [-]  0.13
9    ΔZ/ΔH          [+]  0.06
10   ΔẊ             [+]  1.36
11   ΔẎ             [+]  5.58
12   ΔŻ             [-]  1.24
13 T2 TIG       2/11:09:00              ┌─────────────────────────
17   ΔT             [+] 19.0            │ ORBITER   STATE
18   ΔX             [-] 32.38           │307/19: 0:15.234
19   ΔY             [ ]  0.00           │   X - 16169.205
20   ΔZ             [+]  0.13           │   Y -  8580.648
21 BASE TIME    2/10:23:00              │   Z +11727.738
                                        │   VX -  1.616647
                                        │   VY - 19.382700
                                        │   VZ - 16.416791

   ITEM 27 EXEC
```

Figure 23.31 Results of final VNP2 burn Lambert targeting with adjusted aim point targets (items 18 and 20).

Figure 23.30) represents a trajectory dispersion and was used to adjust the down-track and altitude Lambert targeting constraints in Figure 23.31 (items 18 and 20). Once the delta-velocity vector in Figure 23.31 was executed a coelliptic trajectory was established.

The Flight

STS-39 had a very ambitious flight plan. Some unexpected problems occurred, but all mission objectives during the SPAS-II deployed phase were accomplished. Tables 23.7 and 23.8 provide details for mission events. Figure 23.32 is the actual relative motion of the orbiter with respect to SPAS-II over an approximately 38 hour period.

Launch

The April 23, 1991 launch was scrubbed due to the failure of a transducer in Space Shuttle Main Engine (SSME) #3. *Discovery* launched from pad 39A on April 28 at 6:33:14 am CDT. The launch was delayed by 32 minutes 14 seconds due to a concern with the OPS-2 data recorder. Ascent and orbit insertion performance were nominal.[11]

Replanning of the SPAS-II Detached Phase

STS-39 mission activities were planned with exact times of execution to meet various experiment, SPAS-II, and shuttle orbiter requirements. The slip in launch time required examination of all activities to determine what changes needed to be made to the mission timeline.

The launch delay resulted in celestial observation times that were 2 minutes different than the times determined during pre-mission planning. This was judged to be a minor impact. The IBSS payload had a requirement that all plume observation burns occur on a whole GMT minute. However, liftoff did not occur on a whole GMT minute. The Mission Elapsed Time of all plume burns was moved 14 seconds early to meet the GMT whole minute requirement.[12]

The slip in launch time also impacted the Ti burn time. The time of the Ti burn ensured appropriate orbital lighting during proximity operations and grapple of SPAS-II. In addition, ground assets at Vandenberg Air Force Base required visibility of both the shuttle and CRO-B to support the CRO-B sub-satellite chemical release observation. This event occurred during proximity operations before SPAS-II grapple and retrieval. It was desired for the Vandenberg crossing time to occur before the orbiter reached the plus V Bar (Figure 23.16). This enabled a good communications link between the orbiter and SPAS-II to support SPAS-II fine pointing for CRO-B observation. The Ti burn time was recomputed to be one hour and 29 minutes before the Vandenberg crossing. This constraint would be applied to any subsequent re-planning of Ti, if it were required.[12]

On flight day 2 rendezvous flight controllers were informed that the CIRRIS payload was running out of cryogen. A proposal was made to delay SPAS-II deployment to provide more time for the CIRRIS activity to be completed before the cryogen was depleted. A three hour slip in SPAS-II deploy was judged to be of little impact. However, any slip in the deploy time greater than three hours necessitated a 24 hour delay in SPAS-II deploy. The 24 hour slip was agreed to by the payload customer. In addition, attitude and pointing personnel recommended that the deployment slip 9 minutes early (a slightly less than 24 hour slip) to keep the celestial target observation times close to the times in the current mission plan.[12]

New times for all burns in the rendezvous book were re-published that took into account the 24 hour deployment delay, an additional 9 minute early slip in deployment time, and the 14 second early slip in plume burn times. Orbit designers also re-created the sun angle constraint plots for the plume observations. Rendezvous personnel found the entire re-planning task to be straightforward.[12]

SPAS-II Deploy, Separation and Far Field Arrival

SPAS-II release was on time and the checkout of the SPAS-II attitude control system was successful. At the time of release the orbiter was in a 137 by 134 nm orbit. The separation burn occurred on time at a MET of 2/20:51 with a Mission Control confirmed delta-velocity magnitude of 2.21 feet/second. The orbiter maintained an inertial attitude hold for 10 minutes and then transitioned to a target track attitude.[12]

The MC1FF and MC2FF mid-course correction burns were small, indicating good trajectory performance. Arrival relative position errors at the NFF0 burn point, designed to occur at the desired far field point on the minus V Bar, were 40 feet in X LVLH and 60 feet in Z LVLH. The TFF1 burn was executed one hour and 25 minutes later to transfer the orbiter back to the far field point. The MCFF1 burn was not executed to conserve propellant.

After the NFF0 burn the SPAS-II attitude reference was not accurate enough to support IBSS sensor requirements. The first plume sequence was delayed to provide time to resolve the pointing problem. The NFF1 burn was ground targeted and executed by the crew. NFF1 was designed to return the orbiter to the far field point in one orbital revolution to protect for a possible one revolution delay in execution of the first plume sequence. Execution of the first plume sequence was delayed nine hours due to the time required to resolve the SPAS-II attitude pointing problem. Rendezvous personnel in Mission Control replanned the three far field

Table 23.7 STS-39 Events

Event	Actual ΔV	Planned ΔV	Actual Time	Planned Time	Remarks
Launch	---	---	118/11:33:14	118/11:05	Slipped 32 min 14 sec for OPS recorder malfunction
OMS-2	209.5	209.6	0/00:36:07	0/00/38:00	Ha=139.0, Hp = 137.2
SEP	2.2	2.2	2/20:51:00	1/21:00:00	Delayed 24 hours for CIRRIS OPS
MC1FF	0.2	0.2	2/21:21:00	1/21:30:00	Executed; w/o MC2FF ΔV = 2.5
MC2FF	0.7	0.5	2/22:01:00	1/22:10:00	
NFF0	2.0	2.0	2/22:21:00	1/22:30:00	
TFF1	0.8	0.9	2/23:45:46	1/23:55:00	
MCCF1	---	2.3	3/00:25:46	2/00:35:00	Not executed. OP1 delayed.
NFF1	0.4	0.6	3/00:45:46	2/00:55:00	Football continued for SPAS pointing error.
NCSK1	2.6	2.6	3/05:00:00	2/03:45:00	Biased for energy growth.
NCSK2	3.0	3.0	3/07:20:00	2/06:25:00	Biased for energy growth.
TFF2	2.5	2.6	3/08:49:46	2/08:59:00	
MCFF2	7.5	7.5	3/09:31:45	2/09:39:00	2 min late, large radial component.
NFF2	---	3.1	3/09:49:46	2/09:59:00	ZOE, not confirmed.
OP1	---	16.8	3/09:56:46	2/01:02:00	ZOE, not confirmed.
NOP1	---	15.9	3/10:01:46	2/01:07:00	ZOE, not confirmed.
MCP1	0.6	0.5	3/10:13:46	2/01:19:00	
VNP1	8.1	7.7	3/10:23:46	2/01:29:00	
OP3	17.0	17.0	3/10:30:46	2/10:40:00	
NOP3	15.7	16.1	3/10:35:46	2/10:45:00	
MCP3	0.6	0.5	3/10:47:46	2/10:57:00	
VNP3	5.9	6.1	3/10:57:46	2/11:07:00	
OP2	17.1	17.0	3/11:04:46	2/10:06:00	
NOP2	16.5	16.2	3/11:09:46	2/10:11:00	
MCP2	---	0.1	3/11:21:46	2/10:23:00	Not required, ΔV < 0.2.
VNP2	---	6.1	3/11:31:46	2/10:33:00	ZOE, not confirmed.
RCSP	---	3.0	3/11:38:46	2/11:14:00	ZOE, not confirmed.
NRCS	---	3.0	3/11:39:46	2/11:14:30	ZOE, not confirmed.
CRO-C	~4.0	3.7	3/13:55:20	2/13:57:00	Deploy spring ΔVs, tumbled.
NCSK3	0.8	0.8	3/14:29:00	2/14:19:00	Biased for energy growth.
NHSK4	---	0.3	3/16:45:00	2/15:04:00	Not executed.
NCCNF	1.9	1.4	3/17:58:46	2/18:08:00	
TNF	3.2	3.2	3/18:54:46	2/19:04:00	
MC1NF	1.0	1.1	3/19:19:46	2/19:29:00	Large radial component.
PN1	---	---	---	2/19:40:00	
MC2NF	1.4	1.5	3/19:44:46	2/19:54:00	Large radial component.
MC3NF	2.4	2.1	3/19:59:46	2/20:09:00	Large radial component.
NNF	5.1	4.7	3/20:14:46	2/20:24:00	Large radial component.
OP4	17.1	17.0	3/20:21:46	2/20:31:00	
NOP4	16.3	16.3	3/20:26:46	2/20:36:00	
MCP4	0.4	0.4	3/20:38:46	2/20:48:00	
VNP4	5.9	5.9	3/20:48:46	2/20:58:00	
OP5	17.1	17.0	3/20:55:46	2/21:05:00	
NOP5	16.6	16.3	3/21:00:46	2/21:10:00	
MCP5	0.2	0.4	3/21:12:46	2/21:22:00	
NSR5	5.9	6.0	3/21:22:46	2/21:32:00	

CIRRIS – Cryogenic Infrared Radiance Instrument for Shuttle, **CRO** – Chemical Release Observation, **Ha**– Height of apogee, **Hp** – Height of perigee, **min** – minute, **OPS** – Operations, **sec** – seconds, **SPAS** – Shuttle Pallet Satellite, **w/o** – without, **ZOE** – Zone of Exclusion, no communication with Mission Control, **ΔV** – Delta Velocity

Table 23.8 STS-39 Events (continued)

Event	Actual ΔV	Planned ΔV	Actual Time	Planned Time	Remarks
OPC/NC	8.2	8.2	4/01:32:46	3/01:42:00	1.7 ft/sec retrograde.
NCSK5	1.6	1.5	4/02:10:00		Unscheduled pre-mission, posigrade.
NOPC	8.4	8.1	4/03:46:56	3/03:56:00	1.0 ft/sec posigrade.
CRO-B	~3.7	3.5	4/06:30:15	3/06:46:00	Deploy spring ΔV's, tumbled.
NCC	2.5	2.3	4/08:03:52	3/08:11:10	2.4 ft/sec retrograde.
Ti	3.0	2.9	4/08:59:46	3/09:07:04	1 hour 29 min before CRO-B observation.
MC1	---	0.0	4/09:19:04	3/09:26:22	Not executed.
PN2	---	---	---	3/09:42:00	27 second late TIG slip.
MC2	2.1	2.1	4/09:47:19	3/09:54:10	
MC3	---	0.0	4/09:57:19	3/10:04:10	
MC4	0.3	0.3	4/10:07:19	3/10:14:10	
Manual Phase	---	---	---	3/10:16:10	
V Bar Approach	---	---	---	3/10:39:00	
Grapple Range	---	---	---	3/11:04:00	
Grapple	---	---	4/10:52:00	3/11:11:00	
HA1	9.2	9.0	4/23:05:35	3/12:30:00	Executed for loss of CRO-B tracking.
CIRC1	9.2	9.0	4/23:54:20	3/13:15:00	Executed for loss of CRO-B tracking.
CRO-A		3.3	5/00:36:46	4/21:15:00	Deploy spring ΔV's, tumbled.
HA2	6.2	6.0	5/11:40:00	5/11:30:00	Executed per pre-mission design.
CIRC2	6.1	6.0	5/12:24:52	5/12:15:00	Executed per pre-mission design.
AIS	3.6	3.6	7/04:20:00	---	Not planned pre-mission, burned OOP.
UVPI	1.2	1.3	7/10:53:29	---	Not planned pre-mission, orbit raising.
MPEC	---	2.49	7/20:45:00	---	Posigrade deploy ΔV.
SEP	2.9	3.0	7/21:21:21	7/21:10:00	Retrograde SEP
Deorbit	257.8	257.8	8/06:20:20	8/06:24	KSC selected, EDW no-go for winds.
Landing	---	---	8/07:22:22 (KSC)	8/07:24 (EDW)	Runway KSC15

CRO – Chemical Release Observation, **EDW** – Edwards Air Force Base, **ft/sec** – feet/second, **KSC** – Kennedy Space Center, **OOP** – Out Of Plane, **SEP** – Separation, **TIG** – Time of Ignition, ΔV – delta velocity,

OMS plume sequences and supplied the crew with new timeline pages.

Far Field Station-Keeping Challenges

NCSK burns were designed to be small primary RCS burns to control orbiter down-track (phasing) position relative to SPAS-II. The NCSK1 burn was delayed one revolution to support SPAS-II commanding to regain the attitude reference. The Mission Control targeting of the NCSK1 burn was designed to place the orbiter at a point 10 km (5.4 nm) behind SPAS-II on the minus V Bar for the TFF2 burn after a 3 hour 42 minute transfer. Before NCSK1 about half a revolution of ground tracking with radars was performed. Mission Control orbit determination indicated a significant increase in orbital energy of the shuttle. As a result the vent value used by Mission Control to account for un-modeled accelerations in the Mission Control software was increased. The computed NCSK1 burn solution was computed with ephemeris data that included the new vent value. In addition, the burn delta-velocity in-plane component was biased to account for translational effects from the maneuver to and from the NCSK1 burn attitude. [13,14]

After NCSK1 was performed the predicted trajectory was good and it was believed that a subsequent NCSK2 burn would not be required. However, later trajectory predictions using the latest on-board orbiter and SPAS-II state vectors by Mission Control indicated that the orbiter would approach to within half a nautical mile of SPAS-II. It was also predicted that the TFF2 burn would occur at a range of 3.5 nm behind SPAS-II rather than the desired 5.4 nm. A burn to correct the trajectory dispersion was required before the orbiter reached relative perigee since a posigrade burn could result in an even closer approach to SPAS-II.

The NCSK2 burn was performed 2 hours and 20 minutes after NCSK1 to correct for the trajectory dispersion and to place the orbiter at the desired 10 km (5.4 nm) point. NCSK2 reversed the overall orbiter closing rate on SPAS-II. However, since it was performed approximately 1.2 nm below the V Bar it

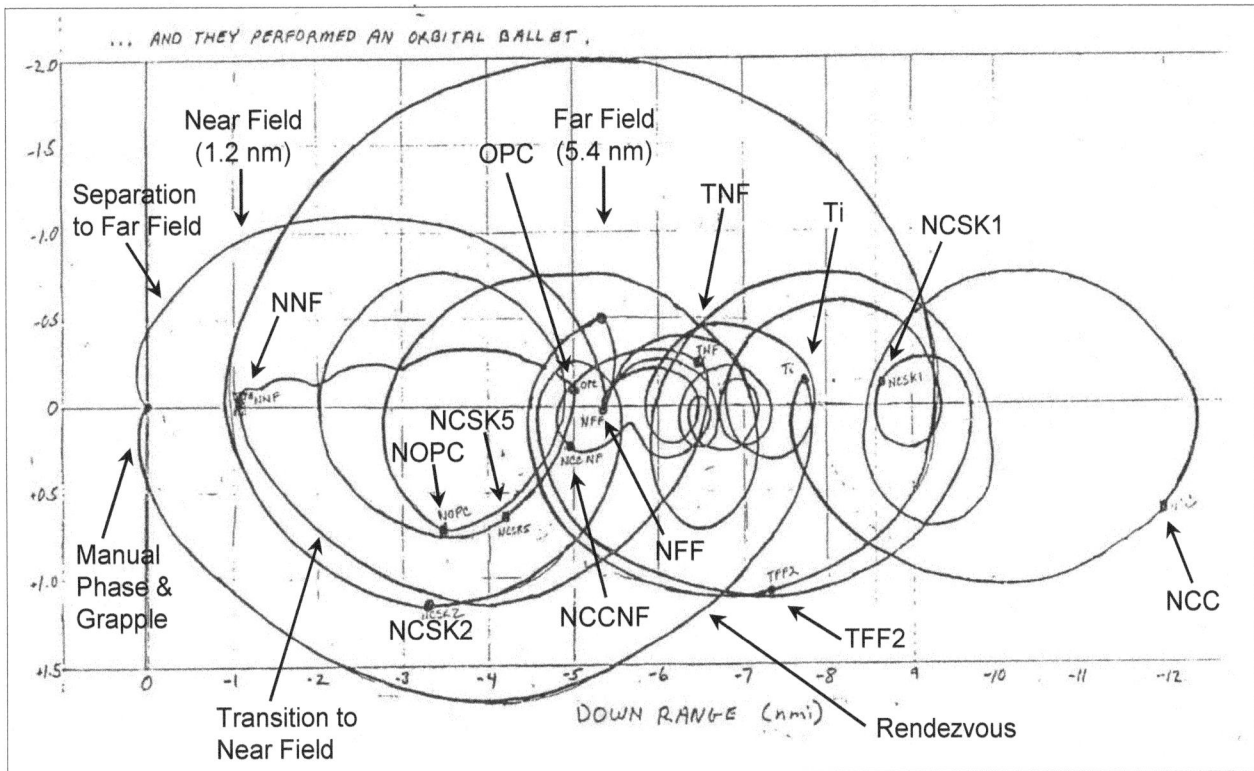

Figure 23.32 Sketch of actual STS-39 relative motion with some burns labeled.

resulted in a large looping trajectory. The orbiter approached to within one nautical mile of SPAS-II, then reached a relative apogee about 1.9 nm above the minus V Bar (Figure 23.32). The trajectory then took the orbiter to a minus V Bar crossing at about 9.1 nm. The subsequent on-board Lambert targeted TFF2 burn took out the opening rate created by NCSK2 and re-established the pre-NCSK2 closing rate.[12] After the TFF2 burn the vent in the Mission Control ephemeris was changed back to the original lower value.[14]

After the mission three possible causes of the trajectory dispersion were investigated. The biases added to NCSK1 and NCSK2 to account for translational effects from the maneuvers to and from the burn attitude were found to be accurate. SPAS-II attitude maneuvers did change the atmospheric drag of SPAS-II but the drag was relatively high after NCSK1. SPAS-II orbital energy did not grow during this time. Post flight analysis indicated that the increased orbiter vent force value used in the Mission Control ephemeris only applied to a short portion of the trajectory. The original lower value of the vent force more accurately represented un-modeled accelerations before and after that short period. Use of the higher vent force value when the actual vent force was lower was found to account for the trajectory dispersion after NCSK1. The granularity of ground tracking data made it difficult to determine vent forces over short periods of time.[14, 15]

Far Field Plume Burns

A problem with SPAS-II pointing delayed the first plume sequence at the far field. The NFF1 burn was targeted by Mission Control as a football to place the orbiter at the 10 km point in 2.5 hours. The OP1 plume sequence was delayed until the next set of plume sequences.[16]

Shortly after TFF2 the radar broke lock on SPAS-II and then re-acquired it. The crew inhibited radar measurements and Mission Control monitored the radar data. The data was good and measurement processing by relative navigation was resumed. However, noisy radar angle residuals were noted during the maneuver to the next plume burn attitude. The noisy angle measurements did not negatively impact the quality of the relative navigation solution.[12]

Before the intermediate on-board Lambert targeting solution for MCFF2, the radar broke lock on SPAS-II. The crew inhibited radar measurement processing. The radar then went through several loss of tracking and re-acquisition cycles. Measurement data during tracking looked good and the intermediate on-board MCFF2 burn solution compared favorably with the Mission Control Flight Dynamics Officer solution based on ground tracking. This gave rendezvous personnel confidence in the on-board relative navigation solution in spite of the radar break tracks. The radar also passed a self test. The

Instrumentation and Communications Officer (INCO) noted that the radar range switch was in the minimum or low power position. This could have explained the break locks around the time of the TFF2 burn. The rendezvous radar was switched to AUTO and immediately re-acquired SPAS-II. [12]

Due to the large size of the MCFF2 delta-velocity and since radar measurements had not been processed for a while the crew was asked to delay the MCFF2 burn and take additional radar data. During the maneuver to the burn attitude radar angle measurements were intentionally not processed. However, the radar broke lock again; indicating that the previous minimum power setting did not cause the previous loss of tracking incidents. [12]

Due to the busy crew timeline Mission Control told the crew to burn the intermediate MCFF2 burn solution and not compute a final burn solution. Normally on-board Lambert targeted burns were executed using Lambert cyclic guidance that closed the loop on the trajectory. However, after the intermediate burn targeting the crew selected the OMS engine on the MNVR EXEC display due to the large size of the burn. This action changed the burn to external delta velocity guidance, per software requirements. Since the crew had been instructed not to perform the final Lambert targeting to save time, MCFF2 was performed open loop with respect to the trajectory. Performing the final Lambert targeting would have reset the burn guidance to closed loop Lambert. [12]

At the end of the post MCFF2 transfer the error in the arrival point was -2,500 feet in X LVLH and -2,000 feet in Z LVLH. Factors that caused this position error were 1) executing the burn two minutes late with open-loop (with respect to the trajectory) external delta velocity guidance, 2) a burn attitude error of 8 degrees, and 3) the intermediate on-board burn solution could not have taken into account translational delta-velocity imparted by the subsequent maneuver to burn attitude (performing the final targeting would have accounted for the maneuver to burn attitude). Although both external delta velocity and Lambert cyclic guidance account for sensed IMU delta-velocity during the burn, use of Lambert cyclic guidance (closed loop with respect to the trajectory) would have improved relative trajectory performance under these conditions. [12]

The NFF2, OP1, and NOP1 burns were performed without communication with Mission Control due to gaps in communications coverage. The crew reported that the OP1 and NOP1 burns were successful, including SPAS-II pointing and data collection. In addition, a playback of telemetry data at this time showed that the previous minimum power setting of the radar had been selected after the radar loss of tracking incidents. This proved that the low power setting was not the cause of the intermittent loss of tracking. The OP3 plume sequence was also successful. The OP2 plume sequence trajectory

performance was good enough that the MCP2 trajectory correction burn was not performed. [12] The plume sequences went well and the crew reported that SPAS-II data collection during the plume burns was successful. The primary RCS (PRCS) plume burn and associated null burn (NRCS) were performed during a time when Mission Control was not in communication with the orbiter due to a gap in communications coverage.

CRO-C was successfully deployed after the far field plume sequences were complete.

In preparation for the orbiter transition to the near field the rendezvous team, Mission Control flight directors, and Mission Operations Directorate management met to review far field relative trajectory performance. The on-board Lambert targeted burns (transfer times less than one revolution) provided better trajectory control than ground targeted station-keeping burns designed to control relative motion over several revolutions. If the trend in predicted relative motion became undesirable a burn could be executed to correct the trend. The Mission Control team and the crew did not have to commit to the transfer to the near field until TNF burn execution. If trajectory control after TNF became an issue a breakout procedure could be executed by the crew.

The NCSK3 burn (phasing adjustment) was ground targeted and designed to place the orbiter at the far field point for the TNF burn. Subsequent far field station-keeping trajectory performance was good enough that the NHSK4 burn (altitude adjustment) was not performed.

SPAS-II was visible to the crew typically between orbital noon and orbital sunset. The attitude of SPAS-II was difficult to determine with binoculars at a range of 5 nm (the far field).

Transition to the Near Field

The NCCNF and TNF burns were performed to initiate the transfer to the near field point (Figure 23.12). The mid-course corrections burns had increasing radial (plus Z LVLH) delta-velocity components, indicating a force pushing the orbiter trajectory up and away from SPAS-II. The orbiter arrived close to the near field aim-point, 110 feet short in X LVLH and 310 feet high in Z LVLH. [12]

Near Field Plume Sequences

The OP4 and OP5 OMS plume sequences went smoothly with good trajectory performance. After the fifth plume observation sequence (at the near field) was complete the NSR5 burn was executed to start football (matched period) relative motion. However, due to orbital energy growth the football trajectory was transformed into a trajectory with an opening rate. Down-range displacement from SPAS-II increased with

Approved for public release via STI DAA
24483. See statement on title page.

188

each revolution. During this period the rendezvous radar went through several loss of lock and re-acquisition cycles.

The crew subsequently maneuvered to the CIV observation attitude. The orbiter was phasing away from SPAS-II and the observations were performed at ranges of between 12,000 feet and 15,000 feet. Orbiter range to SPAS-II was greater than desired for the CIV experiment but the payload customer did not want orbiter propellant expended to change the relative motion.[12]

Transition to Rendezvous and the CRO-B Deploy

The OPC/NC burn was executed at a range from SPAS-II of approximately 5 nm. The burn was targeted by the Mission Control Flight Dynamics Officer so that the orbiter would reach the 8 nm Ti burn point in 5 revolutions. In addition, the out-of-plane burn component was designed to place the orbiter 12 degrees out-of-plane in 1.25 revolutions to support the CRO-C observation. At the end of the burn there was a large -0.7 foot/second delta-velocity residual in the orbiter Z body axis. The residual was not trimmed to avoid waking sleeping crew members with forward primary RCS jet firings.

Post OPC/NC trajectory predictions indicated a closing rate with SPAS-II and a potential intercept in 2 revolutions. To avoid an intercept trajectory the unscheduled NCSK5 burn was targeted by Mission Control and executed by the crew. The unplanned NCSK5 burn was executed 40 minutes after OPC/NC to correct the trajectory and establish a relative motion football. The resulting relative motion was as desired. NOPC was executed one revolution later to null out-of-plane velocity and ensure a 3.5 revolution transfer to the 8 nm Ti point.[15] Mission Control asked the crew to trim the delta-velocity residuals on this burn to ensure good trajectory performance.[12]

Post flight analysis indicated that a -0.43 foot/second OPC/NC over-burn contributed to the undesired closing rate. However, this did not completely account for the all of the undesired relative motion. Post flight analysis also indicated a need for better target spacecraft attitude timeline knowledge and more insight into the exact nature of the SPAS-II helium vent. Another possible factor included uncertainty in orbiter vent force modeling.[15, 17]

CRO-B was scheduled for deployment approximately two hours after NOPC. The radar experienced several break locks and the orbiter relative navigation software Kalman filter rejected 6 range measurements before the deploy. Just before the CRO-B deploy the radar was unable to track SPAS-II. After CRO-B deploy the radar tracked CRO-B for short periods and lost lock.

Rendezvous With SPAS-II

Due to poor radar performance the Mission Control rendezvous team changed the relative sensor plan for the rendezvous with SPAS-II. The radar mode was changed from GPC Acquisition to GPC Designate which was not dependent on radar angle tracking. This mode of radar operation was used for most of the retrieval. However, use of GPC Designate resulted in no radar angle data being made available to the relative navigation filter. Star tracker angle measurements would be processed in parallel with radar range and range rate when orbital lighting permitted a star tracker pass. Before the NCC burn the radar was able to track SPAS-II in the GPC Designate mode. The NCC burn was successfully performed.[12]

After NCC Mission Control supplied the crew with a procedure to fix a problem with the payload bay camera angle digital data. This data would be used to support a back-up ranging technique during proximity operations in the event of a rendezvous radar failure. The recommended fix was to power cycle the cameras.[12]

After the Ti burn was completed to initiate the rendezvous with SPAS-II a star tracker pass was performed. The star tracker shutter remained closed since the Earth horizon was within 16 degrees of the star tracker bright object sensor (which was mounted on the star tracker light shade and pointed along the star tracker bore sight). The shutter later opened just before Mission Control was to ask the crew to open the shutter manually. Relative navigation performance during the star tracker pass was good. Small navigation errors at Ti and the processing of star tracker line-of-sight angle data instead of radar angles after Ti led to a small MC1 burn solution that was not executed.[18] This indicated good trajectory performance. The star tracker pass ended before orbital sunset and the radar lost lock and then re-acquired SPAS-II. The MC-3 and MC-4 burn solutions were small, but they were executed by the crew.[12]

Final Approach and Grapple of SPAS-II

Although the radar experienced periods of loss of tracking and there were periods of noisy state vectors during the SPAS-II detached phase there was confidence in radar quality during the rendezvous (Figures 23.14 and 23.16). Crew plotting of mid-course correction burn locations and relative motion during proximity operations on polar graph paper indicated a nominal approach trajectory.

During the rendezvous sub-tended angle ranging charts provided accurate backup range estimates (Figure 23.33). However, these range estimates were not continuously performed due to the nominal rendezvous trajectory. Ranging ruler overlays on the Closed Circuit Television (CCTV) camera screen were not useful during proximity operations (Figure 23.16) due to CCTV camera blooming caused by sunlight reflected by SPAS-II. Both automatic and manual control of the camera iris did not improve the image. Within a range of 200 feet image blooming was not a problem. This was likely due to the

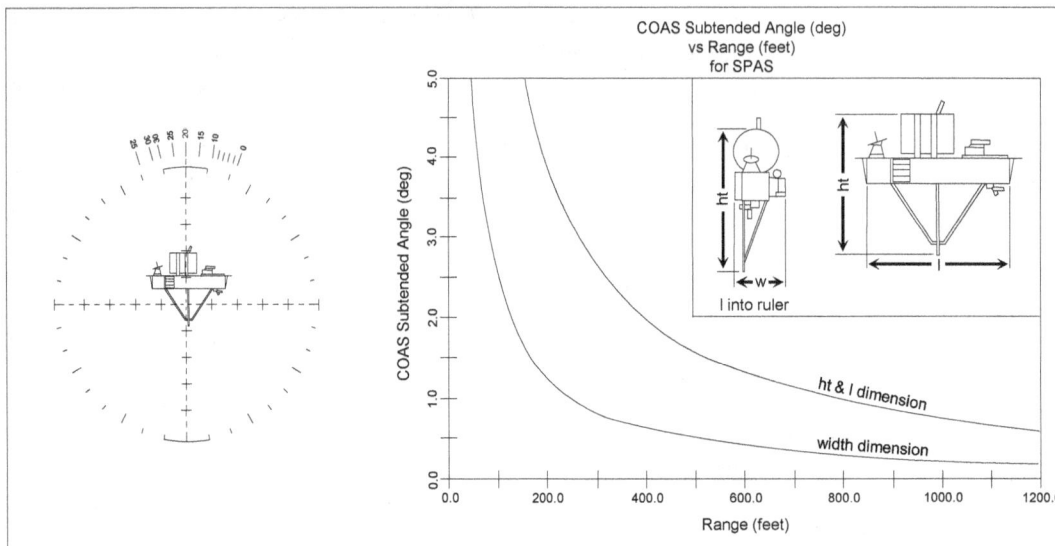

Figure 23.33 COAS (left) and subtended angles range chart (right) from STS-39 (April-May 1991).

SPAS-II being backlit by the sun at this point, rather than reflecting sunlight directly into the orbiter CCTV cameras. By a range of 200 feet out-the-window observations of the SPAS-II above the payload bay provided better situational awareness than the CCTV cameras. However, the proximity operations phase would have been very challenging had a rendezvous radar failure occurred along with the loss of backup ranging rulers due to CCTV camera blooming.*

The laptop computer Payload Bay (PLBAY) program was used occasionally during proximity operations to estimate range using crew entered CCTV payload bay camera angle data. PLBAY was a precursor of the later RPOP program. However, range estimates were not accurate outside of 250 feet, most likely due to inaccuracy in the camera angle encoders. Heavy crew workload inside a range of 250 feet prevented continual range determination using the PLBAY program.[12]

As the orbiter approached the manual phase (post MC-4, proximity operations) the radar frequently lost lock on SPAS-II. The crew changed the radar modes between GPC and AUTO TRACK several times in attempts to track SPAS-II. During tracking the range and range rate measurements were becoming increasingly noisy. However, trajectory performance during the manual piloting phase was excellent.[12]

During the manual phase the crew acquired the CRO-B beacon with the camera on SPAS-II. SPAS-II pointing was accomplished and the CRO-B chemical release observation was successful. The orbiter arrived on the plus V Bar at approximately the 200 foot point. The crew did not place the flight control system in the LOW Z (minimum plume impingement) mode since the propellant remaining was below the LOW Z red line. The radar broke lock inside of 200 feet and the crew transitioned the antenna to the communications mode. This permitted Mission Control to observe video of the final approach and grapple of SPAS-II by the RMS.[12]

During final approach the RMS grapple fixture on SPAS-II was visible by a range of 200 feet. SPAS-II was successfully grappled two minutes before orbital sunset. Had grapple been delayed the payload bay floodlights would have provided sufficient lighting for grapple during orbital night. The plus V Bar approach and grapple phase lasted about 15 minutes. SPAS-II was grappled with the RMS after approximately 38 hours of free flight. [12]

Post Grapple Activities

Difficulty in tracking the CRO-B and CRO-C sub-satellites after the chemical releases resulted in uncertainty about their orbits. While SPAS-II was berthed in the payload bay an orbital altitude adjust burn and a circularization burn was computed by Mission Control and performed by the crew to avoid potential contact with the empty CRO sub-satellites. The posigrade MPEC deployment was successful.

Return to Earth

Discovery landed on KSC runway 15 on May 6, 1991, after a flight of 8 days, 7 hours, 23 minutes, and 17 seconds.[11]

STS-39 Flight Experience Summary

This section provides a summary of the STS-39 SPAS-II detached phase.

---

* The later introduction of the Rendezvous and Proximity Operations Program (RPOP), Hand Held Lidar (HHL), and Trajectory Control Sensor (TCS, another lidar) improved crew situational awareness during proximity operations and provided better sensor redundancy.

SPAS-II deployed activities began with SPAS-II deploy on flight day 4 at a Mission Elapsed Time (MET) of 2/20:51 (days, hours, minutes, seconds) and ended with SPAS-II grapple on flight day 5 at MET 4/10:52. The three far field and two near field plume observation sequences were successfully completed. A total of 16 OMS burns and 41 RCS translational burns were performed. The OMS-2 (orbit insertion) and deorbit burns used both OMS engines. The remaining 14 were seven right OMS burns and seven left OMS burns. RMS performance throughout the flight was nominal.[11, 12, 13]

Requirements – Relative motion and rendezvous design was driven by constraints not normally associated with relative motion trajectories. These included dual shift crew operation, ground site visibility to deployed payloads, on-orbit lighting for experiments, ensuring an adequate communications link between the orbiter and SPAS-II, and other attached payload requirements.

Replanning of the SPAS-II Detached Phase – High cryogenic use by the CIRRIS payload resulted in a delay of SPAS-II deployment by 24 hours. This enabled completion of CIRRIS activities before cryogen depletion. This also delayed the first OMS plume sequence until after the first crew sleep period following SPAS-II deployment. The delay had no impact on the relative motion trajectories.[11] Re-planning of burn times was also required to accommodate the slip in the launch time. Resolution of a SPAS-II attitude pointing problem resulted in a delay of the first far field plume sequence of nine hours. This necessitated replanning of all three far field OMS plume sequences.

Station-Keeping Challenges – Trajectory dispersions during the Mission Control targeted coarse station-keeping phase were higher than expected. This was attributed to inaccuracies in long-term state vector prediction due to uncertainties in orbiter and SPAS-II unmodeled accelerations.[19] The poor performance of ground targeted station-keeping burns raised a concern about the ability of the Shuttle Program to perform close-range station-keeping at the relatively low orbital altitude of approximately 140 nm.[12]

Propellant Consumption – Forward RCS propellant consumption predictions were close to actual consumption. Aft RCS propellant consumption predictions were over 1,000 pounds low. However, mission planners considered the propellant budget to be good even with the unexpected relative trajectory control difficulties.[16]

Lambert Targeting – On-board Lambert burn targeting compared well with Mission Control burn targeting that was performed in parallel. In addition, on-board targeting of coelliptic and football (matched period) burns using crew modification of Lambert targeting trajectory constraints worked well. However, computing these burns required more crew training and more crew time during the flight.[20] These null burns at the far

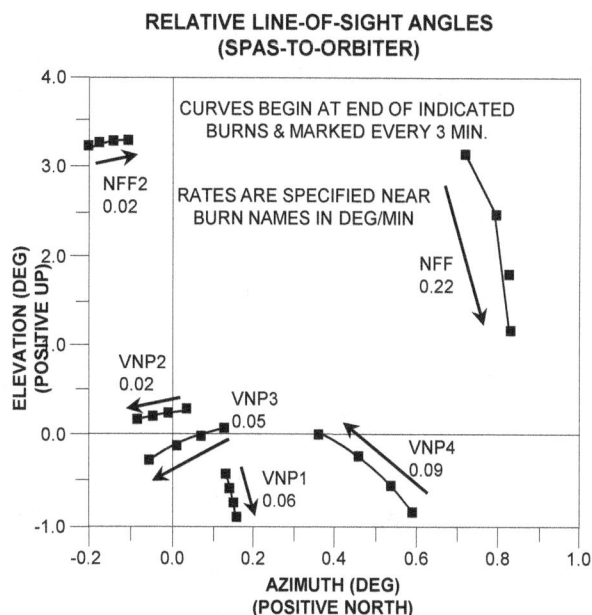

RELATIVE LINE-OF-SIGHT ANGLES (SPAS-TO-ORBITER)

Figure 23.34 Orbiter line-of-sight rates after null burns as seen from SPAS-II.

and near fields successfully minimized relative motion and met requirements of the IBSS payload on SPAS-II. Line-of-sight control was better than expected (Figure 23.34). Re-pointing of SPAS-II to facilitate plume observation was not required.[16] Energy growth observed during the SPAS-II deployed phase had little impact on trajectory performance when on-board targeted burns were executed. This was believed to be due to short transfer times and the use of on-board targeted mid-course correction burns.[20]

Rendezvous Radar – Rendezvous radar performance during the far and near field plume sequences was good. Occasional radar data drop outs introduced noise into the orbiter state vector. However, the relative navigation filter mitigated the noise impact over time and few burns were targeted with noisy state vectors. The frequency of loss of tracking incidences increased during the final approach to SPAS-II grapple. Occurrences of rendezvous radar loss of tracking during the manual piloting phase complicated the piloting task. During periods of loss of lock and reacquisition range rate data became noisy. However, a nominal approach trajectory mitigated the impact of poor radar performance. Some of the occurrences of radar tracking loss were believed to be due to SPAS-II attitude maneuvers and weak returned signal strength.[18, 20]

CRO and MPEC Deployments – All three CRO sub-satellite deployments and chemical release observations were successful. Each of the CRO sub-satellites tumbled after deployment. CRO-C and CRO-B chemical releases occurred approximately 94.6 nm down-range and 43 nm down-range respectively from SPAS-II. The CRO-A chemical release occurred approximately 69 nm down-range from the orbiter. Delta-velocities from CRO-C and

CRO-A chemical releases were higher than predicted. Uncertainty in CRO sub-satellite orbit estimates drove the execution of two burns to ensure the sub-satellites would not come near the orbiter. The MPEC deployment was also successful.

STS-39 Lessons Learned

Successful execution of the STS-39 flight plan was due in part to extensive pre-mission planning and analysis by experienced personnel.

The STS-39 mission stressed the difficulty and importance of determining vent force values to account for un-modeled accelerations that cause trajectory perturbations. Not all sources of orbital acceleration can be mathematically modeled in a straightforward manner, such as gravity be. Hardware design and operation of vents on spacecraft must be investigated and understood. Sources of chaser and target spacecraft un-modeled accelerations that can cause orbital energy growth include, but are not limited to, cross coupling of rotational jet firings into translation, and vents. Long attitude holds during ground tracking can be useful for determining accurate vent force values through precision orbit determination. However, vent force value determination requires long periods of attitude hold without rotational and translational burns.

Accurate knowledge of target spacecraft attitude with respect to time (also known as an attitude timeline) as well as vent forces is useful for accurate target orbit determination and trajectory prediction. Accurate understanding of attitude maneuver cross coupling into translation and resulting trajectory dispersions is necessary.[15]

During long-range station-keeping while within relative sensor range the use of on-board targeted burns and short transfer times (less than 60 minutes) proved useful for mitigating trajectory dispersions resulting from vent and atmospheric drag (spacecraft attitude) uncertainties.[15]

At that time (April/May 1991) relative motion was hand plotted on polar graph paper and transmitted to a large screen in Mission Control using closed circuit television. Plotters, also called "Mr. Hand" (their hands frequently showed up on the closed circuit television image), provided valuable situational awareness for Mission Control personnel during the SPAS-II deployed phase. However, more rapid relative motion graphics using computer displays were highly desirable. The later introduction of the Rendezvous and Proximity Operations Program (RPOP) into Mission Control provided this capability. RPOP was originally intended for use on-board the shuttle during proximity operations but was used on the ground as well.

Additional Mission Control personnel supporting orbit determination and trajectory planning during the SPAS-II deployed phase was found to be beneficial.

The ability to change the target spacecraft atmospheric drag in the on-board navigation software would have improved on-board relative navigation performance. Such a capability (known as a KFACTOR) existed for the orbiter state vector and was eventually added for the target state vector.

The communication link between the orbiter and SPAS-II was never lost, even though the antenna lines-of-sight were close to 90 degrees apart. Personnel believed that the communication range and line-of-sight angle estimates imposed an unrealistic constraint on mission planning, leading to over-design of the SPAS-II detached phase.

STS-37 Mid-Range Station-Keeping Test

The primary purpose of STS-37 was to deploy the Gamma Ray Observatory (GRO). After the deploy Detailed Test Objective (DTO) 822, Mid-Range Station-Keeping, was originally intended to test the long-range station-keeping technique planned for use on STS-39. The DTO involved deployment of a Radar Corner Reflector (RCR) by an Extra Vehicular Activity (EVA) crew member at the end of an EVA. The orbiter was to phase away overnight. The next day station-keeping was to be performed on the minus V Bar at ranges of 8 nm (14.8 km), 1.2 nm (2.2 km), and 5.4 nm (10 km) from the RCR (Figure 23.35).[21]

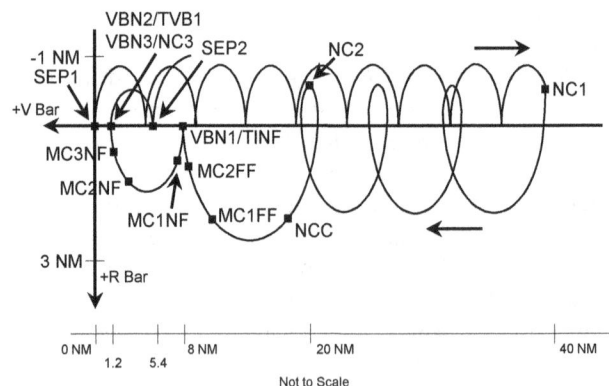

Figure 23.35 Planned relative motion profile for Mid-Range Targeted Station-Keeping DTO before November 1990.

Changes to the DTO

By September of 1990 changes in the Shuttle Program flight manifest moved the STS-37 launch date to after STS-39.[†] The DTO could no longer support STS-39 but several station-keeping techniques were examined for evaluation at a range of 8 nm on the minus V Bar. The motivation for the new DTO was to permit the orbiter to remain at or near the 8 nm point to meet a payload

---

[†] STS-37 ended up flying about 3 and a half weeks before STS-39, in April of 1991.

requirement or provide a delay before a rendezvous was resumed.

A problem on the orbiter or the target spacecraft could prevent completion of a rendezvous. One option would be for the orbiter to execute a break-out at the Ti point and phase away overnight. The rendezvous could be resumed once the problem was resolved. However, this would have a significant impact on the mission timeline, cost additional propellant, and would require repeating much of the rendezvous profile later in the mission. The mid-range station-keeping option, if orbiter systems health permitted it, would permit the orbiter to remain in the vicinity of the 8 nm Ti point until the problem was resolved and the rendezvous could be resumed. In addition, the range might also permit the orbiter to remain within communications range of the target spacecraft. This option would conserve propellant and have less of an impact on the mission timeline. Proving that mid-range station-keeping could be performed with star tracker navigation would make the technique an option for the radar fail case.[21, 22]

DTO options considered were 1) holding at the Ti position, 2) phasing away to a range of 16 nm, 3) flying football relative motion, 4) use of only ground tracking, orbit determination, and burn targeting to support station-keeping, and 5) an out-of-plane station-keeping profile. In October of 1990 Shuttle Program Manager Tommy Holloway approved an out-of-plane station-keeping profile using star tracker measurements for relative navigation. Out-of-plane relative motion would enable the orbiter to stay within communications range of another spacecraft. It could also be executed for any number of integral revolutions or half revolutions.[21]

DTO Profile Details

The new DTO plan involved the same deployment of the RCR at the end of an EVA followed by over-night phasing away from the RCR. The next crew day the orbiter would perform star tracker relative navigation and on-board burn targeting to reach the V Bar Null (VBN1) burn point at 8 nm (Figure 23.36). An out-of-plane

(OOP) delta-velocity would be combined with VBN1 to set-up the out-of-plane motion. Star tracker passes would be performed each revolution to maintain an accurate relative navigation state vector for the orbiter. Ten minutes after the end of each pass a Mission Control targeted phasing burn would be executed to re-target the orbiter for the 8 nm point on the minus V Bar. Ten minutes was considered to be enough time for the Mission Control Flight Dynamics Officer to compute the phasing burn using the on-board orbiter state vector and for a partial burn pad to be voiced up to the crew and entered into the MNVR EXEC display.[21]

Based on Monte Carlo analysis of trajectory dispersions an out-of-plane velocity of minus 5 feet/second was chosen over minus 15 feet/second for the VBN1/OOP burn. The smaller value permitted the out-of-plane motion to degrade quicker, providing a test of how long the out-of-plane station-keeping technique could be used. The larger delta-velocity resulted in more stable out-of-plane motion. The smaller value also consumed less propellant. An additional consideration was performing the star tracker passes on either side of the minus V Bar, or having the pass cross the minus V Bar to provide more geometry change and higher line-of-sight angular rates. Monte Carlo results showed that there was no benefit to permitting the star tracker pass to cross the minus V Bar.[21]

Design of the DTO profile was complete by February of 1991. The orbiter would phase away from the RCR overnight. The NC1 phasing burn would be performed at orbital noon to set-up for the appropriate lighting at VBN1/OOP to permit the orbiter to reach the 20 nm point in three revolutions. NC2 targeted the orbiter for the 8 nm VBN1/OOP point in one revolution.[21]

Addressing Late Issues

Several issues were identified that resulted in a re-design of the DTO profile (Figure 23.37). These were 1) a potential contact hazard between the GRO and the RCR, 2) a potential contact hazard between the GRO and any EVA hardware that could be jettisoned, 3) placing

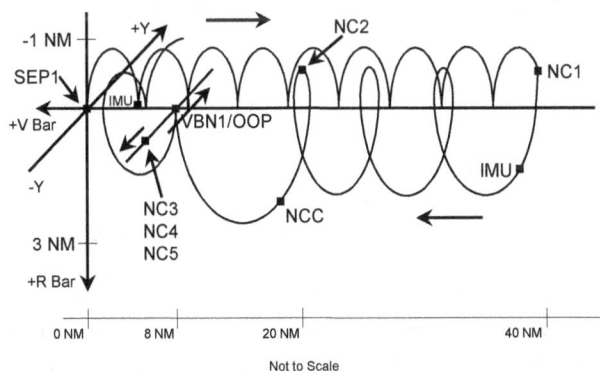

Figure 23.36 Relative motion profile for Mid-Range Targeted Station-Keeping DTO using the Radar Corner Reflector, November 1990.

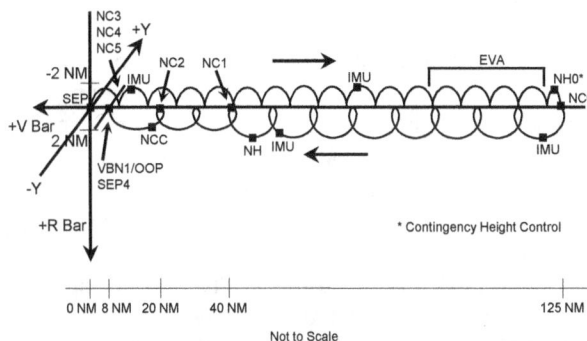

Figure 23.37 Relative motion profile for Mid-Range Targeted Station-Keeping DTO using GRO as the target, February 1991.

orbital debris (the RCR) in the 28.5 inclination orbit, and 4) uncertainty about the orbiter star tracker detecting and tracking the RCR using reflected sunlight.[21]

Only the jettisoned EVA hardware re-contact hazard was not resolved. Changes to the DTO included 1) use of the GRO as the DTO target instead of the RCR, 2) no planning for an EVA hardware jettison, 3) scheduling only one burn after the EVA to deconflict the crew timeline. These changes were required to have minimal impact to planning due to the rapidly approaching launch date.[21]

The GRO separation sequence was not changed. The orbiter was to separate overnight and during the EVA. A NH0 placeholder burn was added to the burn plan in case it was required to counteract orbital energy growth (caused by cross-coupling of rotational RCS jet firings into translation). NC0 would be executed at orbital noon to set-up a phasing rate toward the VBN1/OOP point and establish the appropriate orbital lighting at VBN1/OOP. A NH burn was included to control the height at VBN1/OOP. This was intended to keep the on-board Lambert targeted NCC burn less than 4 feet/second so that it could be executed as a multi-axis primary RCS burn. Such a burn eliminated the need to perform a maneuver to burn attitude that could cause a trajectory dispersion due to rotational jet firings cross coupling into translation. NC1 was targeted for the 20 nm point in two revolutions. The NC2 burn at 20 nm targeted the orbiter for the 8 nm VBN1/OOP point. The out-of-plane component of the VBN1/OOP burn was changed from minus 5 to plus 5 feet/second due to orbiter attitude.[21]

A contingency profile was designed to protect for the possibility of an EVA hardware jettison. If EVA hardware were jettisoned the orbit would be lowered by 3 nm using two orbital adjustment burns separated by half a revolution. The NC0 burn would be used as a hardware separation burn as well as a phasing burn. It would be executed at orbital midnight to set up the proper orbital lighting at VBN1/OOP for the star tracker passes. However, a minimum separation of 11 nm was to be maintained between GRO and the orbiter once the orbiter returned to the GRO orbital altitude.[21]

Mission Execution

Atlantis launched on April 5, 1991 into a post OMS-2 burn orbit of 242 x 244 nm with an inclination of 28 degrees. On flight days 1 and 2 the crew performed on-orbit checkout of GRO and the RMS. GRO was deployed using the RMS on flight day 3, and two crew members performed a contingency EVA to free a stuck GRO high gain antenna boom. On flight day 4 the scheduled EVA was performed. The Mid-Range Station-Keeping DTO was performed on flight day 5, along with the flight control system (OPS 8) checkout and the RCS hot fire test to prepare for landing.

A back-away deploy was performed after GRO release from the RMS on flight day 3. A 0.55 foot/second Low Z primary RCS SEP 1 burn was performed. The LVLH attitude (122 degrees pitch, 17 degrees yaw, and 180 degrees roll) resulted in SEP 1 burn components of -0.4 X (retrograde), 0.0 Y, and +0.4 Z (radial down) feet/second in the LVLH frame. The shuttle rendezvous radar and relative navigation was begun at a range of 155 feet. This was performed to confirm desired relative motion throughout the separation sequence. 22 minutes after SEP 1 the SEP 2 burn was performed with LVLH delta-velocity components of -0.2 X, +2.0 Y, and 0.0 Z feet/second. GRO RCS commanding for attitude control was enabled once the orbiter reached a range of 1,500 feet, a range based on GRO thruster overpressure data. The SEP 3 burn was executed 37 minutes after SEP 2 with LVLH delta-velocity components of +2.9 X, +0.2 Y, and +0.1 Z feet/second. The orbiter was in a –Z body axis target track to facilitate Payload Interrogator communication with GRO. The rendezvous radar was taken to communications mode 51 minutes after GRO release.[23]

Several changes were made to the DTO plan during the flight. One revolution of the DTO was deleted due to the late deployment of GRO and an increase in crew tasks due to the two EVAs. All DTO activities after the NC0 burn were delayed and one revolution of station-keeping data was lost.[21]

NC0 was targeted for 24 nm rather than 20 nm since 24 nm represented the mid-point of a two revolution transfer from 40 nm (the last ground targeted phasing burn) to 8 nm (the Ti point) on a nominal rendezvous profile. This change was made to set-up star tracker pass geometry consistent with a standard stable orbit rendezvous profile. The rationale behind the original 20 nm point for NC2 was to facilitate rendezvous radar acquisition (22 nm maximum range for radar acquisition). The change to 24 nm did not pose a problem for later use of the rendezvous radar. NC1 was deleted and NH was delayed until half a revolution before NC2 to lower the crew work load. These changes resulted in a NC2 range of 28 nm rather than the desired 24 nm.[21]

Three phasing burns were planned during the out-of-plane station-keeping. NC3 was targeted but not executed, NC4 was executed with a delta-velocity of 0.4 feet/second, and NC5 was deleted due to the late GRO deploy.

The down-track position of the orbiter during the out-of-plane station-keeping from VBN1/OOP to the separation burn ranged from -43,000 feet to -49,500 feet. Mission planning predicted dispersions ranged from -39,900 feet to -57,300 feet. Trajectory dispersions and relative navigation performance indicated that out-of-plane (or mid-range) station-keeping using star tracker data was a viable technique in a radar fail case.[21]

The flight day 6 landing was waived off due to weather. Landing at Edwards Air Force Base was on flight day 7.

Summary

STS-39 represented the most complex deploy/retrieve flown by the Shuttle Program. Integration of primary and secondary payloads on the shuttle orbiter and development of mission plans, crew procedures, and ground procedures required extensive coordination by experienced personnel at multiple NASA centers, government contractors, and Department of Defense organizations. Both the STS-39 and STS-37 missions illustrated the complexity of planning and flying missions with relative motion and rendezvous objectives.

References

References with an asterisk at the end may be found in the following compilation document. See the A Note on Sources chapter.

Goodman, John L. (editor), *Space Shuttle Rendezvous and Proximity Operations Overview and Experience Papers*, Volume 2 of 2 (1986-2009), JSC-35050.

1. STS-39 Press Information, Office of Media Relations, Space Systems Division, Rockwell International, April 1991.

2. *STS-39 Flight Requirements Document, Level A Groundrules and Constraints, Appendix A*, NSTS-17462A-39, Close to Launch Cycle, Launch Date April 23, 1991, Revision A, NASA/JSC Flight Design and Dynamics Division, April 1991.

3. Sawin, Dan, "STS-39 Rendezvous Analysis & Operations," Holloway Briefing, Flight Design and Dynamics Department, Rockwell Space Operations Company, January 17, 1991.*

4. Malarkey, John M., *STS-39, Infrared Background Signature Survey (IBSS)*, Rendezvous Flight Data File, JSC-48072-39, Final, January 18, 1991.

5. Sawin, Dan, "STS-39 Information Pack," Flight Design and Dynamics Department, Rockwell Space Operations Company, April 15, 1991. This contains minutes of the January 22, 1991 STS-39 Flight Techniques Splinter Meeting on Radar Failed Operations.*

6. Goodman, John L., *Introduction to Space Shuttle Rendezvous Guidance, Navigation, and Control, Fourth Edition*, JSC-49686, NASA JSC Flight Design and Dynamics Division, November 2009.

7. Goodman, John L., (editor), *Space Shuttle Rendezvous Maneuver Targeting Papers*, JSC-35053, Flight Dynamics Division, Mission Operations Directorate, NASA Johnson Space Center, July 2011.

8. Malarkey, John M., "Targeting Schemes For Onboard Computed Matched Period and Coelliptic Maneuver Capabilities," Orbit Flight Techniques Panel Presentation, Rendezvous Guidance and Procedures Office, Flight Design and Dynamics Division, NASA/JSC Mission Operations Directorate, September 15, 1989.*

9. Goodman, John, "STS-37/39 NSR Burns," informal memo to Larry Bryan of Rockwell International Space Systems Division (Downey, CA), Level 8 Guidance, Navigation, and Control Flight Software Testing Group, Reconfiguration Products Department, Rockwell Space Operations Company, May 24, 1990.*

10. Rahman, Shireen, *IBSS Detached Operations Test Report, Nominal Rendezvous, ONOM2*, Level 8 Test Report, Flight Cycle, STSOC-RT-001287-0015, Reconfiguration Products Department, Rockwell Space Operations Company, Houston, TX, January 28, 1991.

11. Fricke, Robert W., *STS-39 Space Shuttle Mission Report*, JSC-08250, NSTS-08250, NASA/JSC, June 1991.

12. Malarkey, John M., "STS-39 Rendezvous Post Flight Report," Rendezvous Guidance and Procedures Office, Flight Design and Dynamics Division, NASA/JSC Mission Operations Directorate, May 10, 1991.*

13. Britz, William R., "STS-39 Post Flight Review," Orbit Flight Dynamics Group, Flight Design and Dynamics Division, NASA/JSC Mission Operations Directorate, May 17, 1991.*

14. Britz, William R., "Resolution of the Station Keeping Problems During STS-39," Orbit Flight Dynamics Group, Flight Design and Dynamics Division, NASA/JSC Mission Operations Directorate, November 7, 1991.*

15. Burley, P. J., "STS-39 Lessons Learned Ground Targeted Stationkeeping CRO Trajectory Operations," Orbit Flight Dynamics Group, Flight Design and Dynamics Division, NASA/JSC Mission Operations Directorate, August 14, 1991.*

16. Sawin, Dan, "STS-39 Lessons Learned," Presentation to the Orbit Flight Techniques Meeting of the NASA/JSC Mission Operations Directorate, Flight Design and Dynamics Department, Rockwell Space Operations Company, August 14, 1991.*

17. Malarkey, John M., "STS-39 Actual Rendezvous Profile Overview," Rendezvous Guidance and Procedures Office, Flight Design and Dynamics Division, NASA/JSC Mission Operations Directorate, June 18, 1991.*

18. Michels, T. R., "STS-39 Radar Anomaly Study," 660-NAV-640-92-024, Flight Design and Dynamics Department, Rockwell Space Operations Company, April 22, 1992.*

19. Simpson, Roger, "Minutes of Post STS-39 Trench Tag Up," Flight Design and Dynamics Division, NASA/JSC Mission Operations Directorate, May 10, 1991.*

20. Malarkey, John M., "STS-39 Post Flight Lessons Learned Onboard Targeting and Radar Navigation," Orbit Flight Techniques Panel Presentation, Rendezvous Guidance and Procedures Office, Flight Design and Dynamics Division, NASA/JSC Mission Operations Directorate, August 14, 1991.*

21. Gabriel, Dawn M., "STS-37/DTO 822 Post-Flight Summary," SSDD-440-91-059, Flight Design and Dynamics Department, Rockwell Space Operations Company, June 18, 1991.*

22. Goodman, John, "DTO 822 Out-of-Plane Star Tracker," STS-37 Level 8 Software Testing Performance Test Review Presentation, Level 8 Guidance, Navigation, and Control Flight Software Testing Group, Reconfiguration Products Department, Rockwell Space Operations Company, March 7, 1991.*

23. Meyer, Christian K., "STS-37 Post Flight Support," SSDD-F0-730-91-146, Flight Design and Dynamics Department, Rockwell Space Operations Company, July 25, 1991.*

CHAPTER 24 - A CLOSER LOOK AT THE HUBBLE SERVICING MISSIONS

This is an expanded version of a paper originally titled, "Hubble Servicing Challenges Drive Innovation of Shuttle Rendezvous Techniques" by John L. Goodman and Stephen R. Walker. It was presented at the 32nd Annual AAS Guidance and Control Conference in Breckenridge, Colorado, on Saturday, January 31, 2009. This chapter contains seven pages not included in the conference paper.

Introduction

Hubble Space Telescope (HST) servicing performed by Space Shuttle crews contributed to what is arguably one of the most successful astronomy missions ever flown.[1-3] On-orbit servicing performed by five Space Shuttle servicing missions between 1993 and 2009 increased the science return and extended the life of the telescope by correcting performance problems, replacing malfunctioning hardware, and equipping it with more advanced astronomy sensors.[4] Servicing missions involved extensive coordination between specialists in multiple disciplines in both the Shuttle and HST Programs to develop new or adapt existing techniques for HST servicing. These disciplines included trajectory design, robotics, flight control, thermal control, power generation, structures, orbital debris, and Extra-Vehicular Activity (EVA).[5]

HST servicing missions have provided NASA with opportunities to gain insight into servicing mission design and to develop nominal and contingency procedures. HST performance issues have driven new and unanticipated servicing and proximity operations techniques development. Both nominal and contingency procedures and mission plans for rendezvous, proximity operations, jettison, deployment, and tool capture have evolved since HST was deployed on STS-31 in 1990. Although Space Shuttle missions to HST involve human-in-the-loop rendezvous, capture, and servicing, the HST servicing experiences and lessons learned are also applicable to current and future robotic flight programs that involve on-orbit servicing and rendezvous.[6-11] The highly successful Orbital Express robotic servicing demonstration mission illustrated the importance of pre-mission development of contingency procedures to address postulated anomalies, as well as real-time development of contingency procedures in response to unanticipated anomalies.[12] Although HST EVA and robotic activities are outside the scope of this paper, those disciplines have likewise developed and evolved extensive nominal and contingency procedures.

Servicing missions succeeded in part due to the efforts of experienced HST and Shuttle Program personnel (NASA and contractor) from multiple disciplines that had extensive experience planning and flying servicing and assembly missions to a variety of spacecraft. This facilitated application of best practices and lessons learned. These personnel were responsive to unanticipated satellite performance issues that drove late

and significant changes in servicing mission plans. These events drove changes to existing proximity operations, robotic operation, and servicing procedures or required the creation of new procedures and mission plans. HST and Shuttle Program personnel continually learned about emerging HST and shuttle orbiter constraints. Unforeseen constraints and performance limitations drove development of new or changes to existing nominal and contingency plans and procedures. Rendezvous, proximity operations, and other mission techniques from other Space Shuttle missions were successfully applied to mitigate risk to HST servicing mission success.

This chapter provides an overview of HST servicing missions. This is followed by a description of HST design and operations that are pertinent to Space Shuttle rendezvous and proximity operations. Next, relative navigation and shuttle plume impingement challenges are discussed. For the deploy mission and the servicing missions an overview is given of the rendezvous, proximity operations, and deploy procedures that were flown, along with mission results. In addition, contingency procedures to address the HST aperture door failed closed or failed open cases are described. Other contingency proximity operations and hardware jettison procedures are then outlined. Table 24.1 is an overview of HST servicing mission objectives. Table 24.2 is a list of nominal and contingency procedures for each mission that address relative motion. The table lists procedures for rendezvous, proximity operations (approach and grapple), jettison, and deploy and separation.

A rescue mission had been planned if a thermal protection system problem prevented the safe return of the STS-125 crew during the last HST servicing mission in 2009. Since the rescue mission was different in many respects from the HST deployment and servicing missions, nominal and contingency procedures are discussed in a separate section at the end of the chapter.

Early Servicing Concepts

From the beginning of space telescope concept development in the early 1970s, both on-orbit servicing by Space Shuttle EVA crew and ground servicing was included in requirements and operations concepts. On-orbit servicing by Space Shuttle astronauts was to be performed every 2.5 years, and hardware lifetime and reliability requirements were based on this assumption. Every 5 years the shuttle was to return the telescope to

Table 24.1 Space Shuttle Missions Concerning The Hubble Space Telescope

Mission	Orbiter	HST Mission Objectives	Crew	Launch Date, Pad Landing Date, Runway	Remarks
STS-61J	*Atlantis* OV-104	Deploy HST	John Young Charles Bolden Steven Hawley Bruce McCandless Kathryn Sullivan	Mission planned for August, 1986.	Canceled after *Challenger* accident.
STS-31	*Discovery* OV-103	Deploy HST	Loren Shriver Charles Bolden Steven Hawley Bruce McCandless Kathryn Sullivan	4/24/90, 39B 4/29/90, EDW 22	HST successfully deployed. Contingency rendezvous with HST planned but not required.
STS-42	*Discovery* OV-104	Proposed HST photo inspection.	Ronald Grabe Stephen Oswald Norman Thagard David Hilmers William Readdy Roberta Bondar Ulf Merbold	1/22/92, 39A 1/30/92, EDW 22	Proposed inspection was to document solar array tip deflections that could lead to array failure and negatively impact astronomy. Inspection proposal rejected in August 1991. Primary mission objective International Microgravity Laboratory-1.
STS-57	*Endeavour* OV-105	EVA tests of STS-61 HST servicing procedures.	Ronald Grabe Brian Duffy David Low Nancy Sherlock Peter Wisoff Janice Voss	6/21/93, 39B 7/01/93, KSC 33	EVA successful. No rendezvous or prox ops conducted in support of HST. Primary mission objectives EURECA retrieval and SPACEHAB.
STS-61	*Endeavour* OV-105	Servicing Mission 1 (SM1)	Richard Covey Kenneth Bowersox Kathryn Thornton Claude Nicollier Jeffrey Hoffman Story Musgrave Thomas Akers	12/02/93, 39B 12/13/93, KSC 33	Installation of corrective optics. Solar arrays replaced and one old array jettisoned by EVA crew.
STS-82	*Discovery* OV-103	Servicing Mission 2 (SM2)	Kenneth Bowersox Scott Horowitz Joseph Tanner Steven Hawley Gregory Harbaugh Mark Lee Steven Smith	2/11/97, 39A 2/21/97, KSC 15	MECO under-speed. During rendezvous star tracker broke lock on HST, then tracked a star and orbital debris. SEP2 maneuver under-burn. Re-planning and crew procedures executed in response to these issues ensured successful rendezvous and separation.
STS-103	*Discovery* OV-103	Servicing Mission 3A (SM3A)	Curtis Brown Scott Kelly Steven Smith Michael Foale John Grunsfield Claude Nicollier Jean-Francois Clervoy	12/19/99, 39B 12/27/99, KSC 33	Flown in response to HST gyro failures. HST in Hardware Sun Point safe mode at the time of rendezvous due to fourth gyro failure. *Discovery* yaw maneuver due to off nominal HST attitude at the time of grapple.
STS-109	*Columbia* OV-102	Servicing Mission 3B (SM3B)	Scott Altman Duane Carey John Grunsfeld Nancy Currie James Newman Richard Linnehan Michael Massimino	3/01/02, 39A 3/12/02, KSC 33	Rendezvous altitude decayed below insertion altitude, forcing one rendezvous maneuver to be retrograde.
STS-125	*Atlantis* OV-104	Servicing Mission 4 (SM4)	Scott Altman Gregory C. Johnson Michael Massimino Michael Good Megan McArthur John Grunsfeld Andrew Feustel	5/11/09, 39A 5/24/09, EDW 22	Mount passive LIDS docking hardware and laser retro-reflectors on HST for possible missions by future human or robotic spacecraft.
STS-400	*Endeavour* OV-105	Rescue 125 crew if required.	Chris Ferguson Eric Boe Stephen Bowen Robert Kimbrough	Mission not required. Would have been launched from pad 39B.	Rescue orbiter grapples *Atlantis* with RMS. EVA transfer of *Atlantis* crew to rescue orbiter. TCS retro-reflector mounted in *Atlantis* payload bay for use by rescue orbiter TCS.

EDW – Edwards Air Force Base
EURECA – European Retrievable Carrier
EVA – Extra Vehicular Activity
HST – Hubble Space Telescope
KSC – Kennedy Space Center
LIDS – Low Impact Docking System
MECO – Main Engine Cut-Off
OV – Orbiter Vehicle
RMS – Remote Manipulator System
SM – Servicing Mission
STS – Space Transportation System
TCS – Trajectory Control Sensor

Table 24.2 Nominal And Contingency Procedures For HST Servicing Missions

Mission	Rendezvous	Proximity Operations		Jettison	Deploy/Separation
STS-31					
Nominal					• Deploy with RMS
Contingency	• Stable Orbit (2 rev) • Radar Fail • Rndz Breakout • Ti Delay	• Inertial Approach • Fast Flyaround • STS Roll to Align	• Prox Ops Breakout • EVA Rescue • Loss of VRCS		• Emergency RMS Deploy • No RMS Backaway Deploy
STS-61					
Nominal	• Stable Orbit (2 rev)	• Inertial Approach			• Deploy with RMS
Contingency	• Radar Fail • Rndz Breakout • Ti Delay	• Manual Inertial Flyaround Alignment Trim • AUTO Inertial Flyaround • Prox Ops Backoff • Prox Ops Breakout	• Tool chasing • EVA Rescue • Loss of VRCS	• HST Jettison • SAC Jettison • ORUC Jettison • SA Jettison Using Jettison Handle (performed) • SA Jettison Using Portable GF	• RMS Quick Deploy • No RMS Backaway Deploy • Low Propellant Sep (performed)
STS-82					
Nominal	• Stable Orbit (2 rev)	• +R Bar Approach/Inertial Grapple			• Deploy with RMS
Contingency	• Radar Fail • Rndz Breakout • Ti Delay	• Inertial Approach • RBAR Yaw Alignment • Manual Inertial Flyaround Alignment Trim • AUTO Inertial Flyaround • Prox Ops Backoff	• Prox Ops Breakout • Loss of Low Z Braking • Loss of Low Z Breakout • Loss of VRCS • Tool chasing • EVA Rescue	• HST Jettison • EVA Hardware Jettison	• RMS Quick Deploy • No RMS Backaway Deploy • No FRCS Sep
STS-103					
Nominal	• ORBT (2 rev)	• +R Bar Approach/Inertial Grapple			• Deploy with RMS
Contingency	• Stable Orbit (2 rev) • Radar Fail • Rndz Breakout • Ti Delay	• Inertial Approach • RBAR Yaw Alignment • Manual Inertial Flyaround Alignment Trim • AUTO Inertial Flyaround • Prox Ops Backoff • HST R Bar Breakout	• HST Flyaround/Loss of LOW Z Breakout • Loss of VRCS • Loss of Low Z Braking • Tool Chasing • EVA Rescue	• HST Jettison for Rapid Safing • ORUC Jettison • EVA Hardware Jettison	• RMS Quick Deploy • No RMS Backaway Deploy
STS-109					
Nominal	• ORBT (1 rev)	• +R Bar Approach/Inertial Grapple			• Deploy with RMS
Contingency	• Stable Orbit (1 rev) • Radar Fail • Rndz Breakout • Ti Delay	• Inertial Approach • RBAR Yaw Alignment • Manual Inertial Flyaround Alignment Trim • AUTO Inertial Flyaround • Prox Ops Backoff • HST R Bar Breakout	• HST Flyaround/Loss of LOW Z Breakout • Loss of VRCS • Loss of Low Z Braking • Tool Chasing • EVA Rescue	• HST Solar Array Jettison • HST Jettison for Rapid Safing • SAC Jettison • RAC Jettison • EVA Hardware/Solar Array Jettison	• RMS Quick Deploy • No RMS Backaway Deploy
STS-125					
Nominal	• ORBT (1 rev)	• +R Bar Approach/Inertial Grapple			• Deploy with RMS
Contingency	• Stable Orbit (1 rev) • Radar Fail • Rndz Breakout • Ti Delay	• Inertial Approach • RBAR Yaw Alignment • Manual Inertial Flyaround Alignment Trim • AUTO Inertial Flyaround • Prox Ops Backoff • HST R Bar Breakout	• HST Flyaround/Loss of Low Z Breakout • Loss of VRCS • Loss of Low Z Braking • Tool Chasing • EVA Rescue	• HST Jettison for Rapid Safing • SLIC Jettison • ORUC Jettison • EVA Hardware Jettison	• RMS Quick Deploy • No RMS Backaway Deploy
STS-40x					
Nominal	• ORBT (1 rev)	• +R Bar Approach			• Separation
Contingency	• Radar Fail • Rndz Breakout • Ti Delay				

EVA – Extra Vehicular Activity
FRCS – Forward Reaction Control System
GF – Grapple Fixture
HST – Hubble Space Telescope
ORBT – Optimized R Bar Targeted Rendezvous

ORUC – Orbital Replacement Unit Carrier
RAC – Rigid Array Carrier
RMS – Remote Manipulator System
Rndz – Rendezvous
SA – Solar Array

SAC – Solar Array Carrier
SLIC – Super Lightweight Interchangeable Carrier
STS – Space Transportation System
Ti – Transition Initiation
VRCS – Vernier Reaction Control System

to Earth for a more intensive refurbishment. The shuttle would then return the telescope to Earth orbit. However, by the late 1970s, concerns about contamination and structural loads that the telescope could be subjected to during ascent and entry led NASA to limit servicing to on-orbit. It was determined that on-orbit servicing would be adequate to maintain HST during the 15 year design life.

Overview of HST Servicing Missions

Planning for all HST missions involved trade studies, simulations, and extensive technical discussions covering both nominal and contingency mission plans and procedures. Mission preparation included timeline and crew activity planning, procedure development, and trajectory design covering all aspects of the mission. This included ascent, launch aborts, rendezvous, proximity operations, entry and landing, EVA, robotics, etc. Contingency procedures were also developed or adapted to addresses systems anomalies that may occur in the rendezvous, proximity operations, servicing, and deploy phases.

Shuttle rendezvous with HST and grapple, by the Remote Manipulator System (RMS) robotic arm, was normally scheduled for flight day three.* On the morning of flight day three, the shuttle relative navigation sensors (radar and star tracker) obtained relative measurements that were used to improve the estimate of the relative navigation state in the shuttle flight computers. Rendezvous maneuvers were also computed by the shuttle flight computers.

Once the orbiter was within approximately 2000 feet of HST (post MC-4) the proximity operations phase began. The relative motion trajectory was designed to accommodate orbiter and HST constraints such as orbiter Reaction Control System (RCS) jet plume impingent, power generation, and thermal control. The crew grappled HST with the RMS and berthed it in the shuttle payload bay. After several days of servicing by EVA crew members, HST was deployed and eventually resumed the astronomy mission. Deploy procedures were designed to ensure safe separation of the orbiter from HST while also concurrently protecting HST from plume impingement. Deploy procedures also had to meet additional constraints for thermal, lighting, and communications.

The Hubble Space Telescope

Figure 24.1 is an illustration of the HST as it appears on-orbit while conducting the astronomy mission. Two solar arrays provide electrical power. HST attitude and solar array orientation must be carefully managed to ensure that sufficient power is available to recharge HST batteries. In addition, the HST solar arrays, solar array support structure, and rotational mechanisms are sensitive to shuttle RCS jet plume contamination and over-pressure. Significant analysis is required to develop nominal and contingency proximity operations procedures (approach, grapple, deploy) that do not violate HST plume constraints. Furthermore, HST attitude during shuttle proximity operations must be carefully managed to ensure that the HST solar arrays can generate sufficient power, even in the presence of degraded HST attitude control system performance. HST optics are sensitive to plume contamination as well. However, the optics are protected by closing the aperture door during the approach by the shuttle.

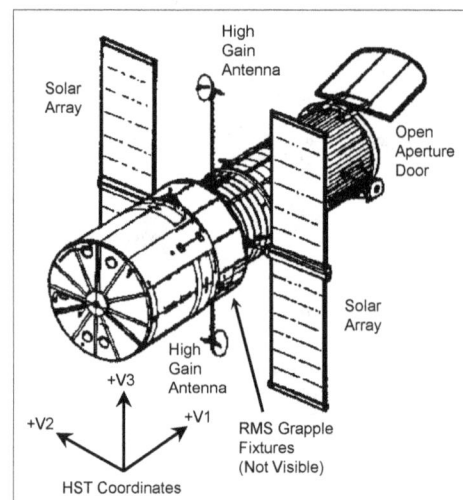

Figure 24.1 Hubble Space Telescope

HST relies on four Reaction Wheel Assemblies (RWAs) for attitude control, rather than using RCS jets. Six Rate Gyro Assemblies (RGAs) provide redundant measurements for attitude control. However, only three RGAs are required for attitude control. The Retrieval Mode Gyro Assembly (RMGA) is a non-redundant set of back-up gyros that are independent of the RGAs. The RMGA can provide course attitude data for limited periods to support shuttle proximity operations and grapple.

In the event of performance anomalies HST has two attitude control safe modes to maintain HST in a power positive configuration. A HST systems anomaly that forces use of one of the safe modes has implications for proximity operations and shuttle robotics procedures. The Hardware Sunpoint (HWSP) safe mode uses RMGA data and points the +V3 axis to the Sun, maintains an inertial attitude hold, aligns the solar arrays with the V1 axis, and closes the aperture door. The Zero Gyro Sun Point (ZGSP) safe mode points the +V3 axis in the general direction of the Sun, maintains a slow spin about the V3 axis, aligns the solar arrays with the V1 axis, and

---

* Flight day one began with crew wakeup at the Kennedy Space Center on the day of launch. Subsequent flight days on-orbit began when the crew wakes up.

closes the aperture door. Coarse rate and Sun position data is obtained from the Coarse Sun Sensor. No RGA data is used by the ZGSP safe mode.

Before the shuttle begins the final approach to grapple HST with the RMS, the HST is placed in a proper systems configuration and attitude. The –V3 High Gain Antenna (HGA) (Figure 24.1) is stowed and latched, and the solar arrays rotated to be parallel with the V1 axis. HST performs a roll maneuver to place the RMS grapple fixture on the north side of the orbital plane. HST continues to maintain an inertial attitude hold during rendezvous and final approach. Two RMS grapple fixtures are mounted on the HST along the –V3 axis (Figure 24.1). The fixtures can be removed and installed by EVA crew, if required. The nominal grapple attitude of HST is not optimal for power generation by the solar arrays. When the roll maneuver completes, a 180 minute Sun pointing timer is started. If HST is not grappled by the orbiter after 180 minutes, HST performs a low rate attitude maneuver to a power optimal attitude. However, this maneuver was not required on the servicing missions flown.

Challenges of HST Servicing Missions

The Space Shuttle was designed in the early 1970s after NASA had successfully demonstrated rendezvous techniques in the Gemini and Apollo Programs. However, technical and mission design challenges emerged from servicing missions to HST, as well as shuttle missions involving other target spacecraft, that were not faced during Gemini and Apollo.[13] This section highlights some of those challenges.

Propulsion, Attitude Control, and Plume Impingement Challenges

The early operational concepts for HST defined in the 1970s included on-orbit servicing by astronauts. HST hardware and systems layout was designed to support servicing. However, the design of both the HST and the Space Shuttle was completed before the potential of HST contamination or structural damage, resulting from over-pressure by shuttle RCS jet plume impingement, was fully understood. As a result, proximity operations design for servicing missions has evolved as insight into plume effects on HST has improved. To minimize risk of plume contamination and over-pressure the shuttle Low Z flight control mode is used for HST and other proximity operations missions, such as Mir and ISS, rather than normal Z-axis firings (Figure 24.2). The Low Z mode provides some RCS braking capability while minimizing RCS plume impingement. The Low Z mode uses X body axis jets that have a small thrust component along the Z-axis, rather than Z-axis jets that direct plumes at the target spacecraft. The X-axis thrust components of the forward and aft-firing jets sum to near-zero, leaving a small Z-axis

component that can be used for braking. Propellant consumption for braking is increased dramatically in the Low Z mode. The Z-axis thrust component of the X-axis jets was not an original Space Shuttle design requirement for proximity operations. The Low Z mode was developed in the 1977-1978 time period, after the shuttle design was finalized and hardware was already under construction.[5,13] Use of the Low Z mode increases propellant consumption on missions that are already propellant limited as the HST orbital altitude is much higher than the orbital altitude of other shuttle missions.

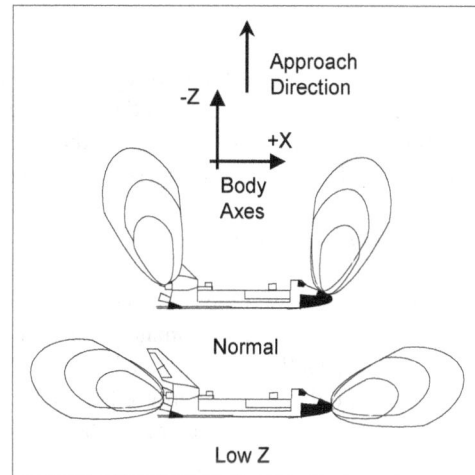

Figure 24.2 Normal and Low Z Primary RCS jet plumes.

HST does not have a propulsion system for orbit maintenance or attitude control. Consequently it is dependent on the shuttle for orbit raising maneuvers to counteract orbital decay due to atmospheric drag. While the HST is in the shuttle payload bay the shuttle may perform a re-boost maneuver to increase the HST orbital altitude. Since years separate servicing missions, HST is placed and maintained at as high an altitude as can be reached by the shuttle. The orbital altitude coupled with the previously mentioned extensive use of the Low Z flight control mode reduces available propellant margins.

Relative Navigation

At the time of the HST design in the mid 1970s shuttle rendezvous sensors were defined as radar without a transponder on the target spacecraft and a star tracker that tracked sunlight reflected by the target.[13] While HST was equipped with a Tracking and Data Relay Satellite System (TDRSS) transponder to allow ground tracking using Doppler and range measurements, HST was not equipped with relative navigation aids. Unlike smaller shuttle rendezvous targets, such as SPARTAN, some rendezvous beam wandering was observed while tracking HST. However, the beam wandering was less than is experienced on International Space Station (ISS) missions. By 1995 the shuttle was flying the Trajectory Control Sensor (TCS) lidar and the Hand Held Lidar (HHL) to provide measurements to the laptop computer Rendezvous and Proximity Operations Program (RPOP). HHL measurements and RPOP data based on TCS and HHL improved crew situational awareness beyond what could be provided by the legacy shuttle computer relative navigation filter and sensors. This led to lower propellant consumption during proximity operations. Proposals to add retro-reflectors to HST to support TCS had not been approved for cost reasons; consequently TCS is not used on servicing missions. Although there was a European Space Agency retro-reflector on the back end of HST, it was not designed to work with TCS. The curved surface of HST makes it a poor target for HHL and causes shuttle payload bay camera blooming, complicating proximity operations piloting. Experience has shown that the RMS grapple fixtures on HST are good targets for the HHL.

Remote Manipulator System

The RMS was used to grapple, maneuver, berth, and deploy the HST. It is an approximately 50 ft long, 6-degree-of-freedom robotic arm equipped with six joints (shoulder yaw, shoulder pitch, elbow pitch, wrist pitch, wrist yaw, and wrist roll). It is located on the port side of the payload bay, and is capable of handling payloads up to 65,000 lb. The RMS end effector on the end of the arm grapples a fixture installed on the payload. An RMS display and control panel, rotational and translational hand controllers, and associated television displays are located in the aft flight deck flight crew station. The RMS is also used as an extension ladder for EVA crews, and for conducting inspections using end effector television camera.

There are two types of RMS end effector techniques used for grappling another spacecraft. The preferred technique, called orthogonal, aligns end effector motion along orbiter +Y body axis, in the direction of the starboard wing. End effector orientation aligns the up, down, left, and right directions observed by the crew with the end effector camera with the orbiter body X and Z axes. This provides the crew with good motion cues. The second technique, non-orthogonal, retains the closing rate cues of the orthogonal technique, but end effector lateral motion (up, down, left, right) is not aligned with the body X and Z axes. Two RMS grapple fixtures are mounted on the HST along the −V3 axis. The fixtures can be removed and installed by EVA crew if required.

Ground Tracking Limitation During Ascent

A Mission Control ground tracking solution is available for comparison with the on-board navigation state after Main Engine Cut-Off (MECO). Ensuring an accurate on-board state vector is important since the

Orbital Maneuvering System 2 (OMS-2) burn helps control orbiter phasing and orbital plane control for rendezvous. In the event that on-board navigation state error in the out-of-plane direction is excessive after MECO, one of two state vectors could be uplinked by Mission Control to the shuttle flight computers. The first is the ground tracking state vector. The second is the GPS state vector downlinked from the orbiter GPS receiver. A new state vector uplink would ensure that the OMS-2 orbit insertion burn will correct for out-of-plane trajectory dispersions. Correcting for such dispersions later in the flight increases propellant consumption.

For 51.6 degree inclination flights to the International Space Station, ground radar tracking is available throughout powered ascent. The elimination of the Bermuda tracking station in 1998 reduced ground tracking coverage during powered ascent for the 28.5 degree inclination HST missions. In the absence of an accurate ground tracking solution, the on-board GPS receiver state vector was used by Mission Control to assess the health of the on-board navigation state on flights STS-103, STS-109, and STS-125. If required, a GPS state vector could have been uplinked to the vehicle before the OMS-2 burn. Mission Control would assess the performance of the GPS receiver before using the GPS receiver as a vector source. However, a GPS update was not required for STS-103, STS-109, or STS-125. This procedure would have also been available for the STS-400 (*Endeavour*) rescue mission of the Atlantis crew, if a rescue had been required. GPS state vectors are normally only used during entry to update the orbiter navigation state.[14]

STS-31 – HST DEPLOY

After a four year delay due to the loss of the Space Shuttle *Challenger*, HST was deployed from the orbiter *Discovery* on April 25, 1990 (flight day two), during the STS-31 mission (Table 24.1).

STS-31 Deploy

After HST was unberthed from the payload bay with the RMS, solar array #2 did not unfurl. Concurrent with crew and ground troubleshooting, preparations began for an unscheduled EVA in the event that a manual unfurl was required. Two EVA crewmembers conducted the in-suit pre-breath activity (required to flush nitrogen from the bloodstream before being exposed to the reduced pressure environment of a spacewalk), and then entered the shuttle airlock. The airlock was then depressurized to 5 psi. However, another pre-planned contingency procedure successfully unfurled the array on the third attempt and the EVA was not required. Solar array #1 and the two HGAs were deployed without incident before HST was released by the RMS on rev. 20 (Figure 24.3).

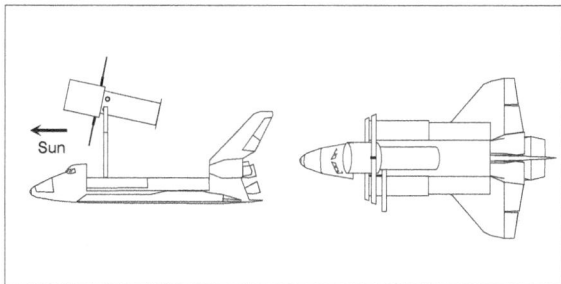

Figure 24.3 STS-31 HST Deployment

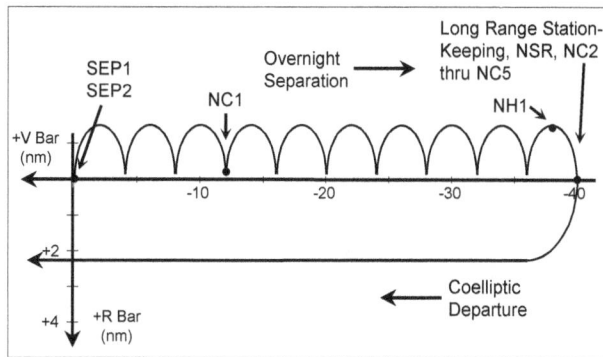

Figure 24.5 STS-31 separation, long range station-keeping, and coelliptic departure profile.

The shuttle rendezvous radar tracked HST from a range of 96 feet to 38,000 feet, when the Ku antenna was taken to the communications mode. Rendezvous radar data was incorporated into on-board navigation during the separation to improve crew and Mission Control knowledge of relative motion (Figure 24.4). Use of the rendezvous radar provided a more accurate relative state solution than could have been obtained with ground radar and TDRSS tracking.

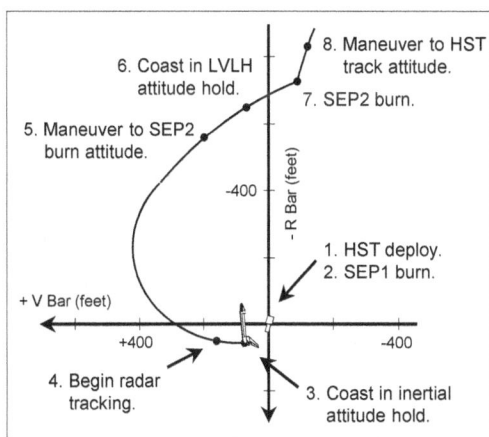

Figure 24.4 STS-31 HST Deploy Profile

The Separation 1 (SEP1) maneuver was performed with the RCS in the Low Z mode to minimize HST contamination by RCS plumes. Separation maneuvers performed by *Discovery* were required to prevent recontact with HST and ensure that the safe separation continued during the crew sleep period following HST deployment (Figure 24.5). The HST inertial deployment was designed to ensure that HST sun sensors would lock onto the Sun after release from the RMS. The time required for HST to acquire and track the Sun to minimize battery discharge and recovery time was also considered.

Continuous communications with *Discovery* was required for pre-defined periods before and after deployment. Once HST was released and *Discovery*

separated to a safe distance, HST mission responsibility was transferred to the HST Director of Orbit Verification at the Space Telescope Operations Control Center (STOCC) at the Goddard Space Flight Center. After HST deploy, *Discovery* separated overnight and conducted long-range station-keeping in the general vicinity of a position 40 nm behind HST on the –V Bar (Figure 24.5). The Shuttle Program was required to maintain a capability to rendezvous with HST for up to 45 hours or until the STOCC verified that the aperture door was open. Long range station-keeping was conducted until HST activation was complete and the aperture door successfully opened by the STOCC. At approximately 1 day and 19 hours after deployment *Discovery* was released from HST operations. A contingency rendezvous was not required. *Discovery* left the long-range station-keeping trajectory using an orbit coelliptic to HST to ensure safe separation (Figure 24.5).

STS-31 Contingency Deploy Planning

A significant amount of planning was performed to ensure that HST could be successfully placed in orbit on STS-31 in the presence of various shuttle anomalies. Although the nominal deploy was on flight day 2, HST could have been deployed on any flight day and orbit if the crew timeline and HST power and thermal constraints permitted it. A flight day 1 contingency deployment could have been performed through the completion of solar array deployment in the event of the following HST failures: 1) Total loss of orbiter power to the HST, and 2) failure to apply orbiter power to HST during HST systems activation on flight day 1. Components within the HST Optical Telescope Assembly (OTA) had a limited life without power from *Discovery* or the HST solar arrays. Extended lack of power could degrade OTA performance and introduce safety risks for returning HST to Earth. A contingency EVA could also have been performed on flight day 2 to make power available to HST from the solar arrays or the orbiter after umbilical disconnect. An EVA would not have been performed on Flight Day 1.

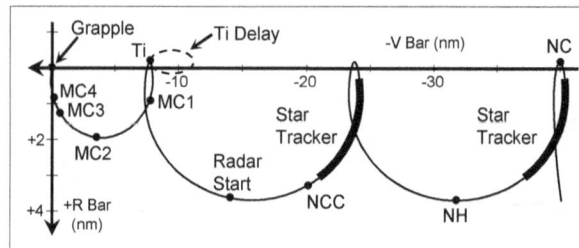

Figure 24.6 Stable orbit rendezvous profile for STS-31 (contingency) and STS-61 and STS-82 (both nominal).

Degraded orbiter systems performance or a component failure might prevent the planned mission duration from being flown, but not require a return to Earth at the next opportunity. In this case a Minimum Duration Flight (MDF) may be declared. The objective of a MDF is to allow high priority mission objectives to be accomplished while minimizing exposure to subsequent failures that could threaten crew safety. The nominal length of a MDF is 72 hours. If a MDF were to be declared before HST deploy, the deployment could have occurred on flight day 2. Entry preparation would occur on flight day 3, with landing on flight day 4. HST could also have been deployed on flight day 3. For a flight day 2 deploy the crew could perform a contingency EVA if required to support the HST deployment. A contingency EVA would not have been performed for a flight day 3 deploy since the entry would be on flight day 4. EVAs for mission success are not performed the day before entry or on entry day. If a landing at the next available Primary Landing Site was required the HST did not have to be deployed for the orbiter and crew to return safely to Earth. A RMS or backaway deploy would be considered in this case if there was sufficient time remaining before the deorbit burn.

An EVA to deploy HST manually using the crew members gloved hands would not have been performed since the size of HST would have prevented EVA crew members from observing each other. Close coordination is required to avoid undesirable re-contact during a by-hand payload deployment. Unscheduled EVAs could also be performed no earlier than flight day 3 to address the following: 1) RMS failures, 2) reattach thermal blankets, 3) umbilical disconnect, 4) HGA deploy, 5) aperture door latch release, and 6) grapple fixture release.

STS-31 Contingency Rendezvous and Inertial Approach

While rendezvous was part of the nominal mission plan for future servicing missions, it was a contingency procedure for STS-31. The only driver, for a contingency rendezvous following deploy, was to open a failed closed HST aperture door (Figure 24.1). The contingency rendezvous timeline was written for a flight day 5 rendezvous and EVA by the crew to open the door, with an additional flight day added to the mission (Figure 24.6). However, HST would have been released even if it was known that an existing orbiter systems problem would prevent a contingency rendezvous from being performed. An orbiter systems problem could require the orbiter to return to Earth sooner than planned. In this case a Minimum Duration Flight (MDF) could be declared, with the orbiter returning to Earth as soon as 72 hours after launch. The MDF mission timeline could not have supported a contingency rendezvous and EVA to open the failed closed aperture door.

The contingency rendezvous profile was the standard stable orbit profile (Figure 24.6) with an inertial approach (Figure 24.7).[13] If a rendezvous and grapple were required, mission responsibility would revert from the STOCC to the Mission Control flight director in Houston. At sunrise, on the grapple orbit, HST would be maneuvered so that the –V1 axis would be pointed into the velocity vector at orbital noon. At this time the –V1 axis would also be pointed at the payload bay of the approaching *Discovery* when it arrived on the +V Bar. At orbital noon HST would then roll about the V1 axis so that the –V3 RMS grapple fixture would be pointed in a specified direction out-of-plane and on the north side of the orbit (Figure 24.8). After the roll maneuver was complete *Discovery* would approach to within 200 feet and the RCS system would be placed in the Low Z mode (Figure 24.2). This HST maneuver sequence was designed to align the HST for capture with the RMS of *Discovery* and to reduce or eliminate the need for *Discovery* to perform additional maneuvers to prepare for capture.

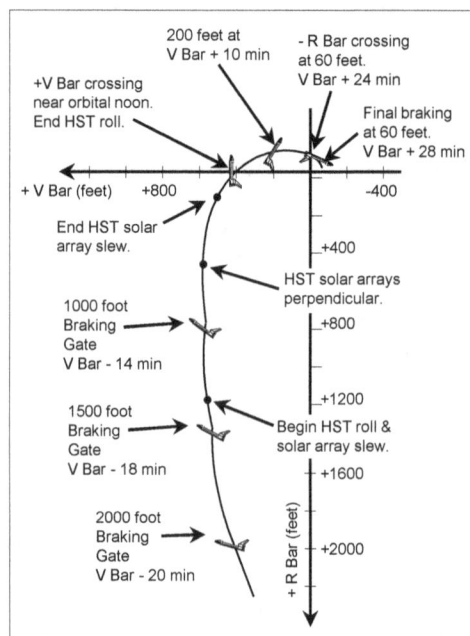

Figure 24.7 Inertial proximity operations approach for STS-31 (contingency) and STS-61 (nominal).

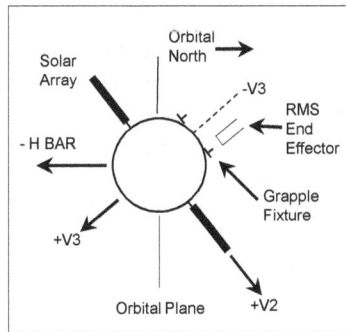

Figure 24.8 Nominal HST attitude as seen from the orbiter at grapple. Aperture door end of HST is pointed away from the orbiter.

STS-42 – The Proposed HST Photo Inspection

In the months following HST deployment, exposure of the HST solar array support mechanisms to thermal gradients when passing through orbital sunrise and sunset resulted in solar array tip deflections as large as +/-1 meter. These sudden deflections could cause image blurring and loss of fine attitude pointing required for astronomical observations. In addition, there was also concern that repeated deflections could lead to failure of the solar arrays and supporting hardware.

By August of 1991 the STS-42 mission was planned for launch in January of 1992 with the International Microgravity Laboratory 1 (IML-1) as the primary payload. Concerns about HST solar array fluctuations and impacts to array structural integrity led NASA to examine the possibility of changing the STS-42 mission to perform a photo inspection of HST. Photographic documentation of solar array deflections during orbital day and night would be used to verify newly developed math models of solar array tip deflection. The math models were critical to predicting how close to failure the solar arrays were. A hand held spotlight would be used to facilitate photography during orbital night.

Flying STS-42 to HST would involve changing the orbital inclination from 57 degrees to 28.5 degrees, and increasing the orbit insertion altitude from 163 nm to 316 nm. The HST rendezvous altitude was 323 nm. IML-1 would remain as the primary payload, but some of the secondary payloads in the payload bay or mid-deck could have been removed to reduce vehicle mass to accommodate the higher mission altitude and propellant budgeting requirements. Proximity operations design and procedures had to minimize plume contamination risk since the aperture door was to remain open. The original concept called for the orbiter to perform four in-plane inertial fly-arounds of HST at orbital rate and a range of 300 feet to allow each solar array to be observed for two orbits. A study reduced the fly-arounds to two due to dual shift crew sleep requirements in support of the IML-1. Use of the Low Z RCS mode could have allowed the

fly-around range to be reduced. There was also a possibility that placing the HST in a LVLH attitude hold with the orbiter station-keeping on the +V Bar would provide propellant savings over the two inertial fly-arounds.

The proposed change of the STS-42 mission to perform the HST inspection was rejected in late August of 1991. Photo documentation of the solar arrays was later accomplished on the first servicing mission before grapple (STS-61, December 1993).

STS-57 – Servicing Procedures Tests During EVA

In preparation for the first HST servicing mission, a series of tests was conducted during an EVA to refine servicing procedures. These tests concerned the use of the foot restraint on the RMS by the EVA crew members. Tests of safety tether management, handling and aligning large objects by hand, and the use of HST servicing tools while the EVA crew member was mounted on the RMS were conducted. RMS handling qualities while the EVA crew member held a large object were also evaluated.

STS-61 – Servicing Mission 1 (SM1)

On June 25, 1990, two months to the day after deployment from *Discovery*, a spherical aberration was discovered in Hubble's primary mirror, significantly reducing the quality of astronomical observations. A major objective of the first servicing mission was to install the Corrective Optics Space Telescope Axial Replacement, or COSTAR. Five corrective mirrors in COSTAR corrected the optical effects of the flawed mirror. Additional upgrades made by the EVA crew included the Wide Field Planetary Camera 2 (WFPC2) to replace WFPC1, new solar arrays and solar array drive electronics, new magnetometers, new coprocessors for the flight computer, two new Rate Sensor Units, two new Gyroscope Electronic Control Units, and a Goddard High Resolution Spectrometer redundancy kit. The new solar arrays reduced the vibration caused by array motion as HST moved from orbital night to day.

STS-61 Rendezvous Burn Targeting Procedural Work-Around

In the months before the mission, a procedural work-around had to be developed to resolve a crew display limitation. Unlike STS-31, a contingency re-rendezvous after HST re-deploy during STS-61 could have occurred when the time since lift-off (Mission Elapsed Time, or MET) had a two digit value for days (10 or higher). The Orbit Targeting display used by the crew to compute on-board targeted Lambert maneuvers (NCC through MC4, Figure 24.6) could only accept a reference MET input by the crew, known as Base Time, for maneuvers with a

Approved for public release via STI DAA 24483. See statement on title page.

205

single digit day. All previous rendezvous profiles executed by the Shuttle Program occurred when the MET Base Time had a single digit day (less than 10). A procedural work-around was developed that manipulated display inputs so that targeting could be performed for maneuver Base Times with double digit days (10 or higher). This procedure was tested and verified on shuttle AP-101S computers in the NASA/Johnson Space Center Software Production Facility. A procedural work-around was preferable to a higher risk software patch. The crew display software was later modified to eliminate the need for the procedural work-around.

HST Roll Maneuver Before Grapple Discussion

During mission planning for STS-61, the HST Project was asked to summarize the risks of HST performing a roll to grapple attitude rather than the orbiter performing a fly-around maneuver to achieve the grapple attitude. The roll maneuver exposed HST to the potential of extended loss of electrical power, increased thermal heating and outgassing concerns, and could lead to battery depletion requiring re-charging later. Normally HST is restricted to power positive attitudes, and could only spend a limited amount of time in the grapple attitude. Although an on-board safing system protects HST against credible failure scenarios, and HST sequences were designed to minimize risk, there were some systems risks that could not be avoided. HST time in the grapple attitude could have been extended to three hours if required by: 1) reducing the electrical load for cases when the Sun was south of the orbital plane, or 2) slewing the solar arrays to an angle not normally flown. Electrical load reduction was undesirable and the solar array orientation was untested, and could expose the arrays to increased plume impingement. The advantage of deleting the roll maneuver in favor of an orbiter fly-around was that HST would remain in a power positive attitude, undesirable procedural work-arounds could be avoided, thermal heating and outgassing concerns would be eliminated, and the need for later battery re-charging would be minimized.

In place of the HST roll, a large orbiter yaw maneuver would have been required, with the magnitude dependent on the position of the solar line-of-sight with respect to the orbital plane. The maneuver would be performed in Low Z to minimize plume impingement, and would take place at a safe distance from HST to avoid contact between the RMS end effector and the HST solar arrays. Disadvantages of the orbiter yaw included magnification of orbital mechanics effects that could cause the two spacecraft to separate due to the larger than normal station-keeping distance, cross coupling from the Low Z mode that would increase the difficulty of piloting, difficulty of performing the yaw maneuver while station-keeping, and loss of RMS end effector camera range-rate cues for piloting. Additional concerns with the orbiter

yaw maneuver included increased propellant consumption, maneuver completion during orbital night, RMS end effector camera pointed at the Sun if the maneuver ended during orbital day, and increased plume impingent and plume contamination of HST. While the yaw maneuver was flyable, and is required as a contingency procedure regardless of the nominal mission plan, it was undesirable as a nominal procedure. The HST roll maneuver was retained (Figure 24.7), but the HST roll angle was selected that facilitated a single joint non-orthogonal capture with the RMS.

STS-61 Proximity Operations Piloting Aids

STS-61 carried three proximity operations piloting aids that were not in the shuttle baseline design in the 1970s. Two were hand held laser rangefinders, the Melios and the LTI. The third was a laptop computer (Payload General Support Computer, or PGSC) running the Payload Bay (PLBAY) program. PLBAY provided the crew with enhanced situational awareness of relative motion. PLBAY could accept laser range, radar range and elevation angle, and shuttle payload bay Closed Circuit Television Camera (CCTV) angle measurements for improving the estimate of relative motion provided the crew. Through the use of paper charts in the crew procedures book the crew could use radar angles, CCTV angles, and SPACEHAB camera angles to determine range out to 240 feet. COAS subtended angles could provide a rough range estimate out to 1500 feet. However angles were backup sources of range measurement to be used only if the rendezvous radar failed.

STS-61 Rendezvous and Proximity Operations

HST was successfully maneuvered to the rendezvous attitude and the aperture door closed on flight day 2. The nominal rendezvous was designed with HST grapple on flight day 3. The on-board targeted phase profile on the day of rendezvous was the standard stable orbit profile (Figure 24.6) that was also carried as a contingency for STS-31. The crew sighted HST near the start of the first star tracker pass (Figure 24.6). Two star tracker passes were performed before the first on-board targeted maneuver, Corrective Combination (NCC).[†] Radar data was incorporated after NCC. The Transition Initiation (Ti) maneuver targeted the orbiter for a HST intercept. Following Mid-course Correction 4 (MC4) the crew began the proximity operations phase and near-continuous manual trajectory control.

[†] "N" originally (1960s) was a Docking Initiation (DKI) targeting program counter variable for the number of the crossing of the chaser line of apsides where the maneuver was performed (as in 1 for first apogee, 1.5 for first perigee, 5 for fifth apogee, etc.). In documentation the burns were named N_C (Catch-up or phasing), N_H (Height), N_{PC} (Plane Change), N_{SR} (Slow Rate or coelliptic), and N_{CC} (Corrective Combination). See Appendix H.

After the orbiter Ti maneuver, HST was configured by the STOCC to reduce the electrical power required, in order to accommodate the HST roll to grapple attitude during proximity operations. At the start of the terminal phase (post MC4), the HST +V3 axis pointed at the sun. The solar arrays were aligned with the V1 axis and the −V3 HGA was stowed to maximize clearance for the RMS grapple. However, HST could have been grappled and berthed with the HGA deployed, if required. Approximately 20 minutes before the orbiter reached the +V Bar, HST began a roll maneuver to place the RMS grapple fixture on the north side of the orbital plane (Figure 24.8).

At a range of 400 feet the crew transitioned the flight control system to the Low Z mode to avoid plume over-pressure on the HST solar arrays (Figures 24.2 and 24.7). Increased fidelity plume impingement analysis resulted in a procedural change to place the flight control system in the Low Z mode at a range of 400 feet, rather than 200 feet as had been called out in the STS-31 contingency rendezvous procedures. This change was made to avoid structural damage to the HST solar array bi-stems.

The HST roll maneuver was completed by the time the orbiter arrived on the +V Bar, at orbital noon, at a range of approximately 350 feet (Figure 24.7). At this time the HST −V1 axis (end of HST opposite the aperture door) was aligned with the +V Bar and pointing at the orbiter (Figure 24.8). The crew continued the inertial approach until reaching the station-keeping range of 35 feet. The grapple was successful, and was scheduled to occur 10 minutes after orbital sunset to minimize shuttle camera blooming and permit completion of photography of solar array deflection during sunset.

During the flight, considerable work was done on possible changes to the solar array jettison procedure to account for the possibility of having to jettison one jammed solar array while the other fragile array was also stuck in a deployed state (Table 24.2). One of the original solar arrays did not retract when commanded, and was subsequently jettisoned by an EVA crew member attached to the end of the RMS. The other array was returned to Earth. After jettison, rotational and translational motion imparted to the solar array by shuttle RCS jet plume impingement was clearly visible to the crew and Mission Control personnel. Some personnel commented that the flapping motion of the array appeared to be like a prehistoric pterodactyl. It was estimated that 3 feet/second of delta-velocity was imparted to the array by RCS jet firings based on radar ground tracking and on-board laser measurements. Solar array motion heightened concerns about plume impingement on HST. The new solar arrays were installed and unfurled successfully. However, the new arrays had a noticeable twist that contributed to increased plume impingement concerns on later missions. Additional work during the flight focused on changes to the tool chasing procedure, and a separation that used the normal Z RCS jets (Figure 24.2).

STS-61 Deploy

Before HST deploy the shuttle performed a re-boost, circularizing the HST orbit at 321 nm. Starting with STS-61, the aperture door was opened before HST was deployed from the shuttle (Figure 24.9). If the door failed to open, the crew could perform an EVA with HST berthed in the payload bay to manually open the door. Unlike STS-31, no contingency re-rendezvous for the crew to manually open the aperture door during an EVA was planned for STS-61 or subsequent servicing missions. However, this did not preclude one from being performed, if required. The HST deploy and separation sequence was designed to be flexible to preserve a re-rendezvous capability.

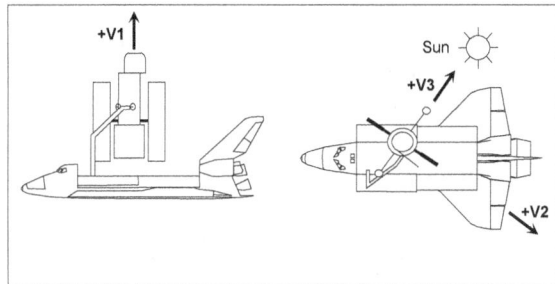

Figure 24.9   HST deployment for STS-61 and subsequent missions.   Note open door.

The nominal separation sequence provided safe post-deploy relative motion and minimized plume impingement, contamination, and propellant consumption (Figure 24.10). Nominal HST deploy was designed to occur at least 20 minutes before sunset. Ground communications with HST was required from before the opening of the deploy window to after deploy. Both HGAs were deployed before HST release from the RMS, with the solar arrays aligned with the V1 axis and the +V3 axis pointed at the sun. This deploy attitude was

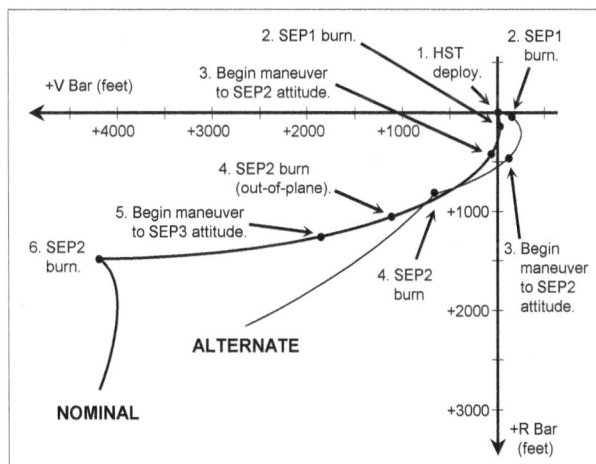

Figure 24.10 Nominal and alternate STS-61 HST deploy profiles.

207

optimal for power generation. An alternate separation sequence was developed late in the mission planning process that required less propellant, but it had a shorter deploy window. HST was successfully re-deployed and the alternate separation sequence was flown.

STS-82 – Servicing Mission 2 (SM2)

Two new science instruments were added to HST during the second servicing mission (Table 24.1). These were the Space Telescope Imaging Spectrograph and the Near Infrared Camera and Multi-Object Spectrometer. Hardware replacements included a refurbished Fine Guidance Sensor, a new Solid State Recorder, one new Reaction Wheel Assembly to replace one of the four original units, and the addition of an Optical Control Electronics Enhancement kit. Other maintenance items included replacement of one of the four Data Interface Units and replacement of one of the two Solar Array Drive Electronics units.

STS-82 Proximity Operations Piloting Aids

STS-82 was the first Hubble mission to carry the Rendezvous and Proximity Operations Program (RPOP). RPOP was originally based on the PLBAY program flown on STS-61, but much improved. Hand Held Lidar (HHL) units were also carried by the crew to obtain range and range rate measurements during the proximity operations phase. By the time of STS-82, RPOP and HHL had been proven on a number of shuttle missions to the Mir space station. The Trajectory Control Sensor (TCS) lidar normally carried in the payload bay was not carried on HST missions since HST does not have retro-reflectors that are compatible with it.

STS-82 Rendezvous and Proximity Operations

At the end of powered ascent, a 6.1 ft/sec Main Engine Cut-Off (MECO) under-speed occurred. This resulted in the re-planning by Mission Control of two burns on the day of rendezvous, before on-board sensor tracking started (Figure 24.6). Similar MECO under-speeds were seen on other HST servicing missions as well. While the under-speeds were within the design margins of the shuttle, mission planning for the later STS-125 and *Atlantis* rescue mission was performed to minimize MECO under-speed and subsequent rendezvous burn impacts.

The rendezvous profile flown by STS-82 was the same stable orbit profile flown by STS-61 (Figure 24.6). During the first star tracker pass, the star tracker lost lock on a dim HST and began tracking what was later determined to be the star Saiph. The relative navigation filter in the shuttle computer rejected two star tracker measurements and then momentarily re-established lock

on a slightly brighter HST. Lock on HST was lost again and the star tracker acquired what was apparently nearby orbital debris. The navigation state was corrupted by three navigation updates during the debris tracking period. A crew command to inhibit navigation processing was not accepted by the shuttle computer due to a known timing issue. The star tracker re-acquired HST and subsequent measurements corrected the error introduced by the spurious measurements. The crew replaced the state vector that had received spurious updates with a backup vector. The star tracker pass continued without incident. Post flight analysis indicated that the HST solar arrays were parallel to the star tracker's line-of-sight and pointed to the Sun. The end of the HST (the V1 axis, Figure 24.1) was pointed to the orbiter. This combination of HST attitude with the sun 90 degrees from the star tracker line-of-sight resulted in a dim target, causing the star tracker to lose lock on HST.

After the first star tracker pass, an additional unplanned out-of-plane correction maneuver was performed based on ground radar tracking data and the results of the first star tracker pass (Figure 24.6). Out-of-plane corrections during the rest of the rendezvous were minor. The remaining rendezvous and grapple activities were nominal.

While the rendezvous profiles for STS-61 and STS-82 were the same, STS-82 flew a different final approach during proximity operations. Just after MC4 the crew transitioned from the inertial approach to a lower energy +R Bar approach (Figure 24.11). The +R Bar approach (Figure 24.12) was developed for the shuttle missions to Mir and the ISS in 1994. It was first flown on STS-66 (November 1994) during the rendezvous with and retrieval of the CRISTA-SPAS deployed payload.[13]

The primary advantage of the new approach was natural orbital mechanics braking. This reduced the risk of plume impingement as fewer RCS jet firings were required. The natural orbital mechanics braking allowed the HST Low Z range constraint to be increased to 1500 ft to provide additional plume protection as the HST slowly rotated above the approaching shuttle (Table 24.3). A +R Bar approach also provided a hands-off separation, that required no RCS jet firings due to orbital mechanics. Once the range to HST was less than 150 feet, the crew would station-keep on the +R Bar and wait for the HST –V1 axis to align with the orbiter –Z axis (Figures 24.12 and 24.13). Once the axes were aligned, the crew would establish an inertial attitude hold and perform an inertial approach to the 35 foot station-keeping range for RMS grapple of HST.

Table 24.3 Comparison of Braking or Range Rate Gates and Low Z Initiation for Nominal and Contingency Rendezvous Profiles

Range (feet)	Inertial Approach from Coelliptic [A] (feet/sec)	STS-31 Inertial Approach from Stable Orbit (feet/sec)	STS-61 Inertial Approach from Stable Orbit (feet/sec)	STS-82 +R Bar Approach from Stable Orbit (feet/sec)	STS-103 +R Bar Approach from ORBT (feet/sec)	STS-103 [B] Inertial Approach from Stable Orbit (feet/sec)	STS-109 +R Bar Approach from ORBT (feet/sec)	STS-109 [B] Inertial Approach from Stable Orbit (feet/sec)	STS-125 +R Bar Approach from ORBT (feet/sec)	STS-125 [B] Inertial Approach from Stable Orbit (feet/sec)
9000	45									
8000	40									
7000	35									
6000	30									
5000	25									
4000	20									
3000	15									
2000	10	4	4	4	3	4	3	4	3	4
1700							2.6		2.6	
1500		2	2	2, LOW Z	2.3, LOW Z	2, LOW Z	2.3, LOW Z	2, LOW Z	2.3, LOW Z	2, LOW Z
1000	5	1	1	1.5	1.5	1	1.5 [C]	1	1.5 [C]	1
900						0.9		0.9		0.9
800						0.8		0.8		0.8
700						0.7		0.7		0.7
600				0.8	0.8	0.6	0.8		0.8	
500		0.5	0.5	0.5	0.5	0.5	0.5	0.5	0.5	0.5
400	4	0.4	0.4, LOW Z	0.4	0.4	0.4	0.4	0.4	0.4	0.4
300	3	0.3	0.3	0.3	0.3	0.3	0.3	0.3	0.3	0.3
200	2	0.2, LOW Z	0.2	0.2	0.2	0.2	0.2	0.2	0.2	0.2
120				0.2 [D]	0.2 [D]	0.2 [D]	0.2	0.2	0.2	0.2
100	0		0.1				0.1	0.1	0.1	0.1
35		0	0	0	0	0	0	0	0	0

[A] 1970s Shuttle baseline dual coelliptic profile with second coelliptic delta height of 10 nm. Low Z initiation range payload dependent.
[B] Contingency procedure.
[C] Gradually reduce approach rate to value at next braking gate.
[D] Gradually reduce to 0.1 feet/sec.

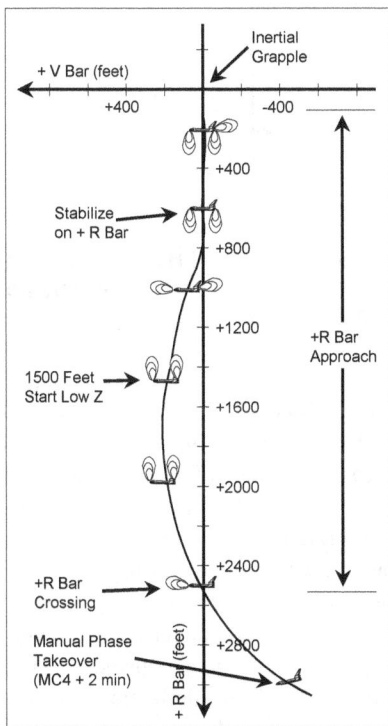

Figure 24.11 STS-82 +R Bar approach from stable orbit profile.

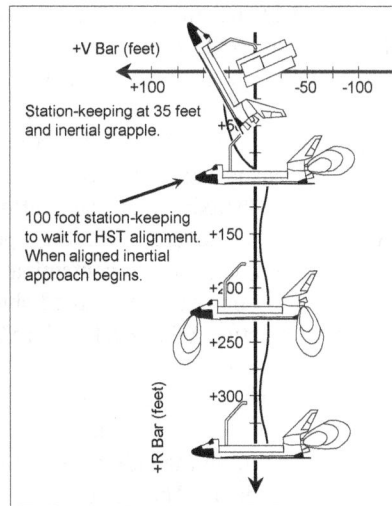

Figure 24.12 Final +R Bar approach and inertial grapple.

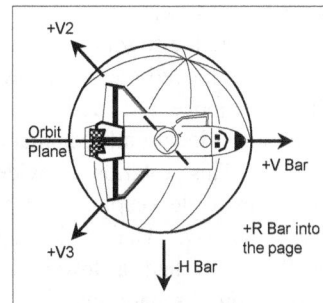

Figure 24.13 Bird's eye view of HST in nominal grapple attitude for the +R Bar approach.

STS-82 Contingency Inertial Approach

If the aperture door failed to close before rendezvous, the unprotected HST optics could be pointed at an orbiter performing a +R Bar approach (Figure 24.12). This would result in unacceptable contamination of the HST optical system. To protect for the aperture door failed open case an inertial proximity operations approach, like that flown on STS-61, would have been performed (Figure 24.7). During an inertial approach the open aperture door would be pointed away from the approaching orbiter during proximity operations. The inertial approach, however, meant increased propellant consumption. For STS-82, the inertial approach procedures were not part of the rendezvous procedures book flown on the orbiter, but they would have been uplinked to the crew, if required.

STS-82 Nominal Deploy

STS-82 included a new deploy requirement as ultraviolet light reflected off of the Earth might enter the telescope when the aperture door was open. The ultraviolet light could cause any contamination that might accumulate on the mirror during the servicing mission to permanently adhere to the mirror. The STS-82 deploy procedure had the same RMS position as on STS-61, but a new orbiter attitude. The new requirement was to point the HST +V1 axis away from the bright Earth limb. HST was to be released in daylight before sunset to allow adequate HST sun sensor acquisition time. The release attitude pointed the +V3 axis at the sun. Both HGAs were deployed. The overall deploy procedures minimized plume impingement, contamination, and propellant use. Deploy design also ensured shuttle crew and ground communication with HST before and after release.

Two deploy and separation profiles were prepared for the mission (Figure 24.14 and 26.15). The appropriate profile was chosen based on the side of the orbital plane where the Sun was located. The initial HST separation burn was changed based on experience from procedures developed for deployments of spacecraft equipped with a solid rocket motor, such as the Inertial Upper Stage. The first separation burn was performed with two forward firing –X RCS jets while the flight control system was in free drift. As this burn moved the orbiter away from the HST, the +Z thrust component of the forward jets caused the orbiter to pitch nose-down until commanded to stop a short while later. This rotation provided adequate clearance to the cabin while keeping HST visible to the crew over the payload bay. The –X jet separation also used less propellant and had a lower risk of plume impingement than a Low Z separation.

HST re-deploy was nominal. However, the second burn in the two burn separation sequence was under-burned (Figure 24.14). While the post-burn

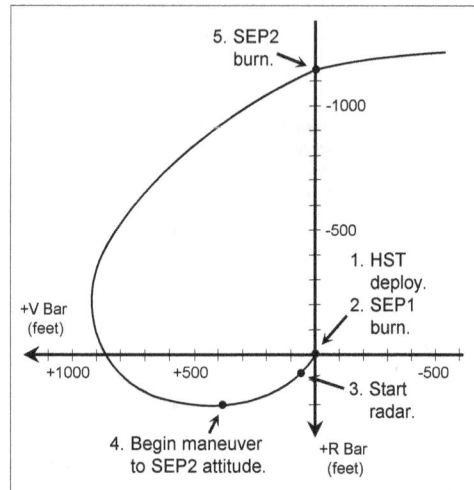

Figure 24.14 Nominal HST deploy profile for Sun north of the orbital plane, missions 82, 103, 109, and 125.

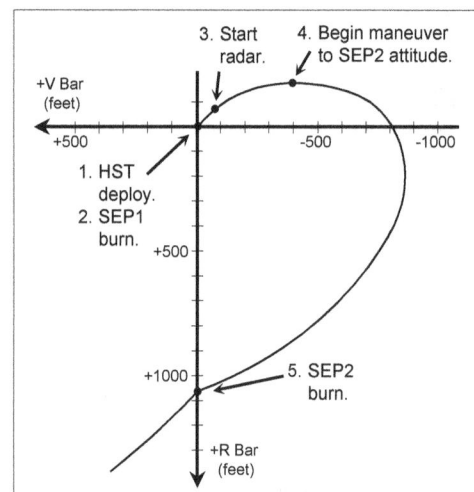

Figure 24.15 Nominal HST deploy profile for Sun south of the orbital plane, missions 82, 103, 109, and 125.

relative motion placed *Discovery* on a safe departure trajectory, the separation rate was less than desired. In addition, cross-coupling, of RCS attitude control firings into translational motion, threatened to further decrease the separation rate of *Discovery*. A third separation burn was computed by Mission Control. Burn data was voiced to the crew and the burn was executed. Post-flight analysis indicated that the under-burn was due to the high rate of Translational Hand Controller (THC) deflection. A restriction on the THC deflection rate was known at one time, but the constraint had not been included in the crew procedure. The procedure was later modified for later flights to replace large numbers of pulses with a single continuous THC deflection, and crew training was improved to increase awareness of the deflection rate limit.

STS-103 – Servicing Mission 3A (SM3A)

The third servicing mission (SM3) was originally planned for June of 2000. However, in February of 1999 a third gyroscope failure occurred. While HST was capable of supporting science activities with no fewer than three gyroscopes, NASA decided to re-schedule the third repair mission to fly before the end of 1999 and replace the failed gyroscopes. Some hardware originally scheduled for the original SM3 mission in 2000 was not ready to support a flight in 1999. As a result, SM3 was split into two missions, SM3A (STS-103) and SM3B (STS-109). Replacement hardware, not available to support the 1999 SM3A mission, was redirected to the newly defined SM3B (STS-109) mission that was later flown in March of 2002.

SM3A servicing objectives included replacement of all six gyroscopes, a new computer, replacement of one of three Fine Guidance Sensors, an aft shroud latches repair, installation of handrail covers, a new outer blanket layer, a new S-Band Single Access Transmitter, shell/shield replacement fabric, and voltage/temperature improvement kits for the batteries.‡ SM3A did not install any new scientific instruments. The failure of a fourth gyroscope on November 19, 1999, a month before the SM3A launch, resulted in HST entering a safe mode. Astronomical observations could not be performed while in safe mode. Significant crew training resources were expended to develop and refine manual piloting techniques to approach and grapple HST in the HWSP and ZGSP safe mode configurations. SM3A mission planning ensured that the deorbit and landing would occur in 1999 to avoid any potential year 2000 rollover computer issues.

STS-103 Rendezvous and Proximity Operations

The previous mission, STS-82, flew a stable orbit profile with a +R Bar final approach (Figure 24.6 and 24.11). However, the stable orbit profile baselined for the shuttle in April of 1983 was originally designed to support inertial approaches with lower energy than inertial approaches from the Apollo legacy coelliptic profile.[13] While the stable orbit/+R Bar combination was successfully flown on a number of missions, starting with STS-66 in November of 1994, stable orbit was not a propellant optimal profile to support a +R Bar approach. A new version of stable orbit rendezvous, Optimized R Bar Targeted (ORBT) rendezvous, was specifically designed to support the +R Bar technique (Figure 24.16 and 24.17). ORBT required fewer jet firings for +R Bar trajectory stabilization and braking than stable orbit. ORBT was first flown on the STS-86 mission to Mir in September-October 1997. STS-103 was the first HST servicing mission to fly the ORBT/+R Bar combination. The change from the stable orbit to the ORBT profile resulted in some differences in STS-82 and STS-103 +R Bar approach procedures.

On the day of rendezvous, during the first star tracker pass, the Moon approached the star tracker line-of-sight to the HST. Anticipating that the bright Moon would cause an automatic closure of the star tracker shutter, flight controllers prepared for the event by providing the crew with times to inhibit star tracker measurements as the Moon passed through the star tracker field of view. However, the star tracker Bright Object Sensor did not close the shutter in response to the Moon until the Moon was well inside the field of view. In response the crew inhibited star tracker measurements for approximately

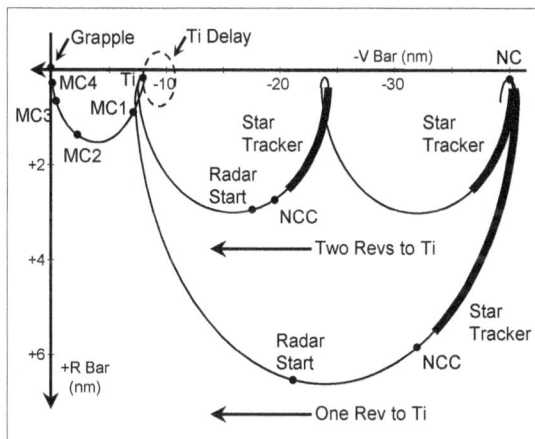

Figure 24.16 Two revs (STS-103) and one rev (STS-109 & -125) to Ti Optimized R Bar Targeted (ORBT) rendezvous profiles.

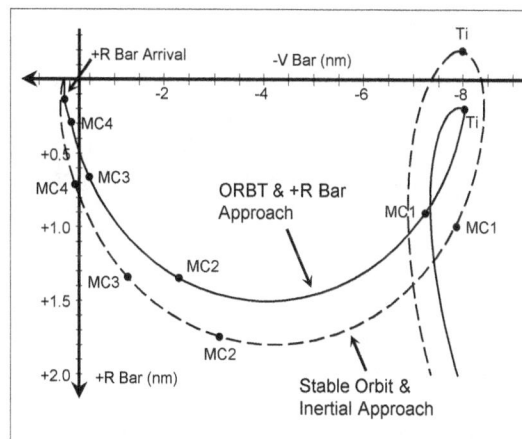

Figure 24.17 Comparison of ORBT and contingency stable orbit rendezvous profiles for STS-103, STS-109, and STS-125.

‡ S. Lee, S. Anandakrishnan, C. Connor, E. Moy, D. Smith, M. Myslinski, L. Markley, and A. Vernacchio, "Hubble Space Telescope Servicing Missions 3A Rendezvous Operations," *2001 Flight Mechanics Symposium*, NASA/CP-2001-209986, NASA Goddard Space Flight Center, June 19-21, 2001, pages 529-543.

seven and a half minutes during the first pass and for approximately eight minutes during the second pass (Figure 24.16). Sufficient navigation data was collected during the two passes. Some noise in the radar angle measurements was noted after the MC4 burn. This was normal and the noise seen on HST missions is much less than that observed on ISS missions.

Due to the fourth gyro failure on November 19, 1999, the HST Program chose to maintain attitude using the Hardware Sunpoint mode and back-up gyros (RMGA). Fortunately, HST entered the Hardware Sunpoint mode with the V1 axis very close to the orbital plane. The crew was able to confirm the Hardware Sunpoint attitude using binoculars about an hour before the grapple. After a nominal +R bar approach the crew executed a 90 degree yaw maneuver on the +R Bar to achieve the grapple attitude. HST was grappled with *Discovery* on the +V1 aperture door end of HST (Figure 24.1). Propellant consumption during proximity operations was higher than expected. Possible causes included noisy radar range rate measurements, and RCS jet cross coupling during the yaw maneuver. Low Z braking starting at 1500 feet also increased propellant consumption. Had the V1 axis not been close to the orbital plane, the grapple alignment maneuver would have been more complicated than a simple yaw and cost even more propellant. In addition, the Sun was close to the orbital plane and above HST, complicating observation of HST and washing out displays in the shuttle cockpit.

STS-103 Contingency Stable Orbit and Inertial Approach

Starting with STS-103, the nominal rendezvous profile was ORBT, supporting a +R Bar approach (Figures 24.12, 24.16, and 24.17). However, should the aperture door fail to close before rendezvous, a +R Bar approach could expose the HST optics to RCS jet plume contamination. Like STS-82, in the event that the HST aperture door failed to close before rendezvous, the orbiter would perform an inertial approach (Figure 24.7).[*] However, this inertial approach would be flown from a legacy stable orbit profile (Figure 24.6, 24.17, and 24.18), instead of the nominal ORBT profile (Figure 24.16, 24.17, and 24.18).

With the proper timing, the failed open HST aperture door could be pointed away from the approaching orbiter throughout an inertial approach, to minimize risk of optics contamination. Execution of a contingency stable orbit would have required re-planning of the last ground targeted maneuver by Mission Control (NC in Figure 24.6 and 26.16). HST attitude would be managed so that the −V1 axis would be pointed at the orbiter at the MC4 + 2 minute point, where the crew took manual control and

---

[*] The exact position and attitude of the orbiter in the LVLH frame at grapple was dependent on the timing of the approach since both the orbiter and HST were in inertial attitude holds.

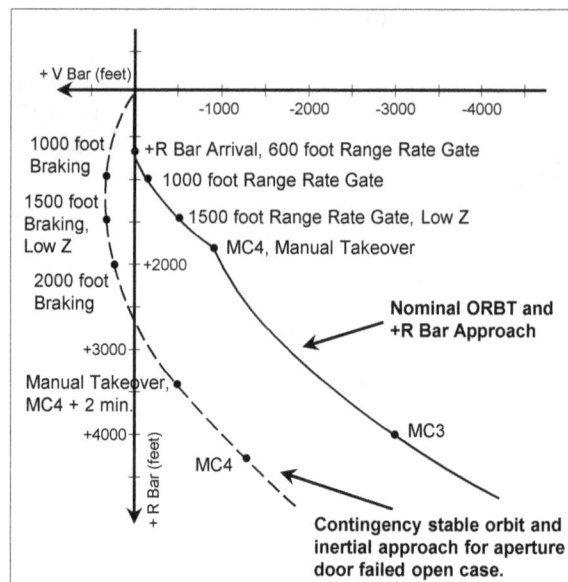

Figure 24.18 Comparison of nominal and contingency approaches to HST for STS-103, STS-109, and STS-125.

placed the orbiter in an inertial attitude hold (Figure 26.18). At +V Bar arrival the −V1 axis would be pointed at the orbiter payload bay. This combination of HST attitude and inertial approach ensured that the failed open aperture door would always be pointed away from the approaching orbiter during proximity operations, minimizing the risk of plume contamination of HST optics. However, as the aperture door had closed at the time of the fourth gyro failure on November 19, there was no need to protect for this contingency.

STS-103 Deploy

STS-103 mission planning included the same nominal deploy sequence options as STS-82 (Figure 24.14 and 26.15). Re-deployment of HST and separation by *Discovery* were nominal (Figure 24.15).

STS-109 – Servicing Mission 3B (SM3B)

Servicing Mission SM3B (March 2002) placed new hardware on HST that was not ready in time to support the SM3A mission flown in December of 1999 (Table 24.1). Maintenance activities included an Advanced Camera for Surveys to replace the Faint Object Camera, replacement of a power control unit, one of four reaction wheel assemblies, and a new cooling system for the Near Infrared Camera and Multi-Object Spectrometer. With the replacement of the Faint Object Camera, none of the optical sensors required the corrective optics installed in HST with COSTAR on STS-61 in December of 1993. In addition, new solar arrays were installed that had more rigidity, produced more power, and were smaller than the arrays installed during STS-61.

STS-109 Rendezvous and Proximity Operations

The rate of orbital decay of HST resulted in a rendezvous altitude that was lower than the orbital insertion altitude. To compensate, the normally posigrade catch-up maneuver (NC in Figure 24.16) before the first star tracker pass on the day of rendezvous was retrograde.

The ORBT rendezvous profile for STS-109 was modified from two revolutions to Ti to one revolution to Ti (Figure 24.16), taking advantage of experience gained in numerous shortened rendezvous profiles flown on missions to the Mir space station and ISS. This eliminated one star tracker pass that had mainly served as a backup relative navigation opportunity, but provided extra time in the crew day in the timeline after HST grapple. Other aspects of the STS-109 ORBT profile were the same as STS-103. STS-109 also carried the same +R Bar approach procedure as STS-103 (Figures 24.12, 24.13, 24.16, and 24.17).

The +R Bar approach proceeded more slowly than in ground simulations. This was consistent with previous missions and likely due to noisy range rate measurements and the difficulty of viewing HST against the Sun. Welding goggles were used by the crew to view HST, but the goggles made it difficult to observe displays in the cockpit. Proximity operations propellant consumption was higher than predicted, but within acceptable margins. HST was successfully grappled.

STS-109 Contingency Stable Orbit and Inertial Approach

STS-109 carried the same contingency stable orbit and inertial approach procedures as STS-103 (Figures 20.17 and 20.18). Unlike the stable orbit profile in Figure 24.6, the STS-109 stable orbit profile would have been one revolution between NC and Ti. However, these procedures were not performed as the HST aperture door was successfully closed before rendezvous.

STS-109 Deploy

STS-109 mission planning included the same nominal deploy procedures as STS-82 and STS-103 (Figures 20.14 and 20.15). HST was successfully deployed (Figure 24.14).

The *Columbia* Accident, Robotic Servicing, and Servicing Mission *4*

At the time of the loss of *Columbia* (February 1, 2003) two further missions to HST were on the Shuttle Program flight planning manifest. SM4 (STS-123) would be flown by *Columbia* on November 18, 2004. Another mission by *Columbia* (STS-144), with a projected launch date of November 19, 2009, was to return HST to Earth if safety and payload bay structural issues could be resolved. However, the loss of *Columbia* and her crew resulted in significant changes to planning for future shuttle missions to the HST.

As a part of the Return to Flight (RTF) activity after the *Columbia* accident, NASA developed a plan to provide Contingency Shuttle Crew Support (CSCS) for a shuttle crew stranded on the ISS due to a compromised shuttle Thermal Protection System (TPS). The crew would remain on the ISS until the shuttle TPS was repaired, or another shuttle was launched to the ISS to return the stranded crew to Earth. However, the ISS would be available to a shuttle crew as a safe haven since the shuttle does not carry enough propellant to transfer from a HST servicing mission orbit to an ISS orbit.[15]

On January 16, 2004, NASA Administrator Sean O'Keefe announced that no more HST servicing missions would be flown since the ISS would not be accessible to the shuttle crew for a safe haven. Future shuttle missions would be limited to those flying to the ISS. However, without another servicing mission, it was expected that HST degraded hardware (gyroscopes, batteries) performance would not permit HST to perform astronomy beyond the year 2010.

On February 20, 2004, NASA issued a Request for Information to industry concerning the feasibility of a robotic servicing and deorbit mission to HST, known as the Hubble Robotic Servicing and De-orbit Mission (HRSDM).[9] This mission would extend the life of the telescope and permit a safe deorbit over the Pacific Ocean once HST was no longer capable of performing the astronomy mission.[16] In September of 2004 NASA awarded a contract to Lockheed Martin to build the de-orbit module for the HST Robotic Vehicle (HRV). The following month MacDonald, Dettwiler and Associates Ltd. was awarded a contract to build the robotic system for the HRV. The NASA Goddard Space Flight Center was to develop the HRV Ejection Module where the robotics would be mounted. HRV faced significant challenges to develop and certify robotic servicing, and automated and autonomous rendezvous, proximity operations, and grapple in time to meet the proposed December 14, 2007 launch date. HRV development was canceled on February 8, 2005, due to cost, schedule, and technical risk concerns.

Successful shuttle missions in 2005 and 2006, along with successful development of TPS inspection and repair methods, led NASA Administrator Michael Griffin to announce in October of 2006 that NASA would fly another servicing mission before the end of the Shuttle Program in 2010. NASA re-examined the risk of an HST mission and the use of existing TPS inspection and repair methods. Shuttle TPS could be inspected and repaired by the crew using only equipment carried on the orbiter. In addition, the concept of using another shuttle to rescue the servicing mission crew was determined to be feasible.

STS-125 – Servicing Mission 4 (SM4)

The primary objective of SM4 (May 2009) was the installation of two new scientific instruments, the Wide Field Camera 3 (WFC3) and the Cosmic Origins Spectrograph (COS).[17, 18] The COSTAR, installed during STS-61 to correct the spherical aberration of the primary mirror, was removed to make room for the COS and returned to Earth. New scientific instruments installed since STS-61 in 1993 had corrective optics and COSTAR was no longer needed. WFC2 was removed from HST as well. The Advanced Camera for Surveys partially failed in 2007 due to an electrical short and it was repaired. The Space Telescope Imaging Spectrograph suffered a power failure in 2004 and was also repaired. In addition, all six gyroscopes and batteries were replaced. One of three Fine Guidance Sensors was replaced and new Outer Blanket Layer insulation was installed. The crew also replaced the Science Instrument Control & Data Handling (SIC&DH) unit. Side A of the HST Control Unit/Science Data Formatter within the SIC&DH failed on September 27, 2008. Side B supported astronomy activities after the failure.

SM4 also mounted a Low Impact Docking System (LIDS) passive interface on the Hubble aft bulkhead. LIDS was developed as the docking hardware for the Constellation Program. LIDS will enable future human or robotic vehicles to dock with HST for servicing or HST deorbit. The Hubble soft capture mechanism, including the LIDS passive interface, was attached to the HST berthing pins that were used to berth Hubble to the Flight Support System (FSS) in the shuttle payload bay. Four retro-reflectors were also be mounted on the assembly to support relative navigation sensors of future human or robotic vehicles.[19, 20] The retro-reflectors are designed to work with the lidar sensors that were commercially available at the time of the Hubble Robotic Servicing and Deorbit Mission design effort.[8] Painted patterns on the target assembly are designed to work with future optical recognition algorithms. The Relative Navigation Sensor (RNS), a NASA Goddard sensor, was also flown for testing during proximity operations.[19, 20]

Impact of HST Hardware Failure

By September of 2008 STS-125 (*Atlantis*) launch was scheduled for October 8. *Endeavour* was assigned the role of the Launch On Need (LON) rescue vehicle, mission STS-400. If the LON mission was required *Endeavour* could launch within 10 days of the launch of *Atlantis*. *Atlantis* was rolled from the Vehicle Assembly Building (VAB) to Pad 39A on Thursday, September 4, 2008. *Endeavour* was rolled from the VAB to Pad 39B on Friday, September 19. This was the first time since July 2001 that two shuttles were on the launch pads at the same time. However, on September 27, side A of the HST Control Unit/Science Data Formatter (CUSDF)

failed, preventing the transmission of science data to Earth. The unit could not be reset. This caused NASA to delay of the launch of STS-125 to no earlier than February of 2009, so that the STS-125 EVA crew could train to replace the malfunctioning hardware. *Atlantis* was rolled back to the VAB. *Endeavour* was moved to pad 39A for the STS-126 mission to ISS (flown November 2008).

The delay resulted in a switch of the LON rescue mission from *Endeavour* to the orbiter *Discovery*. The *Discovery* LON was designated STS-401. After STS-125 flew, no earlier than February 2009, and if no STS-401 mission was required, *Discovery* would fly STS-119. However, the STS-125 launch slipped past the launch date for STS-119. *Discovery* flew the STS-119 mission before STS-125 in March 2009. *Endeavour* was reassigned the role of rescue orbiter and the previous LON designation of STS-400 was re-established.

The original LON crew was the flight deck crew from STS-123 (*Endeavour*, March of 2008). They were Dominic Gorie (commander), Gregory H. Johnson (pilot), Robert Behnken (Mission Specialist 1), and Michael Foreman (Mission Specialist 2). By February of 2009 they had been replaced by the flight deck crew from STS-126 (*Endeavour*, November of 2008). The new LON crew was Chris Ferguson (commander), Eric Boe (pilot), Robert Kimbrough (Mission Specialist 1), and Stephen Bowen (Mission Specialist 2).

STS-125 Propellant, Deorbit, and Landing Challenges

The high HST orbital altitude increased the risk of a collision with orbital debris as the lower atmospheric density does not cause the debris to decay as rapidly as debris at lower orbital altitudes. Increased concern about orbital debris at the HST orbital altitude led the Shuttle Program to reduce the amount of time the orbiter is at that altitude between HST deployment and the deorbit burn. In addition, it is necessary that any maneuvers performed by the orbiter after HST deploy contribute to deorbit. Placing the orbiter in an elliptical orbit before the deorbit burn limits the landing opportunities since deorbit burns near perigee are far more costly in propellant. This is true even for landing sites that are within the orbiter's entry cross-range capability. The post deorbit burn perigee has to be placed at an appropriate latitude for a landing at the Kennedy Space Center or Edwards Air Force Base.

STS-125 Nominal and Contingency Procedures for Rendezvous, Proximity Operations, and Deploy

STS-125 carried the same ORBT rendezvous profile and +R Bar approach procedures as STS-109 (Figures 24.12, 24.13, 24.16, and 24.17). In the event that the HST aperture door failed to close before final approach, STS-125 would have performed the same contingency stable orbit and inertial approach that was prepared for STS-103 and STS-109. The nominal STS-125 deploy sequence was the same as that flown on STS-82, STS-103, and STS-109 (Figures 24.14 and 24.15).

STS-125 Proximity Operations Piloting Aids

The crew would used HHL as a piloting aid during proximity operations, as was done on previous HST missions. TCS was not flown since HST is not equipped with TCS retro-reflectors. However, a TCS retro-reflector would have been installed in the *Atlantis* payload bay on the ODS truss during a contingency EVA to support the rescue orbiter contingency rendezvous.

STS-125 *Atlantis Rendezvous With HST*

STS-125 was launched on Monday, May 11, 2009, from launch pad 39A.[21] The launch window was 66 minutes long. The crew timeline was packed and a slip in the launch time within the 66 minute window would result in a loss of crew timeline equal to the slip. However, the preferred (and actual) launch time was 20 minutes inside the launch window to buy back ascent performance margin.

At MECO there was a 2 foot/second over-speed. The post OMS-2 perigee was changed from 108 nm to 107 nm to account for this.

Pre-mission trajectory analysis indicated that the post Main Propulsion System dump would result in an apogee that was too high. The high apogee could result in a NC-4 burn on the day of rendezvous of less than 2 feet/second. During rendezvous with HST the orbiter's perigee is raised in a controlled manner. If apogee is already high, NC-4 and NH on the day of rendezvous can become very small and possibly retrograde. It is desirable to avoid retrograde maneuvers.

A proposal was made to delete the post External Tank +X RCS burn. Deleting the 5 foot/second burn after ET separation would add 5 feet/second to NC-4. The post External Tank separation +X RCS burn is performed so that the orbiter umbilical well cameras can photograph the entire length of the External Tank. A pitch rate from the maneuver enables the crew to later photograph the External Tank from long range through the aft cockpit overhead windows. However, the proposal was not approved and the +X RCS burn was performed.

If un-repairable TPS damage had been detected during the inspections on FD2 (Flight Day 2), *Atlantis* systems and consumables (such as power and oxygen) could have been managed to keep the crew alive for up to 24 days. If the late TPS inspection on FD10 were to detect un-repairable TPS damage, the crew could have been supported for up to 16.5 days. A STS-400 rescue of *Atlantis* by *Endeavour* could have been conducted no earlier than 15 days and 16 hours after the inspection revealed the damaged TPS.

On FD2 the HST STOCC completed all preparations for rendezvous on FD3. The third rate sensor gyro was activated and added to the control loop, the telescope aperture door was closed, the high gain antennas were stowed, and HST was maneuvered to the rendezvous attitude. The NC-2 burn on FD2 was so small it was not performed.

On FD2 the small predicted size (Delta Velocity, or DV, < 4 feet/second) of NC-4 on FD3 led to the development of multi-axis RCS burn procedures for that maneuver. NC-4 is normally an OMS burn, and modified procedures were uplinked to the crew. The procedures also called for the orbiter to be placed in target track after the FD3 NH maneuver.

A simultaneous supply water (fuel cell water) and waste water dump was completed approximately 30 minutes before the NH burn on FD3. The dump imparted about 0.6 to 0.7 feet/second of DV to the vehicle, increasing the orbital semi-major axis by about 1100 feet. Before the dump the predicted NC-4 DV was 1.5 feet/second posigrade. After the dump it was -0.4 feet/second retrograde.

The NC-4 DV was so small that it was not burned, possibly a first for the Shuttle Program. The star tracker pass was nominal and the crew visually sighted HST during the pass. Star tracker and radar performance during the rendezvous were nominal. The MC-1 and MC-2 DVs were small and not executed.

Between the Ti and MC-1 burns, rendezvous Payload Interrogator (PI) communications was established between *Atlantis* and HST. Between the MC-1 and MC-2 burns the crew reported PI communication problems with HST. The shuttle was unable to lock onto or process HST telemetry. A check by the crew of switch positions and cables associated with HST communications did not resolve the problem. The ground determined that the Bit Synch Assembly (BSA) apparently failed resulting in loss of ability to establish communications with HST on Payload (PL) string 1. However, the crew and STOCC could command over PL string 2, but commands could only be verified by the STOCC using TDRSS communications. The crew could not receive telemetry from HST, and therefore could not verify the results of commanding. The crew could command HST in the blind, but command verification would have to be performed by the STOCC.

HST telemetry reception was restored through TDRSS direct communications with about 32 minutes remaining in the rendezvous. The remaining commands were sent through *Atlantis* with confirmation of commanding performed by the HST STOCC using TDRSS direct communications. The *Atlantis* crew used PL string 2 to command HST. However, this prevented ground commanding of the Relative Navigation Sensor (RNS) payload on HST over PL string 2.[19, 20] The crew was able to command the RNS payload via switch throws in the *Atlantis* cockpit.

Due to periods of no HST data, time spent troubleshooting the communications problem, and a reluctance to command HST in the blind, the execution of the planned pre-flight commands to prepare HST for grapple was delayed by about an hour. Normally the STOCC was to command HST to perform a roll maneuver (~42 degrees) to achieve the grapple attitude. This maneuver takes approximately 10 minutes, and is started at about the nominal MC-4 execution time. If HST was commanded to perform the roll maneuver, the remaining scheduled TDRSS communications time might not have been long enough to allow the roll maneuver to complete while TDRSS was still available. It was critical for the HST STOCC to stop the roll maneuver before TDRSS communications were lost. The shuttle and HST control teams agreed to replace the HST roll maneuver with an *Atlantis* yaw maneuver. This also saved 16 minutes of HST commanding time.

After MC-4 the crew noted that the orbiter was further out-of-plane than expected. RCS jet firings to null out-of-plane velocity resulted in a faster closure rate from a range of about 1,200 feet to about 500 feet. Out-of-plane position was at a nominal value by 400 feet. An additional contributing factor to the faster approach was a lack of HHL measurements between 3,300 and 1,300 feet. The HST was in an inertial attitude hold and appeared to rotate with respect to the orbiter. Good HHL marks could only be obtained when the aperture door was facing the orbiter. Marks were difficult to obtain at other times due to a lack of flat reflective surfaces facing the orbiter. Gaps in HHL marks also occurred during the three previous flights.

Rendezvous radar measurements became noisy within 400 feet, a lower range than on ISS missions due to the smaller size of HST. The payload bay keel camera was not always available for piloting cues due to the Sun in the camera field of view.

The crew performed about 30 minutes of station-keeping on the +R Bar and flew a slower approach due to an early *Atlantis* arrival at the 150 foot point. Station-keeping was necessary to wait for the HST V1 axis to rotate into alignment with the +R Bar. The 150 foot station-keeping was also performed since the HST roll to grapple maneuver had been canceled.

After re-initiating the +R Bar approach from the 150 foot station-keeping point, the crew performed an Auto Inertial Flyaround Alignment contingency procedure to yaw *Atlantis* by ~42 degrees and achieve the grapple attitude. Waiting for the HST V1 axis to align with the +R Bar enabled *Atlantis* to fly a more propellant efficient approach to the grapple relative position and attitude. An orbiter fly-around to the grapple attitude would have consumed more RCS propellant.

Propellant consumption during proximity operations was higher than originally planned. Approximately 100 lbs more forward RCS propellant and 100 lbs more aft RCS propellant was consumed. However, propellant margins in both the forward and aft RCS tanks were still adequate to support the nominal end of mission (HST deploy and separation, deorbit, and entry). The higher than anticipated propellant consumption was caused by the extended +R Bar station-keeping, a faster than normal +R Bar approach to the 150 foot point that required Low Z braking, and the yaw maneuver. Radar and HHL performance during proximity operations were nominal. Due to the small size of HST the radar did not wander as it does on ISS missions.

Later investigation identified the cause of the communications problem. The HST Data Management Unit Communications Module had not been reconfigured from the 1 Mbps science format to the 32 Kbps rate required for downlink through the Shuttle PI. Full HST and RNS telemetry and commanding capability was later restored.

The post HST deploy Orbital Adjustment (OA) burn placed *Atlantis* in an elliptical orbit to reduce the probability that *Atlantis* could encounter orbital debris that do not decay as rapidly at the high orbital altitude of HST. An additional requirement for the OA burn was to preserve two consecutive landing opportunities at KSC, Edwards Air Force Base, and Northrop Strip for the nominal End of Mission (EOM) day, EOM + 1 day, and EOM + 2 days. In addition, these landing opportunities were required to the last and next-to-the-last for each day at each of the three landing sites. These requirements were necessary for an effective crew timeline.

The post HST release OA burn to mitigate orbital debris risk at HST altitude significantly reduced the propellant margin of the vehicle. For this reason a HST re-boost using the shuttle vernier jets was not planned pre-mission. However, a request was made during the mission for a re-boost to reduce the risk of HST encountering orbital debris from the servicing period. The proposal was withdrawn when analysis indicated that raising the orbit of HST would increase the probability of encountering the debris. A re-boost was not performed by *Columbia* on STS-109 (March 2002) due to the heavy structural weight of the orbiter. A re-boost was performed by *Discovery* on STS-103 (December 1999).

A change to the pre-mission timeline was moving the OA burn from FD10 to FD9, one rev after the SEP-2 maneuver. The night before HST release a new SEP-2 DV was uplinked to the crew. The DV was increased to 6 feet/second to ensure safe relative motion after the orbital adjustment maneuver. HST release was delayed by 4 minutes due to intermittent communications. A negative beta angle (Sun to the south of the orbital plane) separation was flown on FD9 after HST was released by the RMS. About an hour and a half after SEP-2 the OA burn was performed to change the orbit of *Atlantis* from 298 x 305 nm to 160 x 305 nm.

During the flight Mission Control personnel assessed the risk of 35 predicted conjunctions of orbital debris with *Atlantis*. Additional Mission Control personnel were called in to help perform conjunction evaluations. However, after extensive analysis no debris avoidance maneuvers were performed. During STS-103 (December 1999) 58 conjunctions were evaluated, while during STS-109 (March 2002) 12 conjunctions were evaluated. After landing an orbital debris hit was found on the nozzle of the right OMS engine.

All five EVAs to upgrade HST were successful. All mission objectives, except for the reboost that was withdrawn, were accomplished.[20, 21] The STS-400 rescue mission (*Endeavour*) was not required and *Atlantis* landed on Edwards Air Force Base runway 22 on Sunday, May 24, 2009. The landing was delayed by two days and moved to Edwards due to unacceptable weather conditions at the Kennedy Space Center.

On May 31, 2009, *Endeavour* was moved from pad 39B to pad 39A to prepare for the STS-127 mission that flew in July of 2009. Pad 39B was then handed over to the Constellation Program for modification to support the Ares I-X test flight that launched on October 28, 2009.

Other Contingency Rendezvous Procedures

There are three contingency rendezvous procedures that were flown on all HST missions (Table 24.2). These are Ti Delay, Radar Fail, and Rendezvous Breakout.

Ti Delay permits the orbiter to fly a relative motion football (Figures 24.6 and 24.16) at the 8 nm Ti point. This delay could provide the crew, Mission Control, and the HST STOCC at the Goddard Space Flight Center with time to resolve a problem before proceeding with the rendezvous. Alternatively, if the problem could not be resolved in time to permit the rendezvous and grapple on that crew day, the orbiter could separate and phase away from the HST overnight. Ti Delay has not been performed on a HST mission. The only Ti Delay flown by the shuttle was on STS-49 (May 1992), in response to a Lambert burn targeting anomaly.[13,22]

The Radar Fail procedure would be used by the crew after the Mid-course Correction (MC3) maneuver, if radar data were not available for relative navigation and proximity operations (Figures 24.6, 24.16, and 24.17). A radar failure did not occur on HST missions.

However, the rendezvous radar did fail before the STS-92 rendezvous with the ISS in October of 2000, and during the STS-131 rendezvous with ISS in April of 2010. The Radar Fail contingency procedure was successfully executed on those missions.[13]

If a shuttle or HST problem prevented the rendezvous and grapple from being completed, the Rendezvous Breakout procedure would enable the orbiter to establish a safe relative motion trajectory that would not come close to the HST. A breakout during the rendezvous phase (Figures 24.6 and 24.16) has not been performed on a HST servicing mission, or on any other shuttle mission.

Other Contingency Proximity Operations Procedures

Like nominal proximity operations procedures, the contingency procedures are heavily influenced by HST and shuttle hardware design. Contingency proximity operations procedures have evolved, but the number of procedures stabilized by the third servicing mission, STS-103 (Table 24.2).

Some contingency procedures are designed to enable the orbiter to safely leave the vicinity of the HST if a problem prevents grapple. The orbiter may station-keep in the vicinity of the HST while Mission Control, the STOCC, and the crew work to resolve the problem. If the problem cannot be resolved in a timely manner, a breakout is performed so the orbiter safely leaves the vicinity of the HST. Prox Ops Backoff allows the orbiter to back away from HST to a safe station-keeping distance. The HST +R Bar Breakout (Table 24.2) was designed for execution starting at a range of 500 feet until the crew initiates the final inertial grapple (Figure 24.12). A backout along the +R Bar may be required to at least 75 feet before the orbiter can leave the +R Bar via a breakout to avoid undesirable contact with HST. Prox Ops Breakout permits the orbiter to safely leave the vicinity of the HST and exit the proximity operations phase.

The EVA Rescue procedure is used to retrieve an EVA crew member that is no longer tethered to the orbiter or EVA tools that are no longer tethered to the crew member. It is desirable for any EVA tools that are lost overboard to be retrieved as they present a collision hazard. The procedures ensure that structural loads imparted by translational RCS jet activity will not cause failure of the connection between the HST and the FSS in the shuttle payload bay.

Some contingency procedures permit grapple to be accomplished or a breakout to be performed in the event of vernier or Low Z RCS jet failures. The Loss of Vernier RCS procedure permits proximity operations to continue if the orbiter 25 pound thrust vernier RCS jets are no longer available for fine attitude control. The Loss of Low Z Braking procedure provides options to use for any loss of or degradation of Low Z capability during the approach. The approach could be continued or a Loss of Low Z breakout performed. The HST Flyaround/Loss of Low Z Breakout is performed between initiation of inertial attitude hold by the crew and grapple. By the time of STS-103, the number of breakout scenarios had increased and a new flow chart was implemented on a cue card to help the crew navigate through the many options.

Other contingency proximity operations procedures listed in Table 24.2 are performed if HST is not in the correct attitude for grapple when the shuttle arrives. These include the STS Roll to Align, Manual Inertial Fly-Around, Auto Inertial Fly-Around, Yaw/Pitch/Yaw Fly-Around, and the R Bar Yaw Alignment.

Contingency Deploy

Contingency deploy procedures have also been carried on all HST missions (Table 24.2). These procedures permit HST deployment if the RMS is not available or if a faster than normal deployment must be accomplished in response to a systems performance anomaly. These anomalies could require the orbiter to perform an emergency deorbit or a perigee adjust.

Contingency procedures were developed to cover partial or complete failures of the RMS (Table 24.2). For a total RMS failure a backaway deployment would have been performed. This procedure has been prepared for all HST missions. The procedure for the STS-31 deploy mission involved releasing HST retaining latches in the payload bay and performing a +Z translation burn (Figure 24.2) by the orbiter to slowly back away from the HST. The procedure for all subsequent flights was designed to allow the HST berthing pins to clear the FSS latches, while avoiding attitude jet firings that could cause the pins to re-contact. The deploy attitude avoids shadowing of the HST solar arrays by orbiter structure.

All HST missions have been equipped with an Emergency RMS Deploy (STS-31) or a RMS Quick Deploy (STS-61, STS-82, STS-103, STS-109, and STS-125) procedure. The RMS Quick Deploy could be performed if a faster than normal release of HST is required in response to an orbiter systems problem. The quick deploy has essentially the same sequence as the nominal deploy, but certain non-mandatory HST crew commanding and orbiter relative navigation procedures are omitted to save time.

Jettison

Jettison procedures are carried to permit the release of payload bay hardware from the orbiter if it cannot be secured in the payload bay or it is stuck in an unsafe configuration (Table 24.2). Jettison procedures are designed to permit the orbiter to safely leave the jettisoned hardware while minimizing risk of re-contact. Some jettison procedures can be executed by the crew from the cockpit, while other procedures may require crew action during EVA. Jettison procedures are not considered nominal, are often payload and payload support hardware specific, and will vary from flight to flight. Jettison procedures for servicing hardware include the Orbiter Replacement Unit Carrier (ORUC), Rigid Array Carrier (RAC), Solar Array Carrier (SAC), and the Super Lightweight Interchangeable Carrier (SLIC).

These procedures require that HST be jettisoned first. The ORUC, SAC, and SLIC jettison procedures require action by EVA crew members.

A HST Jettison would be performed if the orbiter were required to perform a time critical de-orbit in response to problems such as loss of crew cabin pressure or a propellant leak. The jettison procedure can be performed in any attitude. Low Z RCS jet firings are used to back the orbiter away from HST after the FSS latches are opened.

The orbiter payload bay doors must be closed for the orbiter to safely return to Earth. If the RMS or the rendezvous radar cannot be stowed for entry, then they would be jettisoned to enable the payload bay doors to be closed. A generic hardware jettison procedure is available on all flights if the crew has to jettison generic hardware, including EVA hardware.

A solar array jettison procedure was developed for STS-61 and STS-109 in case an array could not be fully retracted and stowed for return to Earth. The power generation side of the array must face away from the Sun when the array electrical lines between HST and the array are disconnected by the EVA crew. The array would be released by an EVA crew member mounted on the RMS with a foot restraint, using either a jettison handle or a portable grapple fixture. One solar array was jettisoned on STS-61. This is the only jettison that has been performed on a HST servicing mission.

Atlantis Rescue, Prepared But Not Flown

After the loss of *Columbia* in 2003, each shuttle mission performed inspection of the Thermal Protection System (TPS) to determine if the TPS sustained damage during ascent from External Tank foam shedding. The primary means of inspection was the Orbiter Boom Sensor System (OBSS) mounted on the end of the RMS. On ISS missions, a +R Bar Pitch Maneuver (RPM) was performed ~600 foot below the ISS to permit ISS crew to photograph the orbiter TPS.[23] Photographs provided an additional source of data on TPS integrity. If TPS damage was detected and was considered to be a safety risk and could not be repaired on-orbit during an EVA, plans were developed to permit a Space Shuttle crew to use the ISS as a safe haven. The next Space Shuttle in the launch preparation flow for an ISS mission would be launched to retrieve the crew from the ISS and return them to Earth. Like ISS missions, the STS-125 crew performed a TPS inspection using the OBSS. The TPS was not compromised during STS-125.

However, had the TPS been compromised and could not be repaired by the crew during an EVA, the STS-125 *Atlantis* crew could not use the ISS as a safe haven as the shuttle did not have sufficient propellant to reach the ISS from the HST orbit. To provide a rescue capability, a Launch On Need (LON) *Atlantis* rescue mission was prepared (Table 24.1). A rescue shuttle flown by the four flight deck crew members from the STS-123 (March 2008) mission to the ISS would have flown the rescue mission, if it were required. The rescue concept required the pre-launch parallel processing of both *Atlantis* and the rescue orbiter at the Kennedy Space Center. The rescue Space Shuttle was on one of the Complex 39 launch pads while *Atlantis* was launched from the other pad. This was a first for the Shuttle Program. Although maximum crew awake time was limited to 18 hours to avoid fatigue, this limit could have been waved in a rescue scenario to ensure the safe retrieval and return of the *Atlantis* crew.

Atlantis Rescue Rendezvous Design

Rendezvous and proximity operations of *Atlantis* and the rescue orbiter would occur a considerable away from the HST. The nominal rendezvous mission plan for the rescue was a flight day 2 grapple of *Atlantis* by the rescue orbiter, with the possibility of a flight day 3 or 4 grapple, if permitted by ample propellant margins. A flight day 2 grapple was preferred so that the rescue orbiter could reach *Atlantis* as quickly as possible and provide maximum on-orbit time for the crew transfer to be completed. This was the first nominally planned flight day 2 rendezvous and grapple in the Shuttle Program and would have been the first rendezvous of one shuttle with another. Ground-up shuttle rendezvous missions to the ISS normally conducted docking/grapple on flight day 3, with a flight day 2 or flight day 4 docking/grapple as a possible contingency. Rendezvous trajectory dispersions were expected to be higher than normal due to the limited amount of time to track out dispersions on flight day 1 in support of rendezvous orbital adjustment burns. The crew rendezvous checklists for both the STS-125 and the rescue mission were combined into one document.

The ORBT rendezvous on flight day 2 (the star tracker pass through the MC4 burn) was similar to that of ISS and HST servicing missions (Figure 24.17 and the one revolution to Ti profile in Figure 24.16). For all shuttle rendezvous missions, in the event of a rendezvous radar failure, a correction burn is performed after the third mid-course correction burn. If *Atlantis* had sufficient propellant and power, contingency night star tracker measurements could have been obtained by the rescue orbiter if the payload bay lights of Atlantis were turned on and the payload bay pointed in the direction of the approaching rescue orbiter. However, if *Atlantis* was not able to perform the procedure, the rendezvous profile timing was adjusted pre-mission to insure *Atlantis* would be lit by the sun to support crew procedures for the radar fail correction burn.

Atlantis Rescue Contingency Rendezvous Procedures

In the nominal rendezvous plan the rescue orbiter performed all maneuvering. Contingency rendezvous recovery plans were also developed in case the rescue orbiter could not execute the nominal rendezvous profile due to an ascent under-speed at MECO or a propellant failure. The rendezvous recovery profile would preserve the flight day 2 grapple, if possible. While it was preferred to fly a rendezvous with the rescue orbiter approaching from behind and below, off-nominal cases could have required a rendezvous with the rescue orbiter ahead and above *Atlantis* for much of the rendezvous. In these contingency cases Atlantis might also have been required to perform orbit adjustments of relative altitude and phasing to enable the rescue orbiter to complete the rendezvous. Propellant margins on both vehicles would have been carefully managed to ensure that the rescue orbiter had sufficient propellant for a safe deorbit.

This technique is known as control box rendezvous, and was performed on STS-49 (INTELSAT VI/F-3 rendezvous, May 1992) and STS-72 (Space Flyer Unit rendezvous, January 1996).[13] The target spacecraft executed a series of maneuvers after the chaser spacecraft was launched. The maneuvers were designed so that the target entered a volume in space, called a control box, at a designated time. This technique reduced chaser vehicle (in this case, the rescue orbiter) propellant consumption. Once the target entered the box, it no longer maneuvered. Rendezvous recovery was planned so that *Atlantis* did not perform orbit adjustments on the day of rendezvous.

The final rendezvous orbit for the rendezvous recovery case impacted landing opportunities for the rescue orbiter. The final orbit must preserve at least one continental United States landing opportunity for the rescue shuttle per day, with two opportunities preferred. If required, a landing could also have been performed at sites outside the continental United States. In addition, achievement of acceptable disposal areas for *Atlantis* was also be factored into rendezvous recovery planning and determination of the final rendezvous orbit. However, protecting the rescue orbiter deorbit propellant margins had a higher priority than *Atlantis* propellant margins for achieving a safe *Atlantis* disposal footprint.

For the on-board targeted phase on the day of rendezvous, the rescue orbiter would have flown three contingency rendezvous procedures flown by other HST servicing and ISS missions. Those were Radar Fail, Rendezvous Breakout, and Ti Delay (Table 24.2).

Atlantis Rescue Nominal Proximity Operations

Atlantis was to maneuver to the grapple attitude just before the rescue orbiter executed the MC4 burn (Figure 24.17). The grapple attitude placed the nose of *Atlantis* out-of-plane toward orbital south and the payload bay pointed at the Earth (Figure 24.19). The flight control system was to maintain this attitude using the 25 pound thrust vernier RCS jets, if available. In the event of a vernier failure the ALT/DAP and primary jets would have been used. *Atlantis* would have also used the Low Z mode to limit plume impingement on the rescue orbiter.

The proximity operations profile (starting at manual crew take-over after MC4) was a +R Bar approach. However, unlike ISS missions, the R Bar Pitch Maneuver would not have been performed.[23] The rescue orbiter flight control system would have been placed in the Low Z mode from a range of 1000 feet through grapple. This range was chosen as the crews from ISS missions are familiar with Low Z operation starting at this range. *Atlantis* and the rescue orbiter would have been at a 90 degree angle to each other (*Atlantis* nose toward orbital south, the rescue orbiter nose pointed along the velocity vector) to minimize plume impingement effects during the Low Z +R Bar approach by the rescue orbiter (Figure 24.19). The rescue orbiter would have carried both Trajectory Control Sensor (TCS) and Hand Held Lidar (HHL) for use during proximity operations.

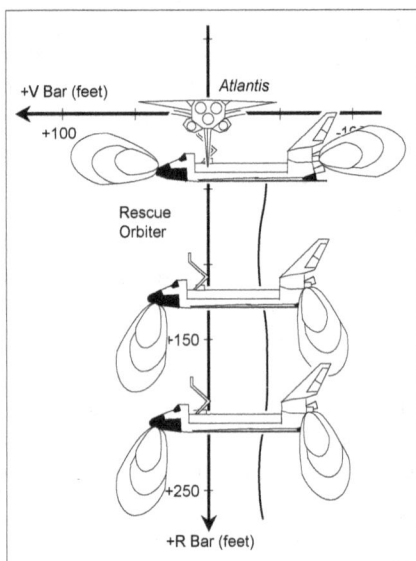

Figure 24.19 Rescue orbiter approach to *Atlantis*.

Capture would have been performed with the RMS of the rescue orbiter grappling the forward grapple fixture on *Atlantis* berthed OBSS. After grapple the OBSS would roll out and the RMS of the rescue orbiter would be used to maneuver *Atlantis* so that both orbiters were nose-to-nose for effective mated attitude control. The rescue orbiter would have then maneuvered the mated stack to a gravity gradient attitude. The RMS of *Atlantis* was not planned for use.

EVA Crew Transfer, Separation, and Deorbit

The rescue involved the transfer by EVA of the seven member *Atlantis* crew to the rescue orbiter on flight days 3 and 4. A total of three EVA transfers from *Atlantis* to the rescue orbiter would have been performed using the white Extra-vehicular Mobility Unit (EMU) suits. Only *Atlantis* crew members were to participate in the EVAs. The four members of the rescue orbiter crew (Table 24.1)

were to remain inside the rescue orbiter. At the start of the first EVA participating crew members were to install a translation rope along the RMS of the rescue orbiter. Astronauts McArthur, Feustel, and Grunsfeld would have transferred to the rescue orbiter during the first EVA. Johnson was to transfer during the second EVA, along with all of the thermal protection system repair hardware.

The third and final EVA would have transferred Altman, Massimino, and Good. Before the last EVA, the remaining crew members on *Atlantis* were to configure the cockpit for the separation and ground commanded deorbit burn. *Atlantis* disposal procedures were based on those developed for damaged orbiter disposal on ISS missions.[24] This included opening allowable attitude error and rate limits so that automatic flight control firings of the RCS jets would not have been performed with the rescue orbiter in close proximity to *Atlantis*. *Atlantis* was to be released by the rescue orbiter on flight day 4 (Figure 24.20). TPS inspection using the OBSS was to be performed on flight day 5, and flight days 6 and 7 were to be used for entry preparation. Rescue orbiter entry and landing was planned for flight day 8.

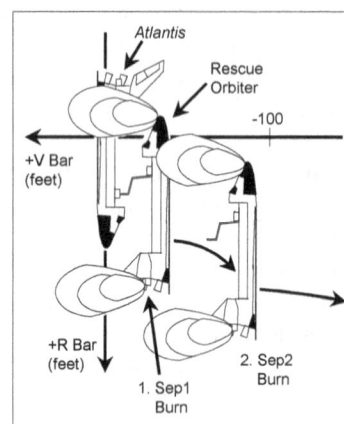

Figure 24.20 Rescue orbiter separation from *Atlantis*.

Observations and Lessons Learned

HST missions succeeded in part due to the efforts of personnel from multiple disciplines that had extensive experience planning and flying servicing missions to a variety of spacecraft. This facilitated application of best practices and lessons learned.[6-11] These personnel are experienced at working in a multi-discipline environment involving multiple NASA organizations and supporting contractors that requires lateral communication. Shuttle Program personnel are experienced in development of contingency procedures, both pre-mission and during a flight, and with interacting with development and operations personnel representing a variety of target spacecraft.

A flight program must be responsive to unanticipated satellite performance issues that may drive late and significant changes in servicing mission plans. These events can drive changes to existing proximity operations, robotic operation, and servicing procedures, or require the creation of new procedures and mission plans. The availability of additional qualified personnel to develop new procedures and operational work-arounds enables a flight program to effectively respond to off-nominal events during real-time operations.

Development and operations personnel continually learn about vehicle systems performance and limitations even after a spacecraft has been built and is in orbit. Unforeseen constraints and performance limitations will emerge that drive development of new or changes to existing nominal and contingency plans and procedures. Over the life of a flight program improvements in analysis and simulation fidelity may reveal additional operational constraints. An example of this was the gradual discovery of HST sensitivity to plume impingement that resulted in the increasing range of Low Z mode initiation from 200 feet out to 1500 feet.

Servicing mission personnel should consider applying rendezvous, proximity operations, and other techniques from other spaceflight missions and flight programs to mitigate risk to mission success. High value missions may drive significant investment in low-probability of occurrence contingency procedures to ensure mission success in the presence of failures and degraded systems performance. However, this may result in an increase in the number of procedures that program personnel must maintain and be prepared to execute over the life of a flight program.

Many nominal and contingency HST procedures were driven by RCS plume impingement overpressure and contamination concerns. Consideration should be given to building spacecraft structures and systems that are not as sensitive to servicing vehicle characteristics such as RCS jet plumes. Furthermore, servicing spacecraft should be designed with RCS and other systems that do not pose a potential hazard to satellites that could be serviced.

The highly reflective surface of HST makes it a poor target for the HHL and causes shuttle payload bay camera blooming, complicating proximity operations piloting. Experience has shown that the RMS grapple fixtures on HST are good targets for the HHL. Proximity operations contingency procedure development for the ZGSP and HWSP HST attitude control safe modes was complicated by a lack of HST retro-reflectors to support the shuttle TCS and HHL. In addition, Mission Control and crew insight into HST attitude during these safe modes was limited, and based primarily on crew observations. Comprehensive telemetry, sensor aids on the vehicle to be serviced, and relative sensors capable of performing relative attitude determination can simplify proximity operations piloting.

In spite of the previously mentioned challenges, ground personnel (HST STOCC, Space Shuttle Mission Control) and shuttle crew members possessed the flexibility, creativity, and situational awareness to analyze unforeseen issues and develop new procedures in a timely manner. Spacecraft and ground support organizations in future robotic or human flight programs should be flexible enough to accommodate late changes in mission requirements. Such responsiveness significantly enhances the probability of mission success.

Summary

The Space Shuttle Program has successfully flown servicing missions that have repaired and upgraded the Hubble Space Telescope. These repair missions increased the science return and extended the life of the telescope by correcting performance problems, replacing malfunctioning hardware, and equipping it with more advanced astronomy sensors. Conducting these missions required the development, adaption, and evolution of numerous crew procedures and flight techniques for performing rendezvous, proximity operations, and deployment. Nominal and contingency procedure development required the efforts of both shuttle and HST Program personnel in disciplines including trajectory design, robotics, flight control, thermal control, power generation, structures, orbital debris, and extra-vehicular activity. Space Shuttle and HST hardware design and limitations placed requirements and constraints on these nominal and contingency techniques. Some constraints were known early in the development of mission techniques in the 1980s, others emerged after HST was placed in orbit in 1990. Particular care was taken to "do no harm" to HST and not impede the ability of HST to perform the science mission. The HST servicing experience and lessons learned are applicable to other programs that perform on-orbit servicing and rendezvous, both human and robotic.

References

1. Doxsey, R., "15 Years of Hubble Space Telescope Science Operations," *AIAA SpaceOps 2006 Conference*, AIAA, Reston, VA, 2006.

2. Zimmerman, R. *The Universe in a Mirror: The Saga of the Hubble Space Telescope and the Visionaries Who Built It*, Princeton University Press, 2008.

3. E. J. Chaisson, *The Hubble Wars: Astrophysics Meets Astropolitics in the Two-Billion-Dollar Struggle over the Hubble Space Telescope*, Harvard University Press, 1998.

4. Dedalis, R. P., and P. L. Mitchell, "Servicing The Hubble – Risk Mitigation, Lessons Learned, And Rewards In Completing Hubble Space Telescope Servicing Missions," *Space 2004 Conference and Exhibit*, AIAA, Reston, VA, 2004.

5. Goodman, J. L., and J. P. Brazzel, "Rendezvous Integration Complexities of NASA Human Flight Vehicles," *32nd Annual AAS Guidance And Control Conference, Guidance and Control 2009, Advances in the Astronautical Sciences*, Univelt, San Diego, CA, 2009.

6. Pfarr, B., W. Ochs, J. Leibee, and C. Fatig, "Operations Lessons Learned From the Hubble Space Telescope First and Second Servicing Missions," *AIAA Defense and Space Programs Conference and Exhibit – Critical Defense and Space Programs for the Future*, AIAA, Reston, VA, 1997.

7. Werneth, R. L., "Lessons Learned from Hubble Space Telescope ExtraVehicular Activity Servicing Missions," *31st International Conference on Environmental Systems*, Orlando, FL, Society of Automotive Engineers, July 2001.

8. Akin, D. L., and B. Roberts, "Robotic Servicing of the Hubble Space Telescope: Lessons Learned from a Short-Lived Program," *AIAA Space 2006 Conference*, AIAA, Reston, VA, September 2006.

9. McGuire, J., and B. Roberts, "Hubble Robotic Servicing and Deorbit Mission: Risk Reduction, Mitigation, and Lessons Learned," AIAA 2007-6255, *AIAA Space 2007 Conference & Exposition*, Reston, VA, September 2007.

10. Cepollina, F. J., "Applying the Secrets of Hubble's Success to Constellation," *ASK Magazine*, Issue 30, Spring 2008, pp. 35-38.

11. Cepollina, F.J., "Lessons from Apollo, Space Shuttle, International Space Station, and the Hubble Space Telescope: Learning From These Great Legacies," *SPACE TIMES*, July/August 2008, pp. 4-9.

12. Kennedy, III, F. G., "Orbital Express: Accomplishments and Lessons Learned," *31st Annual AAS Guidance And Control Conference, Guidance and Control 2008, Advances in the Astronautical Sciences*, Volume 131, Univelt, San Diego, CA, 2008, pp. 575-586.

13. Goodman, J. L., "History of Space Shuttle Rendezvous and Proximity Operations," *AIAA Journal of Spacecraft and Rockets*, Vol. 43, No. 5, Sept.-Oct. 2006, pp. 944-959.

14. Goodman, J. L., and C. A. Propst, "Operational Use of GPS Navigation for Space Shuttle Entry," *IEEE/ION PLANS 2008 Conference*, Monterey, CA, May 5-8, 2008.

15. *NASA's Implementation Plan for Space Shuttle Return to Flight and Beyond*, 10th Edition, NASA, Washington, D.C., June 3, 2005.

16. Lanzerotti, L. J., et al., *Assessment of Options for Extending the Life of the Hubble Space Telescope: Final Report*, National Research Council Of The National Academies, The National Academies Press, Washington, D.C., 2005.

17. Biagetti, C., "Recommissioning Hubble after the 2008 Servicing Mission," *SpaceOps 2008 Conference*, AIAA, Reston, VA, 2008.

18. Walyus, K. D., M. W. Prior, and L. M. Mazzuca, "An Operations Overview of the Hubble Space Telescope Servicing Mission 4," *SpaceOps 2006 Conference*, AIAA, Reston, VA, 2006.

19. Naasz, B. J., et al., "The HST SM4 Relative Navigation Sensor System: Overview and Preliminary Testing Results from the Flight Robotics Lab," *F. Landis Markley Astronautics Symposium, Advances in the Astronautical Sciences*, Univelt, San Diego, CA, 2009.

20. Van Eepoel, J., B. Naasz, S. Queen, "Flight Results from the HST SM4 Relative Navigation Sensor System," *33rd Annual AAS Guidance And Control Conference, Guidance and Control 2010, Advances in the Astronautical Sciences*, Univelt, San Diego, CA, 2010.

21. Jenkins, D. R., and J. R. Frank, *Servicing the Hubble Space Telescope: Space Shuttle Atlantis – 2008*, Specialty Press, North Branch, MN, 2009.

22. Goodman, J. L., *Lessons Learned From Seven Space Shuttle Missions*, NASA Contractor Report NASA/CR-2007-213697, NASA Johnson Space Center, January 2007. See the NASA Technical Reports server at http://ntrs.nasa.gov/, or the Johnson Technical Reports server at http://ston.jsc.nasa.gov/collections/TRS/. Accessed January 20, 2009.

23. Walker, S., J. LoPresti, M. Schrock, and R. Hall, "Space Shuttle Rbar Pitch Maneuver for Thermal Protection System Inspection Overview," *AIAA Guidance, Navigation, and Control Conference*, AIAA, Reston, VA, 2005.

24. Bigonesse, R., and W. R. Summa, "Unmanned Orbiter Undocking: Method for Disposal of a Damaged Space Shuttle Orbiter," *Spaceflight Mechanics 2006, Advances in the Astronautical Sciences*, Univelt, San Diego, CA, 2006.

This page intentionally left blank.

CHAPTER 25 - STS-130 MISSION TO THE ISS

Introduction

The majority of Space Shuttle rendezvous missions were to the International Space Station (ISS). There was some variation in proximity operations procedures for flights to Mir and the ISS, but overall these flights had less mission-to-mission variation in mission plans and procedures as compared to the deploy/retrieve and satellite repair missions of the 1980s and 1990s.

This chapter provides an overview of rendezvous, proximity operations, and separation activities for the February 2010 flight of *Endeavour* to the ISS. Mission activities are discussed by flight day.

Mission Plan

The primary objective of the STS-130 mission (ISS Flight 20A) of the space shuttle *Endeavour* was to transport the ISS Node 3 (*Tranquility*) to the ISS for installation by the STS-130 and ISS Expedition 22 crew members. It was nominally a 13 day mission (during the flight it was extended to 14 days) with a 10 minute launch window.

At docking the mated stack was predicted to weigh, for the first time, in excess of one million pounds. There were no planned hardware jettisons during the docked phase nor any payload deploys from *Endeavour* after undocking and separation.

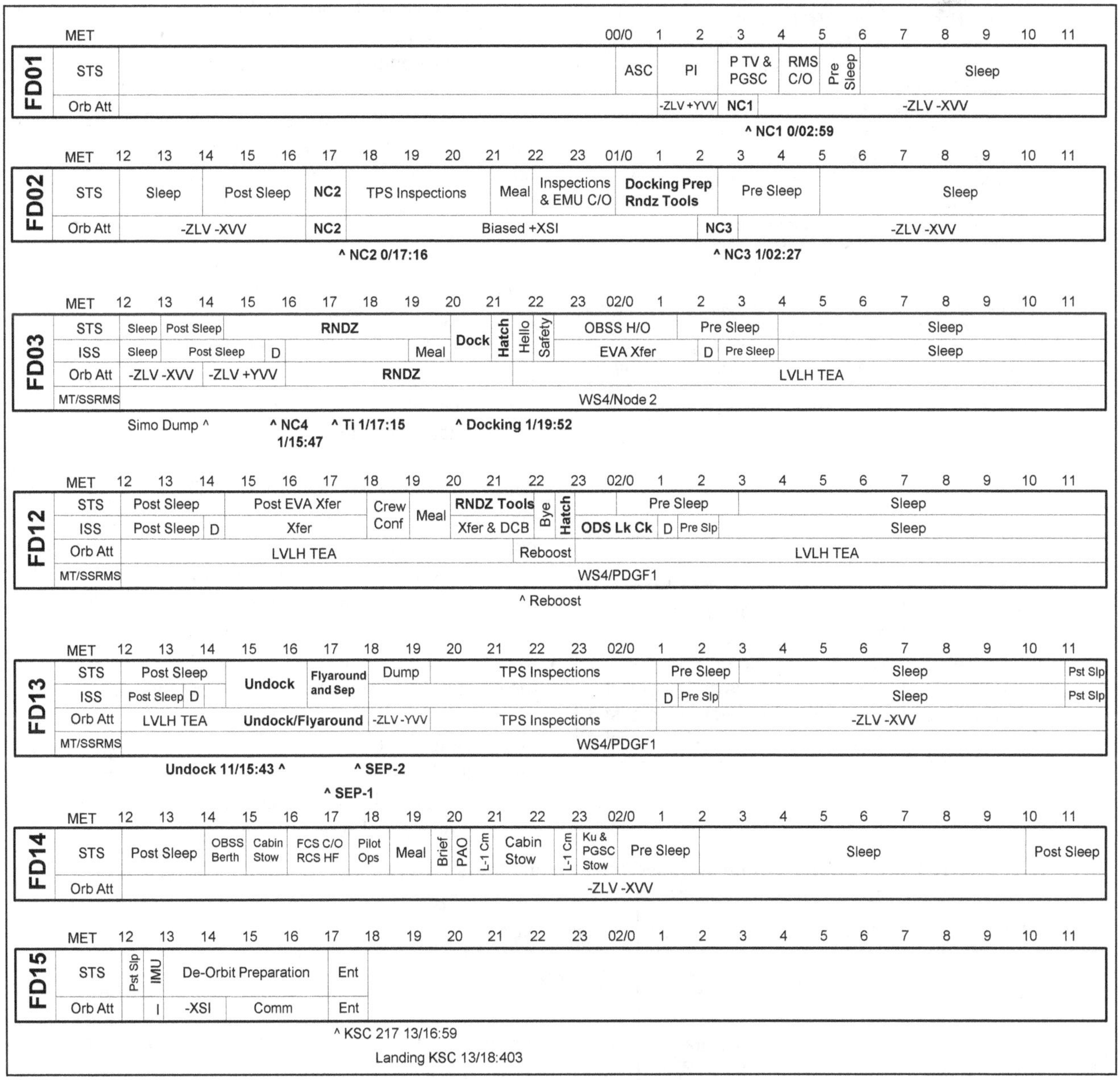

FD01 — MET: 00/0 1 2 3 4 5 6 7 8 9 10 11
- STS: ASC | PI | P TV & PGSC | RMS C/O | Pre Sleep | Sleep
- Orb Att: -ZLV +YVV | NC1 | -ZLV -XVV
- ^ NC1 0/02:59

FD02 — MET: 12 13 14 15 16 17 18 19 20 21 22 23 01/0 1 2 3 4 5 6 7 8 9 10 11
- STS: Sleep | Post Sleep | NC2 | TPS Inspections | Meal | Inspections & EMU C/O | **Docking Prep Rndz Tools** | Pre Sleep | Sleep
- Orb Att: -ZLV -XVV | NC2 | Biased +XSI | NC3 | -ZLV -XVV
- ^ NC2 0/17:16   ^ NC3 1/02:27

FD03 — MET: 12 13 14 15 16 17 18 19 20 21 22 23 02/0 1 2 3 4 5 6 7 8 9 10 11
- STS: Sleep | Post Sleep | **RNDZ** | **Dock** | Hatch | Hello | Safety | OBSS H/O | Pre Sleep | Sleep
- ISS: Sleep | Post Sleep | D | Meal | EVA Xfer | D | Pre Sleep | Sleep
- Orb Att: -ZLV -XVV | -ZLV +YVV | **RNDZ** | LVLH TEA
- MT/SSRMS: WS4/Node 2
- Simo Dump ^   ^ NC4 1/15:47   ^ Ti 1/17:15   ^ Docking 1/19:52

FD12 — MET: 12 13 14 15 16 17 18 19 20 21 22 23 02/0 1 2 3 4 5 6 7 8 9 10 11
- STS: Post Sleep | Post EVA Xfer | Crew Conf | Meal | **RNDZ Tools** | Bye | Hatch | Pre Sleep | Sleep
- ISS: Post Sleep | D | Xfer | Xfer & DCB | ODS Lk Ck | D | Pre Slp | Sleep
- Orb Att: LVLH TEA | Reboost | LVLH TEA
- MT/SSRMS: WS4/PDGF1
- ^ Reboost

FD13 — MET: 12 13 14 15 16 17 18 19 20 21 22 23 02/0 1 2 3 4 5 6 7 8 9 10 11
- STS: Post Sleep | **Undock** | **Flyaround and Sep** | Dump | TPS Inspections | Pre Sleep | Sleep | Pst Slp
- ISS: Post Sleep | D | D | Pre Slp | Sleep | Pst Slp
- Orb Att: LVLH TEA | **Undock/Flyaround** | -ZLV -YVV | TPS Inspections | -ZLV -XVV
- MT/SSRMS: WS4/PDGF1
- Undock 11/15:43 ^   ^ SEP-1   ^ SEP-2

FD14 — MET: 12 13 14 15 16 17 18 19 20 21 22 23 02/0 1 2 3 4 5 6 7 8 9 10 11
- STS: Post Sleep | OBSS Berth | Cabin Stow | FCS C/O RCS HF | Pilot Ops | Meal | Brief | PAO | L-1 Cm | Cabin Stow | L-1 Cm | Ku & PGSC Stow | Pre Sleep | Sleep | Post Sleep
- Orb Att: -ZLV -XVV

FD15 — MET: 12 13 14 15 16 17 18 19 20 21 22 23 02/0 1 2 3 4 5 6 7 8 9 10 11
- STS: Pst Slp | IMU | De-Orbit Preparation | Ent
- Orb Att: I | -XSI | Comm | Ent
- ^ KSC 217 13/16:59
- Landing KSC 13/18:403

Figure 25.1 STS-130 as flown crew overview timeline for orbiter free flight and activities concerning rendezvous and separation. Flight Days 4 through 11 while docked to the ISS are not shown. Rendezvous and separation activities are in bold.

Figure 25.1 provides a detailed overview of crew activities directly concerning ascent, rendezvous, proximity operations, separation, and landing. Table 25.1 provides a summary of major shuttle and ISS crew activities performed on each flight day. Crew members for the STS-130 mission are listed in Table 25.2.

Table 25.1 Primary Crew Activities by Flight Day

Flight Day	Primary Crew Activities
1	• Ascent, Post Insertion, NC-1/NPC, RMS Checkout
2	• NC-2/NPC, Inspection, Rendezvous Tools Checkout, NC-3/NPC
3	• Water Dump, Rendezvous, Docking, Open Hatches, OBSS H/O
4	• EVA Prep, Off Duty
5	• EVA1 (Node 3 Install, OTP Relocate)
6	• Focused Inspection, Node 3 Activation
7	• EVA2 (Node 3 External Outfitting)
8	• Cupola Relocate
9	• PMA3 Relocate, Off Duty
10	• EVA3 (Node 3, Cupola, & PMA Tasks)
11	• ECLSS Rack Transfers, Get Aheads
12	• Transfer, Rendezvous Tools Checkout, Reboost, Hatch Close
13	• Undock, Fly-Around, TPS Inspection
14	• Cabin Stow, FCS Checkout, RCS Hotfire
15	• Deorbit Preparation, Entry & Landing

ECLSS – Environmental Control and Life Support Systems, EVA – Extra-Vehicular Activity, FCS – Flight Control System, NC – Phasing or catch-up burn, NPC – Plane Change burn, OBSS H/O – Orbiter Boom Sensor System Hand-Off, OTP – Orbital Replacement Unit Tool Platform, PMA – Pressurized Mating Adapter, RCS – Reaction Control System, RMS – Remote Manipulator System, TPS – Thermal Protection System

Table 25.2 STS-130 Crew Members

Name	Nickname	Assignment
George D. Zamka	Zambo	Commander
Terry W. Virts Jr.	TV	Pilot
Nicholas J.M. Patrick	Nick	Mission Specialist 1
Robert L. Behnken	Dr. Bob	Mission Specialist 2
Stephen K. Robinson	Stevie Ray	Mission Specialist 3
Kathyrn P. Hire	Kay	Mission Specialist 4

Launch Windows

Launch windows and rendezvous burn plans were periodically revised during the mission planning process and just before launch based on improved predictions of the ISS state vector.

Most Mir and ISS launches were successfully performed at the in-plane launch time, when the targeted orbital plane crossed the launch site. Selection of the launch time was a balance between four considerations, 1) maximize the launch window, 2) provide sufficient East Coast Abort Landing opportunities for multiple Space Shuttle Main Engine (SSME) failure cases, 3) maximize ascent performance to cover SSME failure cases observed during flights (chamber pressure shifts, nozzle leaks, fuel flow meter malfunctions), and 4) achieve the earliest SSME failure time that would permit completion of the mission.

Launch opportunities on all three days met the solar beta angle constraint of $|\beta| < 60$ degrees for thermal considerations during mated operations. For these days a Return to Launch Site (RTLS) abort and nominal end of mission landing would occur during darkness, but a Trans-Oceanic Abort Landing (TAL) would occur in daylight.

Data for the primary launch opportunity (Sunday, February 7) and backup opportunities is given in Table 25.3. Figure 25.2 illustrates the planar launch windows and OMS-2 phasing limits for the February 7, 8, and 9 launch opportunities. The launch window consists of both planar and phase windows. The maximum duration of the planar window is 10 minutes based on thermal limits. The thermal limits and ET disposal footprint constraints end the planar window. The ET disposal limits keep the External Tank from coming too close to national boundaries and landmasses. ET thermal limits are necessary since as the vehicle steers into the desired plane due to a launch slip, the side slip angle of the vehicle increases ET heating. The opening of the planar window occurred approximately 24 minutes earlier each day, and the opening times repeated approximately every 59 days.

Shuttle missions to the ISS typically docked on Flight Day 3 on orbit 30 or 31. The rev number varied based on phase angle at OMS-2, ISS crew sleep shifting, and achieving orbital lighting conditions for the on-board targeted phase and proximity operations. ISS crew sleep

Table 25.3 STS-130 Launch Opportunities Based on Launch Minus 7 Days Flight Dynamics Officer Data

Launch Date	Launch Time	Flight Day 3 Launch Window	Insertion Orbit	Phase Angle	Docking	Undocking
February 7, 2010 Sunday (nominal)	04:39:47 EST	04:34:47 to 04:44:47 EST	123.6 x 84.9 nm	273.3	Tuesday, Feb 9 12:18 AM CST	Wednesday, Feb. 17 6:12 PM CST
February 8, 2010 Monday (actual)	04:14:05 EST	04:09:05 to 04:19:05 EST	123.6 x 109.9 nm	88.1	Tuesday, Feb 9 11:09 PM CST	Thursday, Feb. 18 6:35 PM CST
February 9, 2010 Tuesday	03:51:34 EST	03:46:34 to 03:56:34 EST	123.6 x 84.9 nm	275.4	Wednesday, Feb 10 11:30 PM CST	Friday, Feb. 19 5:25 PM CST

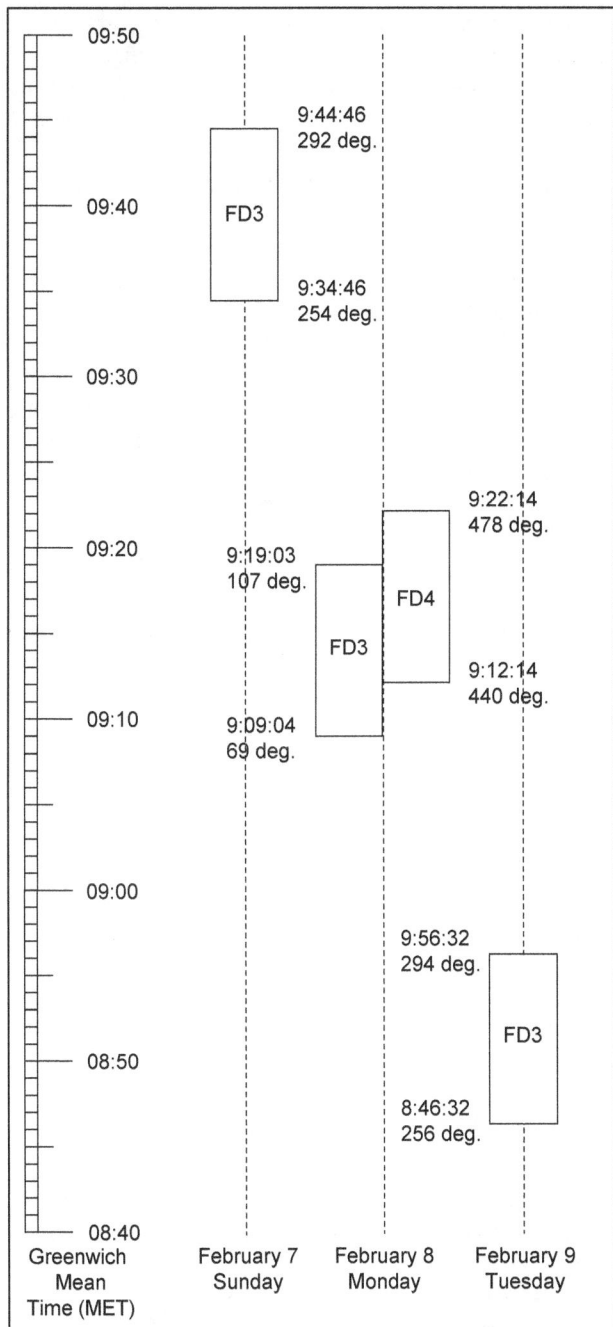

Figure 25.2 Daily planar window opening and closing times with phase angle limits at OMS-2. Note that February 8 has both Flight Day 3 and 4 windows. This plot is based on February 1, 2010 flight design data.

shifting and meeting lighting constraints could shift the time of docking earlier or later in the day.

The Space Shuttle Program has never flown a Flight Day 1 rendezvous and docking/grapple since it results in small launch windows that do not repeat on a daily or weekly basis. Furthermore, Space Adaption Syndrome (SAS) can occur on Flight Days 1 and 2. A Flight Day 3 rendezvous and docking/grapple has been preferred for several reasons: 1) it provides time for the crew to overcome SAS, 2) results in a less challenging crew timeline on Flight Day 1, 3) it permits Flight Day 2 to be used for equipment checkout and troubleshooting, 4) provides greater launch window probabilities than Flight Day 1 and Flight Day 2 rendezvous, and depending on target spacecraft orbital altitude may provide an in-plane launch window everyday within phase angle limits, and 5) provides sufficient opportunities for ground tracking and orbital adjustment burns to correct trajectory dispersions.

However, phasing limits may permit both a Flight Day 2 and a Flight Day 3 option. A Flight Day 2 rendezvous could be flown if a Minimum Duration Flight (MDF) was declared in response to degraded orbiter systems performance or a component failure. The objective of a MDF is to allow high priority mission objectives to be accomplished in the presence of degraded systems performance while minimizing exposure to subsequent failures that could threaten crew safety. In the case of a MDF a Flight Day 2 rendezvous would be acceptable.

Note that in Figure 25.2 for a Monday, February 8 launch there were overlapping planar windows for both Flight Day 3 and Flight Day 4 rendezvous. If the launch was delayed and the predicted launch time slipped out of the Flight Day 3 planar window and into the Flight Day 4 planar window, and if enough propellant was available, a Flight Day 4 rendezvous could have been flown (Figure 25.6). Or the Shuttle Program could have chosen to scrub the launch attempt and wait a day to attempt a launch in a Flight Day 3 launch window.

The phase window was based on a 185 nm average ISS orbital altitude and an 85 nm minimum perigee limit for the orbiter. The phase angle at OMS-2 can vary from 37 degrees to 311 degrees for a Flight Day 3 rendezvous and 211 degrees to 505 degrees for a Flight Day 4 rendezvous. However, a Flight Day 4 phase angle window could go as low as 50 or 60 degrees depending on the rendezvous burn plan. Since a Flight Day 3 docking is preferable to a Flight Day 4 docking, the Flight Day 3 window is used and a transition is made to the Flight Day 4 window once the Flight Day 3 window closes. The portion of the Flight Day 4 window that overlaps Flight Day 3 is not used.

Flight Day 4 planar windows did exist for the Sunday, February 7 and Tuesday, February 9 launch opportunities. However, the Flight Day 4 phasing windows for those dates did not fit into the Flight Day 4 planar windows; hence they are not shown in the Figure 25.2. The Flight Day 4 window is about 372 degrees greater that the Flight Day 3 planar window for all cases. Since the Flight Day 4 windows are limited to a maximum of 522 degrees, adding 372 degrees to February 7th and 9th windows result in a planar window that cannot be supported by phasing.

If the OMS-2 phase angle is low, there is a possibility that the same altitude used in OMS-2 burn targeting can be used for both a Flight Day 2 and a Flight Day 3 rendezvous. This allows the trajectory for both a Flight

Day 2 and Flight Day 3 rendezvous to remain the same up until the NC-1 burns. This provides the Shuttle Program two hours in which to declare a MDF before committing to a Flight Day 3 rendezvous at the NC-1 burn. Propellant is budgeted for a Flight Day 3 rendezvous. Since a Flight Day 2 rendezvous requires less propellant the Flight Day 3 budget covers it.

For a Flight Day 2 docking the orbiter must have a lower orbital altitude that provides a higher phasing (or catch-up) rate (Figure 25.5). For a Flight Day 3 docking the orbiter is at a higher orbital altitude that provides a slower phasing rate (Figure 25.4). The OMS-2 burn controls the altitude of the orbit half an orbital revolution after the burn. This orbital altitude (the OMS-2 burn target altitude) controls the phasing rate until the NC-1 burn. If both Flight Day 2 and Flight Day 3 docking options are to be preserved until the NC-1 burn, the altitude target for the OMS-2 burn is limited by a subset of the orbital altitudes (and therefore phasing rates) required for the Flight Day 2 and Flight Day 3 docking options. In order to accommodate phasing requirements for both options, the launch window is generally broken

up into two segments. The first segment (planar window open to the in-plane launch time) requires a higher OMS-2 target altitude. However, the first segment was not used on the later Mir and all ISS flights since the nominal launch time was the in-plane time. The second segment (in-plane launch time to planar window close) requires a lower OMS-2 target altitude.

The maximum phase angle for a Flight Day 2 rendezvous is about 110 degrees. The minimum phase angle to achieve either a Flight Day 2 or Flight Day 3 rendezvous varies from 36 to 40 degrees. As the launch date approaches ISS orbit determination has less uncertainty and rendezvous planners can be less conservative with phasing angle limits. Note in Figure 25.2 only the Flight Day 3 planar window for Monday, February 8 overlapped with Flight Day 2 phasing limits.

Rendezvous Plans

The pre-launch Flight Day 3 docking rendezvous maneuver plans for each launch opportunity are given in Tables 25.4, 25.5, and 25.6. The data in the tables was

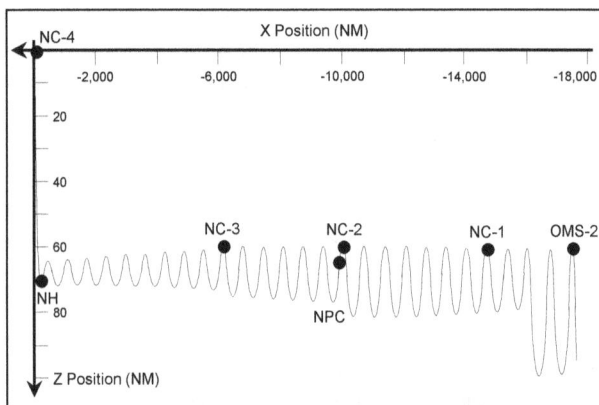

Figure 25.3 February 7 launch opportunity Flight Day 3 profile (not flown), 262 degree phase angle at insertion. The February 9 profile was similar due to ISS orbit design. See Tables 25.4 and 25.6.

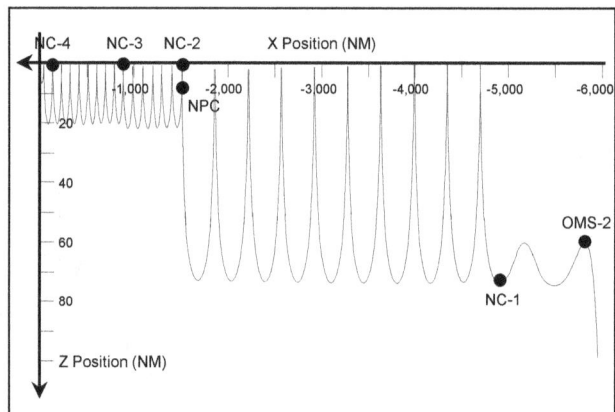

Figure 25.4 February 8 launch opportunity Flight Day 3 docking rendezvous profile, 92 degree phase angle at insertion. This profile was flown on STS-130. See Tables 25.5 and 25.7.

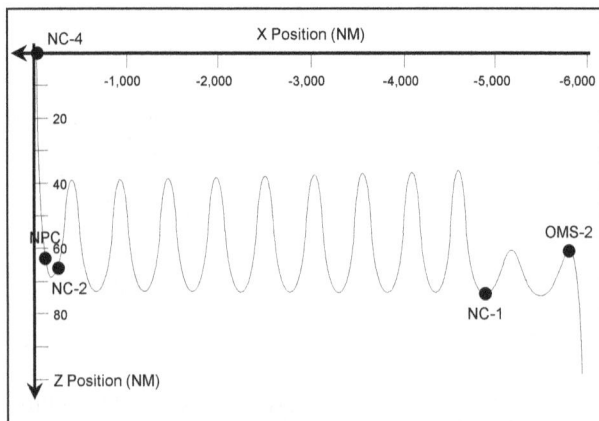

Figure 25.5 February 8 launch opportunity Flight Day 2 docking rendezvous profile (MDF, not flown), 92 degree phase angle at insertion.

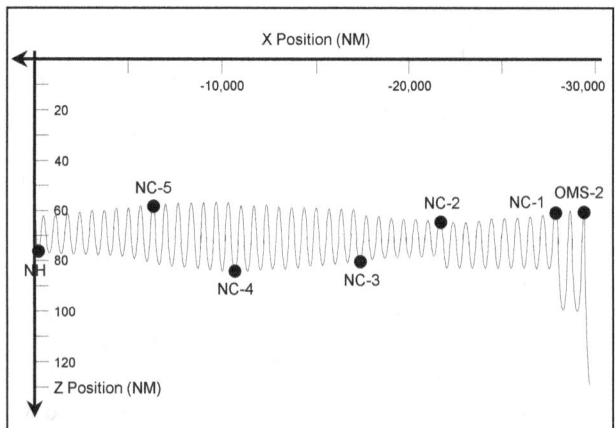

Figure 25.6 February 8 launch opportunity Flight Day 4 docking rendezvous profile (not flown), 464 degree phase angle at insertion.

Table 25.4 Launch minus 7 days rendezvous profile, Flight Day 3 docking, for the February 7 (Sunday) launch opportunity. See also Figure 25.3.

```
CHASER L E1        M 000:00:10:29.000        STS-130 2/7 Launch
TARGET L E3        M 000:00:10:29.000        ISS MDOL 2/7 Launch

  CHASER  DVtot =  390.07  DVx =  388.00  DVy =    1.96  DVz =    1.17
  TARGET  DVtot =    0.00  DVx =    0.00  DVy =    0.00  DVz =    0.00 10  MNVRS
-----------------------------------------------------------------------------
| MNVR  NAME |    GMTIG   IMP  |  DVx  |   HA   |   RANGE   |      Y       |
|  COMMENT   |    METIG        |  DVy  |   HP   |   PHASE   |    Ydot      |
|  DVMag     |    DT           |  DVz  |   DH   | Noon/Mid -|   SR/SS -    |
|------------+-----------------+-------+--------+-----------+--------------|
|   1 HA     | 038:10:18:08.201|  98.18| 123.60 | 4942.9924 |  -173842.6   |
| OMS-2      | 000:00:38:21.201|   0.00|  84.91 |  -86.7611 |       -5.7   |
|   98.2     | 000:02:56:40.840|   0.00|  63.63 | N-00:21:06| SS-00:49:44  |
|------------+-----------------+-------+--------+-----------+--------------|
|   2 NC     | 038:13:14:49.040|  42.97| 123.27 | 5923.8992 |  -153024.5   |
| NC-1       | 000:03:35:02.040|   0.00| 108.85 | -110.8250 |       -4.8   |
|   43.0     | 000:13:19:02.138|   0.00|  63.93 | N-00:27:12| SS-00:55:52  |
|------------+-----------------+-------+--------+-----------+--------------|
|   3 EXDV   | 039:02:33:51.179|   8.00| 123.43 | 7041.9123 |  -106559.5   |
| NC-2       | 000:16:54:04.179|   0.00| 112.80 |  156.3549 |       -3.6   |
|    8.0     | 000:00:12:09.963|   0.00|  64.13 | M-00:05:03| SR-00:22:11  |
|------------+-----------------+-------+--------+-----------+--------------|
|   4 NPC    | 039:02:46:01.142|   0.00| 123.43 | 7024.1573 |   -76378.9   |
| NPC        | 000:17:06:14.142|  -1.88| 112.68 |  155.1690 |       80.5   |
|    1.9     | 000:08:40:59.128|   0.00|  65.20 | N-00:38:36| SR-00:10:02  |
|------------+-----------------+-------+--------+-----------+--------------|
|   5 EXDV   | 039:11:27:00.270|   3.00| 123.14 | 5360.6120 |   -62554.8   |
| NC-3       | 001:01:47:13.270|   0.00| 114.41 |   96.3256 |       15.0   |
|    3.0     | 000:13:56:41.730|   0.00|  64.27 | M-00:20:14| SS-00:02:57  |
|------------+-----------------+-------+--------+-----------+--------------|
|   6 NH     | 040:01:23:42.000| 114.11| 187.20 |  216.7811 |      840.5   |
| NH         | 001:15:43:55.000|   0.00| 114.16 |    3.2787 |        1.5   |
|  114.1     | 000:00:45:51.000|   0.00|  68.89 | M-00:06:04| SR-00:23:30  |
|------------+-----------------+-------+--------+-----------+--------------|
|   7 NC     | 040:02:09:33.000| 110.02| 187.20 |   40.1780 |     -158.8   |
| NC-4       | 001:16:29:46.000|   0.00| 176.62 |    0.6339 |       -0.2   |
|  110.0     | 000:01:29:35.000|   0.00|   0.25 | N-00:05:55| SS-00:34:14  |
|------------+-----------------+-------+--------+-----------+--------------|
|   8 NCC    | 040:03:39:08.000|   8.95| 187.34 |    8.0055 |        0.3   |
| Ti         | 001:17:59:21.000|  -0.03| 181.74 |    0.1263 |       -0.0   |
|    9.0     | 000:01:16:54.000|   0.15|   0.20 | N-00:07:43| SS-00:36:00  |
|------------+-----------------+-------+--------+-----------+--------------|
|   9 SOI    | 040:04:56:02.000|   1.51| 187.64 |    0.3306 |       -1.8   |
| MC-4       | 001:19:16:15.000|  -0.05| 182.44 |    0.0023 |        0.0   |
|    1.7     | 000:00:13:00.000|   0.69|   0.30 | N-00:22:12| SS-00:50:15  |
|------------+-----------------+-------+--------+-----------+--------------|
|  10 SOR    | 040:05:09:02.000|   1.26| 187.67 |    0.0968 |        2.3   |
| Rbar       | 001:19:29:15.000|  -0.01| 183.08 |    0.0000 |        0.0   |
|    1.3     | 000:00:00:00.000|   0.33|   0.10 | N-00:09:12| SS-00:37:27  |
-----------------------------------------------------------------------------
```

```
STS-130/L-7_Feb7    Trajectory Sequence of Events

==============================================================================
| EVENT         |     TIG     |  ORB   |   DV   |   HA   |   HP   |
|               |     MET     |        |   FPS  |   NM   |   NM   |
==============================================================================
| OMS-2         | 00/00:37:47 |    1   |   98.2 |  123.6 |   84.9 |
| NC-1          | 00/03:34:46 |    3   |   42.9 |  123.3 |  108.8 |
| NC-2          | 00/16:53:57 |   12   |    8.0 |  123.4 |  112.8 |
| NPC           | 00/17:06:10 |   12   |    1.9 |  123.4 |  112.7 |
| NC-3          | 01/01:47:06 |   18   |    3.0 |  123.2 |  114.4 |
| NH            | 01/15:43:16 |   27   |  114.3 |  187.3 |  114.2 |
| NC-4          | 01/16:29:08 |   28   |  109.8 |  187.3 |  176.5 |
| Ti            | 01/17:59:21 |   29   |    9.1 |  187.4 |  181.7 |
| MC-4          | 01/19:16:15 |   30   |    1.7 |  187.7 |  182.4 |
| Dock          | 01/20:39:00 |   31   |    0.0 |  187.9 |  183.1 |
| Undock        | 10/14:33:00 |  169   |    0.0 |  189.5 |  180.8 |
| Sep-1         | 10/15:48:00 |  169   |    3.0 |  189.1 |  181.2 |
| Sep-2         | 10/16:16:00 |  170   |    1.5 |  189.8 |  181.3 |
==============================================================================
```

Table 25.5 Launch minus 7 days rendezvous profile, Flight Day 3 docking, for the February 8 (Monday) launch opportunity. See also Figure 25.4.

```
CHASER L E1        M 000:00:10:29.000      STS-130 2/8 Launch
TARGET L E3        M 000:00:10:29.000      ISS MDOL 2/8 Launch

  CHASER  DVtot =  387.67  DVx =  385.98  DVy =    1.59  DVz =   1.08
  TARGET  DVtot =    0.00  DVx =    0.00  DVy =    0.00  DVz =   0.00  9 MNVRS
-----------------------------------------------------------------------------
| MNVR  NAME |    GMTIG  IMP   |  DVx  |   HA   |   RANGE   |      Y     |
|  COMMENT   |    METIG        |  DVy  |   HP   |   PHASE   |    Ydot    |
|  DVMag     |    DT           |  DVz  |   DH   | Noon/Mid -|  SR/SS -   |
|------------+-----------------+-------+--------+-----------+------------|
|  1 HA      | 039:09:52:26.213| 143.22| 123.59 | 5004.3597 |  -53645.2 |
|  OMS2      | 000:00:38:21.213|   0.00| 109.90 |   88.1243 |      11.3 |
|  143.2     | 000:02:13:12.683|   0.00|  63.95 | M-00:23:25| SS-00:06:09|
|------------+-----------------+-------+--------+-----------+------------|
|  2 NC      | 039:12:05:38.895| 109.36| 185.40 | 4270.8262 |   40308.6 |
|  NC1       | 000:02:51:33.895|   0.00| 110.01 |   72.9660 |     -11.2 |
|  109.4     | 000:14:16:51.105|   0.00|  73.48 | N-00:27:18| SS-00:55:36|
|------------+-----------------+-------+--------+-----------+------------|
|  3 NH      | 040:02:22:30.000| 102.14| 186.72 | 1232.2807 |   -7874.9 |
|  NC2       | 000:17:08:25.000|   0.00| 165.81 |   19.5439 |       5.0 |
|  102.1     | 000:00:02:40.501|   0.00|   2.65 | M-00:38:42| SS-00:21:13|
|------------+-----------------+-------+--------+-----------+------------|
|  4 NPC     | 040:02:25:10.501|   0.00| 186.72 | 1231.7574 |   -7715.0 |
|  NPC       | 000:17:11:05.501|  -1.52| 165.78 |   19.5402 |       7.0 |
|  1.5       | 000:08:38:54.499|   0.00|   3.73 | M-00:36:00| SS-00:18:31|
|------------+-----------------+-------+--------+-----------+------------|
|  5 EXDV    | 040:11:04:05.000|   3.00| 187.17 |  726.8121 |     772.1 |
|  NC3       | 001:01:50:00.000|   0.00| 167.19 |   11.5018 |     -12.0 |
|  3.0       | 000:13:57:13.000|   0.00|   4.55 | N-00:19:44| SS-00:47:44|
|------------+-----------------+-------+--------+-----------+------------|
|  6 NC      | 041:01:01:18.000|  16.62| 187.51 |   40.0213 |    -172.9 |
|  NC4       | 001:15:47:13.000|   0.00| 176.44 |    0.6314 |      -0.2 |
|  16.6      | 000:01:28:25.000|   0.00|   0.31 | N-00:05:00| SS-00:33:06|
 ------------+-----------------+-------+--------+-----------+------------
|  7 NCC     | 041:02:29:43.000|   8.87| 187.66 |    8.0002 |      -6.8 |
|  Ti        | 001:17:15:38.000|  -0.02| 181.55 |    0.1262 |      -0.0 |
|  8.9       | 000:01:16:54.000|   0.11|   0.21 | N-00:07:58| SS-00:36:02|
|------------+-----------------+-------+--------+-----------+------------|
|  8 SOI     | 041:03:46:37.000|   1.51| 187.94 |    0.3305 |      -2.2 |
|  MC4       | 001:18:32:32.000|  -0.04| 182.13 |    0.0023 |       0.0 |
|  1.6       | 000:00:13:00.000|   0.64|   0.30 | N-00:22:27| SS-00:50:17|
|------------+-----------------+-------+--------+-----------+------------|
|  9 SOR     | 041:03:59:37.000|   1.26| 187.99 |    0.0969 |       2.2 |
|  Rbar      | 001:18:45:32.000|  -0.01| 182.83 |    0.0000 |       0.0 |
|  1.3       | 000:00:00:00.000|   0.33|   0.10 | N-00:09:27| SS-00:37:29|
-----------------------------------------------------------------------------


STS-130/L-7_Feb8    Trajectory Sequence of Events


=============================================================================
| EVENT          |   TIG      | ORB  |   DV   |  HA   |  HP   |
|                |   MET      |      |   FPS  |  NM   |  NM   |
=============================================================================
| OMS2           | 00/00:37:33|   1  |  143.2 | 123.6 | 109.9 |
| NC1            | 00/02:50:56|   2  |  109.4 | 185.4 | 110.0 |
| NC2            | 00/17:07:50|  12  |  102.0 | 186.7 | 165.8 |
| NPC            | 00/17:11:02|  12  |    1.5 | 186.7 | 165.8 |
| NC3            | 01/01:49:53|  18  |    3.0 | 187.2 | 167.1 |
| NC4            | 01/15:47:06|  27  |   16.7 | 187.5 | 176.4 |
| Ti             | 01/17:15:38|  28  |    8.9 | 187.6 | 181.6 |
| MC4            | 01/18:32:32|  29  |    1.7 | 187.9 | 182.1 |
| Dock           | 01/19:55:00|  30  |    0.0 | 188.1 | 182.9 |
| Undock         | 10/15:21:00| 169  |    0.0 | 189.6 | 180.5 |
| Sep-1          | 10/16:36:00| 169  |    3.0 | 189.2 | 180.9 |
| Sep-2          | 10/17:04:00| 170  |    1.5 | 190.0 | 181.0 |
|===========================================================================|
```

Table 25.6 Launch minus 7 days rendezvous profile, Flight Day 3 docking, for the February 9 (Tuesday) launch opportunity. See Figure 25.3.

```
CHASER L E1        M 000:00:10:29.000        STS-130 2/9 Launch
TARGET L E3        M 000:00:10:29.000        ISS MDOL 2/9 Launch

  CHASER  DVtot =  389.95  DVx =  387.89  DVy =    2.02  DVz =    1.03
  TARGET  DVtot =    0.00  DVx =    0.00  DVy =    0.00  DVz =    0.00 10  MNVRS
--------------------------------------------------------------------------------
| MNVR  NAME |    GMTIG  IMP   |   DVx  |   HA   |   RANGE    |      Y      |
|  COMMENT   |    METIG        |   DVy  |   HP   |   PHASE    |    Ydot     |
|  DVMag     |    DT           |   DVz  |   DH   | Noon/Mid - |  SR/SS -    |
|------------+-----------------+--------+--------+------------+-------------|
|  1 HA      | 040:09:29:55.220 |  98.18 | 123.60 | 4844.7793  | -175934.1  |
| OMS-2      | 000:00:38:21.220 |   0.00 |  84.91 |  -84.6318  |      -6.2  |
|   98.2     | 000:02:56:40.843 |   0.00 |  64.06 | N-00:22:30 | SS-00:50:30 |
|------------+-----------------+--------+--------+------------+-------------|
|  2 NC      | 040:12:26:36.063 |  40.78 | 123.27 | 5846.1166  | -154955.0  |
| NC-1       | 000:03:35:02.063 |   0.00 | 107.63 | -108.6828  |      -4.7  |
|   40.8     | 000:13:18:49.812 |   0.00 |  64.35 | N-00:28:35 | SR-00:00:29 |
|------------+-----------------+--------+--------+------------+-------------|
|  3 EXDV    | 041:01:45:25.875 |   8.00 | 123.41 | 7058.7115  | -107359.4  |
| NC-2       | 000:16:53:51.875 |   0.00 | 111.57 |  157.7307  |      -4.2  |
|    8.0     | 000:00:13:46.348 |   0.00 |  64.57 | M-00:06:33 | SR-00:24:13 |
|------------+-----------------+--------+--------+------------+-------------|
|  4 NPC     | 041:01:59:12.223 |   0.00 | 123.40 | 7038.8957  |  -67911.9  |
| NPC        | 000:17:07:38.223 |  -1.87 | 111.45 |  156.3991  |      89.2  |
|    1.9     | 000:08:39:14.291 |   0.00 |  66.26 | N-00:38:29 | SR-00:10:28 |
|------------+-----------------+--------+--------+------------+-------------|
|  5 EXDV    | 041:10:38:26.514 |   3.00 | 123.11 | 5397.0059  |  -63388.5  |
| NC-3       | 001:01:46:52.514 |   0.00 | 113.19 |   97.1957  |      14.9  |
|    3.0     | 000:13:57:02.486 |   0.00 |  64.73 | M-00:21:50 | SS-00:04:03 |
|------------+-----------------+--------+--------+------------+-------------|
|  6 NH      | 042:00:35:29.000 | 114.67 | 187.53 |  217.7958  |     867.0  |
| NH         | 001:15:43:55.000 |   0.00 | 112.94 |    3.2919  |       1.5  |
|  114.7     | 000:00:45:37.000 |   0.00 |  69.77 | M-00:07:15 | SR-00:25:06 |
 --------------------------------------------------------------------------
|  7 NC      | 042:01:21:06.000 | 111.57 | 187.64 |   39.9993  |    -153.3  |
| NC-4       | 001:16:29:32.000 |   0.00 | 176.29 |    0.6311  |      -0.3  |
|  111.6     | 000:01:30:35.000 |   0.00 |   0.22 | N-00:07:19 | SS-00:35:13 |
|------------+-----------------+--------+--------+------------+-------------|
|  8 NCC     | 042:02:51:41.000 |   8.92 | 187.80 |    8.0136  |      13.7  |
| Ti         | 001:18:00:07.000 |   0.07 | 181.36 |    0.1264  |      -0.0  |
|    8.9     | 000:01:16:54.000 |   0.02 |   0.20 | N-00:08:07 | SS-00:36:00 |
|------------+-----------------+--------+--------+------------+-------------|
|  9 SOI     | 042:04:08:35.000 |   1.51 | 188.06 |    0.3303  |      -2.5  |
| MC-4       | 001:19:17:01.000 |  -0.07 | 181.95 |    0.0023  |       0.0  |
|    1.7     | 000:00:13:00.000 |   0.69 |   0.30 | N-00:22:36 | SS-00:50:15 |
|------------+-----------------+--------+--------+------------+-------------|
| 10 SOR     | 042:04:21:35.000 |   1.26 | 188.09 |    0.0970  |       2.3  |
| Rbar       | 001:19:30:01.000 |  -0.01 | 182.60 |    0.0000  |       0.0  |
|    1.3     | 000:00:00:00.000 |   0.33 |   0.10 | N-00:09:36 | SS-00:37:28 |
 --------------------------------------------------------------------------


STS-130/L-2_Feb9    Trajectory Sequence of Events


==================================================================================
| EVENT         |    TIG     |  ORB  |   DV   |   HA   |   HP   |
|               |    MET     |       |   FPS  |   NM   |   NM   |
==================================================================================
| OMS-2         | 00/00:37:47 |   1   |  98.2  | 123.6  |  84.9  |
| NC-1          | 00/03:34:47 |   3   |  40.7  | 123.3  | 107.6  |
| NC-2          | 00/16:53:45 |  12   |   8.0  | 123.4  | 111.6  |
| NPC           | 00/17:08:29 |  12   |   1.9  | 123.4  | 111.5  |
| NC-3          | 01/01:46:45 |  18   |   3.0  | 123.2  | 113.2  |
| NH            | 01/15:43:16 |  27   | 114.9  | 187.6  | 113.0  |
| NC-4          | 01/16:28:54 |  28   | 111.3  | 187.7  | 176.1  |
| Ti            | 01/18:00:07 |  29   |   9.2  | 187.9  | 181.4  |
| MC-4          | 01/19:17:01 |  30   |   1.7  | 188.1  | 181.9  |
| Dock          | 01/20:39:00 |  31   |   0.0  | 188.3  | 182.6  |
| Undock        | 10/14:34:00 | 169   |   0.0  | 189.8  | 180.2  |
| Sep-1         | 10/15:49:00 | 169   |   3.0  | 189.4  | 180.6  |
| Sep-2         | 10/16:17:00 | 170   |   1.5  | 190.1  | 180.7  |
==================================================================================
```

Table 25.7 STS-130 Rendezvous and Separation Burns As Executed

Burn	MET d/hh:mm:ss	GMT d/hh:mm:ss	Local (CST) d, hh:mm:ss	DV Total ft./sec.	LVLH X DV ft./sec.	LVLH Y DV ft./sec.	LVLH Y DZ ft./sec.
OMS-2	0/00:38:35	39/09:52:40	Mon, 03:52:40	141.96	+141.9	0.0	0.0
NC-1	0/02:59:20	39/12:13:25	Mon, 06:13:25	102.94	+102.7	-5.2	+4.2
NC-2	0/17:15:58	40/02:30:03	Mon, 20:30:03	104.28	+104.2	0.0	+4.1
NC-3	1/02:27:59	40/11:42:04	Tue, 05:41:64	1.40	+1.4	0.0	0.0
NC-4	1/15:42:09	41/00:56:14	Tue, 18:56:14	22.60	+22.6	0.0	0.2
NCC	1/16:16:36	41/01:30:43	Tue, 19:30:43	0.36	+0.4	0.0	-0.1
Ti	1/17:14:18	41/02:28:25	Tue, 20:28:25	6.65	+6.6	-0.3	+0.8
MC-1	1/17:34:18	41/02:48:25	Tue, 20:48:25	0.44	+0.1	+0.3	-0.3
MC-2	1/18:02:23	41/03:16:30	Tue, 21:16:30	0.51	+0.1	0.0	-0.5
MC-3	1/18:19:23	41/03:33:30	Tue, 21:33:30	0.25	-0.2	-0.1	-0.1
MC-4	1/18:29:23	41/03:43:30	Tue, 21:43:40	1.93	+1.7	-0.2	+0.9
Undock	11/15:39:45	51/00:53:50	Fri, 18:53:50				
SEP-1	11/16:47:47	51/02:01:52	Fri, 20:01:52	1.50	0.0	0.0	-1.5
SEP-2	11/17:16:50	51/02:30:55	Fri, 20:30:55	3.30	-3.3	0.0	0.0

Docking occurred on Tuesday, February 9, 2010. Undocking occurred on Friday, February 19, 2010.

computed at launch minus 7 days (L-7). Data in Table 25.5 does not necessarily reflect the as-flown rendezvous profile. Table 25.7 provides actual burn data for the rendezvous and separation burns.

Rendezvous maneuver plans were re-computed after orbit insertion and changed based on vehicle performance. The ground targeted phase of shuttle rendezvous profiles is designed to place the Transition Initiation (Ti) maneuver on the day of rendezvous at the appropriate time relative to sunset to achieve the proper lighting during the R Bar Pitch Maneuver (RPM).

Figure 25.3 is the Flight Day 3 profile for the February 7 launch opportunity. The actual profile flown after the launch on February 8 is illustrated in Figure 25.4. The February 8 launch opportunity also had options for Flight Day 2 and Flight Day 4 rendezvous and docking. Figure 25.5 illustrates the Flight Day 2 profile and Figure 25.6 the Flight Day 4 profile. The Flight Day 3 and Flight Day 2 profiles (Figures 25.4 and 25.5) were the same up until the NC-1 burn. This preserved the Flight Day 2 and Flight Day 3 docking options up to the NC-1 burn.

Flight Day One – Launch and Orbit Insertion

The launch attempt on Sunday, February 7 was scrubbed due to cloud conditions that violated constraints for a Return to Launch Site abort. The scrub was

declared during the poll of the launch team to come out of the T-9 minute hold. The Mission Management Team later decided to conduct a 24 hour launch turn-around.

Endeavour launched at 4:14:05 AM EST from Pad 39A on Monday, February 8, 2010. Ascent was nominal. The 141.96 feet/second OMS-2 burn was performed and no RCS trim burn was required (Table 25.7). The shuttle Ku band antenna used for both TDRSS communications and rendezvous radar was activated. The 102.9 feet/second NC-1 burn was also successful and no RCS trim was required. The NC-1 burn was earlier than on previous flights due to the short phase angle at insertion (92 degrees). Since a nodal crossing occurred near the planned NC-1 time NC-1 was moved to the nodal crossing and a NPC burn was combined with NC-1.

Flight Day Two – Rendezvous Tools and TPS Checkout

Rendezvous related activities performed on Flight Day 2 (Figure 25.1) included the NC-2 and NC-3 phasing burns, centerline camera installation, docking ring extension, laptop computer set-up, and the Trajectory Control Sensor (TCS) and Hand Held Laser (HHL) checkouts.

Table 25.8 STS-130 Prime Orbit Timeline For Nominal Approach and Docking

Event	PET d/hh:mm:ss	MET d/hh:mm:ss	GMT d/hh:mm:ss	Local (CST) d, hh:mm:ss	DMT d/hh:mm:ss
Ti	-0/02:38:06	1/17:14:18	41/02:28:25	Tue, 20:28:25	41/05:28:25
US Solar Arrays Feathered for Docking	-0/02:07:31	1/17:44:53	41/02:59:00	Tue, 20:59:00	41/05:59:00
Range = 33000 ft	-0/02:04:07	1/17:48:17	41/03:02:24	Tue, 21:02:24	41/06:02:24
ISS Tracking Light On	-0/02:02:24	1/17:50:00	41/03:04:07	Tue, 21:04:07	41/06:04:07
Sunset	-0/02:02:13	1/17:50:11	41/03:04:18	Tue, 21:04:18	41/06:04:18
ISS Maneuver Start	-0/01:48:31	1/18:03:53	41/03:18:00	Tue, 21:18:00	41/06:18:00
ISS In Docking Attitude	-0/01:43:31	1/18:08:53	41/03:23:00	Tue, 21:23:00	41/06:23:00
Range = 10000 ft	-0/01:39:30	1/18:12:54	41/03:27:01	Tue, 21:27:01	41/06:27:01
Range = 5000 ft	-0/01:30:51	1/18:21:33	41/03:35:40	Tue, 21:35:40	41/06:35:40
ISS In Prox Ops Mode	-0/01:27:31	1/18:24:53	41/03:39:00	Tue, 21:39:00	41/06:39:00
Sunrise	-0/01:26:55	1/18:25:29	41/03:39:36	Tue, 21:39:36	41/06:39:36
ISS Tracking Light Off	-0/01:26:24	1/18:26:00	41/03:40:07	Tue, 21:40:07	41/06:40:07
Range = 3000 ft	-0/01:25:22	1/18:27:02	41/03:41:09	Tue, 21:41:09	41/06:41:09
MC4	-0/01:21:12	1/18:31:12	41/03:45:19	Tue, 21:45:19	41/06:45:19
Range = 1500 ft	-0/01:17:12	1/18:35:12	41/03:49:19	Tue, 21:49:19	41/06:49:19
RPM Start Window Open	-0/01:14:51	1/18:37:33	41/03:51:39	Tue, 21:51:39	41/06:51:39
Range = 1000 ft	-0/01:12:12	1/18:40:12	41/03:54:19	Tue, 21:54:19	41/06:54:19
KU to LO (800 ft)	-0/01:09:12	1/18:43:12	41/03:57:19	Tue, 21:57:19	41/06:57:19
+Rbar Arrival (725 ft)	-0/01:08:12	1/18:44:12	41/03:58:19	Tue, 21:58:19	41/06:58:19
Range = 600 ft	-0/01:03:00	1/18:49:24	41/04:03:31	Tue, 22:03:31	41/07:03:31
Start Pitch Maneuver	-0/01:01:06	1/18:51:18	41/04:05:25	Tue, 22:05:25	41/07:05:25
Noon	-0/00:58:53	1/18:53:31	41/04:07:37	Tue, 22:07:37	41/07:07:37
End Pitch Maneuver	-0/00:53:06	1/18:59:18	41/04:13:25	Tue, 22:13:25	41/07:13:25
RPM Full Photo Window Close	-0/00:50:56	1/19:01:28	41/04:15:35	Tue, 22:15:35	41/07:15:35
Initiate TORVA (575 ft)	-0/00:50:30	1/19:01:54	41/04:16:01	Tue, 22:16:01	41/07:16:01
RPM Start Window Close	-0/00:42:32	1/19:09:52	41/04:23:59	Tue, 22:23:59	41/07:23:59
Russian Solar Arrays Feathered	-0/00:42:24	1/19:10:00	41/04:24:07	Tue, 22:24:07	41/07:24:07
+Vbar Arrival (310 ft)	-0/00:39:00	1/19:13:24	41/04:27:31	Tue, 22:27:31	41/07:27:31
Range = 300 ft	-0/00:38:10	1/19:14:14	41/04:28:21	Tue, 22:28:21	41/07:28:21
Range = 250 ft	-0/00:34:00	1/19:18:24	41/04:32:31	Tue, 22:32:31	41/07:32:31
Sunset	-0/00:30:52	1/19:21:32	41/04:35:39	Tue, 22:35:39	41/07:35:39
Range = 200 ft	-0/00:29:50	1/19:22:34	41/04:36:41	Tue, 22:36:41	41/07:36:41
Range = 170 ft	-0/00:27:20	1/19:25:04	41/04:39:11	Tue, 22:39:11	41/07:39:11
Range = 150 ft	-0/00:25:40	1/19:26:44	41/04:40:51	Tue, 22:40:51	41/07:40:51
Range = 100 ft	-0/00:21:30	1/19:30:54	41/04:45:01	Tue, 22:45:01	41/07:45:01
Range = 75 ft	-0/00:18:30	1/19:33:54	41/04:48:01	Tue, 22:48:01	41/07:48:01
Range = 50 ft	-0/00:14:20	1/19:38:04	41/04:52:11	Tue, 22:52:11	41/07:52:11
Range (30 ft) SK Start	-0/00:11:00	1/19:41:24	41/04:55:31	Tue, 22:55:31	41/07:55:31
SK (30 ft) End (Push To Dock)	-0/00:06:00	1/19:46:24	41/05:00:31	Tue, 23:00:31	41/08:00:31
Range = 10 ft	-0/00:01:40	1/19:50:44	41/05:04:51	Tue, 23:04:51	41/08:04:51
Contact	+0/00:00:00	1/19:52:24	41/05:06:31	Tue, 23:06:31	41/08:06:31
Sunrise	+0/00:04:27	1/19:56:51	41/05:10:58	Tue, 23:10:58	41/08:10:58
Noon	+0/00:32:28	1/20:24:52	41/05:38:59	Tue, 23:38:59	41/08:38:59

CST – Central Standard Time, **DMT** – Decreed Moscow Time, **GMT** – Greenwich Mean Time, **ISS** – International Space Station, **LO** – Low, **MC4** – Mid-course Correction 4, **MET** – Mission Elapsed Time, **PET** – Phase Elapsed Time, **RPM** – R Bar Pitch Maneuver, **SK** – Station Keeping, **Ti** – Transition Initiation, **TORVA** – Twice Orbital Rate V Bar Approach

NC-2 and NC-3 Burns

The NC-2 burn on Flight Day 2 was successful and no RCS trim was required (Table 25.7). The 50 foot long Orbiter Boom Sensor System (OBSS) was grappled by the Remote Manipulator System (RMS), un-berthed, and thermal protection surveys of the starboard wing, port wing, and nose cap were completed. Imagery analysis by ground personnel began. The OBSS was re-berthed before the NC-3 burn. Both Extra-vehicular Mobility Units (EMUs) were checked out. On the ISS PMA-2 pressurization was performed in preparation for the docking.

Rendezvous Tools Checkout

The crew installed and aligned the centerline camera that would be used on Flight Day 3 during the docking. The crew reported that the installation went well and that camera performance at the 10 and 40 degree zoom settings was satisfactory. No shimming was required. The APAS Docking ring was extended and the docking hardware was then powered down.

The crew set up the two laptop computers (primary and backup) that ran Rendezvous and Proximity Operations Program (RPOP) and TCS software. A third laptop that ran WinDECOM software to provide orbiter Pulse Code Modulation Master Unit (PCMMU) data to the RPOP laptops was also set up. This hardware was left set up overnight for use on Flight Day 3.

TCS checkout was good. The crew tested the HHL with the night scope by using an Orbital Maneuvering System (OMS) pod as a target since the aft bulkhead was obscured by Node 3 (*Tranquility*) in the payload bay. HHL marks indicated that the OMS pod was not moving relative to the crew cabin. Quick thinking Mission Control personnel concurred that this was desirable.

At MET 0/17:30 (Monday, February 8, 8:44 pm CST) a star/Inertial Measurement Unit (IMU) alignment was performed using data from stars 93 (Iota Centauri) and 83 (Theta Aurigae).

Rendezvous Event and Lighting Information

The crew was provided with a rendezvous event and lighting information summary by Mission Control (Table 25.8). The Transition Initiation (Ti) maneuver was planned to occur at MET 1/17:14:18, about 36 minutes before orbital sunset. This provided appropriate lighting for the RPM photo session. The RPM was a 0.75 degree/second attitude maneuver that permitted the ISS crew to photograph the thermal protection system on the underside of the orbiter.

The solar beta angle during the approach would be minus 23 degrees (Sun south of the orbital plane). The Sun was not expected to enter the centerline camera field of view but could skirt the edges of the cockpit overhead windows during proximity operations near the +R Bar and again near the +V Bar when the Sun set behind the ISS.

The ISS was to begin a 5 minute maneuver to the docking attitude approximately 50 minutes after the Ti burn, at an MET of 1/18:04. If a post Ti night star tracker pass were required due to a rendezvous radar failure the ISS attitude maneuver would occur during the pass. The US solar arrays were to be feathered about 10 minutes after the MC-1 burn, at an MET of 1/17:45. Russian Service Module solar arrays were to be feathered before orbiter arrival on the +V Bar at an MET of 1/19:10.

The RPM window was predicted to open at MET 1/18:37:33 and close at 1/19:09:52 (Figure 25.7). The RPM window is driven by the Sun and camera lines-of-sight to the orbiter and ISS shadowing of the orbiter. The latest time the crew could start the RPM and allow the ISS crew to obtain two complete sets of RPM photos within the prescribed lighting constraints was MET 1/19:01:28. Optimum RPM lighting occurred +/- 4 minutes with respect to orbital noon. For the February 8 launch date no +R Bar station-keeping was required before the opening of the RPM start window for a nominal trajectory. Had the launch occurred on February 7 no station-keeping would have been required either. However, if the launch had slipped to February 9 +R Bar station-keeping would have been required before the RPM.

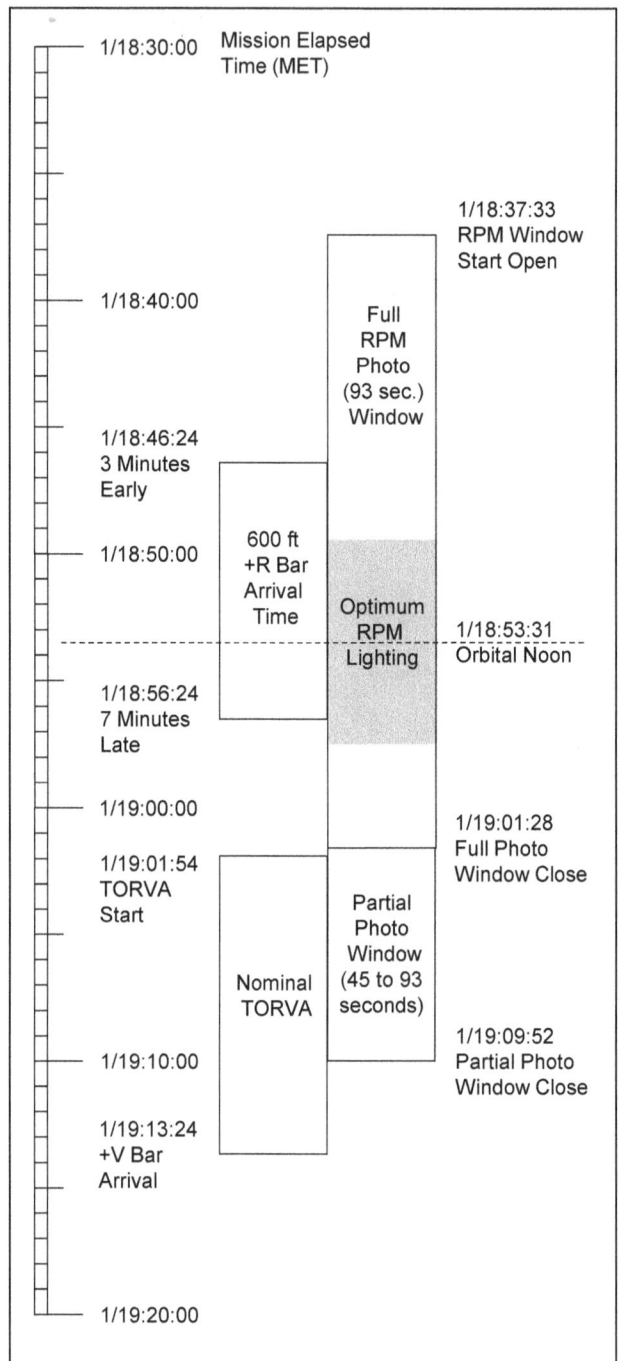

Figure 25.7 Full and partial RPM photo windows.

Sunset was predicted to occur approximately half an hour before docking, at a range on the +V Bar of about 200 feet, at an MET of 1/19:22. The predicted docking time was 1/19:52 MET, approximately four minutes before sunrise. This time assumed 5 minutes of +V Bar station-keeping for an auto-angular flyout at a range of 30 feet. However, if the relative alignment determined by the *Endeavour* crew from reading the ISS docking target was within limits the station-keeping and auto-angular flyout would not be performed.

Table 25.9 Rendezvous Team Audio Communication (DVIS) Loops

Loop	Function
AFD CONF	Shift handover briefings and special topics.
A/G	Communication with orbiter crew.
FD	Flight Director loop.
GNC Coord	Rendezvous tools and GNC discussions with the MER.
Landing Support	Additional loop for rendezvous team.
Nav Support	Communication between FDO/TRAJ and ONAV and NAV.
MOCR DYN1	RGPO and FDO with personnel other than Flight Dynamics.
MOCR GNC/PROP	Front room console discussions with GNC and PROP.
MPSR DYN A	FDO loop for communicating with MARS and PROFILE.
MPSR DYN B	RGPO loop for communicating with TARGET, ONAV, RPS, and PROX OPS.

AFD CONF – Alternate Flight Director Conference, **DVIS** – Digital Voice Intercommunications System, **DYN** – Dynamics, **FD** – Flight Director, **FDO** – Flight Dynamics Officer, **GNC** – Guidance, Navigation, and Control, **MARS** – Maneuver and Rendezvous Specialist, **MER** – Mission Evaluation Room, **MOCR** – Mission Operations Control Room, **MPSR** – Multi-Purpose Support Room, **NAV** – Navigation, **ONAV** – On-board Navigation, **PROFILE** – Rendezvous Trajectory Profile, **PROP** – Propellant Officer, **PROX OPS** – Proximity Operations, **RGPO** – Rendezvous Guidance and Procedures Officer, **RPS** – Rendezvous Procedures Support, **TRAJ** – Trajectory

Flight Day Three – Rendezvous and Docking

The crew was awakened at approximately 1/13:00 MET (4:14 pm CST, Tuesday, February 9, 2010) to begin Flight Day 3, the day of rendezvous (Figure 25.1). After the Group B power-up IMUs 2 and 3 were aligned to IMU 1. Stars 88 (Zeta Canis Majoris) and 80 (Epsilon Persei) were acquired for the alignment at MET 1/13:40. A retrograde waste water dump was begun at an MET of approximately 01/13:53 (5:07 pm CST). The retrograde dump ensured that the water would remain below the ISS orbital altitude, avoiding contamination of the ISS.

Handover to the Rendezvous Execute Shift

The handover briefing for the Mission Control Orbit 1 team, also known as the execute shift, started at MET 01/14:04 (5:18 pm CST) on the AFD CONF DVIS loop (Table 25.9). All Mission Control positions reported that their systems were performing well. The Flight Activities Officer (FAO) reported that the waste water dump was in progress.

The Flight Dynamics Officer (FDO) reported that the orbiter was in a 187x165 nm orbit, trailing ISS by 119 nm, closing at a rate of 85 nm per orbit. The NC-3 burn performed yesterday (Flight Day 2) was nominal with a total delta-velocity of 1.4 ft/sec. There was no NH (altitude adjust burn) in the rendezvous profile today. NC-4 TIG was MET 1/15:42 and at a range of 33 nm

behind the ISS. Although the nominal NC-4 range was 40 nm, the shorter 33 nm range was within the shuttle flight experience base.

The Rendezvous Guidance and Procedures Officer (RGPO) saw no issues with the 33 nm range at NC-4. The Flight Day 2 centerline camera checkout was good, as was the rendezvous tools checkout. A message on proximity operations lighting and an event timeline were sent to the crew on Flight Day 2. Some of the predicted proximity operations times could change due to MC-2 TIG slip and manual piloting. The nominal docking time was predicted to be 11:06 pm CST. Some TDRSS communications gaps had been closed by Ground Control (GC). It was expected that the crew would enter the rendezvous procedures book at a MET of 1/14:30.

GC reported a 6 minute TDRSS communications gap starting 16 minutes after the nominal docking time. GC was working hard to close it. The Guidance, Navigation, and Control (GNC) controller reported that GNC systems were performing well. A star of opportunity IMU alignment had just completed. The Assembly Checkout Officer (ACO) reported that the ISS was making progress in preparing for docking later in the day. The Payload Deployment and Retrieval System (PDRS) officer reported that the OBSS was berthed. The orbiter RMS was cradled and both Manipulator Positioning Mechanisms (MPMs) were deployed.

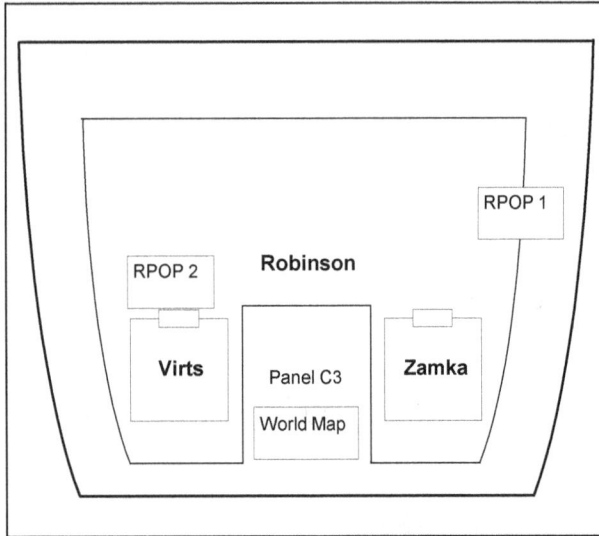

Figure 25.8 Diagram of orbiter flight deck showing approximate positions of crew members during rendezvous thru Ti. Crew positioning varied in zero gravity.

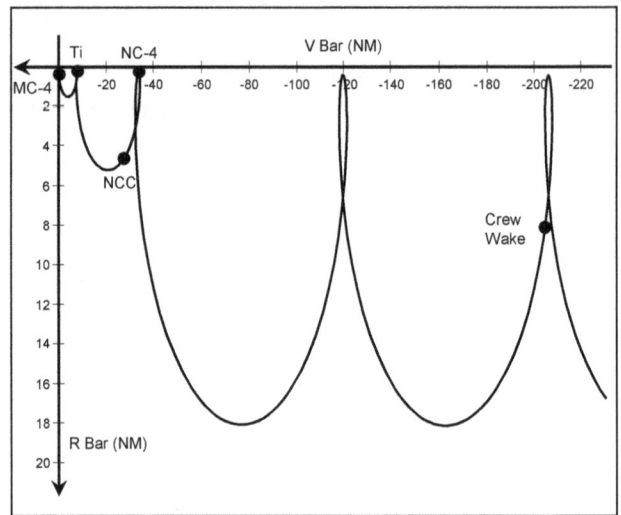

Figure 25.9 Flight Day 3 rendezvous profile. The MC-1, MC-2, and MC-3 burns between Ti and MC-4 are not shown.

Table 25.10 STS-130 Task Assignments for Rendezvous Thru MC-4

Task	Zamka	Virts	Robinson	Hire	Patrick	Behnken
Targeting/Burns/DAP (Pre & Post Ti)	Primary/Backup	Backup/Primary				
Navigation (sensors)	Backup	Primary				
Master Checklist		Primary	Backup			
Communications/PADS		Primary		Backup		
CCTV Configuration			Primary	Backup		
TRAD/PGSC Configuration			Primary			Backup
TRAD/PGSC Ops (RPOP/TCS)			Primary	Backup		
Systems/Reference Data		Backup		Primary		
Photo/TV OPS (Recorder, V10)		Backup		Primary		

CCTV – Closed Circuit Television, DAP – Digital Auto Pilot, OPS – Operations, PADS – Burn data read to the crew by Mission Control and recorded in the crew procedures, PGSC – Payload General Support Computer (laptop), RPOP – Rendezvous and Proximity Operations Program, TCS – Trajectory Control Sensor, TRAD – Tools for Rendezvous and Docking, Ti – Transition Initiation, TV - Television

Table 25.11 Flight Day 3 Rendezvous Package Preliminary Burn TIGs

Burn	Mission Elapsed Time (day:hr:min:sec)
NC4	001:15:42:17.648
NCC	001:16:16:36.000
TI	001:17:14:18.000 PET = 0:0, SS – 36 MIN
MC1	001:17:34:18.000
MC2	001:18:04:12.000 PET = 0:0
MC3	001:18:21:12.000 MC2 + 17 MIN
MC4	001:18:31:12.000 MC2 + 27 MIN
DOCK	001:19:52:00.000

MC – Mid-course Correction, MIN – Minute, NC – Phasing burn, NCC – Corrective Combination burn, PET – Phase Elapsed Time, SS – Sunset, Ti – Transition Initiation

NC-4 to Radar Acquisition

Figure 25.8 shows approximate crew positions from pre-NC-4 thru the Ti burn. Table 25.10 lists rendezvous task assignments for each crew member thru the MC-4 burn. Figure 25.9 illustrates the Flight Day 3 rendezvous profile and Table 25.11 lists the burn Times of Ignition (TIGs) provided to the crew by Mission Control.

The waste water dump was completed at MET 01/14:33 (5:47 pm CST). The crew entered the rendezvous book before the NC-4 burn. Orbiter and target state vectors and a target KFACTOR were uplinked to the vehicle. The Ku (rendezvous radar) self test was successful. The crew was informed that after NC-4 a period of ratty communications would begin at about 1/15:51 MET and end at about 1/16:03 MET. NC-4 was performed using both OMS engines. The crew performed one -X RCS pulse for trim. After the burn the crew reported that Payload General Support Computer (PGSC) set-up for the rendezvous and docking was complete.

The crew was not able to get through the entire initial star tracker relative navigation pass procedures before the period of ratty communications began. Rather than being a period of intermittent communications, the period was actually a black-out. The crew performed the intermediate Lambert targeting of the NCC burn in the blind. Continuous communications were re-established before the end of the star tracker pass due to the handover from the TDRS at 171 degrees west longitude to the TDRS at 41 degrees east longitude.

Overall the star tracker pass was successful and the crew inhibited star tracker measurement processing after 115 marks had been obtained. The brightness of the ISS could cause the star tracker shutter to close shortening the pass. However, that did not occur. The star tracker pass had the lowest FILT-PROP position difference of the ISS flights so far, 349 feet. One Mir flight had a lower number, and STS-109 (HST servicing mission 3B) had the record for lowest FILT-PROP position difference at 187 feet. NCC, the first on-board targeted maneuver, was successful. Radar tracking began at a range of approximately 138,500 feet.

Radar measurement processing began once the orbiter was within 135,000 feet of the ISS. The first position update based on radar was 410 feet. Subsequent position updates were 294, 204, and 80 feet. Radar processing was continued until the Ku antenna was transitioned to communications mode during final approach, except for during burns (Ti, MC-1, MC-2, MC-3, and MC-4) and the maneuvers to and from the Ti burn attitude.

Ti Through MC-2

Mission Control gave the crew a go to perform the Ti burn at an MET of 01/16:44 (7:58 pm CST). The crew later reported that before Ti the ISS was extremely bright and that much ISS detail was visible. Ti was successfully performed with the left OMS engine. The crew performed one +Y RCS pulse to trim the residual delta-velocity. Approximate crew positions on the orbiter flight deck from post-Ti thru MC-4 are shown in Figure 25.10. After Ti the proximity operations covariance matrix was uplinked. MC-1 was successfully performed.

After MC-1 the TCS unit was activated and it passed the self test. Between MC-1 and MC-2 the ISS crew took sunset photos of the orbiter (Figure 25.11). The preliminary, intermediate, and final MC-2 on-board Lambert targeting all had a TIG slip of minus 1 minute 49 seconds. Normally variation in TIG slip is observed. MC-2 was successfully executed.

After MC-2, at MET 01/18:05 (9:19 pm CST) ISS was maneuvering to the docking attitude and the U.S. solar arrays had been feathered. U.S. solar array feathering is required before the orbiter can proceed inside of 600 feet during the Twice Orbital Rate +V Bar Approach (TORVA).

HHL and TCS Acquisition, MC-3, and MC-4

The first HHL range measurement of 15,620.3 feet was acquired at MET 01/18:03:53 (Figures 25.12 and 25.13). At MET 01/18:11 (9:25 pm CST) the range to the ISS passed below 10,000 feet and the TCS shutter opened. At MET 01/18:12 (9:26 pm CST) ISS reported that it was in the docking attitude and that the ISS crew was go for the shuttle RPM. The ISS is required to be in the docking attitude (0 degrees in pitch, yaw, and roll in a Local Vertical Local Horizontal or LVLH frame) by the time the orbiter reached a range of 620 feet before the RPM. However, the solar arrays did not have to be

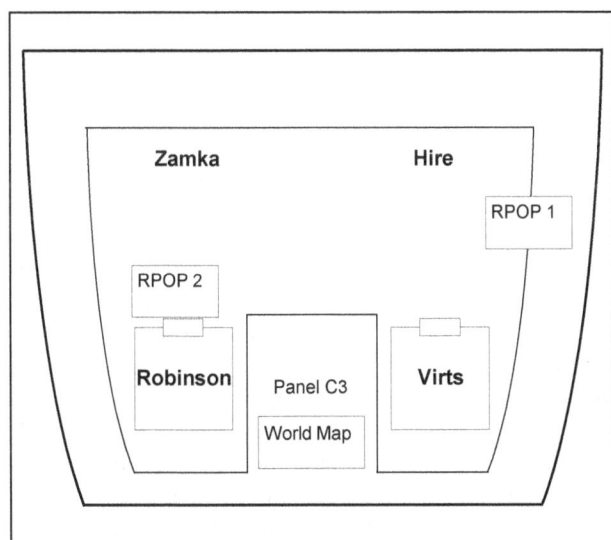

Figure 25.10 Diagram of orbiter flight deck showing approximate positions of crew members from post Ti thru MC-4. Crew positioning varied in zero gravity.

Figure 25.11 Photo of *Endeavour* taken by the ISS crew at orbital sunset. At this point (30:56 after Ti and 11:15 before MC-2) the range from *Endeavour* to the ISS was ~27,763 feet.

Figure 25.12 RPOP1 (primary) laptop photo soon after the first HHL mark of 15,620.3 feet was taken. Paper to the left of the laptop screen, for easy reference, is a chart (TRAD FAIL RANGE AND RANGE RATE DETERMINATION) that details how the crew should determine range and range rate during proximity operations in the event of sensor failures.

Table 25.12 Day of Rendezvous Predictions for TCS Acquisition Range*

Range (feet)	Name	Function
4200	Ray Bigonesse	RGPO Console
4900	Jessica LoPresti-Bellock	PROX Ops Console
5267	Steve Gauvain	Crew Training
5280	Andrzej Stewart	Observer
6429	Jorge Frank	Crew Training
7000	Alan Fox	Crew Training
7001	Dave Dannemiller	Observer
7002	John Goodman	SPAN DM

*Actual TCS acquisition range was 4,595 feet.
DM – Flight Dynamics Division, **PROX** – Proximity Operations, **RGPO** – Rendezvous Guidance and Procedures officer, **SPAN DM** – Flight Dynamics Division representative in the SPacecraft ANalysis control room, **TCS** – Trajectory Control Sensor

feathered by this time for the RPM to proceed. MC-3 was successfully performed.

Before TCS acquisition rendezvous personnel performed the customary ritual of estimating the TCS acquisition range (Table 25.12). At a range of about 4,600 feet TCS went through several loss of lock/reacquisition cycles before continuous tracking with the pulse laser began at a range of 4,595 feet. Normally TCS did not exhibit loss of lock/reacquisition cycles. Two minutes before MC-4 the crew reported HHL range was within 6 feet of TCS range, and HHL derived range

Figure 25.13 Mission Specialist Steve Robinson photographed on the port side of the aft flight deck. The primary RPOP laptop is to the right of him and he holds an HHL unit. Above left of Robinson is a CCTV screen with a ranging ruler overlay. Above that is another CCTV screen for the centerline camera, also equipped with an overlay. Robinson used the LED headlamp to read procedures in the low-light environment on the flight deck. Photo taken before MC-3 and around the time the first HHL mark was taken.

rate computed by RPOP was within 0.2 ft/sec of TCS range rate.

Mission Control was happy with the on-board Lambert targeted MC-4 burn solution. However, the RGPO asked the crew to ignore the Y body axis (the Y body axis corresponded to the Y LVLH axis at this point) delta velocity burn solution component due to radar noise. The RPOP relative state that included filtered TCS measurements indicated that the orbiter's out-of-plane relative motion was as desired and further tweaking was not necessary. Ground personal noted that the vehicle executed MC-4 when it was inside the MC-4 circle on the ground RPOP display, indicating very low trajectory dispersions (Figure 25.14).

RPM, TORVA, and TCS Performance

After MC-4 the crew transitioned to the approach cue card and Mission Control gave them a go to perform the RPM. Approximate crew positions on the flight deck from post-MC-4 thru docking are illustrated in Figure 25.15 (see also Figure 25.16). Crew tasks are listed in Table 25.13. Two crew members were assigned to Attitude Flyout since the commander considered this to be a two person task. Planned proximity operations relative motion and events are depicted in Figure 25.17. At this time the crew reported that TCS range and HHL range were within a few feet of each other and within 0.4 feet/second of each other in range rate (Figure 25.18). The crew transitioned the Digital Auto Pilot (DAP) to the Low Z mode at a range of 1,000 feet. Mission Control gave the crew a go to proceed inside of 600 feet.

RPOP – STS-130 ISS Rendezvous (Rev C)

File Edit Control Views Display Sensors Help

Prop Age   3

R 1716   Ṙ -2.61

X	-696	Ẋ 2.26
Y	12	Ẏ -0.01
Z	1569	Ż -1.85

Orb CG to Tgt CG   1000

Raw TCS 2(Pls)   MET: 1/18:30:45
Refl   7   Pitch   64
Age   3   Alt   184
Rng   1729
Rdot   -2.78
Elv   -1.38
Azi   1.02

x
TGT
LVLH z

1000 —

	RESID	RATIO	ACPT	REJ
RNG	-2.3	0.01	136	0
RDOT	-0.12	0.03	136	0
ELV	0.07	0.10	136	0
AZI	0.02	0.03	136	0

TCS NAV   PCM

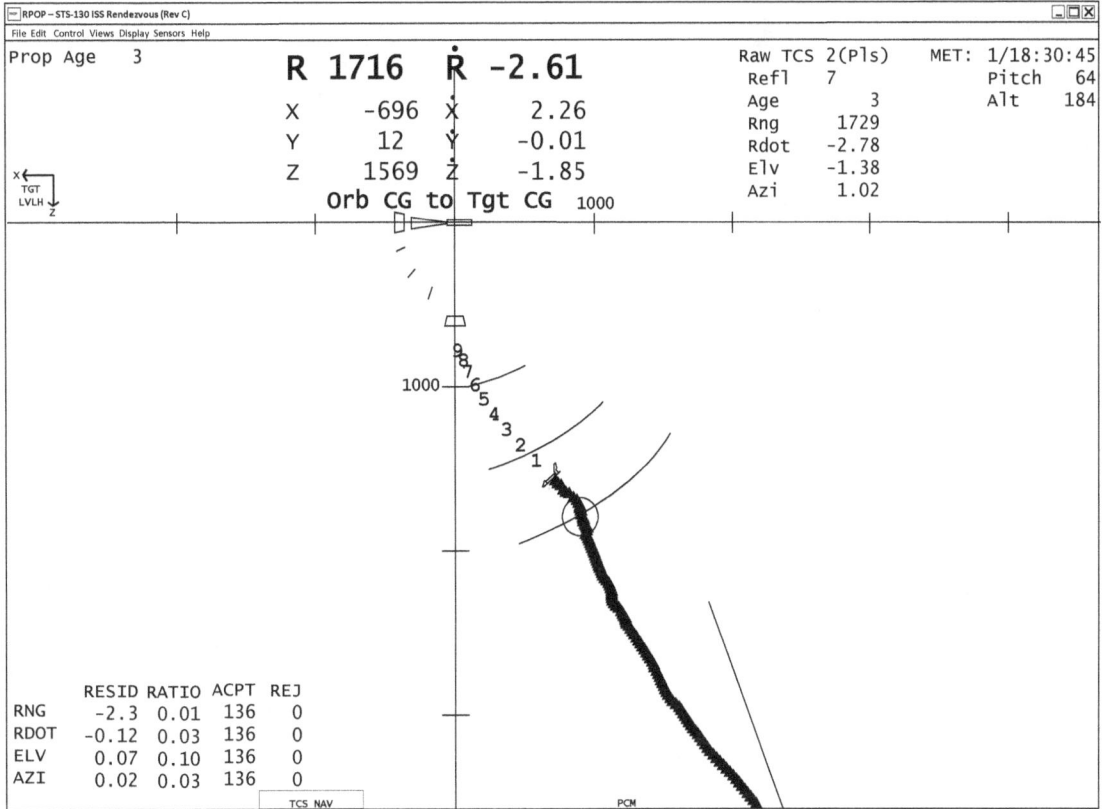

Figure 25.14 Re-created STS-130 RPOP display. Oval in center of display represents relative position of the MC-4 burn under nominally expected dispersions.

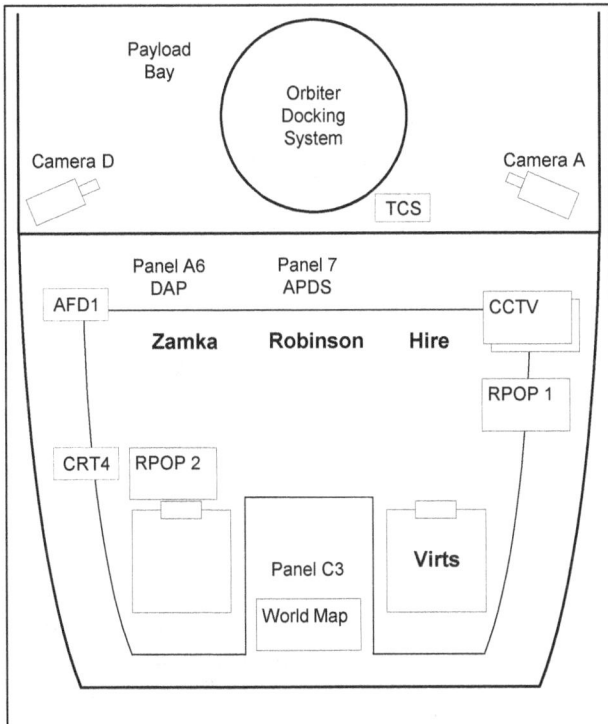

Payload Bay

Orbiter Docking System

Camera D   Camera A

TCS

Panel A6   Panel 7
DAP   APDS

AFD1   CCTV

Zamka   Robinson   Hire

RPOP 1

CRT4   RPOP 2

Panel C3

World Map

Virts

Figure 25.15 Diagram of orbiter flight deck showing approximate positions of crew members during proximity operations (post MC-4) and docking. Crew positioning varied in zero gravity.

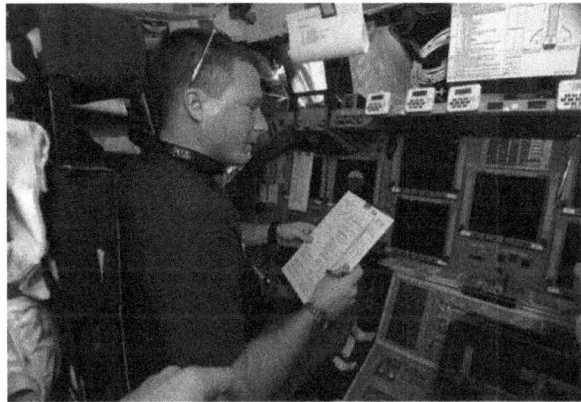

S130E006543   NASA

Figure 25.16 Pilot Terry Virts at the commander's station between MC-4 and +R Bar arrival. He is holding the APPROACH cue card. The RNDZ TIGs flight note and the RCS FAILURE/RESPONSE cue card are clamped to the overhead panel in front of him for easy reference. At the lower right of the photo is a laptop running a centerline camera repeater.

239

Table 25.13 STS-130 Task Assignments for Proximity Operations (Post MC-4)

Task	Zamka	Virts	Robinson	Hire	Patrick	Behnken
Piloting (aft station)	Primary	Backup				
Hand Held Lidar (HHL) Operations			Backup	Primary		
APDS Configuration (Docking System)					Primary	Backup
Range Ruler Callouts (TCS & Range Ruler)		Backup	Primary			
Attitude Flyout		Backup	Primary	Primary		

APDS – Androgynous Peripheral Docking System, **TCS** – Trajectory Control Sensor

TERMINAL PHASE, RPM, AND TORVA

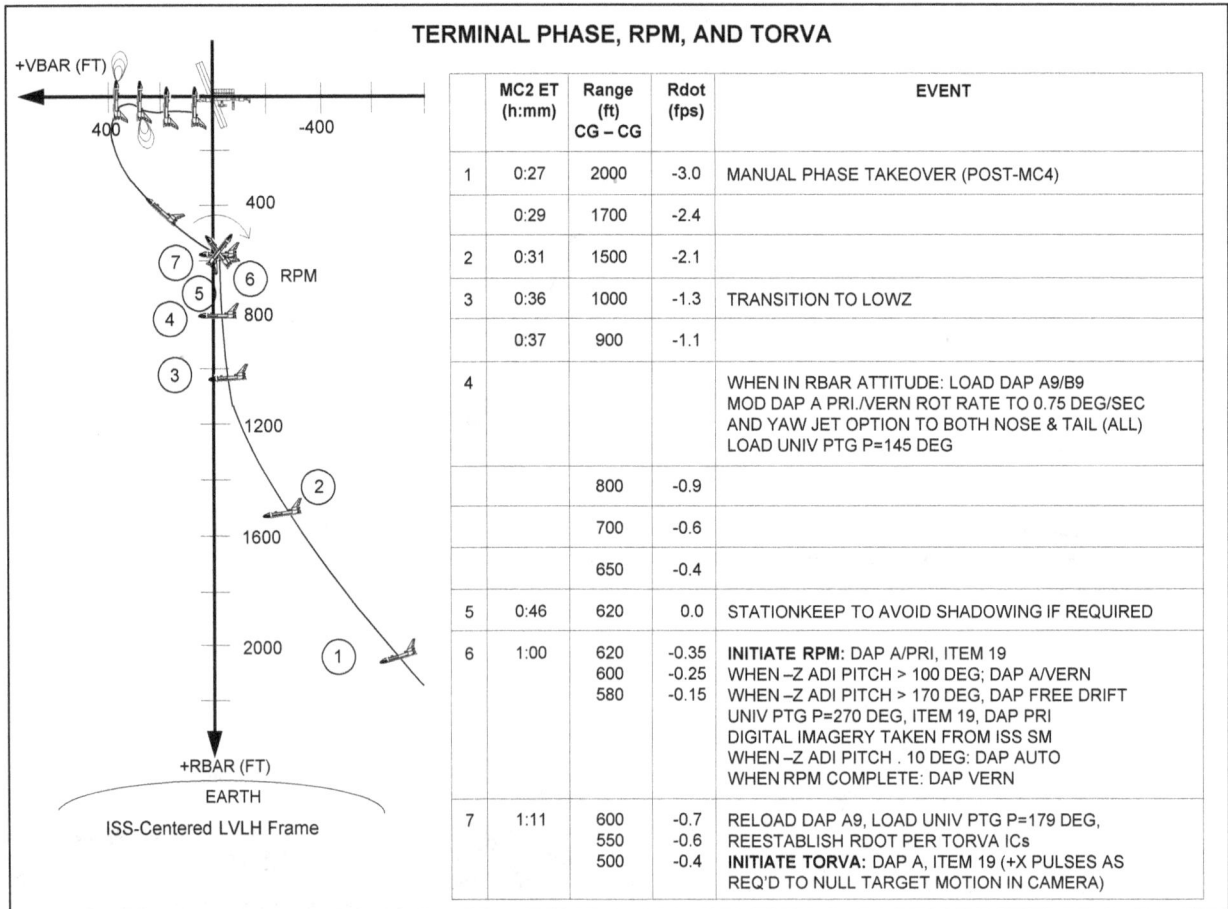

	MC2 ET (h:mm)	Range (ft) CG – CG	Rdot (fps)	EVENT
1	0:27	2000	-3.0	MANUAL PHASE TAKEOVER (POST-MC4)
	0:29	1700	-2.4	
2	0:31	1500	-2.1	
3	0:36	1000	-1.3	TRANSITION TO LOWZ
	0:37	900	-1.1	
4				WHEN IN RBAR ATTITUDE: LOAD DAP A9/B9 MOD DAP A PRI./VERN ROT RATE TO 0.75 DEG/SEC AND YAW JET OPTION TO BOTH NOSE & TAIL (ALL) LOAD UNIV PTG P=145 DEG
		800	-0.9	
		700	-0.6	
		650	-0.4	
5	0:46	620	0.0	STATIONKEEP TO AVOID SHADOWING IF REQUIRED
6	1:00	620 600 580	-0.35 -0.25 -0.15	**INITIATE RPM:** DAP A/PRI, ITEM 19 WHEN –Z ADI PITCH > 100 DEG; DAP A/VERN WHEN –Z ADI PITCH > 170 DEG, DAP FREE DRIFT UNIV PTG P=270 DEG, ITEM 19, DAP PRI DIGITAL IMAGERY TAKEN FROM ISS SM WHEN –Z ADI PITCH . 10 DEG: DAP AUTO WHEN RPM COMPLETE: DAP VERN
7	1:11	600 550 500	-0.7 -0.6 -0.4	RELOAD DAP A9, LOAD UNIV PTG P=179 DEG, REESTABLISH RDOT PER TORVA ICs **INITIATE TORVA:** DAP A, ITEM 19 (+X PULSES AS REQ'D TO NULL TARGET MOTION IN CAMERA)

Figure 25.17 Planned proximity operations profile and events.

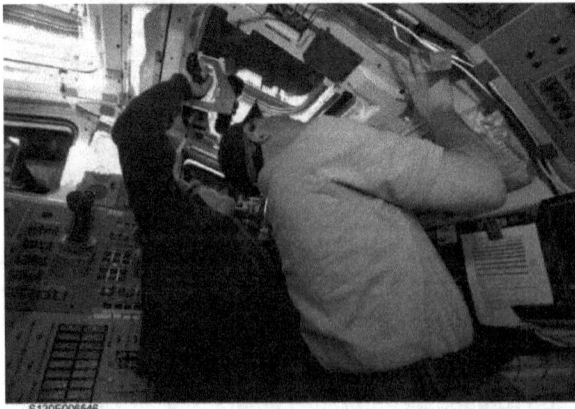

Figure 25.18 Kay Hire (left) taking HHL marks through the aft port overhead window while Steve Robinson (right) looks at the ISS. This photo was taken when the range was just outside of 1,000 feet between MC-4 and +R Bar arrival.

The crew reported that the TCS range rate measurements were noisy. Mission Control personnel observed that the raw TCS range rate value fluctuated between 0.7 feet/second and 5 feet/second. Normally TCS transitioned from the pulse laser to the CW laser at a range of around 1,000 feet. TCS remained mostly in pulse laser tracking with occasional and temporary transitions to CW laser tracking. TCS did not perform continuous tracking of ISS with the CW laser. However, the TCS filter edited the noisy measurements and the filtered state provided flight controllers with expected range and range rate values.

Before the RPM flight controllers discussed options to achieve continuous CW laser tracking during the TORVA to support +V Bar acquisition and the final approach. If TCS re-acquired pulse mode tracking after

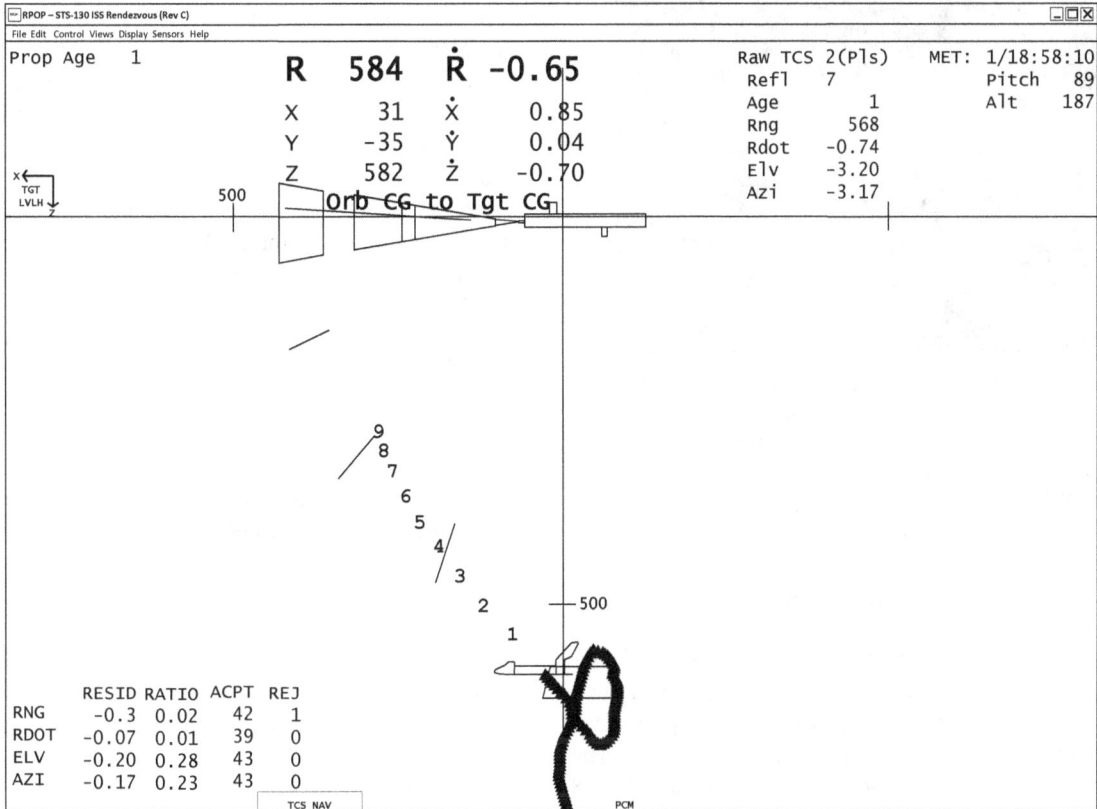

Figure 25.19 Re-created STS-130 RPOP display showing relative motion during the RPM and the start of the TORVA.

completion of the RPM and did not transition to CW mode tracking, the crew could force the TCS to perform CW tracking as a test. This might result in continuous CW laser tracking. If TCS failed to acquire in CW mode it would transition back to pulse. If pulse laser data was acceptable it could be used by the crew up until a range of approximately 65 feet, the minimum TCS tracking range in pulse mode.

The RPM was nominally initiated at a range of ~600 feet (center-of-gravity to center-of-gravity) with a closing rate of ~0.3 feet/second. 80 degrees into the maneuver the orbiter DAP was placed in free drift to prevent RCS firings that could cause RCS jet plume loads on the ISS. The DAP remained in free drift until ~80 degrees of rotation were left in the maneuver. No station-keeping was performed before the RPM and the crew successfully initiated the rotation.

Before and after the RPM (at a range of approximately 600 feet) CW laser measurements were noisy but usable. The RPM was successfully completed and the TORVA fly-around from the +R Bar to the +V Bar was begun. Figure 25.19 is a re-creation of the Mission Control RPOP display depicting relative motion before, during, and after the RPM using flight data. Figure 25.20 is a photo of the aft flight deck taken during the RPM. Figure 25.21 is a photo of *Endeavour* taken by the ISS crew during the TORVA. The RGPO elected to delay the pulse mode override until the orbiter was stable

Figure 25.20 Commander George Zamka (left) and Steve Robinson looking at the ISS through the Orbiter overhead windows just before RPM completion. Zamka piloted the orbiter throughout proximity operations and docking. Zamka was wearing his hat backwards so that the bill would not limit his ability to get close to the overhead window. The hat was from the II Marine Expeditionary Force Air Ground Logistics Team at Camp LeJeune, North Carolina. This photo was taken from the aft flight deck floor.

on the +V Bar. However, even without the pulse override to force CW tracking TCS personnel in the Mission Evaluation Room (MER) were able to determine that the CW laser was not functioning properly and that there were jumps in range and range rate.

Figure 25.21 Photo of *Endeavour* during the TORVA taken by the ISS Expedition 22 crew from the forward and port window in the Docking Compartment. In the foreground is Soyuz TMA-17/21S docked to the FGB nadir port.

+V Bar Acquisition Through Docking

Planned relative motion and events from +V Bar acquisition through docking are illustrated in Figure 25.22. By +V Bar acquisition the TCS had performed solid tracking with the CW laser for 8 minutes. The crew reported that HHL and TCS range were within 8 feet of each other and that docking mechanism power-up was complete.

Mission Control gave the crew a go for docking. The Ku sub-system was transitioned from radar to communications mode on the +V Bar at a range of 283

feet. At a range of approximately 250 feet on the +V Bar the crew overrode the pulse laser to force CW laser tracking. A MER recommendation to override the CW laser to re-establish pulse laser tracking was vetoed since additional TCS troubleshooting could distract the crew and good CW tracking data would provide the crew with a backup source of range and range rate data. The RGPO elected to maintain the present configuration of TCS for the rest of the approach.

CW laser tracking was not an improvement over the earlier pulse mode performance. The RPOP TCS filter edited bad data, but the filter re-initialized several times due to consecutive edits of bad measurements. Jumps in position and changes in range rate occurred. At one point the Mission Control RPOP display showed the orbiter approaching the ISS at a rate of 5 feet/second.

Figure 25.23 illustrates the poor TCS performance that resulted in relative motion on the RPOP display that did not reflect actual relative motion during the corridor approach. Note that the range rate at this point was -5.15 feet/second, an unrealistic value for this point in the approach. Figure 25.24, from STS-126 (November 2008), illustrates actual relative motion during an approach with good TCS performance. The rendezvous team lost confidence in TCS and recommended that the crew use HHL as the primary source of range data, in conjunction with the centerline camera vertical angle and the ranging ruler overlay on the CCTV camera display. RPOP computed range rate based on HHL range measurements.

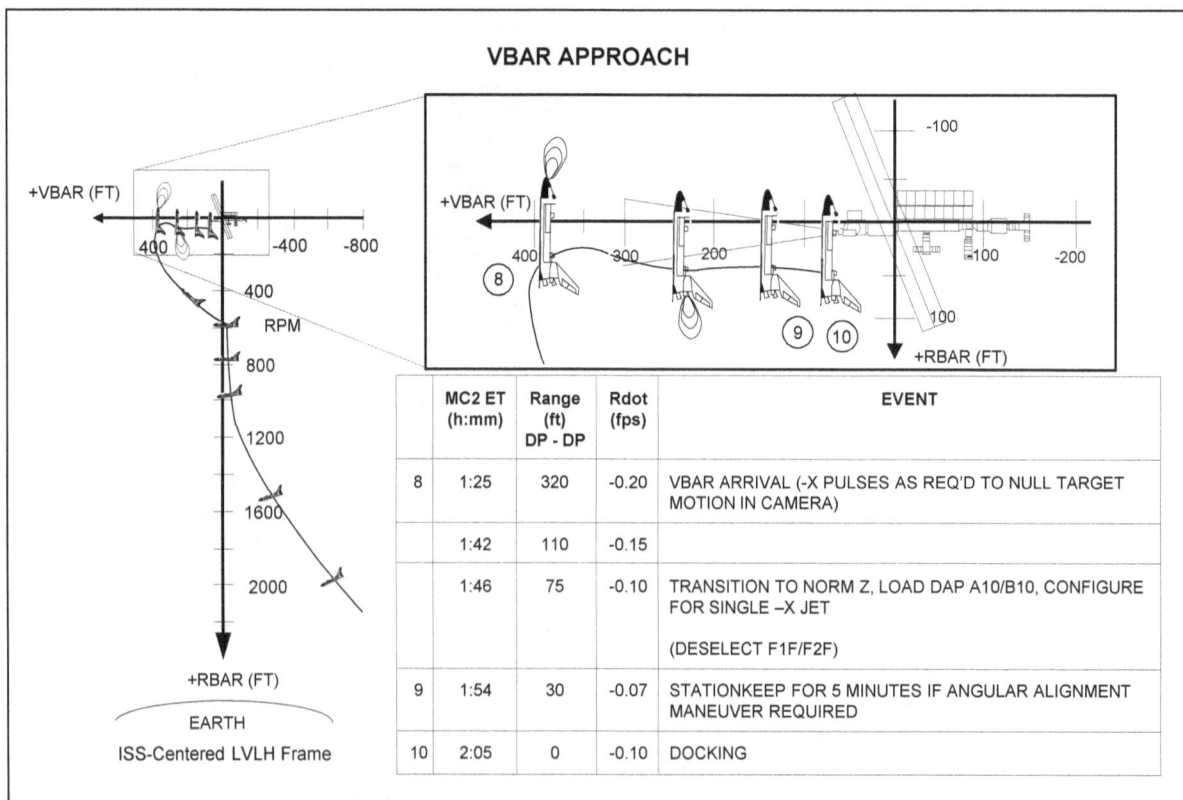

VBAR APPROACH

	MC2 ET (h:mm)	Range (ft) DP - DP	Rdot (fps)	EVENT
8	1:25	320	-0.20	VBAR ARRIVAL (-X PULSES AS REQ'D TO NULL TARGET MOTION IN CAMERA)
	1:42	110	-0.15	
	1:46	75	-0.10	TRANSITION TO NORM Z, LOAD DAP A10/B10, CONFIGURE FOR SINGLE –X JET (DESELECT F1F/F2F)
9	1:54	30	-0.07	STATIONKEEP FOR 5 MINUTES IF ANGULAR ALIGNMENT MANEUVER REQUIRED
10	2:05	0	-0.10	DOCKING

Figure 25.22 Planned final approach profile and events.

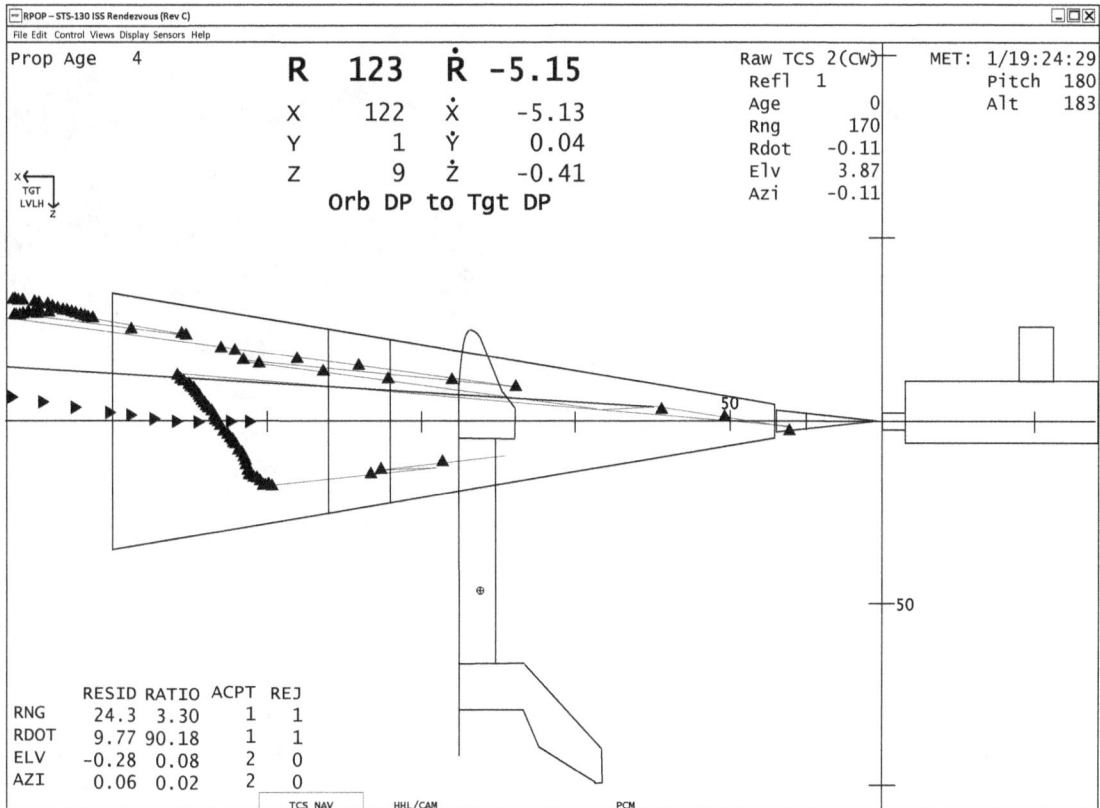

Figure 25.23 Re-created STS-130 RPOP display during final approach. Erratic relative motion indicators are due to erroneous TCS measurements. Note unrealistic range rate of -5.15 feet/second. Digital future relative motion indicators are off the screen due to the high range rate.

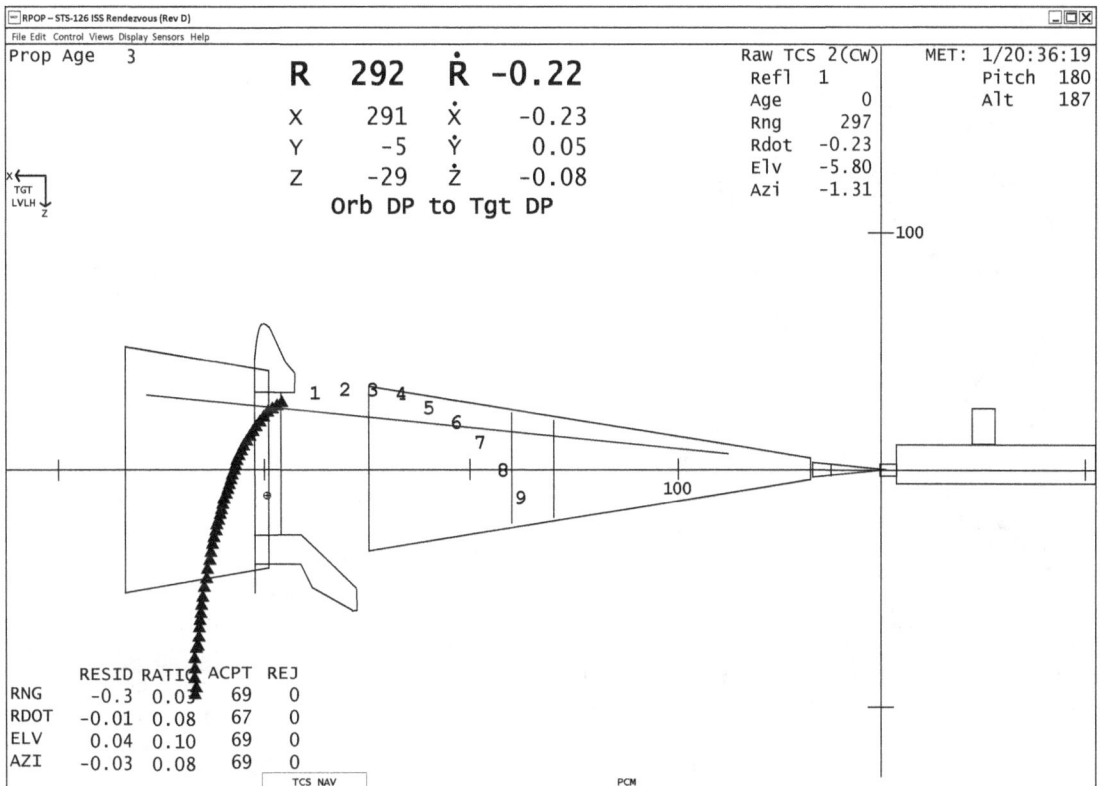

Figure 25.24 Re-creation of Shuttle RPOP display from STS-126 (November 2008). Numbers in front of the orbiter represent future predicted relative position at one minute intervals.

Use of HHL as the primary source of range data was preferable since attempts to recover acceptable TCS performance during the final approach could have distracted the crew and reduced situational awareness. Mission Control also recommended that the crew keep TCS and RPOP running in case good data was available to compare with HHL. Good TCS measurements were occasionally available and it was obvious to the crew when the measurements were good or bad. During periods of good TCS the measurements were compared to RPOP range and range rate measurements as a sanity check on HHL data.

The crew had been taking HHL measurements and inputting centerline camera vertical angle measurements (with horizontal measurements locked at zero degrees) into RPOP throughout the manual piloting phase even when TCS performance was acceptable. This resulted in a seamless transition from TCS to HHL as the primary range sensor. The crew had practiced this back-up procedure extensively during training to prepare for a TCS failure scenario. Although the crews normally take marks by placing the HHL against the window, acceptable marks were taken with the HHL further inside the cockpit, as long as the crew member was braced and not moving. Node 2 Forward on the ISS was used as the HHL aim point during the approach.

The crew checked the range to the ISS twice using the angle subtended by specific parts of the ISS structure on the centerline camera image in conjunction with a chart. These range values agreed with RPOP range determined by filtering HHL range and vertical centerline camera

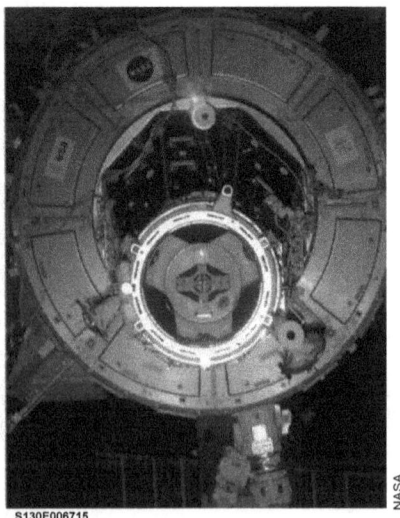

S130E006715

Figure 25.25 Digital still camera photo of docking target and PMA-2 during orbital night about 6 minutes before reaching the 30 foot point. The continuous starboard red ISS moding light on PMA-2 indicates that ISS is still in attitude hold. A flashing light indicated that ISS attitude control was in free drift. One of three methods of informing the orbiter crew of the ISS attitude control status is required. These are the moding light, Mission Control, or the ISS crew.

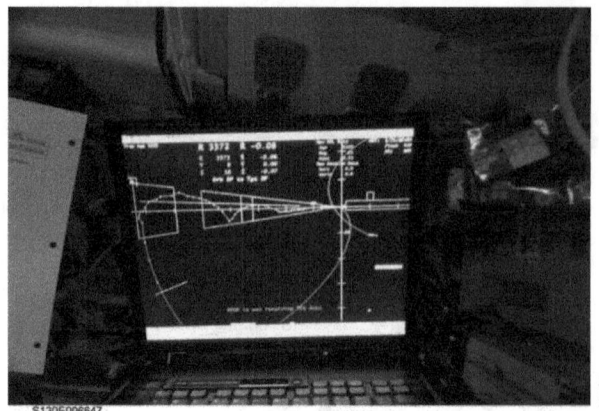

S130E006647

Figure 25.26 Post hard dock on-board photo of primary RPOP showing purple relative motion trajectory based on HHL range marks and centerline camera vertical angle measurements.

angle measurements. The crew did not use range determined by centerline camera sub-tended angles during the approach.

The crew had planned to start reading the docking target on the ISS hatch at range of 50 feet, but was able to read the target on the PGSC repeater outside of 50 feet. The crew determined relative misalignment using the repeated centerline camera image (Figure 25.25). The centerline camera had to be re-focused throughout the final approach.

The crew continued to provide RPOP with HHL range measurements and centerline camera angles (Figure 25.26). At 50 feet the vertical centerline camera angles input into RPOP were frozen at zero degrees. At a range of 35 feet the crew reported that no auto-angular fly-out was required. At 12 feet the crew switched to the ranging ruler overlay on the CCTV screen for range data. The minimum range for HHL is 12 feet (docking interface to docking interface range at this point is 6 feet). Range rate computed from ranging ruler measurements was consistent with the HHL derived range rate computed by RPOP. In addition, the TCS remained in CW mode and TCS data just before docking was good. The crew armed Post Contact Thrusting inside 15 feet. Docking hardware capture and transition of the shuttle and ISS attitude control systems to free drift were confirmed. The contact velocity determined by post-flight analysis was 0.113 feet/second. The successful docking was facilitated by intensive crew training in backup procedures developed to enable the crew to dock if TCS failed.

The docking occurred during orbital night therefore there were no solar lighting issues. The crew later reported that the Orbiter Docking System (ODS) truss lights were adequate for the final approach. Sunlight (the solar beta angle was -23 degrees) during manual piloting was not a problem and solar lighting did not degrade camera images used for piloting cues.

The ISS is visually complex with many elements that move and reflect differently, presenting a potential

Figure 25.27 After capture, hard dock was delayed until relative motion between the two docking mechanisms was dampened and alignment was achieved. This photo, taken from an orbiter aft flight deck payload bay window, shows the large angular misalignment between the mechanisms after capture.

distraction. Identifying specific elements of the complex ISS structure station required effort due to its constantly changing appearance. This was caused by spacecraft relative motion coupled with variations in solar lighting.

Digital still and video photography of the ISS was performed throughout proximity operations and the final approach (Figure 25.25 is one example). Before the flight the two shuttle crew members assigned to the Photo/TV task practiced in simulations with the other crew members assigned to proximity operations and docking tasks. This ensured that all tasks on the flight deck were performed in a seamless manner.

Docking mechanism alignment was lost during retraction. Approximately 35 minutes was required to permit relative motion to dampen and regain alignment due to the gravity gradient before continuing retraction and hook drive (Figure 25.27). A similar signature was seen after the STS-126 (November 2008) docking. In addition it was noted that petal 3 drove slightly slower than expected. This had been seen on previous flight of *Endeavour* (STS-127, July 2009).

Both the orbiter and the ISS were successfully hard docked. Rendezvous personnel in Mission Control remained on console until hard docking was confirmed in the event they were needed to support a contingency undock and separation. After hard dock the ISS initiated attitude control of the mated stack in the Torque Equilibrium Attitude. The orbiter controled mated stack attitude for large attitude maneuvers, dumps, and orbital re-boost burns. Before the ISS hatch was opened the crew of *Endeavour* removed the stand-off cross from the docking target and placed a protective cover over it.

Post Docking

After docking the RGPO informed the FAO that downlinking files with RPOP and TCS data was a high

priority to support the TCS failure investigation. All RPOP and TCS data files were downlinked from *Endeavour* and provided to investigators. The team was eager to resolve the TCS performance issue and develop a sensor plan to support the undocking and separation then scheduled for Flight Day 12 (Thursday, February 18). Questions to be answered were, 1) Why did the TCS have trouble transitioning to the CW laser mode? and 2) Which laser provided better data, Pulse mode or CW mode?

Flight Day Five – TCS Investigation

On Friday, February 12, results of the NASA/JSC Engineering Directorate TCS failure investigation were presented to STS-130 rendezvous and proximity operations personnel. One of the three electronic tone boards that controlled the CW laser, the intermediate board that generated the range signal, was intermittently emitting a bad bit resulting in range measurement spikes of approximately +/- 30 feet. The problem could not be overcome with a TCS power cycle since it was a hardware problem. A review of data from the previous flight of TCS unit 1007 on STS-127 did not show this poor performance. STS-130 was the first time a loss of a TCS tone board had occurred during flight.

However, pulse laser performance was not impacted by the CW laser problem. Pulse laser performance seen during the rendezvous (measurement noise) matched that seen on previous missions. This was confirmed during a playback of the Flight Day 3 rendezvous and proximity operations RPOP data on Flight Day 4 (Thursday). The questionable pulse laser lock-on signature had also been seen on previous missions.

The TCS unit flown on *Endeavour* (1007) had been designated to fly on the next flight of *Endeavour*, STS-134, in support of the ISS docking and the Sensor Test for Orion RelNav Risk Mitigation Detailed Test Objective (STORRM DTO). After the mission NASA/JSC Engineering personnel planned to pull the TCS unit from *Endeavour* for further investigation. TCS unit 1010 would be shipped to KSC for installation on *Endeavour* to support STS-134. It had recently completed between flight maintenance.

Modified TCS Procedures for Undocking and Separation

RGPO and GNC personnel had reviewed and redlined modified TCS procedures to be used by the crew during the separation and fly-around later in the mission (Figure 25.28). Since the CW laser was not recoverable the pulse laser would be used during the separation and fly-around, even though it was less accurate than the CW laser. Although the pulse laser was noisier than the CW laser it was deemed acceptable for separation since the RPOP navigation filter filtered the noise and provided good relative state information during the approach on Flight

UNDOCKING, TORS/TORF, AND FINAL SEPARATION

	UNDOCK ET (h:mm)	Range (ft) DP - DP	EVENT
	-0:03	0	ORBITER AND ISS IN FREE DRIFT TO BEGIN UNHOOKING (ISS LVLH PYR 0, 0, 0, ATTITUDE)
1	0:00	0 2	**UNDOCKING** AT MIDNIGHT-2 MIN; DAP B/ALT MODE TO LVLH; MAINTAIN CORRIDOR
	0:01		SELECT VERNS; PERFORM DAP B +Z NORMZ BURNS AT 10 SEC INTERVALS TO BUILD OPENING RATE TO 0.15 FPS
	>0:03	>30	DAP B +Z NORMZ BURNS AT 10 SEC INTERVALS TO BUILD OPENING RATE TO 0.20 FPS
		50	RESELECT −Z JETS (F1F, F2F)
2	0:07	75	TRANSITION TO LOW Z
3	0:29 [1:15]*	>400 (CG-CG)	**SEP1**: 1.5 FPS +X RADIAL BURN [IF PROP AVAILABLE, PERFORM 1/4 LAP TORS BETWEEN 400 AND 600 FT (CG-CG); NULL OPENING RATE OUTSIDE 600 FEET; PERFORM 3/4 LAP TORF BETWEEN 600 AND 700 FT; THEN PERFORM SEP1]
4	0:57 [1:43]*	>2000 (CG-CG)	**SEP2**: 3.0 FPS +X, NORMZ RETROGRADE BURN

* Alternate times are for Flyaround Case.

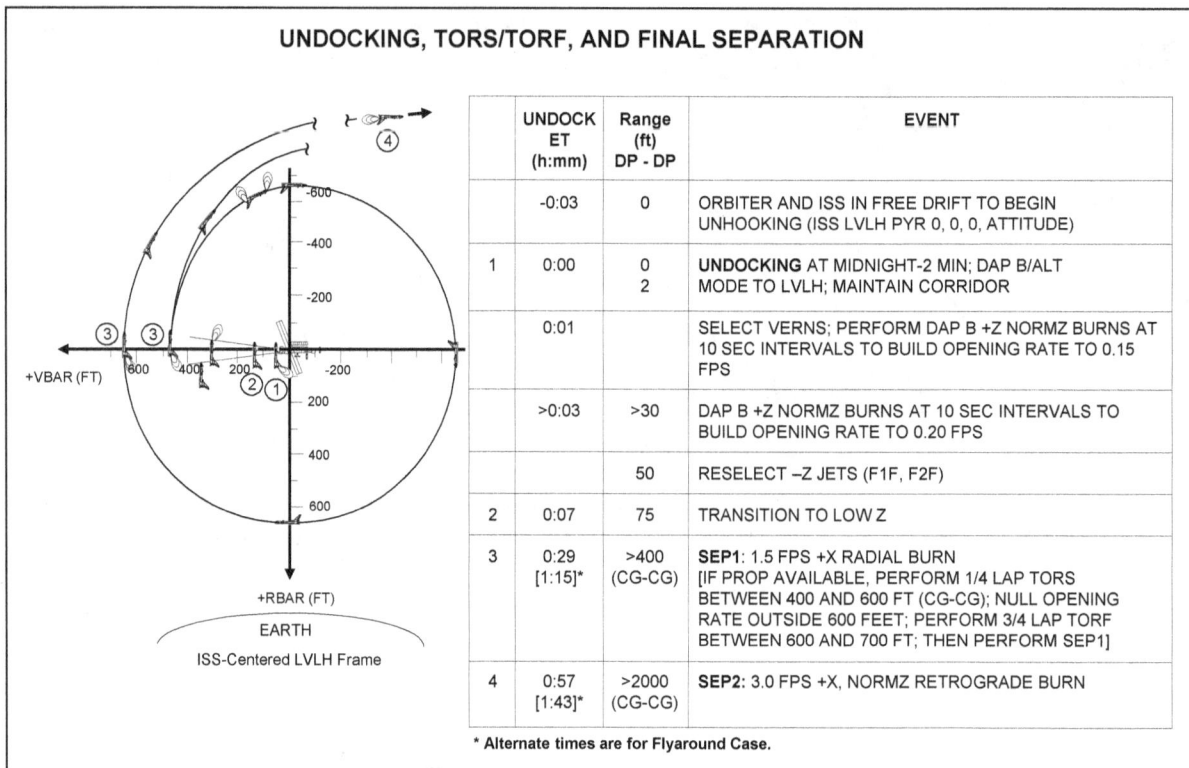

Figure 25.28 Planned separation profile and events.

Day 3. The pulse laser was only capable of tracking outside a minimum range of 66 feet.

However, HHL had always been the primary proximity operations sensor during fly-arounds, and HHL must be working before an ISS fly-around could be initiated. The ranging ruler would provide the initial range and range rate data at undocking for the first five feet of the separation, until HHL data became usable. The APDS ring would not be extended as it was for docking and therefore would not match the cockpit television screen overlay. The crew would have to point payload bay cameras A and D so that the image matched the overlay. Range was expected to be in error by one foot therefore ranging ruler data could only be used for calculating range rate by hand.

Flight Day Eight

On Monday, February 15, starting at 1 pm RGPO and crew training personnel verified the modified TCS procedure for use of pulse laser tracking for separation and fly-around in the Shuttle Mission Simulator.

Flight Day Twelve

The crew successfully performed the rendezvous tools checkout in preparation for the undocking and separation on Flight Day 13. The CW laser self-test was good during the checkout. However, this did not change the undocking procedure to not use the CW laser. The HHL was checked by firing it at the ISS S0 truss. The centerline camera was aligned and did not require shimming. Transfer of the shuttle crew to *Endeavour* was complete by 1 am CST on Friday, February 19. The crew re-installed the stand-off cross on the ISS docking target. The hatch was closed at 2:08 am CST.

Flight Day Thirteen – Undocking and Separation

Undocking was originally planned for Flight Day 12. However, the Shuttle Program decided to add one docked day to provide more time for working Node 3 installation issues and other activities.

Change to Separation-2 Burn

Soon after docking the ISS Program decided to perform an 11.8 foot/second posigrade re-boost using the Russian Service Module engines. The burn was scheduled to be performed on February 20, after the undocking of *Endeavour*. This orbit raising maneuver of 25 minute 57 seconds duration, in conjunction with the orbiter vernier RCS jet re-boost burn while docked, was designed to set up the appropriate phasing to support the upcoming Soyuz TMA-16/20S undocking and landing, the Soyuz TMA-18/22S launch and docking, and the STS-131/19A launch and docking. However, this conflicted with the planned 1.5 foot/second posigrade Separation-2 (SEP-2) burn to be performed by *Endeavour* after undocking. Both vehicles would have performed orbit raising maneuvers within 24 hours of each other and risked the possibility of unsafe relative motion. The

Table 25.14 STS-130 Prime Orbit Timeline For Nominal Undock

Event	MET d/hh:mm:ss	PET hh:mm:ss	GMT d/hh:mm:ss	Local (CST) hh:mm:ss	DMT d/hh:mm:ss
ISS in Prox Ops Mode	11/12:44:53	-0/02:55:07	50/21:59:00	Fri, 15:59:00	51/00:59:00
Sunrise	11/12:57:14	-0/02:42:46	50/22:11:21	Fri, 16:11:21	51/01:11:21
Noon	11/13:25:14	-0/02:14:46	50/22:39:21	Fri, 16:39:21	51/01:39:21
US Arrays Feathered	11/13:30:53	-0/02:09:07	50/22:45:00	Fri, 16:45:00	51/01:45:00
Sunset	11/13:53:14	-0/01:46:46	50/23:07:21	Fri, 17:07:21	51/02:07:21
Start Maneuver to Undock Attitude	11/14:24:54	-0/01:15:06	50/23:39:01	Fri, 17:39:01	51/02:39:01
Sunrise	11/14:28:37	-0/01:11:23	50/23:42:44	Fri, 17:42:44	51/02:42:44
Orbiter/ISS in Undock Attitude	11/14:52:53	-0/00:47:07	51/00:07:00	Fri, 18:07:00	51/03:07:00
Noon	11/14:56:38	-0/00:43:22	51/00:10:45	Fri, 18:10:45	51/03:10:45
Sunset	11/15:24:39	-0/00:15:21	51/00:38:46	Fri, 18:38:46	51/03:38:46
Russian Segment Arrays Feathered	11/15:30:00	-0/00:10:00	51/00:44:07	Fri, 18:44:07	51/03:44:07
Undocking	11/15:40:00	+0/00:00:00	51/00:54:07	Fri, 18:54:07	51/03:54:07
DAP:B/LVLH/ALT, Maintain Corridor	11/15:40:05	+0/00:00:05	51/00:54:12	Fri, 18:54:12	51/03:54:12
Initial Separation Pulses	11/15:41:00	+0/00:01:00	51/00:55:07	Fri, 18:55:07	51/03:55:07
ISS Snaps and Holds Current Attitude	11/15:41:40	+0/00:01:40	51/00:55:47	Fri, 18:55:47	51/03:55:47
Range = 50 feet (DP-DP), Reselect –X jets	11/15:45:00	+0/00:05:00	51/00:59:07	Fri, 18:59:07	51/03:59:07
Range = 75 feet (DP-DP), DAP: LOW Z	11/15:47:00	+0/00:07:00	51/01:01:07	Fri, 19:01:07	51/04:01:07
Sunrise	11/16:00:11	+0/00:20:11	51/01:14:18	Fri, 19:14:18	51/04:14:18
RS Arrays Resume Tracking	11/16:00:12	+0/00:20:12	51/01:14:19	Fri, 19:14:19	51/04:14:19
Range = 400 ft (CG-CG), Start Flyaround	11/16:09:00	+0/00:29:00	51/01:23:07	Fri, 19:23:07	51/04:23:07
Range = 600 feet (CG-CG)	11/16:18:30	+0/00:38:30	51/01:32:37	Fri, 19:32:37	51/04:32:37
US Arrays Resume Tracking	11/16:18:31	+0/00:38:31	51/01:32:38	Fri, 19:32:38	51/04:32:38
ISS Maneuver to TEA (Range > 600 ft)	11/16:18:54	+0/00:38:54	51/01:33:01	Fri, 19:33:01	51/04:33:01
–Rbar Crossing	11/16:20:30	+0/00:40:30	51/01:34:37	Fri, 19:34:37	51/04:34:37
Noon	11/16:28:08	+0/00:48:08	51/01:42:15	Fri, 19:42:15	51/04:42:15
–Vbar Crossing	11/16:32:00	+0/00:52:00	51/01:46:07	Fri, 19:46:07	51/04:46:07
+Rbar Crossing	11/16:43:30	+0/01:03:30	51/01:57:37	Fri, 19:57:37	51/04:57:37
Sep 1 Burn on +Vbar (1.5 fps radial burn)	11/16:55:00	+0/01:15:00	51/02:09:07	Fri, 20:09:07	51/05:09:07
Sunset	11/16:56:05	+0/01:16:05	51/02:10:12	Fri, 20:10:12	51/05:10:12
Sep 2 Burn (3.0 fps +X retrograde burn)	11/17:23:00	+0/01:43:00	51/02:37:07	Fri, 20:37:07	51/05:37:07

ALT – Alternate DAP, **CG** – Center of Gravity, **CST** – Central Standard Time, **DAP** – Digital Auto Pilot, **DMT** – Decreed Moscow Time, **DP** – Docking Port, **GMT** – Greenwich Mean Time, **ISS** – International Space Station, **LVLH** – Local Vertical Local Horizontal, **MET** – Mission Elapsed Time, **PET** – Phase Elapsed Time, **RS** – Russian Solar arrays, **Sep** – Separation, **TEA** – Torque Equilibrium Attitude

shuttle flight control team determined that SEP-2 could be changed to a 3.0 foot/second retrograde burn to maintain safe relative motion while preserving enough propellant to raise *Endeavour's* orbit for a contingency re-rendezvous with the ISS. Re-rendezvous and docking with the ISS would be required if the late thermal protection system inspection found damage that ruled out a safe re-entry.

The separation plan was for a standard +V Bar undocking and corridor separation followed by one complete fly-around of the ISS (Figure 25.28). Planned event times are in Table 25.14. Nominal undocking was planned to occur two minutes before orbital midnight to ensure lighting for the entire fly-around. The mated stack was to be in a 0,0,0 LVLH attitude, also known as +XVV +ZLV. The ISS automatically re-established ISS attitude control 100 seconds after undocking and snapped and held the current LVLH attitude. Mission Control read the attitude to the *Endeavour* crew for entry into RPOP. At MET 11/16:19, or when the orbiter was at a range of more than 600 feet, the ISS automatically began a 5 minute maneuver to the Torque Equilibrium Attitude (TEA) of -2.2 degrees pitch, +1.0 degrees yaw, and +0.7 degrees roll in the LVLH frame. Mission Control read the TEA to the *Endeavour* crew for entry into RPOP.

RPOP used the ISS attitude to perform accurate TCS reflector identification.

The US solar arrays were to be feathered by 11/13:00 MET. US array sun tracking would resume once the orbiter range increased beyond 600 feet. The Russian solar arrays were to be feathered by 11/15:10 MET, 30

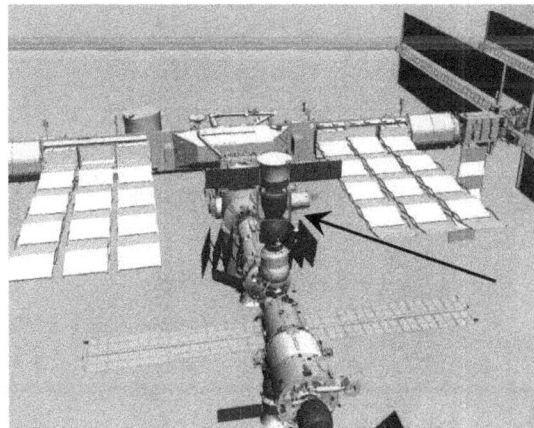

Figure 25.29 ISS graphic transmitted to the crew in preparation for the fly-around. The location of the loose MLI blanket on Soyuz TMA-16 to be photographed during the fly-around is indicated by the arrow and box.

minutes before undocking. Sun tracking would resume at MET 11/16:00 as the orbiter approached the fly-around range.

Sunrise would occur after undocking at an MET of 11/16:00, about 10 minutes before the start of the fly-around. The next sunset would occur at about MET 11/16:56, at about the scheduled SEP-1 burn time.

During the fly-around the crew was to take photos of the ISS and a loose Multilayer Insulation (MLI) blanket on the Soyuz TMA-16 (20S) that was docked to Mini-Research Module 2 (MRM 2, also known as the *Poisk* docking module, located on the *Zvezda* module's zenith Port, Figure 25.29). The photos were to be taken while the orbiter traversed from the –R Bar to the –V Bar.

The solar beta angle was 22.8 degrees placing the sun in the overhead widows of the orbiter crew compartment near the end of the fly-around. The sun was not expected to enter the centerline camera field of view. SEP-1 was planned to be a 1.5 foot/second radial up burn. SEP-2, a 3 foot/second retrograde burn, would follow 28 minutes later.

Pre-Undocking Activities

The crew completed an IMU alignment at MET 11/14:15. At MET 11/14:23 ISS attitude control was in FREE DRIFT and *Endeavour* assumed automated attitude control of the mated stack. A 27 minute 48 second attitude maneuver to the undocking attitude was performed. TCS was initialized at MET 11/14:56. The undocking attitude was achieved. Mission Control stated that *Endeavour* was go for undocking at MET 11/15:22 and ISS reported that they were go for undocking at MET 11/15:28. The Russian solar arrays were reported feathered one minute later. Mated stack attitude and rates were acceptable and within the desired limits. The undocking attitude was the same as the docking attitude. Before undocking the orbiter was placed in attitude control of the stack and the ISS was in free drift. The orbiter flight control system was placed in free drift 3 minutes before undocking.

Undocking Through Fly-Around Start

Approximate crew positions on the flight deck are illustrated in Figure 25.30. Crew task assignments for undocking and separation are listed in Table 25.15. Undocking occurred at MET 11/15:39:45 (6:53 pm, Friday, February 19). As soon as the docking system

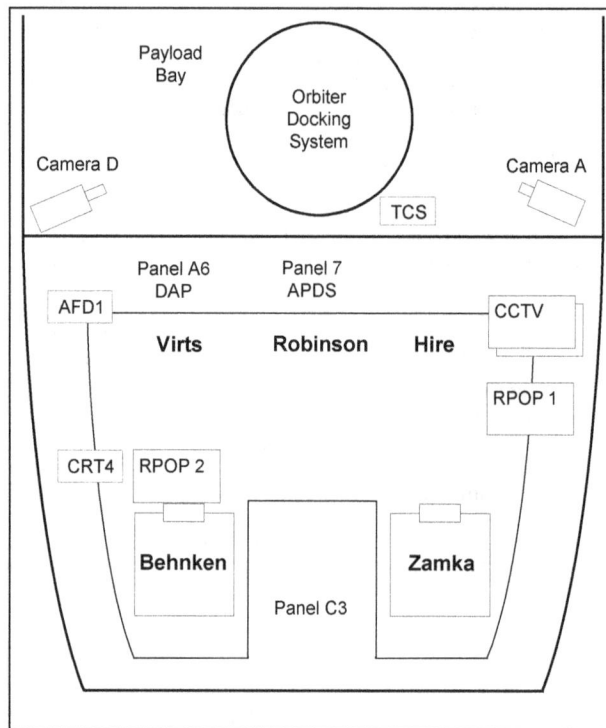

Figure 25.30 Diagram of orbiter flight deck showing approximate positions of crew members during the undocking and fly-around. Crew positioning varied in zero gravity.

Table 25.15 STS-130 Task Assignments for Undocking and Fly-Around

Task	Zamka	Virts	Robinson	Hire	Patrick	Behnken
Piloting (aft station)	Backup	Primary				
Navigation (sensors)	Primary	Backup				
Guidance & Control (targeting/burns)	Backup	Primary				
Master Checklist	Primary		Backup			
Communications/PADS	Primary			Backup		
APDS Operations (docking flights)					Backup	Primary
CCTV Operations			Primary	Backup		
PGSC Configuration			Primary			
TRAD/PGSC Operations			Primary	Backup		
Hand Held Lidar (HHL) Operations			Backup	Primary		
Systems/Reference Data	Backup			Primary		
Photo/TV Operations		Backup		Primary		

APDS – Androgynous Peripheral Docking System, **CCTV** – Closed Circuit Television, **PADS** – Burn data read to the crew by Mission Control and recorded in the crew procedures, **PGSC** – Payload General Support Computer (laptop), **TRAD** – Tools for Rendezvous and Docking, **TV** – Television

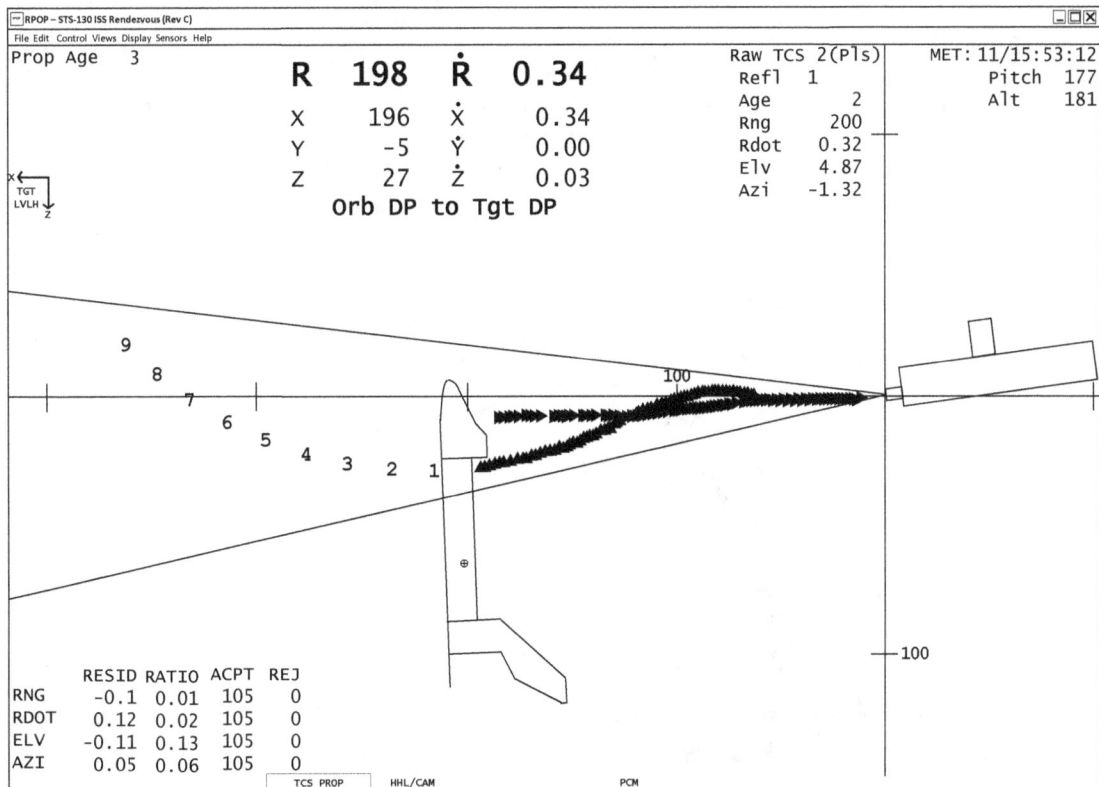

```
RPOP – STS-130 ISS Rendezvous (Rev C)                                    _ □ X
File Edit Control Views Display Sensors Help
Prop Age    3              R  198   Ṙ  0.34      Raw TCS 2(Pls)│ MET: 11/15:53:12
                                                  Refl   1    │    Pitch  177
                            X   196   Ẋ   0.34    Age     2    │    Alt    181
                            Y    -5   Ẏ   0.00    Rng   200    │
x ←                         Z    27   Ż   0.03    Rdot  0.32   │
TGT                         Orb DP to Tgt DP      Elv   4.87   │
LVLH ↓                                            Azi  -1.32   │
   Z



      9
        8
         7                                        100
          6
            5
              4   3   2   1


                      ⊕                                        — 100


          RESID RATIO ACPT REJ
RNG       -0.1  0.01  105   0
RDOT      0.12  0.02  105   0
ELV      -0.11  0.13  105   0
AZI       0.05  0.06  105   0
              TCS PROP    HHL/CAM         PCM
```

Figure 25.31 Re-created RPOP display showing relative motion during the corridor backout before the start of the fly-around. The trajectory that starts at undocking is based on HHL range and range rate measurements. The slow variation in the HHL trajectory is due to orbiter attitude dead-banding. The trajectory that starts at ~66 feet is based on TCS.

petals were clear, a few seconds after physical separation, the crew of *Endeavour* placed the DAP in an LVLH attitude hold with attitude control provided by the vernier RCS jets (Figure 25.28). ISS automatically resumed attitude control 100 seconds after physical separation by snapping and holding the current LVLH attitude.

The crew used the ranging ruler for range data and then transitioned to the HHL. The crew had trained for a TCS fail separation extensively, just as they had trained for the TCS fail approach. During the +V Bar separation the front of Node 2 (Figure 25.25), at the 1 o'clock position opposite the NASA logo, was used as the HHL aim point. The crew transitioned to ISS Node 3 as the HHL aim point during the fly-around. The base of the Cupola was also used. Airlock range data, formerly considered a good HHL aim point, was less consistent than Node 3 range data due to the presence of tanks on the airlock.

TCS began tracking in pulse laser mode at a range of approximately 66 feet, as expected. The *Endeavour* DAP was placed in LOW Z mode at a range of 75 feet (MET 11/15:47). The ISS crew rang the bell to signal that *Endeavour* had departed the ISS. The *Endeavour* crew completed the docking mechanism power down at MET 11/15:47. Figure 25.31 illustrates relative motion during the corridor back-out. Approximately 100 seconds after docking the ISS snapped the current LVLH attitude and the ISS attitude control system held that attitude.

Fly-Around and Separation

Figure 25.32 illustrates relative motion during the corridor back-out and the initial part of the twice orbital rate fly-around. The crew began the fly-around approximately eight minutes early. The orbiter reached the 600 foot range approximately 7 minutes ahead of schedule, with sunrise occurring as both vehicles flew over the Himalayas. The orbiter crossed the –R Bar about 7.5 minutes early at an MET of 11/16:13. The fly-around range of 600 feet is nominally reached about the time the orbiter reaches the –R Bar. Both the fly-around start and the arrival at a range of 600 feet were early since the crew flew the +V Bar separation slightly faster than the published 0.2 foot/second separation rate. This had no impact on the mission.

Two minutes later TCS tracking was temporarily lost while the ISS was maneuvering back to the Torque Equilibrium Attitude. ISS achieved the TEA at MET 11/16:20. At MET 11/16:27 the Rendezvous Guidance and Procedures Officer noted that RPOP did a good job of accurately propagating the orbiter state between the –R Bar and -V Bar when TCS measurements were not available due to ISS structural blockage. On the RPOP display (Figure 25.33) orbiter relative position during TCS tracking is indicated by a triangle, while propagated relative position when TCS data is not available is indicated by a cross. While the mission commander flew

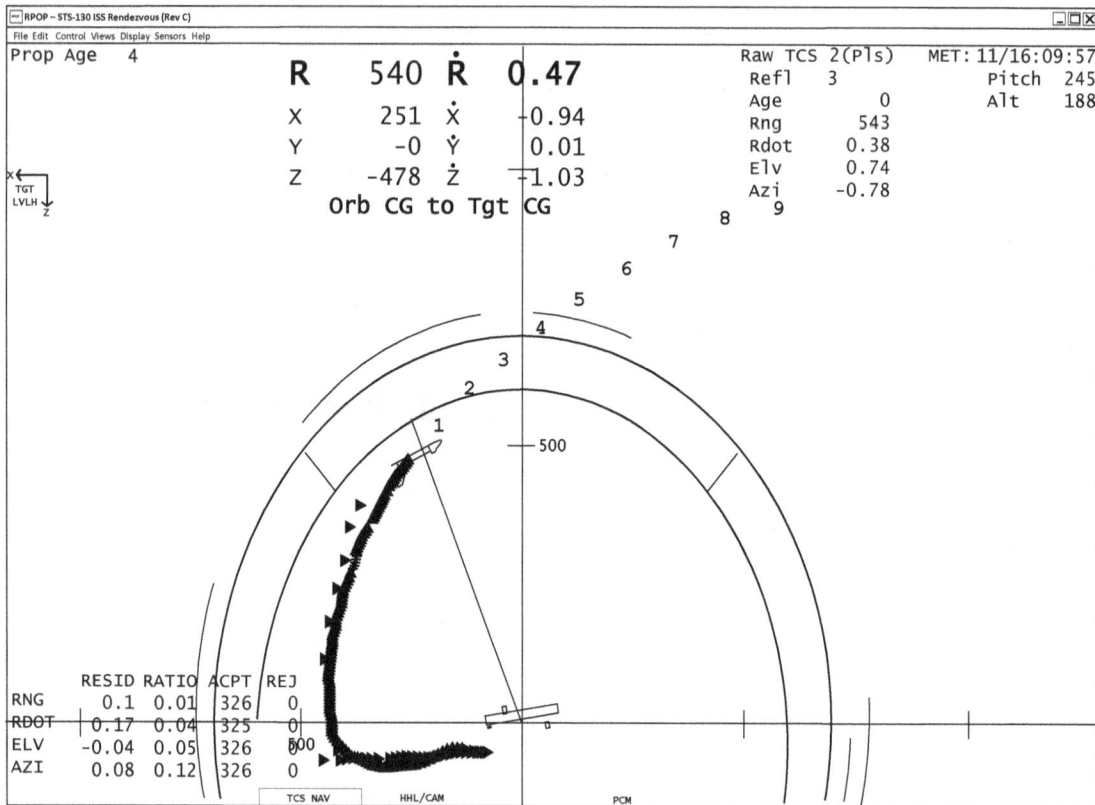

Figure 25.32 Re-created Shuttle RPOP display showing the backout and part of the twice orbital rate fly-around.

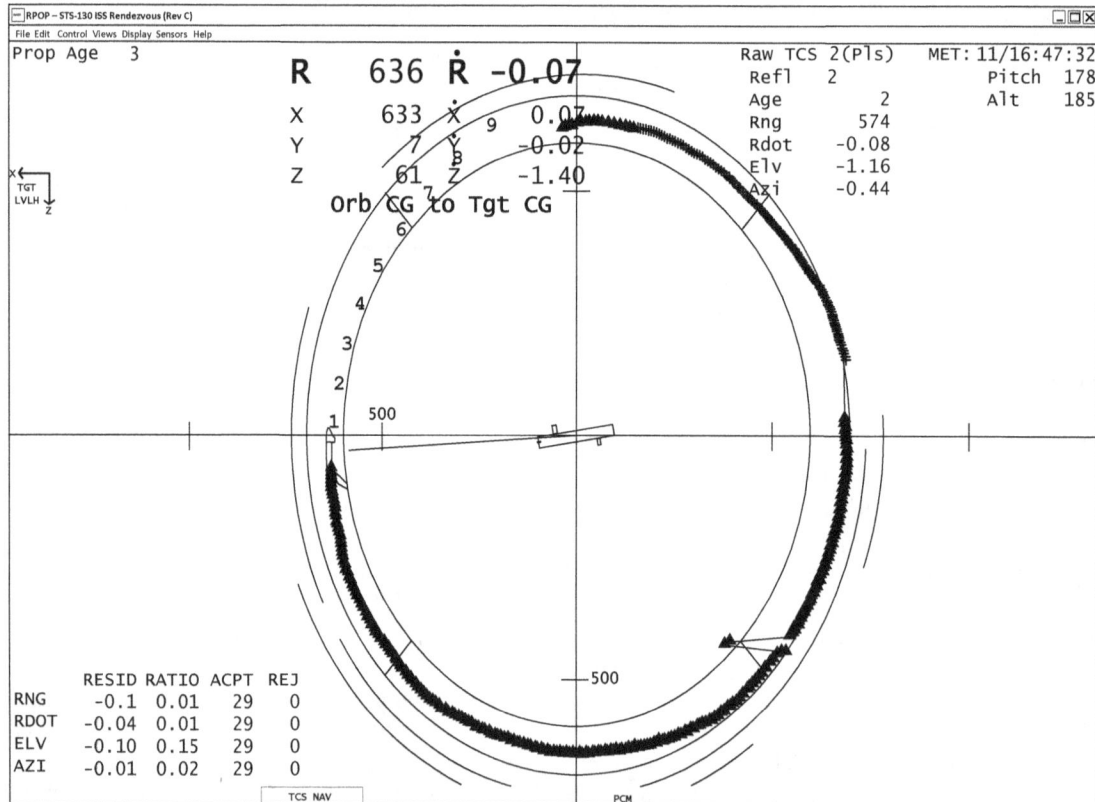

Figure 25.33 Re-created RPOP display illustrating the fly-around from the –R Bar (top) to the SEP-1 burn on the +V Bar (left). The crosses in the upper right hand quadrant indicate that TCS measurements were not available. This occurred due to structural blockage of the TCS reflectors by the ISS structure.

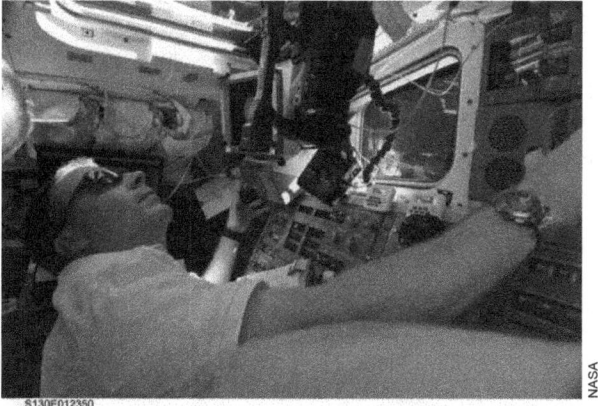

S130E012350 · NASA

Figure 25.34 Pilot Terry Virts flying the orbiter during the later part of the fly-around, before the SEP-1 burn. Virts manipulated the Translational Hand Controller with his left hand (as shown) to control relative motion during the fly-around. His right hand is guarding the Rotational Hand Controller (not normally used) from accidental deflection by other crew members taking photos through the other overhead window. Sunglasses were worn to improve visibility of the ISS in the presence of direct sunlight while the Sun skirted the edge of the overhead window field of view.

the approach and docking, the pilot flew the undocking, fly-around, and separation (Figure 25.34). This was standard practice on ISS flights. The crew obtained the

S130E013111 · NASA

Figure 25.35 Soyuz TMA-16 photographed during the fly-around, close to the –V Bar, to check for loose MLI.

requested photos of Soyuz TMA-16 (Figure 25.35). They did not observe any loose MLI.

Approximately 8 minutes before fly-around complete (+V Bar arrival) and the SEP-1 burn the crew began the rendezvous radar acquisition procedure. After radar converged the first position update was 1,176 feet. The crew executed the 1.5 foot/second radial up burn at MET 11/16:47:47. At MET 11/17:02 the orbiter was in the SEP-2 burn attitude. At MET 11/17:10 (range from the ISS approximately 5,000 feet) the on-board navigation console reported that radar tracking was good with some beam wandering about the ISS and large measurement ratios. However, the relative navigation filter had not

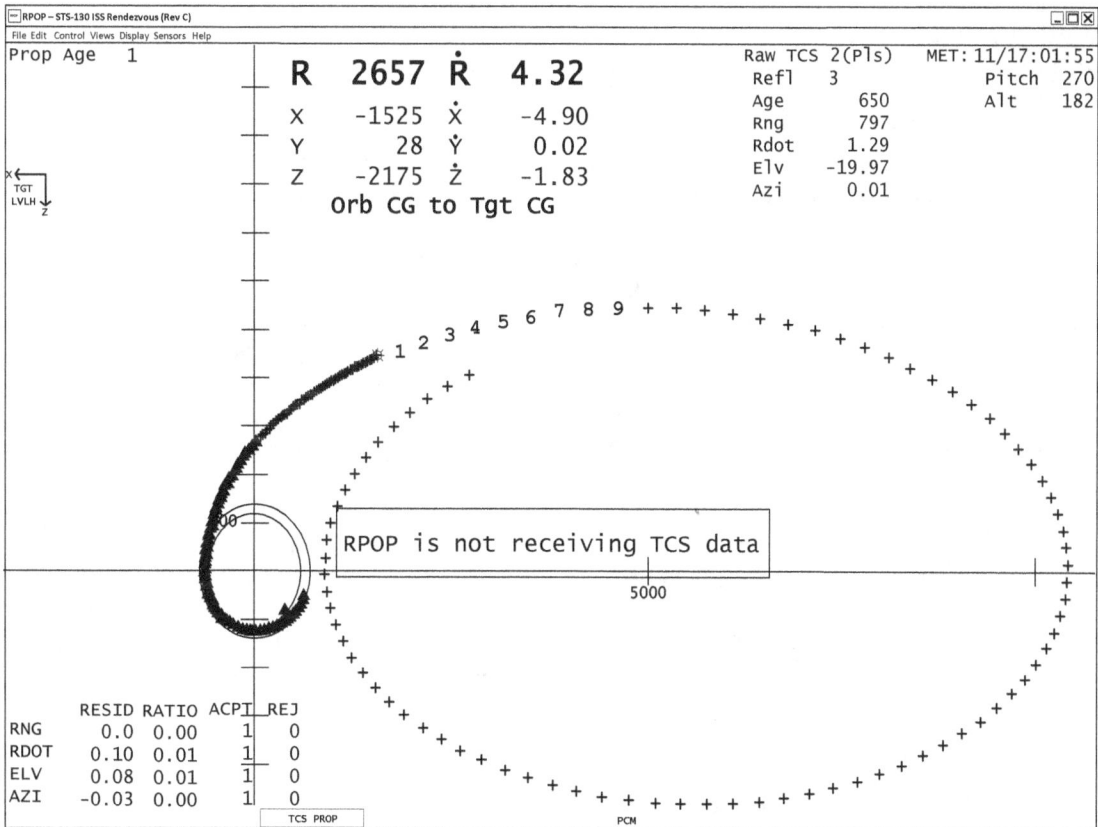

Figure 25.36 Re-created RPOP display depicting orbiter relative motion during the latter half of the fly-around and after the SEP-1 burn.

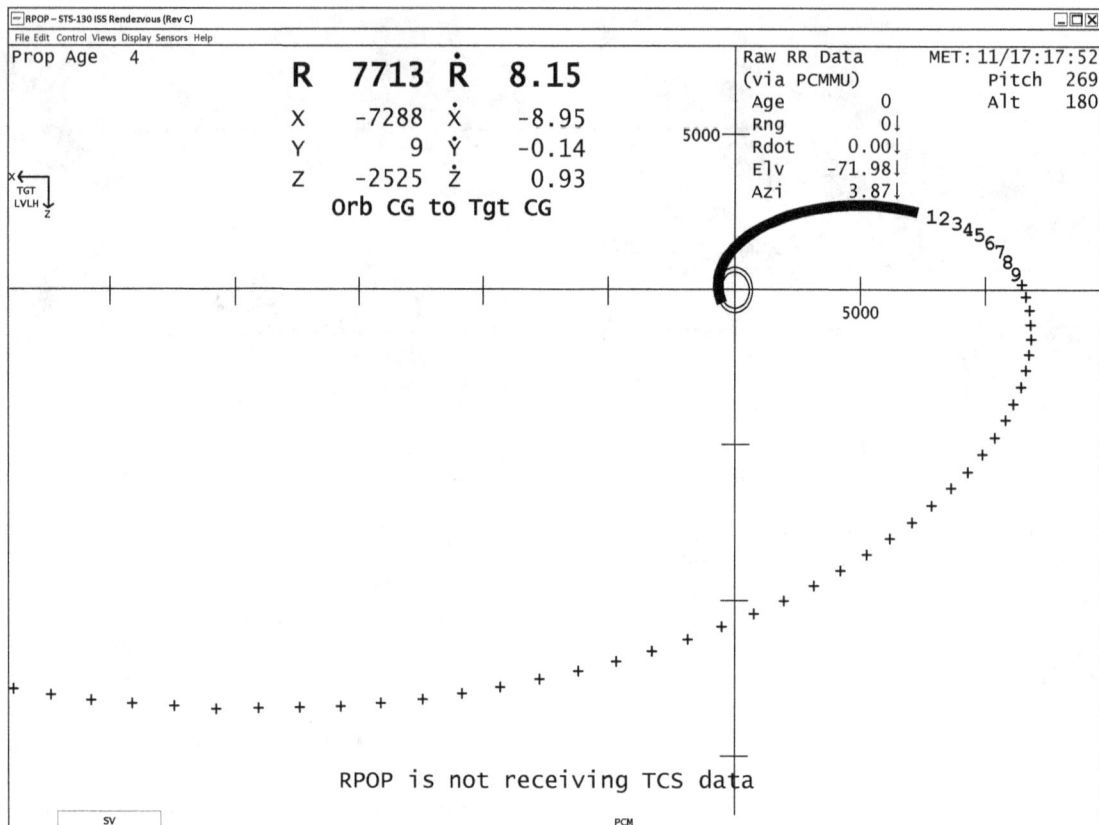

Figure 25.37   Re-created RPOP display approximately one minute after the SEP-2 burn was performed. Relative state data is derived from rendezvous radar measurements.

edited any radar measurements. At MET 11/17:14 (range approximately 6,500 feet) the crew deactivated the TCS. Figure 25.36 illustrates relative motion during the latter part of the fly-around and the initial part of the separation after the SEP-1 burn. Note that the radial SEP-1 burn results in the orbiter eventually returning to the vicinity of the ISS if the SEP-2 burn were not performed.

Just before the SEP-2 burn was performed the relative navigation filter edited a radar measurement. This was expected and the crew inhibited radar data. The SEP-2 burn began at MET 11/17:16:50 and a range from the ISS of approximately 7,500 feet. The Flight Dynamics Officer reported that a retrograde burn of 3.3 feet/second was executed by the crew. The crew transitioned the Ku antenna to communications mode and by MET 11/17:20 (8:36 pm CST) the crew had completed all separation procedures. Predicted relative motion after the SEP-2 burn is depicted in Figure 25.37.

The next day, Saturday, February 20, the ISS executed the re-boost burn at 3:15 pm CST. The burn was successful and a delta-velocity of 12.3 feet/second was achieved.

Flight Day Fifteen

Endeavour landed on the first KSC opportunity on runway 15 at 10:20 pm EST on Sunday, February 21, 2010.

Summary

STS-130 was the 32nd Space Shuttle mission to the International Space Station. Most rendezvous and proximity operations activities on ISS missions were standard and exhibited little flight-to-flight variation, particularly after TORVA and +V Bar approaches were introduced on STS-102 (March 2001).

After the loss of *Columbia* much of Flight Day 2 was devoted to thermal protection system inspection. The rendezvous tools checkout, docking ring extension, and centerline camera installation continued to be conducted on Flight Day 2. The R Bar Pitch Maneuver was added to proximity operations to permit the ISS crew to photograph the underside of the orbiter to help check thermal protection system integrity.

What made STS-130 unusual was the partial failure of the TCS lidar during proximity operations. This had not occurred on a shuttle mission since the first operational flight of TCS on STS-71 (June-July 1995). Extensive crew training with backup procedures using the HHL, RPOP, and the centerline camera was the key to a smooth approach, docking, undocking, and fly-around.

CHAPTER 26 - WHY WAS SHUTTLE RENDEZVOUS AND DOCKING NOT FULLY AUTOMATED?

Frequent questions concerning NASA human flight vehicles are, "Why do the astronauts always manually pilot rendezvous and docking?" or "Why hasn't rendezvous been automated?" Automated rendezvous and docking is not new. The Soviets accomplished the first automated rendezvous and docking with the Kosmos 186 and 188 spacecraft on October 1967.[1,2] Since that time there have been numerous successful automated rendezvous and dockings flown by Soviet/Russian Soyuz and Progress vehicles. Automated rendezvous and docking has been flown to add modules to Salyut-7 (three TKS vehicles), Mir (two Kvant, Kristall, Spektr, Priroda), and the ISS (Zvezda, Pirs, Poisk).[3-6] At the time of this writing the European Space Agency has successfully flown two automated missions to the ISS (Automated Transfer Vehicle, or ATV) and the Japanese have flown two automated missions to the ISS as well (H-II Transfer Vehicle, or HTV).[7-9] More ATV and HTV missions are planned. Four American demonstration missions have been flown to demonstrate various activities of automated rendezvous, proximity operations, and docking. These are DART, XSS-10, XSS-11, and Orbital Express.[10] The Orion and Altair vehicles in the Constellation Program had requirements for automated rendezvous and docking.[11]

This chapter first appeared as an appendix in a 2007 AIAA conference paper titled "Challenges of Orion Rendezvous Development." It was written to provide insight into what has, and what has not, been automated on the Mercury, Gemini, Apollo, and Space Shuttle vehicles. The text has been updated since it was first published in 2007.[11]

Introduction

An understanding of the rationale behind decisions on automation in previous NASA human flight programs is necessary to understand the technical, programmatic, and cultural challenges faced by rendezvous and proximity operations personnel working to meet future spacecraft automation requirements.[1,2]

The design philosophy behind the Gemini and Apollo spacecraft was to keep systems and system interfaces simple by using manual sequencing and manual systems management whenever practical.[13-15] This took advantage of the human ability to recognize, analyze, and diagnose performance anomalies and take corrective action. In addition, the safety critical nature of Apollo rendezvous in lunar orbit, coupled with 1950s and 1960s experiences with autopilots in high performance aircraft, led to a requirement to accomplish rendezvous in the event of a computer failure.

Balancing system complexity with technical, cost, and schedule risk was an important consideration associated with meeting President Kennedy's goal of reaching the Moon and returning to Earth by 1970. The Soviets took a different approach during development of the Soyuz spacecraft in the 1960s. Soyuz systems management and rendezvous were highly automated. The possibility that the high level of automation designed into Soyuz had a negative impact on the progress of Soyuz development during the 1960s has been a topic of discussion since then.[1, 2, 16]

Attitude control of all U.S. human flight vehicles while on-orbit has been performed using automated, semiautomatic, and manual modes. Each new spacecraft has possessed increasingly sophisticated on-orbit automated attitude control, resulting in lower crew workload and increased attitude and pointing flexibility to meet mission requirements. This chapter concerns automated versus manual control of translational dynamics.

Mercury

The three primary objectives of the Mercury Project were: 1) Place a manned spacecraft in orbital flight around the earth, 2) Investigate man's performance capabilities and his ability to function in the environment of space, and 3) Recover the man and the spacecraft safely.[17] The Project Mercury Summary lists the philosophy behind manual and automatic control for the Mercury Spacecraft:

"Redundancy probably increased the complexity of the systems more than any other requirement. Because the spacecraft had to be qualified by space flight first without a man onboard and then because the reactions of man and his capabilities in the space environment were unknown, provisions for a completely automatic operation of the critical spacecraft functions were provided. To insure reliable operation, these automatic systems were backed up by redundant automatic systems." [17]

"The pilot must be given the capability of manually controlling spacecraft attitude." [17]

Mercury ascent was automated. The on-orbit and re-entry phases of flight were used to test manual, automated, and semi-automated means of attitude control. Mercury translational control on-orbit was limited to booster separation and deorbit maneuvers.

The planned primary mode of attitude control for the deorbit burn and re-entry was automatic. However, due to systems problems, only one Mercury orbital flight (MA-8, October 1962) used automated attitude control exclusively for the deorbit burn and re-entry. The other three orbital missions used a combination of automatic, semi-automatic, and manual control.[17]

Gemini

The short development schedule of Gemini (3.3 years from contract award to first human flight) necessitated careful decisions concerning what tasks to automate and what tasks to perform manually. Project Mercury flight experience proved that a human was capable of efficient operation of a spacecraft in weightlessness. The Gemini design philosophy was to increase the level of human participation in piloting and spacecraft operation.[18]

"In the Mercury Program, automatic systems were used almost exclusively in the Guidance and Control System with manual operations used solely as optional back-up modes. The experience gained proved that man is fully capable of making decisions and executing control during most phases of a space mission. There are obvious advantages to exploiting this proven capability, such as savings in equipment and providing mission flexibility, therefore, the Gemini Project had attempted to fully integrate into the system such manual abilities as decision making, navigation, mode selection, and flight control." [19]

"Although manual systems are emphasized, automatic control modes still afford many of the same advantages to a space vehicle that they do to a high performance aircraft. For instance, solutions to problems of launch guidance, re-entry, and long term attitude control are better suited to automatic systems." [19]

Gemini Ascent

Ascent during Gemini was flown automated. Since Gemini had an on-board computer and IMU, both were used to provide a back-up automated ascent guidance capability in the event of a Titan II radio guidance failure. The switch from the primary Titan II guidance to the automated backup was performed by the crew based on predefined criteria.

Gemini Orbit and Rendezvous

One goal of the Gemini Program was to determine what activities could be accomplished by a crew in zero gravity. Gemini rendezvous and docking was performed manually by the crew using data from a computer, a rendezvous radar, an IMU, visual observations, charts and graphs, and hand calculations.

On-orbit pitch and roll attitude could be controlled automatically while yaw control was manual.[18]

Gemini Re-Entry

Another programmatic objective of the Gemini Program was to successfully demonstrate both automated and manual re-entry of a lifting capsule to a target splashdown point in preparation for Apollo. The on-board computer could perform automated entry as well as provide cues for manual piloting. Back-up manual procedures were also available in the event of a computer failure. The Gemini III crew initially flew the manual back-up procedure, then used computer generated piloting cues. A computer failure on Gemini IV forced the crew to fly the manual backup procedure. Entries for Gemini missions V through X were flown manually using computer cues. Automated entry was successfully flown on Gemini missions XI and XII.[20]

Apollo, Skylab, and Apollo-Soyuz

Apollo spacecraft development was subjected to a short development schedule like Gemini. Minimizing risk to cost, schedule, development, and certification required careful choices concerning automation versus manual control, and on-board versus Mission Control roles and responsibilities.

Apollo Ascent Into Low Earth Orbit

Ascent into Earth orbit on the Saturn IB and Saturn V was flown in an automated mode. Starting with Apollo 10, backup guidance was available in the event of a Saturn Instrumentation Unit (IU) Inertial Measurement Unit (IMU) failure. During the first stage, the crew could switch to automatic backup guidance provided by the Command Module Primary Guidance, Navigation, and Control System (PGNCS). For backup guidance during the second and third stages the crew provided hand controlled steering inputs via the PGNCS to the Saturn IU.

Apollo Rendezvous, Docking, and In-Space Operations

On many missions the separation and transposition maneuver, carried out before the Apollo CSM docked with the LM, was executed in an automated mode using the digital autopilot. An upgrade to the digital autopilot on Apollo 10 eliminated the need for a crew member to monitor and periodically adjust the thermal control roll mode performed to and from the Moon. This automation enabled all three crew members to sleep at the same time. The Gemini missions proved that rendezvous and docking could be performed using manual piloting procedures. An excerpt from a 1966 NASA memo by Bill Tindall, the Chief of Apollo Data Priority Coordination, concerning

Approved for public release via STI DAA 24483. See statement on title page.

254

Apollo terminal phase guidance requirements, best illustrates the rationale behind automatic versus manual approach and docking decisions during the Apollo and Shuttle Programs.

> "Based on Gemini experience, the crew has emphasized that there is no requirement for automatic execution of the braking maneuvers by the G&N system. As previously reported, it is felt that this task can be carried out just as well, if not better, by the crew if they are provided the proper information; namely, the range and range rate data.......................Recognizing that procedures are available for utilizing the remaining computer processors to carry out the G&N controlled braking maneuvers by proper pilot manipulation of the computer, we deleted the requirement for automatic computer logic for this task. The point is, we felt that there was insufficient justification to carry out the extra programming, debugging, verification, and documentation, as well as using some 50 to 100 words of precious computer storage, for a program which was not needed, except in rather remote contingency situations, as long as procedures were available to handle all situations. And, they are." [21]

Although the Apollo CSM was nominally the passive vehicle during rendezvous, the CSM pilot performed relative navigation and targeting tasks using the PGNCS lest a LM systems problem kept the LM from completing the rendezvous. CSM procedures for both the passive and active vehicle roles were complex and labor intensive. The first test of Apollo CSM single piloting for rendezvous occurred during the Apollo 7 rendezvous with it's own S-IVB stage (October 1968).

After the return of Apollo 11, comments made by CSM pilot Michael Collins regarding the heavy single piloting workload (approximately 850 keystrokes) prompted the Apollo Program to automate some CSM rendezvous procedures in the fall of 1969. This resulted in the MINKEY program that was flown on the last three lunar missions, as well as the Skylab and Apollo-Soyuz missions. The level of automation in the CSM passive and active rendezvous procedures was limited by on-board computer capacity and the need for the CSM pilot to manually take sextant marks. When available, ground monitoring of CSM systems reduced the workload of the pilot so he could concentrate on guidance, navigation, and control procedure execution and performance monitoring. The automation of some crew tasks did not reduce the flexibility already inherent in the CSM rendezvous procedures.

Apollo Lunar Landing and Ascent Into Lunar Orbit

The Apollo lunar landing profile was flown automatically from 50,000 feet to an altitude under 500 feet above the lunar surface. During this phase the crew could re-designate the landing point based on visual observation of boulders, craters, or sloping terrain. This feature was used on missions 12, 14, 15, 16, and 17 as surface conditions differed from what was expected, based on photographs taken from low lunar orbit.[22, 23] At an altitude under 500 feet the crew transitioned to a semi-automatic control mode (manual attitude control with automated descent engine throttle control) for the rest of the descent. Completely automatic or completely manual modes (manual attitude and throttle control) were also available, but the semi-automatic mode reduced crew workload through automatic throttling while manual attitude control provided precise control over the landing location.[24-27] LM ascent was flown automatically, although a manual procedure was available in the event of certain hardware failures.

Apollo Re-entry

All Apollo atmospheric entries were flown automatically by the PGNCS, though the initiation altitude of automated control varied on the early missions. Backup manual piloting options were available to support a PGNCS failure or failures of other sensors. In the event of a digital computer or IMU failure (PGNCS), the Entry Monitoring System (EMS) provided piloting cues for manual flight of the entry profile.[27-28] Entry was flown automatically once cross checks between the PGNCS and the EMS verified PGNCS performance as acceptable. Much of the entry profile was flown automatically on each Apollo mission, although Apollo 7 did not begin automatic control until an altitude of ~202,000 feet.

Space Shuttle

The Space Shuttle was more automated than the Mercury, Gemini, and Apollo vehicles. However, there was a requirement for manual backup for all automated flight modes.[15] Additional automation was introduced over the life of the program. Shuttle Program emphasis has been placed on applying automation to flight phases where manual or semiautomatic control is difficult, thereby increasing safety. Over the life of the Shuttle Program priority has been given to upgrades that provide return in terms of improved safety, correction of software anomalies, reduction of life cycle costs, or the ability to support a new mission requirement. In the context of an operational system, automation of a task for which a proven, procedural work-around exists is considered to be a "nice to have." The benefit such automation would provide may not justify the cost and the risk associated with making changes to safety-critical software. The practice of carefully considering where to apply program resources is illustrated by an automation decision made early in the Shuttle Program. In 1974 a proposal was drafted to develop an automated flight capability that would allow the shuttle to be test flown without a crew, in the same manner that the Soviet Buran shuttle was later tested in 1988. After careful consideration, Shuttle Program management decided that the budget and resources needed to develop such an automated capability would be better spent on enhancing safety of flight for piloted shuttle missions.[29]

The Space Shuttle was the first human space flight vehicle that flew with a crew on the first mission. All previous vehicles were first tested without a crew. One consideration that led to the decision to fly the first Space Shuttle flight with a crew was uncertainty about hypersonic flight control system performance during the 13 minute communications blackout during re-entry. Flight control system stability derivatives had been determined through wind tunnel testing but there was some uncertainty about the accuracy of the derivatives. If a flight control system problem occurred Mission Control had no insight into vehicle performance nor could it take action to resolve the issue and save the shuttle orbiter during the communications blackout. If a crew flew the first mission they could recognize a stability problem, change flight control system gains, and handle dynamic instabilities to ensure that the orbiter successfully returned to a runway landing.[30]

Shuttle Ascent and Aborts

To date all Space Shuttle ascents have been flown automatically, although a manual backup option, called Control Stick Steering (CSS), is available in the event of a contingency. However, procedures do not permit the crew to engage CSS in the event of a guidance problem until 1 minute 30 seconds after lift-off. This restriction is in place since the dynamics of aerodynamic load relief during the high dynamic pressure (q-bar) region are too complex for the crew to fly manually. In the event of a guidance problem between liftoff and T+1:30 procedures call for the crew to engage the Back-up Flight System (BFS).

The Space Shuttle mission profile includes intact abort modes for loss of a single main engine during ascent that would prevent the vehicle from reaching the desired orbit. These aborts, which permit safe return to a runway landing, originally included Return to Launch Site, Abort Once Around, and Abort to Orbit.[31] However, a runway landing was not possible for some failure scenarios. Training simulations led the STS-2 crew (Engle and Truly) to recommend the development of manual procedures to enable a runway landing in Spain as an alternative to a crew ejection procedure. Cue cards with the manual Transoceanic Abort Landing (TAL) procedures were flown on STS-2 (November 1981). TAL

procedures were automated for STS-3 (March 1982).

Contingency aborts are performed in response to failures other than the loss of one main engine. These failures included loss of thrust from multiple engines or multiple failures in other orbiter systems. In the twenty years since the loss of *Challenger*, contingency procedures during ascent and entry have been continuously developed and refined. Time critical ascent and entry contingency procedures that are difficult for the crew to execute have been automated to improve safety. These include flight computer software changes made to automate challenging procedures for emergency landings due to multiple ascent engine failures. One example of automation of manual procedures is the East Coast Abort Landing (ECAL). Automated ECAL was first available on STS-102 (March 2001) and permits landings at sites on the eastern coast of the United States and Canada for high inclination missions.

In addition to TAL (1982) and ECAL (2001), other conversions of manual piloting procedures to automation were the Main Propulsion System (MPS) Dump Capability (1983), Automation of Normal Acceleration (Nz) Hold Maneuver (1989), Single Engine (2 failed engines) Contingency Abort Procedure Automation (1992), Low Energy TAL Automation (1993), and Three Engine Out Automation (1997). Software improvements to existing ascent abort automation included the Abort Sequencing Redesign (1992), Trans-Atlantic Abort Landing (TAL) Droop Capability (1993), and Return to Launch Site (RTLS) External Tank (ET) Separation Improvements (2007). Crew monitoring of automated flight performance during powered ascent and hypersonic re-entry was improved with new Ascent/Entry Bearing Displays (2007). [50]

Shuttle Rendezvous, Docking, and On-Orbit

The Remote Manipulator System (RMS, the robotic arm) has been used by the shuttle for grappling and deploying payloads during proximity operations. It has both manual and supervised automation modes of operation. Rendezvous was defined as a shuttle capability at the beginning of the Shuttle Program and early in the Program some studies of automated rendezvous and docking were performed. However, the success of rendezvous and docking during Gemini and Apollo led to rendezvous being of lower priority (an "optional service"), than ascent and entry, during the development and flight phases of the Shuttle Program. The shuttle had to overcome other significant technical challenges in rendezvous and proximity operations in areas such as plume impingement and variation in target spacecraft characteristics. However, these challenges were not as safety critical as those in the ascent and entry phases of flight.

A proximity operations autopilot was tested on two missions in 1985 (STS-51G and STS-61B), but was not certified for regular use during missions.[31, 32] At that time the shuttle did not have proximity operations sensors to enable effective automation of guidance, navigation, and control during proximity operations. A programmatic requirement driving certification of a proximity operations autopilot for routine use during missions did not exist. Furthermore, the proximity operations autopilot was added to a special version of flight software that was only to be used for the test. It was not added to and certified for use in the on-orbit flight software normally used during the on-orbit phase of missions.

The Universal Pointing function provided a higher level of on-orbit attitude maintenance automation than was available on previous vehicles. Attitude control during rendezvous, proximity operations, and docking is performed automatically with the crew performing systems management and oversight. Therefore shuttle rendezvous, proximity operations, and docking is more accurately described as semi-automatic rather than manual.

One example of automated attitude control is the final alignment of the orbiter with the ISS for docking. At 30 feet, the crew checks the alignment of the orbiter with the docking target on the ISS. Attitude errors are determined visually by the crew and input into the Universal Pointing display for automatic attitude adjustment. This method is preferred over the Rotational Hand Controller (RHC) as the Digital Auto Pilot can more accurately remove small attitude errors while minimizing ISS plume impingement. This procedure was also performed on the missions to Mir.

Normally the RHC was not used during the on-orbit phase of a shuttle mission due to automated attitude control. However, there were several procedures that required manual attitude control:

• Any attitude separation from another spacecraft.

• Calibration and alignment of the Crew Optical Alignment Sight (COAS) or Heads Up Display.

• The manual attitude alignment contingency procedure during before grapple of the HST.

• Small payload fly-arounds during proximity operations.

• During the Shuttle-Mir missions there was a contingency separation procedure for the no forward RCS jet case that started with a RHC deflection in pitch.

At times during rendezvous and proximity operations automated attitude control is performed to track the target spacecraft to facilitate relative sensor measurements. This in turn requires that an on-board relative navigation solution be maintained. The advantage of a manual attitude control capability is that it permits the crew to keep another spacecraft under visual observation even if relative sensor measurements and a relative navigation solution are not available to support automated attitude control. Continuous out-the-window observation of another spacecraft is required to detect off-nominal relative motion and reduce the risk of collision.

In 2004, Shuttle Program personnel examined the integration of TCS and RPOP into the flight-critical avionics system. In addition, an automated proximity operations capability using RPOP guidance for translational control was successfully demonstrated in a high fidelity simulator for ISS approach and docking. However, the crew still would have been required to perform a visual check of alignment with the ISS docking hardware, and command the automatic attitude adjustment. These shuttle modification proposals were not adopted due to the decision to retire the shuttle fleet in 2010.

Due to the wide variety of target vehicles and associated constraints placed on shuttle proximity operations, proximity operations procedures varied significantly from flight to flight.[32] Flight to flight variation of procedures would have made automation of shuttle rendezvous and proximity operations for all target vehicles expensive in terms of safety critical software life cycle costs. While automated translational control was never implemented for rendezvous, attitude control during all phases of shuttle rendezvous, proximity operations, target capture, target berthing, and docking is performed in an automated manner with the crew performing a systems management and oversight function.

There are two examples of automation proposals that were not approved due to higher priorities for software development resources and budget. The first involved automating the set-up of attitudes to be flown.

The Universal Pointing function in the on-orbit guidance, navigation, and control (GNC) flight software provided desired attitude and rate commands to the Digital Auto Pilot (DAP). The DAP in turn automatically maneuvered to and maintained the commanded attitude. The crew performed keyboard inputs to the Universal Pointing display to set up the attitude maneuvers. The ASTRO-1 astronomy payload flown on STS-35 (December 1990) required over 200 different attitudes to facilitate astronomical observations. This amounted to about 8,930 crew keystrokes.

After the mission the heavy crew workload and potential for input errors led Mission Control personnel to devise an automated attitude table concept that eventually became a GNC flight software requirements change request (CR 90702, OI-24 candidate, 1992). The table would have contained a sequence of attitude commands as a function of time that would have been uplinked to the orbiter GNC software by Mission Control. The capability would have eliminated the need for crew item entries to define and command new attitudes to be maintained automatically by the DAP. Addition of the automated

Approved for public release via STI DAA
24483. See statement on title page.

257

attitude maneuver set-up capability to the GNC software was not approved due to other software changes that had a higher priority for budget and software development resources. Only one shuttle mission, STS-35, has required such a large number of attitude maneuvers.

The second example concerns automation of translational burns performed with the 870 pound thrust primary Reaction Control System (RCS) jets. Burns executed with the Orbital Maneuvering System (OMS) engines were performed automatically by the GNC software after the crew gave an authority to proceed. Smaller burns executed with the primary RCS jets were performed manually by the crew using a Translational Hand Controller (THC). The STS-51A mission (November 1984) performed rendezvous with two different communications satellites, PALAPA-B2 and WESTAR-VI. Both had been deployed on STS-41B (February 1984) but did not reach their assigned orbits due to malfunctions of the Payload Assisted Modules on each satellite. In addition, two satellites were deployed, TELESAT-H and SYNCOM IV-I. The satellite deployments and both rendezvous profiles required a total of 44 OMS and RCS burns.

After the mission the crew recommended that the GNC software be modified so that the crew could chose to automatically execute a translational RCS burns in the same manner that OMS burns were automated. Automating RCS burns could reduce crew work-load, eliminate the need to maneuver to a burn attitude for some RCS burns, and lower the risk of errors during burn execution that could result in trajectory dispersions and higher propellant consumption. However, this would not have automated execution of burns during proximity operations and docking.

This proposal was seriously considered several times over the life of the Shuttle Program, the last time during the Cockpit Avionics Upgrade (CAU, canceled in December of 2004). However, it was never approved for implementation in the orbiter GNC software since propellant savings could not be quantified and there were other software upgrades that provided more payback (risk reduction, necessary to meet new mission requirements) and therefore had a higher priority for the limited software development budget and resources. Most shuttle missions did not have as many OMS and RCS burns as STS-51A (44 total).

Shuttle Re-Entry and Landing

Shuttle entries are typically flown in an automated mode until just below Mach 1. Mach 1 typically occurs between 47,000 to 55,000 feet Above Ground Level (AGL). At this point the crew transitions to Control Stick Steering (manual piloting via the shuttle computers). All Space Shuttle landings, including the five Approach and Landing Tests (ALT) conducted at Edwards Air Force Base in 1977, have been flown manually.

The shuttle was designed and built with an automatic landing (autoland) capability for flight between approximately 10,000 feet and wheels stop on the runway. The autoland requirement developed during shuttle design in the 1970s was driven by Department of Defense shuttle reference Missions 3A and 3B (never flown) that could encounter fog while landing at Vandenberg Air Force Base.[51, 52] Use of autoland required the presence of a Microwave Landing System (MLS) at the runway. One possible advantage of autoland was that it could permit a landing at a runway with a lower cloud ceiling than would be permissible for a manually piloted landing. Manual piloting capability was provided as a backup in the event of performance issues and systems failures.[33] Not all shuttle contingency runways around the world were equipped with MLS, therefore autoland was not an option for landings at these runways. Autoland guidance and use of MLS data for navigation was only in the primary computer software. The backup computer software did not support autoland or MLS navigation. Several tests of autoland during parts of the landing phase were conducted early in the Shuttle Program. During the third ALT free flight (September 1977) autoland was flown down to an altitude of 900 feet AGL.[34, 35] More ALT autoland test flights had been proposed but the total number of free flights was reduced to five due to budget issues. This reduced the amount of testing performed.

On STS-2 (November 1981) autoland was flown from 5,000 feet to 1,300 feet AGL in roll and yaw and to 300 feet AGL in pitch. STS-3 (March 1982) flew autoland from 10,000 feet to 120 feet AGL, and STS-4 (June/July 1982) flew autoland from 10,000 feet to 2,500 feet AGL.[33] However, manual takeover and crew monitoring issues, coupled with successful manual landings, led the Shuttle Program to designate manual landing as the primary method. The orbiter autoland capability was never certified and only to be used for contingencies.[36]

MLS accuracy was also an issue for autoland. There were two configurations of the runway based MLS. In the senior configuration the elevation transmission antennas were 3,350 feet from the runway threshold on the approach end of the runway. Senior azimuth and range transmissions were from a transmitter at a different station at the far end of the runway. The lower elevation angle of MLS Senior resulted in higher multi-path errors close to touchdown. A low elevation angle cut-out to mitigate the worst multipath resulted in no MLS data for approximately the last 100 feet of the approach. This in turn could result in unacceptable errors at touchdown. The MLS junior configuration, the configuration used during much of the Shuttle Program, had the elevation, azimuth, and range transmitters at a point 1,500 from the runway threshold at the approach end of the runway. The MLS Junior configuration did not supply measurements during rollout while the MLS Senior configuration still supplied range and azimuth data during rollout.

Over the next 10 years a number of flight control

system changes were made to improve performance of both manual and automated flight below 10,000 feet. This included automation of the speed brake function during manual piloting.[31]

Another test of autoland, including an automated landing, was planned for STS-53 (December 1992) in support of the Long Duration Orbiter (LDO) project (missions of longer than 30 days).[37] The LDO concept would have also required automated deployment of the air data probes, landing gear, the drag chute, and automatic braking on the runway. The proposed STS-53 Detailed Test Objective (DTO) required the use of the MLS Junior configuration. In the Junior configuration all MLS signals were transmitted from the MLS hardware to the orbiter from a location near the touchdown point. MLS Junior would have avoided the multipath and associated navigation errors of the MLS Senior configuration (azimuth antenna at the far end of the runway, therefore low elevation angles and greater multipath). The Junior configuration permitted MLS data use all the way to touchdown. There was also a special ground monitoring alarm limit requirement for MLS. However, MLS Junior azimuth antenna locations did not support automated steering during rollout and manual steering was required after nose gear touchdown. Limits (placards) were placed on orbiter weight and center-of-gravity location for automated landings. Autoland was certified for the STS-53 DTO and as a backup for LDO missions. Certification for the full range of orbiter center-of-gravity and weight values would have required more work on MLS antenna locations and tighter MLS monitoring alarm limits. In addition, incorporation of radar altimeter data may have been necessary.[38, 39]

The STS-53 commander and pilot underwent extensive autoland and late manual takeover training in the Vertical Motion Simulator at NASA/Ames in California. The crew trained for the DTO with the serious attitude of test piloting professionals. The autoland DTO had the support of Shuttle Program Manager and former astronaut Robert Crippen. However, a new Shuttle Program Manager canceled the STS-53 autoland DTO due to safety concerns based on his experience with legacy automated landings on aircraft carriers. Shuttle autoland personnel believed that these concerns were not relevant to shuttle autoland. The Space Shuttle never flew an LDO mission.

Astronauts were occasionally asked why automated landings are not flown as a nominal flight technique. A summary of their responses follows. [40-44] *

1. Since the shuttle is a glider, it does not have the ability, like a powered airplane, of calling off an approach and making another landing attempt in the event of a problem.

2. Autoland touchdown airspeed dispersions using the Shuttle Training Aircraft (the STA, a Gulfstream II modified to fly like the shuttle orbiter) have been very close to the airspeed limit of the landing gear. Manual piloting could control the touchdown airspeed slightly better than the autoland system. †

3. Unlike a powered aircraft that performs automatic landings, the shuttle orbiter did not have redundant airspeed measurements nor was there an independent means for the crew to cross check autoland performance during the approach.

4. In the event of poor autoland performance, a crew member who has been in a weightless environment for several days or weeks might have difficulty making the necessary corrections during the dynamic and time critical final approach phase. An earlier manual takeover, well before the final approach, provides the crew member with more time to become acclimated to stick feedback and vehicle response.

After the loss of *Columbia* (February 2003), a Remote Control Orbiter (RCO) capability was developed to return an uncrewed orbiter with thermal protection system damage to Earth for a runway landing. The crew of the damaged orbiter would use the ISS as a safe haven, and would be returned to Earth by a later shuttle mission. Return of an uncrewed orbiter in this manner requires the use of autoland.[46] STS-121 (July 2006) was the first mission with RCO capability.

On STS-126 (November 30, 2008 landing) the orbiter *Endeavour* was to perform a 340 degree left overhead turn on the Heading Alignment Cone (HAC) before landing on Edwards Air Force Base runway 04. HAC turns this large were rare for the shuttle. A large turn coupled with high winds required that the commander precisely follow the guidance commands on the Heads Up Display during Control Stick Steering (CSS). If the guidance commands were not precisely followed it could result in an undesirable low energy and low altitude

---

* Autoland was the subject of many studies and debates in 1981-1983 and during the planning for STS-53 (1991-1992). It should be noted that interviewees are relying on memories of meetings and events many years in the past. The STS-2 and STS-3 landings had a strong influence on people's opinions about autoland. For a technical discussion about the STS-3 landing from a control system perspective see pages 41 and 42 of reference 45.

† It is possible that the inconsistent autoland performance during STA landings was caused by the MLS Senior configuration. The STA eventually incorporated radar altimeter data to stabilize errors during autoland. Although the orbiters have radar altimeter data, the data was never incorporated into the GNC system. On several early shuttle missions the altimeter locked onto the orbiter nose gear.

scenario with headwinds. In addition, too many g's could be pulled as the commander attempted to catch-up with the guidance commands. Normally the commander engaged CSS before HAC intercept. However, after analysis by landing personnel and discussion with the crew, the commander chose to let autoland fly the orbiter onto the HAC. Autoland performed an automatic energy dump/pull-up maneuver to achieve a subsonic HAC intercept at Mach 0.95. The commander engaged CSS 22 seconds after HAC intercept at Mach 0.81.

In summary, autoland was a contingency capability only to be used if the crew were incapacitated or incapable of landing the orbiter using Control Stick Steering. Since autoland was only a contingency capability the Shuttle Program never identified a firm requirement to demonstrate it and accept the additional risk of a demonstration. Return to Launch Site (RTLS) abort and Trans-oceanic Abort Landing (TAL) were never demonstrated for the same reason. ‡

Although the Space Shuttle did not fly an automated landing, the shuttle autoland development and certification effort did contribute to successful automated landing of another space vehicle. While the X-37 project was run by NASA, before it was transferred to the Department of Defense, Boeing based the X-37 autoland capability on that of the shuttle. After seven months in orbit the X-37B performed a successful automated and autonomous re-entry and landing at Vandenberg Air Force Base on December 3, 2010. Although some within the Shuttle Program had reservations about the safety and performance aspects of shuttle autoland, the successful X-37B experience should be taken into account. Shuttle autoland specialists believed that shuttle autoland would have worked and the success of the shuttle based X-37B autoland capability was proof.

Historical Automation Summary

Choosing complete automation, semi-automatic control, or completely manual control is done on a case-by-case basis. Considerations include cost, technical risk, schedule risk, safety, and the ability of the crew to accomplish the task.

All powered flight ascents flown by the Mercury (Redstone, Atlas), Gemini (Titan II), Apollo (Saturn IB, Saturn V, Apollo Lunar Module ascent stage), Skylab (Saturn IB), Apollo-Soyuz (Saturn IB), and the Space Shuttle vehicles have been flown in an automated mode. The X-15 and the Scaled Composites SpaceShipOne are the only U.S. vehicles that reached space while being flown manually during powered flight.[47, 48, **] Most entries have been flown in an automated mode as well, although the crew manually flies most of the shuttle landing profile below Mach 1.

During flights of the Gemini Program (1965-1966) it was proven that manual control of close proximity spacecraft translation and docking using sensor measurements and out-the-window piloting cues was intuitive once skills were developed through training; much like flying an airplane, driving a car, or skiing. Manual piloting during powered flight (ascent) and hypersonic re-entry was not intuitive and entailed greater risk and fewer margins for error. Therefore automation development and application during the Gemini, Apollo, and Space Shuttle Programs focused on powered flight and hypersonic re-entry. Unlike powered ascent and hypersonic re-entry, the risk posed by manual piloting of rendezvous and docking/capture was not great enough to warrant development of fully automated rendezvous and docking. The Gemini, Apollo, and Space Shuttle vehicles were not equipped to perform fully automatic rendezvous and docking since programmatic requirements for such a capability were not needed to meet mission objectives with an adequate safety margin. Automation of attitude control during rendezvous and docking was performed to reduce the crew workload.

Successful Space Shuttle automated powered flight and hypersonic re-entry was demonstrated throughout the Shuttle Program starting with STS-1 (1981). Automated abort modes were also developed for the shuttle and were available (if needed) starting with STS-1. Over the life of the Shuttle Program new powered flight contingency abort flight techniques were developed to lower the risk of loss of vehicle and crew. These manual piloting procedures were eventually automated since they were difficult to manually fly and required very precise timing. Software improvements to existing ascent abort automation were also performed. [50]

When evaluating the use of manual piloting techniques versus automation for rendezvous and docking, it is important to note that between June 1965 and July 2011 a total of 101 U.S. human flight missions have included at least one objective related to rendezvous, docking, capture and berthing, or proximity operations. Only one of those missions (Gemini IV, June 1965) failed to achieve the relative motion required to accomplish mission objectives due to difficulty in controlling relative motion.

---

‡ See Appendix B - NASA Response to March 1993 Annual Report, Finding and Recommendation #9, on pages B-10 and B-11 of reference 39.

** The Scaled Composites SpaceShipTwo built for Virgin Galactic was also planned to be manually flown from release from the WhiteKnightTwo aircraft through landing. See reference 49.

References

1. Siddiqi, Asif A., *Challenge to Apollo – The Soviet Union and the Space Race*, 1945-1974, NASA SP-2000-4408, NASA, Washington, DC, 2000.

2. Chertok, Boris, *Rockets and People, Volume III, Hot Days of the Cold War*, NASA SP-2009-4110, NASA, Washington, DC, 2009.

3. Suslennikov, V., "Radio System for Automatic Rendezvous and Docking of Soyuz, Progress Spacecraft and Mir Space Station," *Proceedings of the Third European In-Orbit Operations Technology Symposium*, European Space Research and Technology Centre (Estec), Noordwijk, The Netherlands, June 22-24, 1993.

4. Hall, R. D., and D. J. Shayler, *Soyuz – A Universal Spacecraft*, Springer-Praxis, London, 2003.

5. Legostaev, Victor P., "Russian Space Programs: Achievement and Prospects of Automatic Control Applications," *16th IFAC Symposium on Automatic Control in Aerospace*, St. Petersburg, Russia, June 14-18, 2004.

6. Murrain, Dr. Rafail, and Sergey Budylov, "Short Rendezvous Missions for Advanced Russian Human Spacecraft," *60th International Astronautical Congress*, Daejeon, Republic of Korea, October 12-16, 2009.

7. Baize, Lionel, Martial Vanhove, Pascale Flagel, and Alberto Novelli, "The ATV "Jules Verne" Supplies the ISS," *SpaceOps 2008 Conference*, AIAA, Reston, VA, 2008.

8. Bruno Cavrois, Bruno, Les Mureaux, Stéphane Reynaud, Grégory Personne, Siegfrid Chavy, and Stein Strandmoe, "ATV GNC and Safety Functions Synthesis: Overall Design, Main Performances and Operations," *AIAA Guidance, Navigation and Control Conference and Exhibit*, AIAA, Reston, VA, 2008.

9. Blarre, L., J. Ouaknine, C. Moussu, K. Michel, B. Moebius, P. Da Cunha, and S. Strandmoe, "Description and In-Flight Performance of Rendezvous Sensor for the ATV," *32nd Annual AAS Guidance And Control Conference, Guidance and Control 2009*, Advances in the Astronautical Sciences, Univelt, San Diego, CA, 2009.

10. Kennedy, III, F. G., "Orbital Express: Accomplishments and Lessons Learned," *31st Annual AAS Guidance And Control Conference, Guidance and Control 2008*, Advances in the Astronautical Sciences, Volume 131, Univelt, San Diego, CA, 2008, pp. 575-586.

11. Goodman, J. L., J. P. Brazzel, and D. A. Chart, Ph.D., "Challenges of Orion Rendezvous Development," *AIAA Guidance, Navigation, and Control Conference*, AIAA, Reston, VA, 2007.

12. Goodman, J. L., *Knowledge Capture and Management for Space Flight Systems*, NASA Contractor Report NASA/CR-2005-213692, NASA Johnson Space Center, October 2005. See the NASA Technical Reports server at http://ntrs.nasa.gov/,or the Johnson Technical Reports server at http://ston.jsc.nasa.gov/collections/TRS/.

13. *Gemini Midprogram Conference*, NASA SP-121, NASA, Washington, D.C., 1966.

14. Low, G. M., "Apollo Spacecraft," *AIAA 6th Annual Meeting and Technical Display*, AIAA, Reston, VA, 1969.

15. North, W. J., "The Pilot's Role In Manned Space Flight," *Behavioral Objectives in Aviation Automated Systems Symposium; Proceedings of the Aerospace Congress and Exposition*, Society of Automotive Engineers, Inc., 1982, p. 29-38.

16. Gerovitch, S., "Human-Machine Issues in the Soviet Space Program," in *Critical Issues in the History of Spaceflight*, edited by S. J. Dick and R. D. Launius, NASA SP-2006-4702, NASA, Washington, DC, 2006, pp. 107-140.

17. *Mercury Project Summary, Including Results of the Fourth Manned Orbital Flight*, NASA SP-45, NASA, Washington, D.C., 1963.

18. Blatz, W. J., et al., "Gemini Design Features," *AIAA 3rd Manned Space Flight Meeting*, Houston, TX, November 4-6, 1964.

19. Clausen, John T., "The Role of Simulation in the Development of Gemini Guidance and Control," *AIAA 2nd Manned Space Flight Meeting*, Dallas, Texas, April 22-24, 1963.

20. *Gemini Summary Conference*, NASA SP-138, NASA, Washington, D.C., 1967.

21. Tindall, W. H., "Rendezvous Terminal Phase Guidance Program in the Apollo Spacecraft Computer," 66-FM1-94, Mission Planning and Analysis Division, Manned Spacecraft Center, July 18, 1966. Part of the unpublished Tindallgrams collection.

22. Engle, M., "Operational Considerations for Manned Lunar Landing Missions – Lessons Learned From Apollo," *AIAA Space 2004 Conference and Exhibit*, AIAA, Reston, VA, 2004.

23. Klumpp, A. R., "Apollo Lunar Descent Guidance," *Automatica*, Vol. 10, No. 2, March 1974, pp. 133-146.

24. Nevins, J. L., "Man-Machine Design For The Apollo Navigation, Guidance, and Control System -- Revisited," *Proceedings of the Third International Conference on Automatic Control in Space*, Instrument Society of America, Pittsburgh, PA, 1970.

25. Eyles, D., "Apollo LM Guidance Computer Software for the Final Lunar Descent," *Automatica*, Vol. 9, 1973, pp. 243-250.

Approved for public release via STI DAA 24483. See statement on title page.

261

26. Baird, A., "How to Land Next to a Surveyor: Bill Tindall and the Apollo Pinpoint Lunar Landing," *Quest: The History of Spaceflight Quarterly*, Vol. 14, No. 2, May 2007, pp. 19-27.

27. Mindell, David A., *Digital Apollo: Human and Machine in Spaceflight*, MIT Press, Cambridge, MA, 2008.

28. Frank, A. J., Johnson, B. C., and Knotts, E. F., "An Entry Monitor System For Maneuverable Vehicles," *AIAA Journal of Spacecraft and Rockets*, Vol. 3, No. 8, 1966, pp. 1229-1234.

29. Heppenheimer, T. A., *Development of the Space Shuttle 1972-1981*, Smithsonian Institution Press, Washington, DC, 2002, p. 328.

30. "Human Rating for Future Spaceflight," *Aerospace America*, Vol. 48, No. 7, AIAA, Reston, VA, July-August 2010, pages 26-41.

31. Schmidgall, R. A., "Space Shuttle Ascent Aborts," *Aerospace Technology Conference and Exposition*, Society of Automotive Engineers, Warrendale, PA, 1989.

32. Goodman, J. L., "History of Space Shuttle Rendezvous and Proximity Operations," *AIAA Journal of Spacecraft and Rockets*, Vol. 43, No. 5, Sept.-Oct. 2006, pp. 944-959.

33. Tsikalas, G. M., "Space Shuttle Autoland Design," *AIAA Guidance and Control Conference*, AIAA, Reston, VA, 1982.

34. Jenkins, D. R., *Space Shuttle – The History of the National Space Transportation System – The First 100 Missions*, Specialty Press Publishers, North Branch, MN, 2001.

35. Merlin, P. W., "Free Enterprise: Contributions of the Approach and Landing Test (ALT) Program to the Development of the Space Shuttle Orbiter," *AIAA Space 2006 Conference*, AIAA, Reston, VA, 2006.

36. McWhorter, L., and Law, H., "Orbiter Autoland," *AIAA Space Programs and Technologies Conference*, AIAA Reston, VA, 1992.

37. Glynn, P., and Barrera, T., "Space Shuttle Orbiter Extended On-Orbit Duration Implementation And Benefits," *AIAA Space Programs and Technologies Conference and Exhibit*, AIAA, Reston, VA, 1993.

38. Email communication to the author from Scott Murray, March 17, 2011.

39. Aerospace Safety Advisory Panel, Annual Report, NASA, Washington, DC, March 1994.

40. Williams, Donald E., Oral History Transcript, Interviewed By Rebecca Wright, NASA Johnson Space Center Oral History Project, Houston, TX, July 19, 2002.

41. Bolden, Charles F., Oral History Transcript, Interviewed By Sandra Johnson, NASA Johnson Space Center Oral History Project, Houston, TX, January 6, 2004.

42. Engle, Joe H., Oral History Transcript, Interviewed By Rebecca Wright, NASA Johnson Space Center Oral History Project, Houston, TX, May 27, 2004.

43. Covey, Richard O., Oral History Transcript, Interviewed By Jennifer Ross-Nazzal, NASA Johnson Space Center Oral History Project, Houston, TX, November 1, 2006.

44. Cockrell, Kenneth, "Ask the Crew: STS-98, Question #2," http://spaceflight.nasa.gov/feedback/ expert/answer/crew/sts-98/, accessed March 31, 2010.

45. McWhorter, Larry B., Milt Reed, et al., *Space Shuttle Entry Digital Autopilot*, NASA/SP–2010–3408, NASA Johnson Space Center, Houston, TX, February 2010.

46. Bigonesse, R., and Summa, W. R., "Unmanned Orbiter Undocking: Method for Disposal of a Damaged Space Shuttle Orbiter," *Spaceflight Mechanics 2006, Advances in the Astronautical Sciences*, Univelt, San Diego, CA, 2006.

47. Jenkins, D., *X-15: Extending the Frontiers of Flight*, NASA SP-2007-562, NASA Washington, DC, 2007.

48. Dornheim, M. A., "Flying in Space For Low Cost," *Aviation Week and Space Technology*, April 20, 2003.

49. Morring, Frank, "Help Wanted, As Its First Spaceflight approaches, Virgin Galactic is Recruiting Pilots," *Aviation Week and Space Technology*, Volume 173, No. 13, April 11, 2011, pages 34-35.

50. Hickey, Christopher J., James B. Loveall, James K. Orr, and A. L. Klausman, "The Legacy of Space Shuttle Flight Software," *AIAA SPACE 2011 Conference & Exposition*, AIAA, Long Beach, CA, September 27-29, 2011.

51. Henderson, E., and T. Nguyen, "Space Shuttle Abort Evolution," *AIAA SPACE 2011 Conference & Exposition*, AIAA, Long Beach, CA, September 27-29, 2011.

52. Email communication to the author from Edward M. Henderson, September 19, 2011.

PART IV – APPENDICIES

This page intentionally left blank.

A NOTE ON SOURCES

This history was originally written for an *AIAA Journal of Spacecraft and Rockets* article. Publication considerations prevented all of the material from being published in the journal. The first edition of this JSC document was published in October of 2006 to preserve all of the material.

Many of the references in this work are available in the open literature. These references will provide the reader with additional detail and insight.

Other sources were internal memos, presentations, reports, crew procedures, console procedures, and training documents. Some of these sources were found in the mission binders maintained by NASA/JSC Flight Dynamics Division Rendezvous Guidance and Procedures Office (Code DM34). Others were obtained from United Space Alliance Flight Design and Dynamics personnel that supported the NASA/JSC Flight Dynamics Division (Code DM). Other sources were found in the NASA/JSC Scientific and Technical Information Center (STIC) Library in Building 45. Still other sources were collected by the author over the years. At the end of the Shuttle Program many of those sources were preserved by the author in the following JSC documents as compilations. These are available from the STIC Library in Building 45.

• Goodman, John L. (editor), *Space Shuttle Rendezvous and Proximity Operations Overview and Experience Papers*, Volumes 1 (1970-1985) and 2 (1986-2009), JSC-35050, NASA Johnson Space Center, July 2011.

• Goodman, John L. (editor), *Space Shuttle Rendezvous Profile Papers*, Volumes 1 (1969-1983) and 2 (1984-2007), JSC-35051, NASA Johnson Space Center, July 2011.

• Goodman, John L. (editor), *Space Shuttle Proximity Operations Papers*, Volumes 1 (1970-1979) and 2 (1980-2009), JSC-35052, NASA Johnson Space Center, July 2011.

• Goodman, John L. (editor), *Space Shuttle Rendezvous Maneuver Targeting Papers*, JSC-35053, NASA Johnson Space Center, July 2011.

• Goodman, John L. (editor), *Space Shuttle Relative Navigation Papers*, Volumes 1 (1969-1979) and 2 (1980-2010), JSC-35054, NASA Johnson Space Center, July 2011.

• Goodman, John L. (editor), *Gemini Rendezvous Papers*, JSC-35055, NASA Johnson Space Center, July 2011.

• Goodman, John L. (editor), *Apollo Rendezvous Papers*, Volumes 1 (Lunar Missions) and 2 (Skylab and Apollo/Soyuz), JSC-35056, NASA Johnson Space Center, July 2011.

Two documents by James Oberg provide detail on rendezvous development and shuttle missions through 1986. *The History of Orbital Rendezvous* contains excerpts from some internal memos and reports.

• Oberg, James E., *Rendezvous and Proximity Operations Handbook, Appendix: STS Rendezvous And Proximity Operations Experiences 1981-1986 – Basic*, JSC- 10589, Flight Design and Dynamics Division, Mission Operations Directorate, NASA/JSC, May 16, 1988.

• Oberg, James E., *The History Of Orbital Rendezvous*, 660-FO-730-91-225, Flight Design and Dynamics Department, Rockwell Space Operations Company, Houston, TX, October 1, 1991. Accession number 92T-10780.

Additional detail on shuttle rendezvous may be found in the following reports.

• Goodman, John L., and Kelli S. Wiuff, *Space Shuttle Rendezvous and Proximity Operations Experience Report*, JSC-49626, Flight Design and Dynamics Division, Mission Operations Directorate, NASA Johnson Space Center, February 2003.

• Goodman, John L., *Lessons Learned From Seven Space Shuttle Missions*, NASA Contractor Report NASA/CR-2007-213697, NASA Johnson Space Center, January 2007. See the NASA Technical Reports server at http://ntrs.nasa.gov/.

Technical details on shuttle rendezvous and rendezvous concepts in general may be found in the following training books.

• Goodman, John L., *Introduction to Space Shuttle Rendezvous Guidance, Navigation, and Control, Fourth Edition,* JSC-49686, NASA JSC Flight Design and Dynamics Division, November 2009.

• Goodman, John L., *Introduction to Relative Navigation Concepts for Visiting Vehicles Officers, First Edition,* JSC-36583, Flight Dynamics Division, Mission Operations Directorate, NASA Johnson Space Center, September 2010.

• Goodman, John L., *Introduction to Rendezvous Burn Targeting and Guidance Concepts for Visiting Vehicles Officers, First Edition,* JSC-36584, Flight Dynamics Division, Mission Operations Directorate, NASA Johnson Space Center, September 2010.

RENDEZVOUS PERSONNEL

It is impossible to assemble an accurate list of all personnel that supported some aspect of rendezvous, proximity operations, docking, or berthing from the initial Phase A studies in 1969 through the end of the Shuttle Program over 40 years later. However, the author felt it would be prudent to acknowledge in some way the work performed by NASA and contractor personnel. Appendices A through F give the names of those who performed real-time support of rendezvous and proximity operations or trained the crews for this flight phase. The author did not attempt to expand the appendices beyond real-time support and crew training due to the complexity of the task and the limited time available.

Unfortunately assembling even these records was not easy. Finding the names of trajectory backroom personnel that supported rendezvous and proximity operations before the Space Transportation System Operations Contract (STSOC) was particularly challenging. Rendezvous and proximity operations missions flown during that period spanned STS-7 (June 1983) to STS-61B (November 1985). Errors and omissions likely exist in the appendices.

This page intentionally left blank.

APPENDIX A – RENDEZVOUS GUIDANCE AND PROCEDURES OFFICERS

This appendix lists the flight assignments for Rendezvous Guidance and Procedures Officers (RGPO) and their associated Trajectory Multi-Purpose Support Room (TRAJ MPSR, the backroom) support position, Rendezvous Procedures Support (RPS).

Starting with STS-7 the Rendezvous Procedures Group of the Operations Division (Mission Operations Directorate, or MOD) supported proximity operations and rendezvous missions with three Mission Control positions. The front room position was the Rendezvous Phase Specialist, call sign Rendezvous. There were two backroom support positions called Rendezvous Phase Support 1 and 2 (RPS-1 and RPS-2). In 1985 when part of the Mission Planning and Analysis Division (MPAD) became part of MOD the Rendezvous Procedures Group became part of the Flight Design and Dynamics Division (DM). In 1987-1988 the Rendezvous Phase Specialist position was combined with the Orbit Guidance Officer (GUIDO) position to become the Rendezvous Guidance and Procedures Officer (RGPO).

Two other potions are listed, Mir Approach and Docking Support (MADS) and Station Approach and Docking Support (SADS). The Lead RGPO was the primary point-of-contact for a specific mission, the Backup RGPO was the secondary point-of-contact. GR stands for Ground-up Rendezvous, D-R stands for Deploy-Retrieve.

The source of the flight assignments was a webpage maintained by the Rendezvous Guidance and Procedures Office (Code DM34). Names of personnel in the following flight assignment tables are:

Ray Bigonesse
Mick Chang
Sally Davis
Andy Dougherty
Scott Dunham
Tom Erkenswick
Malise Fletcher
Sarah Graybeal (Ruiz)
Craig Gross
Dustin Hamm
Dave Harshman
Rick Heib
Gary Johnson
Mike Machula
Joe Malarkey
Chris Meyer
Todd Miller
Randy Moon
Duane Mōsel
Bill Ober
Jim Oberg
Nick O'Dosey
Sean O'Rourke
Ted Rickerl
Mark Rowles
Jose Ruiz
Barbara Schwartz
Lynda Slifer (Gavin)
Paul Snow
Jeannette Spehar
Mark Thomas
Michael Veres
Steve Walker
Joe Williams
Jerry Yencharis

Mission	Launch Date	Rndz Type	Mission Payload	Lead Backup	Console Position	Orbit 1 Shift	Orbit 2 Shift	Orbit 3 Shift	Orbit 4 Shift
STS-7 *Challenger*	6/18/83	D-R	SPAS-01	Mosel	RGPO RPS	**Mosel** Rolwes	Heib Oberg		
STS-41B *Challenger*	2/3/84	D	GAS-IRT	Rowles	RGPO RPS	**Rolwes**	Mosel Gross	Rolwes	
STS-41C *Challenger*	4/6/84	GR	Solar Max	Heib	RGPO RPS OJT	**Heib** Veres Thomas			
STS-51A *Discovery*	11/8/84	GR GR	PALAPA-B2 WESTAR-VI	Thomas Heib	RGPO RPS	**Heib** Thomas			
STS-51D *Discovery*	4/12/85	D-R	LEASAT-3	Heib	RGPO RPS	**Heib** Mosel/Gross		Thomas Veres	
STS-51G *Discovery*	6/17/85	D-R	Spartan-1	Gross Rolwes	RGPO RPS	**Gross** Rolwes		Thomas	
STS-51F *Challenger*	7/29/85	D-R	PDP	Veres	RGPO RPS	**Veres**	Heib	Thomas	
STS-51I *Discovery*	8/27/85	GR	LEASAT-3	Thomas	RGPO RPS	**Thomas**			
STS-61B *Atlantis*	11/26/85	D	OEX DAP Target	Veres	RGPO RPS	**Veres**			
STS-51L *Challenger*	1/28/86	D D-R	TDRS-2 Spartan Halley	Dougherty	RGPO RPS	**Dougherty** Harshman			
STS-32 *Columbia*	1/9/90	GR D	LDEF SYNCOM IV	Oberg	RGPO RPS1 RPS2	**Oberg** Meyer Johnson	Malarkey Thomas Schwartz	Dougherty Harshman	
STS-31 *Discovery*	4/24/90	D	HST	Dougherty	RGPO RPS	**Dougherty**			
STS-37 *Atlantis*	4/5/91	D	GRO	Meyer	RGPO RPS	**Meyer** Dougherty	Malarkey	Oberg	
STS-39 *Discovery*	4/28/91	D-R	IBSS-SPAS-II	Malarkey	RGPO RPS1 RPS2	Thomas	Malarkey Dougherty Schwartz	Meyer Oberg Slifer	
STS-48 *Discovery*	9/12/91	D	UARS	Slifer	RGPO RPS	**Slifer** Oberg	Thomas	Oberg	
STS-49 *Endeavour*	5/7/92	GR	Intelsat VI	Meyer Schwartz	RGPO RPS1 RPS2	Meyer Miller Ober	Slifer Schwartz	Thomas	

Mission	Launch Date	Rdnz Type	Mission Payload	Lead Backup	Console Position	Orbit 1 Shift	Orbit 2 Shift	Orbit 3 Shift	Orbit 4 Shift
STS-46 Atlantis	7/31/92	D	TSS 1	Malarkey Gavin	RGPO RPS Tether Profile	Malarkey Miller Snow Williams	Gavin	Meyer Ober Chang Hamm	
		D-R	EURECA	Meyer	RGPO RPS	Malarkey Miller	Gavin	Meyer Ober	
STS-52 Columbia	10/22/92	D D	Lageos-II CANEX-2	Miller	RGPO RPS	Miller Thomas			
STS-56 Discovery	4/8/93	D-R	Spartan 201-01	Miller	RGPO RPS	Miller Schwartz	Meyer	Thomas Williams	
STS-57 Endeavour	6/21/93	GR	EURECA Spacehab	Meyer Williams	RGPO RPS	Meyer Williams	Gavin Ober	Thomas	
STS-51 Discovery	9/12/93	D-R D	ORFEUS-SPAS ACTS/TOS	Ober Thomas	RGPO RPS	Thomas Ober	Gavin	Miller	
STS-61 Endeavour	12/2/93	GR	HST SM-01	Gavin Harshman	RGPO RPS	Gavin Harshman	Thomas	Ober Williams	
STS-60 Discovery	2/3/94	D-R	WSF-01 Spacehab	Miller Ober	RGPO RPS	Thomas Harshman	Miller Williams	Ober	
STS-64 Discovery	9/9/94	D-R	Spartan 201-02	Ober Fletcher	RGPO RPS OJT	Miller Davis	Ober Fletcher	Harshman	
STS-66 Atlantis	11/3/94	D-R	CRISTA-SPAS	Davis Harshman	RGPO RPS OJT	Harshman Thomas Davis	Williams	Gavin Ober	Thomas
STS-63 Discovery	2/3/95	GR	Mir-00 Spacehab	Williams Gavin	RGPO RPS MADS	Davis Harshman	Williams Fletcher Dunham	Gavin Ober	Miller
		D-R	Spartan-204	Harshman Davis	RGPO RPS	Davis Harshman	Williams Fletcher	Gavin Ober	
STS-71 Atlantis	6/27/95	GR	Mir-01	Gavin Fletcher	RGPO RPS MADS	Gavin Fletcher Spehar	Williams Dunham	Miller Moon	
STS-69 Endeavour	9/7/95	D-R	WSF-02	Ober Walker	RGPO RPS OJT	Ober Thomas	Miller Moon Hamm	Davis	
		D-R	Spartan 201-03	Harshman Moon	RGPO RPS OJT	Harshman Moon	Miller Hamm	Davis	
STS-74 Atlantis	11/12/95	GR	Mir-02 Docking Module ODS	Miller Fletcher	RGPO RPS MADS	Miller Fletcher Dunham	Gavin Erkenswick	Williams Moon	

Mission	Launch Date	Rndz Type	Mission Payload	Lead Backup	Console Position	Orbit 1 Shift	Orbit 2 Shift	Orbit 3 Shift	Orbit 4 Shift
STS-72 Endeavour	1/11/96	GR	SFU	Harshman Hamm	RGPO RPS	**Harshman** Ober	Moon Davis	Fletcher	
		D-R	OAST-Flyer	Moon Harshman	RGPO RPS RGPO RPS	Harshman Ober **Moon** Harshman	Moon Davis Davis	Fletcher Fletcher	
STS-75 Endeavour	2/22/96	D-R	TSS 1-R	Williams Davis	RGPO TPS Tether	**Williams** Snow Staas	Davis Hamm Le	Ober Moon Hamilton	
STS-76 Atlantis	3/22/96	GR	Mir-03 Spacehab	Fletcher Miller	RGPO RPS MADS	**Fletcher** Harshman Dunham	Miller Erkenswick	Moon Ober	
STS-77 Endeavour	5/19/96	D-R	Spartan-207 IAE-Inflatable Antenna Exp	Harshman Walker	RGPO RPS OJT	Ober Thomas Snow	**Harshman** Walker	Miller Williams	
		D-R	PAMS-STU	Harshman Snow	RGPO RPS OJT	Ober Thomas Snow	**Harshman** Walker	Miller Williams	
STS-79 Atlantis	9/16/96	GR	Mir-04 Spacehab	Davis Miller	RGPO RPS MADS	**Davis** Ober Erkenswick	Miller Spehar	Fletcher Walker	
STS-80 Columbia	11/19/96	D-R	ORFEUS-SPAS-02	Williams Snow	RGPO RPS	**Williams** Snow	Harshman Hamm	Walker Gavin	Thomas Ober
		D-R	WSF-03	Williams Hamm	RGPO RPS	**Williams** Hamm	Harshman Snow	Walker Gavin	Thomas Ober
STS-81 Atlantis	1/12/97	GR	Mir-05 Spacehab D	Ober Harshman	RGPO RPS MADS	**Ober** Snow Spehar	Harshman Erkenswick	Fletcher Hamm	
STS-82 Discovery	3/13/97	GR	HST SM-02	Walker Thomas	RGPO RPS	**Walker** Thomas	Ober Hamm	Williams Snow	Fletcher
STS-84 Atlantis	5/15/97	GR	Mir-06 Spacehab D SMOE	Harshman Snow	RGPO RPS MADS	**Harshman** Williams Dunham	Walker Erkenswick	Snow Fletcher	
STS-85 Discovery	8/7/97	D-R	CRISTA-SPAS-02	Hamm Williams	RGPO RPS	Fletcher	**Williams** Hamm	Thomas	
STS-86 Atlantis	9/25/97	GR	Mir-07	Snow Walker	RGPO RPS MADS	**Snow** Walker Erkenswick	Harshman Spehar	Williams Thomas Dunham	
STS-87 Columbia	11/19/97	D-R	Spartan 201-04	Harshman Spehar	RGPO RPS	**Harshman** Spehar	Snow Hamm	Thomas Walker	

Mission	Launch Date	Rndz Type	Mission Payload	Lead Backup	Console Position	Orbit 1 Shift	Orbit 2 Shift	Orbit 3 Shift	Orbit 4 Shift
STS-89 *Endeavour*	1/22/98	GR	Mir-08	Walker Hamm	RGPO RPS MADS	**Walker** Dunham Erkenswick	Harshman Thomas Spehar	Fletcher Hamm	
STS-91 *Discovery*	6/2/98	GR	Mir-09 Spacehab	Dunham Snow	RGPO RPS	**Dunham** Fletcher	Harshman Yencharis	Snow Thomas	
STS-95 *Discovery*	10/29/98	D-R	Spartan 201-05	Harshman Walker	RGPO RPS	**Harshman** Yencharis	Thomas	Walker Spehar	
STS-88 *Endeavour*	12/4/98	GR	ISS-2A Node 1 PMA-1&2	Williams Yencharis	RGPO RPS	**Williams** Yencharis	Walker Snow	Dunham Fletcher	Harshman Spehar
STS-96 *Discovery*	5/27/99	GR	ISS-2A 1 Starshine	Snow Rickerl	RGPO RPS	**Snow** Rickerl	Fletcher Walker	Yencharis Harshman	
STS-103 *Discovery*	12/19/99	GR	HST SM-03A	Walker O'Rourke	RGPO RPS	**Walker** O'Rourke	Harshman Rickerl	Yencharis Spehar	
STS-101 *Atlantis*	5/19/00	GR	ISS-2A 2A	Yencharis Rickerl	RGPO RPS	**Yencharis** Fletcher	Dunham Rickerl	Spehar Walker	
STS-106 *Atlantis*	9/8/00	GR	ISS-2A 2B	Harshman O'Dosey	RGPO RPS	**Harshman** O'Dosey	Fletcher O'Rourke	Walker Yencharis	
STS-92 *Discovery*	10/11/00	GR	ISS-3A Z1 Truss PMA-3	Spehar Harshman	RGPO RPS	**Spehar** Harshman	Dunham O'Dosey	Yencharis Rickerl	
STS-97 *Endeavour*	11/30/00	GR	ISS-4A P6 Truss	Walker O'Rourke	RGPO RPS	**Walker** O'Rourke	Fletcher O'Dosey	Rickerl Spehar	
STS-98 *Atlantis*	2/7/01	GR	ISS-5A US Lab	Yencharis O'Dosey	RGPO RPS	**Yencharis** O'Dosey	O'Rourke Dunham	Walker Bigonesse	
STS-102 *Discovery*	3/8/01	GR	ISS-5A 1 MPLM-Leonardo	Rickerl Harshman	RGPO RPS	**Rickerl** Harshman	Fletcher	Dunham Spehar	
STS-100 *Endeavour*	4/19/01	GR	ISS-6A SSRMS MPLM-Raffaello	O'Rourke Yencharis	RGPO RPS	**O'Rourke** Walker	Yencharis	Rickerl Bigonesse	
STS-104 *Atlantis*	7/12/01	GR	ISS-7A Airlock	Harshman Bigonesse	RGPO RPS	**Harshman** Bigonesse	O'Rourke	O'Dosey Walker	
STS-105 *Discovery*	8/10/01	GR	ISS-7A 1 EAS MPLM-Leonardo	Rickerl Walker	RGPO RPS	**Rickerl** Walker	Yencharis	Harshman O'Dosey	

Mission	Launch Date	Rndz Type	Mission Payload	Lead Backup	Console Position	Orbit 1 Shift	Orbit 2 Shift	Orbit 3 Shift	Orbit 4 Shift
STS-108 Endeavour	12/5/01	GR	ISS-UF1 MPLM-Raffaello Starshine 2	O'Dosey Yencharis	RGPO RPS	**O'Dosey** Yencharis	Fletcher	Bigonesse Dunham	
STS-109 Columbia	3/1/02	GR	HST SM-03B	Walker O'Rourke	RGPO RPS	**Walker** O'Dosey	O'Rourke Fletcher	Yencharis Rickerl	
STS-110 Atlantis	4/8/02	GR	ISS-8A S0 Truss	Bigonesse Harshman	RGPO RPS	**Bigonesse** Harshman	Dunham	O'Dosey Spehar	
STS-111 Endeavour	6/5/02	GR	ISS-UF2 MPLM-Leonardo	Yencharis Rickerl	RGPO RPS	**Yencharis** Spehar	O'Dosey	Rickerl Machula	
STS-112 Atlantis	10/7/02	GR	ISS-9A S1 Truss	O'Rourke Spehar	RGPO RPS	**O'Rourke** Spehar	Dunham Yencharis	Walker Rickerl	
STS-113 Endeavour	11/23/02	GR	ISS-11A P1 Truss	Walker Bigonesse	RGPO RPS	**Walker** Machula	Fletcher O'Dosey	Bigonesse Harshman	
STS-114 Discovery	7/26/05	GR	ISS-LF1 MPLM-Raffaello	O'Dosey O'Rourke	RGPO RPS	**O'Dosey** O'Rourke	Machula Yencharis	Rickerl Bigonesse	Walker
STS-121 Discovery	7/4/06	GR	ISS-ULF1 1 MPLM-Leonardo	Machula Bigonesse	RGPO RPS	**Machula** Bigonesse	Spehar Harshman	Rickerl O'Dosey	Yencharis O'Rourke
STS-115 Atlantis	9/9/06	GR	ISS-12A P3/P4 Truss	Harshman Yencharis	RGPO RPS	**Harshman** Yencharis	O'Rourke Walker	O'Dosey Rickerl	Bigonesse Machula
STS-116 Discovery	12/9/06	GR	ISS-12A 1 P5 Truss Spacehab	Rickerl O'Dosey	RGPO RPS	**Rickerl** O'Dosey	Yencharis Bigonesse	Machula O'Rourke	Harshman Spehar
STS-117 Atlantis	6/22/07	GR	ISS-13A S3/S4 Truss	Bigonesse Machula	RGPO RPS	**Bigonesse** Machula	Harshman Spehar	Walker Yencharis	Rickerl O'Dosey
STS-118 Endeavour	8/8/07	GR	ISS-13A 1 S5 Truss Spacehab	O'Rourke Harshman	RGPO RPS	**O'Rourke** Harshman	O'Dosey Machula	Rickerl Spehar	Bigonesse Yencharis
STS-120 Discovery	10/23/07	GR	ISS-10A Node 2 P6 Relocate	Yencharis J Ruiz	RGPO RPS	**Yencharis** J Ruiz	Bigonesse Rickerl	O'Dosey Walker	O'Rourke Harshman
STS-122 Atlantis	2/7/08	GR	ISS-1E Columbus Lab	Machula O'Dosey	RGPO RPS	**Machula** O'Dosey	Harshman O'Rourke	Bigonesse J Ruiz	Yencharis Walker
STS-123 Endeavour	3/11/08	GR	ISS-1J/A JEM-PS SPDM (Dexter)	Rickerl Bigonesse	RGPO RPS	**Rickerl** Bigonesse	O'Rourke Yencharis	O'Dosey Walker	Machula Spehar

Mission	Launch Date	Type of Rndz	Mission Payload	Lead Backup	Console Position	Orbit 1 Shift	Orbit 2 Shift	Orbit 3 Shift	Orbit 4 Shift
STS-124 *Discovery*	5/31/08	GR	ISS-1J Kibo Module J-RMS	Harshman Graybeal	RGPO RPS	**Harshman** Graybeal	Machula J Ruiz	Yencharis Spehar	Rickerl Bigonesse
STS-126 *Endeavour*	11/14/08	GR	ISS-ULF2 MPLM-Leonardo	Bigonesse Yencharis	RGPO RPS	**Bigonesse** Yencharis	Harshman J Ruiz	Machula Graybeal	O'Rourke O'Dosey
STS-119 *Discovery*	3/15/09	GR	ISS-15A S6 Truss	Rickerl Graybeal	RGPO RPS	**Rickerl** Graybeal	O'Dosey Yencharis	O'Rourke Spehar	Bigonesse
STS-125 *Atlantis*	5/11/09	GR	HST SM-04	O'Dosey O'Rourke	RGPO RPS	**O'Dosey** O'Rourke	Walker Graybeal	Bigonesse Rickerl	Harshman Yencharis
STS-400 *Endeavour*	N/A	GR	*Atlantis*	400 Team	RGPO RPS	**O'Dosey** O'Rourke	Harshman Graybeal	Yencharis Machula	N/A N/A
				125 Team	RGPO RPS	**Bigonesse** Rickerl	Walker Spehar	N/A N/A	N/A N/A
STS-127 *Endeavour*	7/15/09	GR	ISS-2J/A JEM-ES ELM-ES	O'Rourke Harshman	RGPO RPS	**O'Rourke** Harshman	Bigonesse Yencharis	Rickerl S Ruiz	Machula
STS-128 *Discovery*	8/28/09	GR	ISS-17A MPLM-Leonardo TriDAR	Yencharis O'Dosey	RGPO RPS	**Yencharis** O'Dosey	S Ruiz Bigonesse	Machula O'Rourke	Harshman
STS-129 *Atlantis*	11/16/09	GR	ISS-ULF3 ELC 1&2	S Ruiz Bigonesse	RGPO RPS	**S Ruiz** Bigonesse	Harshman Rickerl	O'Dosey O'Rourke	Yencharis
STS-130 *Endeavour*	2/8/10	GR	ISS-20A Cupola/ Node 3	Bigonesse Rickerl	RGPO RPS	**Bigonesse** Rickerl	Yencharis Harshman	O'Rourke	S Ruiz
STS-131 *Discovery*	4/5/10	GR	ISS-19A MPLM	O'Dosey S Ruiz	RGPO RPS	**O'Dosey** S Ruiz	Rickerl O'Rourke	Harshman	Bigonesse
STS-132 *Atlantis*	5/14/10	GR	ULF4 MRM1	Rickerl O'Rourke	RGPO RPS	**Rickerl** O'Rourke	O'Dosey Harshman	Bigonesse	S Ruiz
STS-133 *Discovery*	2/14/11	GR	ULF5 PLM	S Ruiz O,,Rourke	RGPO RPS	**S Ruiz** Rickerl	O'Dosey O'Rourke	Bigonesse	Harshman
STS-134 *Endeavour*	5/16/11	GR	ULF6 AMS STORRM DTO	Harshman O'Dosey	RGPO RPS	**Harshman** O'Dosey	O'Rourke Bigonesse	S Ruiz	Rickerl
STS-135 *Atlantis*	7/08/11	GR	ULF7 MPLM	O'Dosey Rickerl	RGPO RPS	**O'Dosey** Rickerl	Harshman Ruiz	O'Rourke Bigonesse	Bigonesse

This page intentionally left blank.

APPENDIX B – RENDEZVOUS AND ORBIT FLIGHT DYNAMICS OFFICERS

This appendix lists the flight assignments for Rendezvous and Orbit Flight Dynamics Officers (FDO). The source of the flight assignments was a webpage maintained by the NASA/JSC Flight Dynamics Division Orbit Flight Dynamics Office (Code DM32). Names of personnel in the following flight assignment tables are:

> Marc Abadie
> Dan Adamo
> Mark Anderson
> Roger Balettie
> Bill Britz
> Phil Burley
> Bill Clarke
> Ron Cohen
> Chris Edelen
> Ed Gonzalez
> Mark Haynes
> Rebecca Cutri-Kohart
> Bill Jacobs
> Brian Jones
> Darrin Leleux
> Bryan Lowman
> David Mayhew
> Jen Mendeck
> Roger Rojas
> Roger Simpson
> Jason Smith
> Bob Stein
> Steve Stich
> Dick Theis
> Bill Tracy

Mission	Launch Date	Rndz Type	Mission Payload	Console Position	Orbit 1 Shift	Orbit 2 Shift	Orbit 3 Shift	Orbit 4 Shift
STS-7 *Challenger*	6/18/83	D-R	SPAS-01	FDO TRAJ	l'Anson Cohen		Oliver González	
STS-41B *Challenger*	2/3/84	D	GAS-IRT	FDO TRAJ		B. Jones	González	
STS-41C *Challenger*	4/6/84	GR	Solar Max	FDO TRAJ	Epp Combs	Cohen Perry	B.Jones W.Jones	
STS-51A *Discovery*	11/8/84	GR GR	PALAPA-B2 WESTAR-VI	FDO TRAJ	B.Jones Combs	Epp Soileau	Cohen Rask	
STS-51D *Discovery*	4/12/85	D-R	LEASAT-3	FDO TRAJ	Burley W.Jones	González Stewart	Rask Lancaster	
STS-51G *Discovery*	6/17/85	D-R	Spartan-1	FDO TRAJ	Epp Stewart	B.Jones Lancaster	W.Jones Hilty	
STS-51F *Challenger*	7/29/85	D-R	PDP	FDO TRAJ	Lancaster Stewart	Rask Brown	González Haynes	
STS-51I *Discovery*	8/27/85	GR	LEASAT-3	FDO TRAJ	B.Jones Haynes	W.Jones Hilty	Lancaster Brown	
STS-61B *Atlantis*	11/26/85	D	OEX DAP Target	FDO TRAJ	W.Jones Haynes	Stewart Sims	Rask Fletcher	
STS-51L *Challenger*	1/28/86	D D-R	TDRS-2 Spartan Halley	FDO TRAJ	N/A N/A	N/A N/A	N/A N/A	
STS-32 *Columbia*	1/9/90	GR D	LDEF SYNCOM IV	FDO TRAJ	Haynes Fletcher	Burley Brown	Rask Kessler	
STS-31 *Discovery*	4/24/90	D	HST	FDO TRAJ	Burley Tracy	Fletcher Langan	Theis Adamo	
STS-37 *Atlantis*	4/5/91	D	GRO	FDO TRAJ	Theis Balettie	Burley Stich	Kessler Haynes	
STS-39 *Discovery*	4/28/91	D-R	IBSS-SPAS-II	FDO TRAJ	Haynes Britz	Burley Shore	Brown Theis (Stich)	
STS-48 *Discovery*	9/12/91	D	UARS	FDO TRAJ	Stich Adamo	Theis Shore	Burley Riggio	
STS-49 *Endeavour*	5/7/92	GR	Intelsat VI	FDO TRAJ	Haynes Stich	Burley Britz	Theis Brown (Riggio)	

Mission	Launch Date	Rndz Type	Mission Payload	Console Position	Orbit 1 Shift	Orbit 2 Shift	Orbit 3 Shift	Orbit 4 Shift
STS-46 *Atlantis*	7/31/92	D D-R	TSS 1 EURECA	FDO TRAJ	Adamo Shore	Tracy Simpson	Brown Britz	
STS-52 *Columbia*	10/22/92	D D	Lageos-II CANEX-2	FDO TRAJ	Burley Simpson	Shore Brown	Tracy Stich	
STS-56 *Discovery*	4/8/93	D-R	Spartan 201-01	FDO TRAJ	Stich Adamo	Theis Riggio	Tracy Balettie	
STS-57 *Endeavour*	6/21/93	GR	EURECA Spacehab	FDO TRAJ	Theis Adamo	Britz Stich	Burley Shore	
STS-51 *Discovery*	9/12/93	D-R D	ORFEUS-SPAS ACTS/TOS	FDO TRAJ	Burley Shore	Balettie Tracy	Adamo Theis	
STS-61 *Endeavour*	12/2/93	GR	HST SM-01	FDO TRAJ	Stich Britz	Theis Shore	Tracy Adamo	Burley Simpson
STS-60 *Discovery*	2/3/94	D-R	WSF-01 Spacehab	FDO TRAJ	Adamo Tracy	Stich Theis	Britz Balettie	
STS-64 *Discovery*	9/9/94	D-R	Spartan 201-02	FDO TRAJ	Burley Balettie	Tracy Theis	Stich Stein	
STS-66 *Atlantis*	11/3/94	D-R	CRISTA-SPAS	FDO TRAJ	Britz Adamo	Theis McCraw	Balettie Tracy	Stein
STS-63 *Discovery*	2/3/95	GR D-R	Mir-00 Spacehab Spartan-204	FDO TRAJ	Britz Stich	Adamo Balettie	Burley Theis	
STS-71 *Atlantis*	6/27/95	GR	Mir-01	FDO TRAJ	Burley Balettie	Stich Hammer	Theis Shore	
STS-69 *Endeavour*	9/7/95	D-R D-R	WSF-02 Spartan 201-03	FDO TRAJ	Tracy Balettie	Theis Edelen	Adamo McCraw	
STS-74 *Atlantis*	11/12/95	GR	Mir-02 Docking Module ODS	FDO TRAJ	Stich Stein	Balettie Edelen	Adamo Hammer	
STS-72 *Endeavour*	1/11/96	GR D-R	SFU OAST-Flyer	FDO TRAJ	Theis McCraw	Simpson Britz	Adamo Edelen	
STS-75 *Endeavour*	2/22/96	D-R	TSS 1-R	FDO TRAJ	McCraw Theis	Tracy Hammer	Stein Stich	
STS-76 *Atlantis*	3/22/96	GR	Mir-03 Spacehab	FDO TRAJ	Balettie Simpson	Stich Edelen	Adamo McCraw	
STS-77 *Endeavour*	5/19/96	D-R D-R	Spartan-207 IAE-Inflatable Antenna Exp PAMS-STU	FDO TRAJ	Burley Simpson	Tracy Britz	Theis Hammer	Simpson Edelen

Mission	Launch Date	Rndz Type	Mission Payload	Console Position	Orbit 1 Shift	Orbit 2 Shift	Orbit 3 Shift	Orbit 4 Shift
STS-79 *Atlantis*	9/16/96	GR	Mir-04 Spacehab	FDO TRAJ	Stich Stein	Balettie Edelen	Adamo Hammer	
STS-80 *Columbia*	11/19/96	D-R D-R	ORFEUS-SPAS-02 WSF-03	FDO TRAJ	Adamo McCraw	Britz Edelen	Theis Schaf	Burley Stein
STS-81 *Atlantis*	1/12/97	GR	Mir-05 Spacehab D	FDO TRAJ	Simpson Adamo	Balettie Hammer	Edelen Theis	
STS-82 *Discovery*	3/13/97	GR	HST SM-02	FDO TRAJ	Tracy Hammer	McCraw Stich	Stein Britz	
STS-84 *Atlantis*	5/15/97	GR	Mir-06 Spacehab D SMOE	FDO TRAJ	Adamo Balettie	Simpson Schaf	McCraw Tracy	
STS-85 *Discovery*	8/7/97	D-R	CRISTA-SPAS-02	FDO TRAJ	Edelen Tracy	Theis Schaf	Britz Clarke	
STS-86 *Atlantis*	9/25/97	GR	Mir-07	FDO TRAJ	Balettie McCraw	Tracy Adamo	Stich Jones	
STS-87 *Columbia*	11/19/97	D-R	Spartan 201-04	FDO TRAJ	Schaf Adamo	Burley Clarke	Theis Tran	Gonzalez Rask
STS-89 *Endeavour*	1/22/98	GR	Mir-08	FDO TRAJ	Tracy Clarke	Stein Jones	Adamo Tran	Edelen McCraw
STS-91 *Discovery*	6/2/98	GR	Mir-09 Spacehab	FDO TRAJ	Stein Tracy	Edelen Spencer	Jones Theis	
STS-95 *Discovery*	10/29/98	D-R	Spartan 201-05	FDO TRAJ	Edelen Tracy	Clarke Burley	Theis Tran	
STS-88 *Endeavour*	12/4/98	GR	ISS-2A Node 1 PMA-1&2	FDO TRAJ	Adamo Tran	Theis Barrett	Tracy Spencer	
STS-96 *Discovery*	5/27/99	GR	ISS-2A 1 Starshine	FDO TRAJ	Theis Barrett	Adamo Clarke	Tracy Tran	Spencer Theis
STS-103 *Discovery*	12/19/99	GR	HST SM-03A	FDO TRAJ	Tracy Clarke	Stein Spencer	Adamo Schaf	Barrett Clarke
STS-101 *Atlantis*	5/19/00	GR	ISS-2A 2A	FDO TRAJ	Adamo Clarke	Theis Schaf	Tracy Tran	
STS-106 *Atlantis*	9/8/00	GR	ISS-2A 2B	FDO TRAJ	Burley Barrett	Tran Tracy	Edelen McDonald	
STS-92 *Discovery*	10/11/00	GR	ISS-3A Z1 Truss PMA-3	FDO TRAJ	Tracy McDonald	Adamo Schaf	Stein Edelen	
STS-97 *Endeavour*	11/30/00	GR	ISS-4A P6 Truss	FDO TRAJ	Theis Barrett	Burley McCraw	Spencer Stein	

Mission	Launch Date	Rndz Type	Mission Payload	Console Position	Orbit 1 Shift	Orbit 2 Shift	Orbit 3 Shift	Orbit 4 Shift
STS-98 *Atlantis*	2/7/01	GR	ISS-5A US Lab	FDO TRAJ	Edelen McDonald	Adamo Clarke	Tran Stein	
STS-102 *Discovery*	3/8/01	GR	ISS-5A 1 MPLM-Leonardo	FDO TRAJ	Burley Tran	Barrett Adamo	Stein Rojas	
STS-100 *Endeavour*	4/19/01	GR	ISS-6A SSRMS MPLM-Raffaello	FDO TRAJ	Tracy McDonald	Spencer Adamo	Clarke Edelen	
STS-104 *Atlantis*	7/12/01	GR	ISS-7A Airlock	FDO TRAJ	Stein-Tran Tran-Clarke	Burley-Tracy Rojas-McDonald	Theis Schaf	
STS-105 *Discovery*	8/10/01	GR	ISS-7A 1 EAS MPLM-Leonardo	FDO TRAJ	Clarke Edelen	Adamo Schaf	McDonald Tracy	
STS-108 *Endeavour*	12/5/01	GR	ISS-UF1 MPLM-Raffaello Starshine 2	FDO TRAJ	Theis Rojas	Edelen Tran	Burley Barrett	
STS-109 *Columbia*	3/1/02	GR	HST SM-03B	FDO TRAJ	Tracy Stein	McDonald Theis	Adamo Rojas	
STS-110 *Atlantis*	4/8/02	GR	ISS-8A S0 Truss	FDO TRAJ	Edelen Barrett	Tran Adamo	Theis Mayhew	
STS-111 *Endeavour*	6/5/02	GR	ISS-UF2 MPLM-Leonardo	FDO TRAJ	Burley Rojas	Tracy Mayhew	Stein Leleux	
STS-112 *Atlantis*	10/7/02	GR	ISS-9A S1 Truss	FDO TRAJ	Stein Mayhew	Barrett Adamo	Spencer Tracy	
STS-113 *Endeavour*	11/23/02	GR	ISS-11A P1 Truss	FDO TRAJ	Theis Leleux	Adamo McDonald	Burley Rojas	
STS-114 *Discovery*	7/26/05	GR	ISS-LF1 MPLM-Raffaello	FDO TRAJ	Tracy Rojas	Edelen Tran	Adamo Mendeck	
STS-121 *Discovery*	7/4/06	GR	ISS-ULF1 1 MPLM-Leonardo	FDO TRAJ	Burley Mendeck	Stein Mayhew	Rojas Leleux	
STS-115 *Atlantis*	9/9/06	GR	ISS-12A P3/P4 Truss	FDO TRAJ	Theis Barrett	Rojas Jacobs	Mayhew Gruber	
STS-116 *Discovery*	12/9/06	GR	ISS-12A 1 P5 Truss Spacehab	FDO TRAJ	Stein Mayhew	Adamo Cutri-Kohart	Mendeck Leleux	
STS-117 *Atlantis*	6/22/07	GR	ISS-13A S3/S4 Truss	FDO TRAJ	Rojas Jacobs	Burley Cutri-Kohart	Leleux Mayhew	

Mission	Launch Date	Rndz Type	Mission Payload	Console Position	Orbit 1 Shift	Orbit 2 Shift	Orbit 3 Shift	Orbit 4 Shift
STS-118 Endeavour	8/8/07	GR	ISS-13A 1 S5 Truss Spacehab	FDO TRAJ	Theis Leleux	Adamo Tracy	Gruber Cutri-Kohart	
STS-120 Discovery	10/23/07	GR	ISS-10A Node 2 P6 Relocate	FDO TRAJ	Tracy Cutri-Kohart	Mayhew Adamo	Mendeck Jacobs	Tran Abadie
STS-122 Atlantis	2/7/08	GR	ISS-1E Columbus Lab	FDO TRAJ	Leleux Barrett	Rojas Adamo	Stein Jacobs	
STS-123 Endeavour	3/11/08	GR	ISS-1J/A JEM-PS SPDM (Dexter)	FDO TRAJ	Mendeck Mayhew	Rojas Abadie	Tracy Cutri-Kohart	Tran Smith
STS-124 Discovery	5/31/08	GR	ISS-1J Kibo Module J-RMS	FDO TRAJ	Stein Cutri-Kohart	Mayhew Abadie	Jacobs Smith	Burley Rojas
STS-126 Endeavour	11/14/08	GR	ISS-ULF2 MPLM- Leonardo	FDO TRAJ	Rojas Smith	Mendeck CutriKohart Abadie	Jacobs Abadie	Leleux Burley
STS-119 Discovery	3/15/09	GR	ISS-15A S6 Truss	FDO TRAJ	Mayhew Tracy	Jacobs Abadie	Stein Anderson	Cutri-Kohart Burley
STS-125 Atlantis	5/11/09	GR	HST SM-04	FDO TRAJ	Tracy Abadie	Stein Smith	Cutri-Kohart Burley	
STS-400 Endeavour	Not Flown	GR	Atlantis	FDO TRAJ				
STS-127 Endeavour	7/15/09	GR	ISS-2J/A JEM-ES ELM-ES	FDO TRAJ	Mendeck Abadie	Cutri-Kohart Smith	Leleux Anderson	Mayhew Jacobs
STS-128 Discovery	8/28/09	GR	ISS-17A MPLM- Leonardo TriDAR	FDO TRAJ	Stein Jacobs	Abadie Anderson	Rojas Smith	Tracy Barrett
STS-129 Atlantis	11/16/09	GR	ISS-ULF3 ELC 1&2	FDO TRAJ	Rojas Anderson	CutriKohart Stein	Smith Burley	
STS-130 Endeavour	2/8/10	GR	ISS-20A Cupola/ Node 3	FDO TRAJ	Tracy Mayhew	Smith Jacobs	Abadie Rojas	Burley Leleux
STS-131 Discovery	4/5/10	GR	ISS-19A MPLM	FDO TRAJ	Mayhew Smith	Jacobs Anderson	Stein Lowman	Mendeck Burley
STS-132 Atlantis	5/14/10	GR	ULF4 MRM1	FDO TRAJ	Cutri-Kohart Anderson	Smith Lowman	Abadie Burley	

Mission	Launch Date	Rndz Type	Mission Payload	Console Position	Orbit 1 Shift	Orbit 2 Shift	Orbit 3 Shift	Orbit 4 Shift
STS-133 *Discovery*	2/24/11	GR	ULF5 PLM	FDO TRAJ	Stein Lowman	Mayhew Smith	Anderson Burley	
STS-134 *Endeavour*	5/16/11	GR	ULF6 AMS STORRM DTO	FDO TRAJ	Rojas Mendeck	Anderson Abadie	Cutri-Kohart Tracy	Smith Leleux
STS-135 *Atlantis*	7/08/11	GR	ULF7 MPLM LMC	FDO TRAJ	Smith Cutri-Kohart	Anderson Mayhew	Lowman Abadie	

This page intentionally left blank.

APPENDIX C – RENDEZVOUS CREW TRAINERS

This appendix lists the flight assignments for rendezvous crew training personnel. The crew trainers created this list at the request of the author in Feburary of 2010. Names of personnel in the following flight assignment tables are:

Rob Banfield
Steve Clark
Rick Davis
Alan Fox
Jorge Frank
Steve Gauvain
Tim Hagin
Gail Hennington Barnett
Bob Mahoney
Lisa Martignetti
Todd Miller
Chuck Moede
Val Murdock
Jim Pendergast
Dave Rose
Dan Sedej
Jeff Tuxhorn

Mission	Launch Date	Type of Rndz	Mission Payload	Trainer
STS-7 *Challenger*	6/18/83	D-R	SPAS-01	Dan Sedej Steve Clark
STS-41B *Challenger*	2/3/84	D	GAS-IRT	Alan Fox Dan Sedej
STS-41C *Challenger*	4/6/84	GR	Solar Max	Dan Sedej Alan Fox
STS-41G *Challenger*	10/5/84	D	Earth Radiation Budget Satellite (ERBS)	Alan Fox Dan Sedej
STS-51A *Discovery*	11/8/84	GR GR	PALAPA-B2 WESTAR-VI	Todd Miller Dan Sedej
STS-51D *Discovery*	4/12/85	D-R	LEASAT-3 (contigency rndz with Flyswatter)	Alan Fox
STS-51G *Discovery*	6/17/85	D-R	Spartan-1	Alan Fox Dan Sedej
STS-51F *Challenger*	7/29/85	D-R	PDP	Dan Sedej
STS-51I *Discovery*	8/27/85	GR	LEASAT-3	Alan Fox
STS-61B *Atlantis*	11/26/85	D	OEX DAP Target	
STS-51L *Challenger*	1/28/86	D D-R	TDRS-2 Spartan Halley	Alan Fox
STS-32 *Columbia*	1/9/90	GR D	LDEF SYNCOM IV	Alan Fox
STS-31 *Discovery*	4/24/90	D	HST	Rob Banfield
STS-37 *Atlantis*	4/5/91	D	GRO	Chuck Moede
STS-39 *Discovery*	4/28/91	D-R	IBSS-SPAS-II	Gail Hennington
STS-48 *Discovery*	9/12/91	D	UARS	Rob Banfield
STS-49 *Endeavour*	5/7/92	GR	Intelsat VI	Alan Fox
STS-46 *Atlantis*	7/31/92	D	TSS 1	Rob Banfield Bob Mahoney
		D-R	EURECA	Chuck Moede

Mission	Launch Date	Type of Rndz	Mission Payload	Trainer
STS-52 *Columbia*	10/22/92	D D	Lageos-II CANEX-2	Alan Fox
STS-56 *Discovery*	4/8/93	D-R	Spartan 201-01	Rick Davis
STS-57 *Endeavour*	6/21/93	GR	EURECA Spacehab	Gail Hennington
STS-51 *Discovery*	9/12/93	D-R D	ORFEUS-SPAS ACTS/TOS	Alan Fox
STS-61 *Endeavour*	12/2/93	GR	HST SM-01	Alan Fox
STS-60 *Discovery*	2/3/94	D-R	WSF-01 Spacehab	Rick Davis
STS-64 *Discovery*	9/9/94	D-R	Spartan 201-02	Alan Fox
STS-66 *Atlantis*	11/3/94	D-R	CRISTA-SPAS	Gail Hennington
STS-63 *Discovery*	2/3/95	GR D-R	Mir-00 Spacehab Spartan-204	Bob Mahoney
STS-71 *Atlantis*	6/27/95	GR	Mir-01	Alan Fox
STS-69 *Endeavour*	9/7/95	D-R D-R	WSF-02 Spartan 201-03	Jim Pendergast
STS-74 *Atlantis*	11/12/95	GR	Mir-02 Docking Module ODS	Gail Hennington
STS-72 *Endeavour*	1/11/96	GR D-R	SFU OAST-Flyer	Rick Davis
STS-75 *Endeavour*	2/22/96	D-R	TSS 1-R	Bob Mahoney Dave Rose
STS-76 *Atlantis*	3/22/96	GR	Mir-03 Spacehab	Alan Fox
STS-77 *Endeavour*	5/19/96	D-R D-R	Spartan-207 IAE-Inflatable Antenna Exp PAMS-STU	Rick Davis
STS-79 *Atlantis*	9/16/96	GR	Mir-04 Spacehab	Alan Fox
STS-80 *Columbia*	11/19/96	D-R D-R	ORFEUS-SPAS-02 WSF-03	Bob Mahoney

Mission	Launch Date	Type of Rndz	Mission Payload	Trainer
STS-81 *Atlantis*	1/12/97	GR	Mir-05 Spacehab D	Dave Rose
STS-82 *Discovery*	3/13/97	GR	HST SM-02	Lisa Martignetti
STS-84 *Atlantis*	5/15/97	GR	Mir-06 Spacehab D SMOE	Jorge Frank
STS-85 *Discovery*	8/7/97	D-R	CRISTA-SPAS-02	Bob Mahoney Lisa Martignetti
STS-86 *Atlantis*	9/25/97	GR	Mir-07	Alan Fox
STS-87 *Columbia*	11/19/97	D-R	Spartan 201-04	Jorge Frank
STS-89 *Endeavour*	1/22/98	GR	Mir-08	Lisa Martignetti
STS-91 *Discovery*	6/2/98	GR	Mir-09 Spacehab	Val Murdock
STS-95 *Discovery*	10/29/98	D-R	Spartan 201-05	Lisa Martignetti
STS-88 *Endeavour*	12/4/98	GR	ISS-2A Node 1 PMA-1&2	Alan Fox
STS-96 *Discovery*	5/27/99	GR	ISS-2A 1 Starshine	Jorge Frank
STS-103 *Discovery*	12/19/99	GR	HST SM-03A	Tim Hagin
STS-101 *Atlantis*	5/19/00	GR	ISS-2A 2A	Val Murdock
STS-106 *Atlantis*	9/8/00	GR	ISS-2A 2B	Tim Hagin
STS-92 *Discovery*	10/11/00	GR	ISS-3A Z1 Truss PMA-3	Alan Fox
STS-97 *Endeavour*	11/30/00	GR	ISS-4A P6 Truss	Jorge Frank
STS-98 *Atlantis*	2/7/01	GR	ISS-5A US Lab	Tim Hagin Val Murdoc
STS-102 *Discovery*	3/8/01	GR	ISS-5A 1 MPLM-Leonardo	Tim Hagin

Mission	Launch Date	Type of Rndz	Mission Payload	Trainer
STS-100 *Endeavour*	4/19/01	GR	ISS-6A SSRMS MPLM-Raffaello	Jorge Frank
STS-104 *Atlantis*	7/12/01	GR	ISS-7A Airlock	Alan Fox
STS-105 *Discovery*	8/10/01	GR	ISS-7A 1 EAS MPLM-Leonardo	Val Murdoc
STS-108 *Endeavour*	12/5/01	GR	ISS-UF1 MPLM-Raffaello Starshine 2	Jorge Frank
STS-109 *Columbia*	3/1/02	GR	HST SM-03B	Tim Hagin
STS-110 *Atlantis*	4/8/02	GR	ISS-8A S0 Truss	Jeff Tuxhorn
STS-111 *Endeavour*	6/5/02	GR	ISS-UF2 MPLM-Leonardo	Alan Fox
STS-112 *Atlantis*	10/7/02	GR	ISS-9A S1 Truss	Val Murdoc
STS-113 *Endeavour*	11/23/02	GR	ISS-11A P1 Truss	Jorge Frank
STS-114 *Discovery*	7/26/05	GR	ISS-LF1 MPLM-Raffaello	Jeff Tuxhorn
STS-121 *Discovery*	7/4/06	GR	ISS-ULF1 1 MPLM-Leonardo	Jorge Frank
STS-115 *Atlantis*	9/9/06	GR	ISS-12A P3/P4 Truss	Alan Fox
STS-116 *Discovery*	12/9/06	GR	ISS-12A 1 P5 Truss Spacehab	Jeff Tuxhorn
STS-117 *Atlantis*	6/22/07	GR	ISS-13A S3/S4 Truss	Jorge Frank
STS-118 *Endeavour*	8/8/07	GR	ISS-13A 1 S5 Truss Spacehab	Steve Gauvain

Mission	Launch Date	Type of Rndz	Mission Payload	Trainer
STS-120 *Discovery*	10/23/07	GR	ISS-10A Node 2 P6 Relocate	Alan Fox
STS-122 *Atlantis*	2/7/08	GR	ISS-1E Columbus Lab	Jeff Tuxhorn
STS-123 *Endeavour*	3/11/08	GR	ISS-1J/A JEM-PS SPDM (Dexter)	Jorge Frank
STS-124 *Discovery*	5/31/08	GR	ISS-1J Kibo Module J-RMS	Steve Gauvain
STS-126 *Endeavour*	11/14/08	GR	ISS-ULF2 MPLM-Leonardo	Jeff Tuxhorn
STS-119 *Discovery*	3/15/09	GR	ISS-15A S6 Truss	Jorge Frank
STS-125 *Atlantis*	5/11/09	GR	HST SM-04	Alan Fox
STS-400 *Endeavour*	Not Flown	GR	STS-125 Rescue	Alan Fox Jorge Frank Jeff Tuxhorn
STS-127 *Endeavour*	7/15/09	GR	ISS-2J/A JEM-ES ELM-ES	Alan Fox
STS-128 *Discovery*	8/28/09	GR	ISS-17A MPLM-Leonardo TriDAR	Jeff Tuxhorn
STS-129 *Atlantis*	11/16/09	GR	ISS-ULF3 ELC 1&2	Steve Gauvain
STS-130 *Endeavour*	2/8/10	GR	ISS-20A Cupola/Node 3	Jorge Frank
STS-131 *Discovery*	4/5/10	GR	ISS-19A MPLM	Alan Fox
STS-132 *Atlantis*	5/14/10	GR	ULF4 MRM1	Steve Gauvain
STS-133 *Discovery*	2/24/11	GR	ULF5 PLM	Alan Fox
STS-134 *Endeavour*	5/16/11	GR	ULF6 AMS STORRM	Jorge Frank
STS-135 *Atlantis*	7/08/11	GR	ULF7	Alan Fox Steve Gauvain

APPENDIX D – ON-BOARD NAVIGATION PERSONNEL

This appendix lists the flight assignments for on-board rendezvous navigation personnel who supported the Rendezvous Guidance and Procedures Officer (RGPO) from the Trajectory Multi-Purpose Support Room (TRAJ MPSR, the backroom) Michele Kocen provided this information to the author in February of 2010.

The author was unable to locate names of on-board navigators that supported rendezvous and proximity operations before the Space Transportation System Operations Contract (STSOC). Pre-STSOC rendezvous and proximity operations missions flown spanned STS-7 (June 1983) to STS-61B (November 1985).

Names of personnel in the following flight assignment tables are:

> Mark Biggs
> Steve Carothers
> Wayne Hensley
> Michele Kocen
> Bryan Lowman
> Todd Michaels
> Todd Miller
> Darrel Monroe
> Valerie Murdock
> Carolyn Propst
> Jerry Yencharis
> Patrick Zimmerman

Mission	Launch Date	Rndz Type	Mission Payload	Rendezvous	Undock	Deploy	Observer
STS-7 *Challenger*	6/18/83	D-R	SPAS-01				
STS-41B *Challenger*	2/3/84	D	GAS-IRT				
STS-41C *Challenger*	4/6/84	GR	Solar Max				
STS-51A *Discovery*	11/8/84	GR GR	PALAPA-B2 WESTAR-VI				
STS-51D *Discovery*	4/12/85	D-R	LEASAT-3				
STS-51G *Discovery*	6/17/85	D-R	Spartan-1				
STS-51F *Challenger*	7/29/85	D-R	PDP				
STS-51I *Discovery*	8/27/85	GR	LEASAT-3				
STS-61B *Atlantis*	11/26/85	D	OEX DAP Target				
STS-51L *Challenger*	1/28/86	D D-R	TDRS-2 Spartan Halley				
STS-32 *Columbia*	1/9/90	GR D	LDEF SYNCOM IV	Kocen Miller			Biggs
STS-31 *Discovery*	4/24/90	D	HST			Miller Biggs	
STS-37 *Atlantis*	4/5/91	D	GRO			Kocen	
STS-39 *Discovery*	4/28/91	D-R	IBSS-SPAS-II	Miller Michaels		Kocen Biggs	
STS-48 *Discovery*	9/12/91	D	UARS			Kocen Michaels	
STS-49 *Endeavour*	5/7/92	GR	Intelsat VI	Kocen Murdock			

Mission	Launch Date	Rndz Type	Mission Payload	Rendezvous	Undock	Deploy	Observer
STS-46 *Atlantis*	7/31/92	D	TSS 1			Biggs (1) Murdock (1) Kocen (2) Miller (3)	
		D-R	EURECA				
STS-52 *Columbia*	10/22/92	D D	Lageos-II CANEX-2			Michaels	
STS-56 *Discovery*	4/8/93	D-R	Spartan 201-01	Kocen Hensley		Kocen Hensley	
STS-57 *Endeavour*	6/21/93	GR	EURECA Spacehab	Murdock Kocen			
STS-51 *Discovery*	9/12/93	D-R D	ORFEUS-SPAS ACTS/TOS	Kocen Hensley		Kocen Hensley	
STS-61 *Endeavour*	12/2/93	GR	HST SM-01	Hensley Murdock		Hensley	
STS-60 *Discovery*	2/3/94	D-R	WSF-01 Spacehab	Miller Biggs		Miller Biggs	
STS-64 *Discovery*	9/9/94	D-R	Spartan 201-02	Zimmerman Hensley		Zimmerman	
STS-66 *Atlantis*	11/3/94	D-R	CRISTA-SPAS	Kocen Propst		Kocen	
STS-63 *Discovery*	2/3/95	GR	Mir-00 Spacehab	Biggs Hensley			
		D-R	Spartan-204	Propst Biggs		Propst	
STS-71 *Atlantis*	6/27/95	GR	Mir-01	Hensley	Hensley		
STS-69 *Endeavour*	9/7/95	D-R	WSF-02	Zimmerman		Zimmerman	
		D-R	Spartan 201-03	Kocen		Kocen	
STS-74 *Atlantis*	11/12/95	GR	Mir-02 Docking Module ODS	Propst	Propst		

Mission	Launch Date	Rndz Type	Mission Payload	Rendezvous	Undock	Deploy	Observer
STS-72 *Endeavour*	1/11/96	GR	SFU			Hensley	
		D-R	OAST-Flyer	Biggs		Kocen	
STS-75 *Endeavour*	2/22/96	D-R	TSS 1-R			Hensley	
STS-76 *Atlantis*	3/22/96	GR	Mir-03 Spacehab	Zimmerman	Zimmerman		
STS-77 *Endeavour*	5/19/96	D-R	Spartan-207 IAE-Inflatable Antenna Exp	Propst		Propst	
		D-R	PAMS-STU	Kocen		Kocen	
STS-79 *Atlantis*	9/16/96	GR	Mir-04 Spacehab	Hensley	Hensley		
STS-80 *Columbia*	11/19/96	D-R	ORFEUS-SPAS-02	Kocen		Kocen	Yencharis
		D-R	WSF-03	Zimmerman		Zimmerman	Santos Yencharis
STS-81 *Atlantis*	1/12/97	GR	Mir-05 Spacehab D	Kocen	Kocen		Yencharis
STS-82 *Discovery*	3/13/97	GR	HST SM-02	Zimmerman		Zimmerman	
STS-84 *Atlantis*	5/15/97	GR	Mir-06 Spacehab D SMOE	Yencharis	Yencharis		Propst
STS-85 *Discovery*	8/7/97	D-R	CRISTA-SPAS-02	Propst		Propst	Propst
STS-86 *Atlantis*	9/25/97	GR	Mir-07	Zimmerman	Zimmerman		Santos
STS-87 *Columbia*	11/19/97	D-R	Spartan 201-04	Kocen		Kocen	

Mission	Launch Date	Rndz Type	Mission Payload	Rendezvous	Undock	Deploy	Observer
STS-89 Endeavour	1/22/98	GR	Mir-08	Propst	Propst		
STS-91 Discovery	6/2/98	GR	Mir-09 Spacehab	Zimmerman	Zimmerman		Carothers
STS-95 Discovery	10/29/98	D-R	Spartan 201-05	Zimmerman	Zimmerman		Carothers
STS-88 Endeavour	12/4/98	GR	ISS-2A Node 1 PMA-1&2	Kocen	Kocen		Carothers
STS-96 Discovery	5/27/99	GR	ISS-2A 1 Starshine	Kocen	Kocen		Carothers
STS-103 Discovery	12/19/99	GR	HST SM-03A	Carothers		Carothers	Kocen
STS-101 Atlantis	5/19/00	GR	ISS-2A 2A	Kocen	Kocen		
STS-106 Atlantis	9/8/00	GR	ISS-2A 2B	Zimmerman	Zimmerman		
STS-92 Discovery	10/11/00	GR	ISS-3A Z1 Truss PMA-3	Carothers	Carothers		
STS-97 Endeavour	11/30/00	GR	ISS-4A P6 Truss	Kocen	Kocen		
STS-98 Atlantis	2/7/01	GR	ISS-5A US Lab	Zimmerman	Zimmerman		
STS-102 Discovery	3/8/01	GR	ISS-5A 1 MPLM-Leonardo	Carothers	Carothers		
STS-100 Endeavour	4/19/01	GR	ISS-6A SSRMS MPLM-Raffaello	Kocen	Kocen		
STS-104 Atlantis	7/12/01	GR	ISS-7A Airlock	Zimmerman	Zimmerman		
STS-105 Discovery	8/10/01	GR	ISS-7A 1 EAS MPLM-Leonardo	Kocen	Kocen		

Mission	Launch Date	Rndz Type	Mission Payload	Rendezvous	Undock	Deploy	Observer
STS-108 Endeavour	12/5/01	GR	ISS-UF1 MPLM-Raffaello Starshine 2	Zimmerman	Zimmerman		
STS-109 Columbia	3/1/02	GR	HST SM-03B	Zimmerman		Zimmerman	
STS-110 Atlantis	4/8/02	GR	ISS-8A S0 Truss	Zimmerman	Zimmerman		
STS-111 Endeavour	6/5/02	GR	ISS-UF2 MPLM-Leonardo	Kocen	Kocen		
STS-112 Atlantis	10/7/02	GR	ISS-9A S1 Truss	Zimmerman	Zimmerman		
STS-113 Endeavour	11/23/02	GR	ISS-11A P1 Truss	Kocen	Kocen		
STS-114 Discovery	7/26/05	GR	ISS-LF1 MPLM-Raffaello	Zimmerman	Zimmerman		
STS-121 Discovery	7/4/06	GR	ISS-ULF1 1 MPLM-Leonardo	Kocen	Kocen		
STS-115 Atlantis	9/9/06	GR	ISS-12A P3/P4 Truss	Zimmerman	Zimmerman		
STS-116 Discovery	12/9/06	GR	ISS-12A 1 P5 Truss Spacehab	Kocen	Kocen		
STS-117 Atlantis	6/22/07	GR	ISS-13A S3/S4 Truss	Monroe Kocen	Monroe Kocen		
STS-118 Endeavour	8/8/07	GR	ISS-13A 1 S5 Truss Spacehab	Zimmerman	Zimmerman		
STS-120 Discovery	10/23/07	GR	ISS-10A Node 2 P6 Relocate	Monroe	Monroe		
STS-122 Atlantis	2/7/08	GR	ISS-1E Columbus Lab	Kocen	Kocen		
STS-123 Endeavour	3/11/08	GR	ISS-1J/A JEM-PS SPDM (Dexter)	Zimmerman	Zimmerman		

Mission	Launch Date	Rndz Type	Mission Payload	Rendezvous	Undock	Deploy	Observer
STS-124 *Discovery*	5/31/08	GR	ISS-1J Kibo Module J-RMS	Monroe	Monroe		
STS-126 *Endeavour*	11/14/08	GR	ISS-ULF2 MPLM-Leonardo	Zimmerman	Zimmerman		
STS-119 *Discovery*	3/15/09	GR	ISS-15A S6 Truss	Lowman Kocen	Lowman Kocen		
STS-125 *Atlantis*	5/11/09	GR	HST SM-04	Kocen		Kocen	
STS-400 *Endeavour*	Not Flown	GR	*Atlantis*	Kocen	Kocen		
STS-127 *Endeavour*	7/15/09	GR	ISS-2J/A JEM-ES ELM-ES	Monroe	Monroe		
STS-128 *Discovery*	8/28/09	GR	ISS-17A MPLM-Leonardo TriDAR	Kocen	Kocen		
STS-129 *Atlantis*	11/16/09	GR	ISS-ULF3 ELC 1&2	Zimmerman	Zimmerman		
STS-130 *Endeavour*	2/8/10	GR	ISS-20A Cupola/ Node 3	Kocen	Kocen		
STS-131 *Discovery*	4/5/10	GR	ISS-19A MPLM	Monroe	Monroe		
STS-132 *Atlantis*	5/14/10	GR	ULF4 MRM1	Kocen	Kocen		
STS-133 *Discovery*	2/24/11	GR	ULF5 PLM	Monroe	Monroe		
STS-134 *Endeavour*	5/16/11	GR	ULF6 AMS STORRM DTO	Zimmerman	Zimmerman		
STS-135 *Atlantis*	7/08/11	GR	ULF7	Monroe	Monroe		

This page intentionally left blank.

APPENDIX E – BURN TARGETING AND PROXIMITY OPERATIONS PERSONNEL

This appendix lists the flight assignments for Trajectory Multi-Purpose Support Room (TRAJ MPSR, the backroom) who supported the Flight Dynamics Officer with ground burn targeting and the Rendezvous Guidance and Procedures Officer (RGPO) with on-board burn targeting or proximity operations. Names in the tables were assembled by USA Flight Design and Dynamics Department Orbit personnel at the request of the author in March of 2010. The tables may contain errors and omissions and are not a complete record of flight assignments.

The author was not able to find accurate flight assignment records for TRAJ MPSR Orbit personnel that supported rendezvous and proximity operations before the 1986 Space Transportation System Operations Contract (STSOC). Pre-STSOC rendezvous and proximity operations missions flown spanned STS-7 (June 1983) to STS-61B (November 1985).

Personnel that supported these positions were:

NASA/DM, RSOC, and USA (1988-2011)

Bill Atkins
Jim Bacher
Charlie Barrett
Greg Bartz
Lynda Bermudez (Slifer/Gavin)
Brian Bertrand
Ray Bigonesse
Colleen de Bont
Bill Britz
Dana Brownfield
Jay Chadwell
Rick Christian
Anthony Foti
Jorge Frank
Dawn Gabriel
Rick Gavin
Barton Gibson
Eduardo Guevara
John Hallstrom
Doug Hamilton
Paul Lane
Jessica LoPresti-Bellock
Scott McKeel
Tom Meissen
Chris Meyer
Mark Miller
Raymundo Moreno
Sean O'Rourke
Joe Pascucci
Don Pearson
Kris Pettinger
Jared Renshaw
Ted Rickerl
Bill Roberts
Dan Sawin
Greg Schrage
Mark Schrock
Kurt Seidensticker
Katie Simons Spotz
Megan Sip

Paul Snow
Susan Stultz Snyder
Tom Snyder
Matt Steinmueller
Kyle Stovall
Tim Stuit
Bill Summa
Farhad Teymurian
Bill Tracy
Hung Tran
Steve Walker
Brian Yarbrough
Jerry Yencharis

NASA Mission Planning and Analysis Directorate (1983-1986)

Bob Becker
Al DuPont
Don Pearson
Ken Young

McDonnell Douglas (1983-1986)

Norm Alexander
Palmer Chiu
Dave Dannemiller
Paul Dowty
Rick Gavin
M. Dan Johnston
Don Pearson
Greg Schrage
Steve Staas

Mission	Launch Date	Rndz Type	Mission Payload	Profile Support	Maneuver and Rndz Specialist	Target	Prox Ops / Tools Checkout
STS-7 *Challenger*	6/18/83	D-R	SPAS-01				
STS-41B *Challenger*	2/3/84	D	GAS-IRT	Pearson (O1)			
STS-41C *Challenger*	4/6/84	GR	Solar Max	Pearson (O1)			
STS-51A *Discovery*	11/8/84	GR GR	PALAPA-B2 WESTAR-VI			Pearson (O1)	
STS-51D *Discovery*	4/12/85	D-R	LEASAT-3				
STS-51G *Discovery*	6/17/85	D-R	Spartan-1				
STS-51F *Challenger*	7/29/85	D-R	PDP				
STS-51I *Discovery*	8/27/85	GR	LEASAT-3				
STS-61B *Atlantis*	11/26/85	D	OEX DAP Target				
STS-51L *Challenger*	1/28/86	D D-R	TDRS-2 Spartan Halley				
STS-26 *Discovery*	9/29/88	D	TDRS-3/IUS				Gibson Meyer
STS-32 *Columbia*	1/9/90	GR D	LDEF SYNCOM IV	Bermudez Gavin Yarbrough		Sawin Britz Snow Chadwell	Schrock Schrage
STS-31 *Discovery*	4/24/90	D	HST	Roberts			Walker
STS-37 *Atlantis*	4/5/91	D	GRO	Atkins Roberts de Bont		Sawin (T1/T2) Snyder (T1) Simons (T2) Chadwell (T1/T2)	
STS-39 *Discovery*	4/28/91	D-R	IBSS-SPAS-II	de Bont (O1) Atkins (O2) Roberts (O3)		Sawin (T1) Snyder (T1) Simons (T2) Chadwell (T2)	
STS-48 *Discovery*	9/12/91	D	UARS				Schrock Summa
STS-49 *Endeavour*	5/7/92	GR	Intelsat VI	Roberts Lane deBont O'Rourke (O3)		Sawin Snyder (T1) Simons (T2)	

Approved for public release via STI DAA 24483. See statement on title page.

300

Mission	Launch Date	Rndz Type	Mission Payload	Profile Support	Maneuver and Rndz Specialist	Target	Prox Ops / Tools Checkout
STS-46 Atlantis	7/31/92	D D-R	TSS 1 EURECA				 Bertrand Brownfield
STS-52 Columbia	10/22/92	D D	Lageos-II CANEX-2				Rickerl Stovall
STS-56 Discovery	4/8/93	D-R	Spartan 201-01	Lane Roberts O'Rourke (O3)		Snyder (T1) Spotz (T2)	Walker Seidensticker
STS-57 Endeavour	6/21/93	GR	EURECA Spacehab	Roberts		Gabriel Spotz	Bertrand Brownfield
STS-51 Discovery	9/12/93	D-R D	ORFEUS-SPAS ACTS/TOS	O'Rourke (O1) Lane Roberts		Gabriel (T1) T. Snyder (T2)	Walker Rickerl Frank
STS-61 Endeavour	12/2/93	GR	HST SM-01	Lane Spotz O'Rourke		Gabriel Rickerl (T2)	Walker Brownfield
STS-60 Discovery	2/3/94	D-R	WSF-01 Spacehab	Lane Roberts Spotz		Snyder Rickerl	Walker Summa (WSF) Rickerl (ODERACS)
STS-64 Discovery	9/9/94	D-R	Spartan 201-02	Spotz Lane Pettinger (O3)		Tran Snyder	
STS-66 Atlantis	11/3/94	D-R	CRISTA-SPAS	O'Rourke (O1) Lane (O2) Pettinger (O3)		Snow (T1) Gabriel (T2)	Rickerl
STS-63 Discovery	2/3/95	GR D-R	Mir-00 Spacehab Spartan-204	Pettinger (O1) Spotz Lane		Sawin (T1) Snyder (T2) Gabriel (T1) Rickerl (T2)	Bertrand Michaux
STS-71 Atlantis	6/27/95	GR	Mir-01	Lane (O1) O'Rourke (O2) Spotz (O3)		Snyder Gabriel	Frank Bertrand
STS-69 Endeavour	9/7/95	D-R D-R	WSF-02 Spartan 201-03	Spotz O'Rourke Lane		Gabriel	McKeel
STS-74 Atlantis	11/12/95	GR	Mir-02 Docking Module ODS	Spotz (O1) Pettinger (O2) O'Rourke (O3)		Tran	Rickerl Summa

Mission	Launch Date	Rndz Type	Mission Payload	Profile Support	Maneuver and Rndz Specialist	Target	Prox Ops / Tools Checkout
STS-72 *Endeavour*	1/11/96	GR D-R	SFU OAST-Flyer	Spotz (O1) Pettinger (O1) O'Rourke (O2) Lane (O3)		Snyder Yencharis	Schrock Brownfield
STS-75 *Endeavour*	2/22/96	D-R	TSS 1-R			Hamilton	
STS-76 *Atlantis*	3/22/96	GR	Mir-03 Spacehab	Christian Lane Pettinger		Gabriel	McKeel
STS-77 *Endeavour*	5/19/96	D-R D-R	Spartan-207 IAE-Inflatable Antenna Exp PAMS-STU	O'Rourke Christian Lane Pettinger		Tran Snyder Yencharis	Bertrand Moreno
STS-79 *Atlantis*	9/16/96	GR	Mir-04 Spacehab	Christian Bacher Lane		Yencharis	Rickerl Brownfield
STS-80 *Columbia*	11/19/96	D-R D-R	ORFEUS-SPAS-02 WSF-03	Bacher (O1) Pettinger (O1) O'Rourke Christian		Snyder Gabriel	Rickerl Guevara
STS-81 *Atlantis*	1/12/97	GR	Mir-05 Spacehab D	Christian O'Rourke Yarbrough		Miller	McKeel Moreno
STS-82 *Discovery*	3/13/97	GR	HST SM-02	Christian Bacher O'Rourke		Gabriel	Summa Teymurian
STS-84 *Atlantis*	5/15/97	GR	Mir-06 Spacehab D SMOE	Yarbrough Bacher Lane		Miller	Rickerl Guevara
STS-85 *Discovery*	8/7/97	D-R	CRISTA-SPAS-02	Yarbrough O'Rourke Bacher		Hamilton	McKeel Brownfield
STS-86 *Atlantis*	9/25/97	GR	Mir-07	Bacher (O1) O'Rourke (O2) Pettinger (O3)		Gabriel	Summa Teymurian
STS-87 *Columbia*	11/19/97	D-R	Spartan 201-04	O'Rourke (O1) Lane (O2) Bacher (O3)		Miller	McKeel Moreno

Mission	Launch Date	Rndz Type	Mission Payload	Profile Support	Maneuver and Rndz Specialist	Target	Prox Ops / Tools Checkout
STS-89 *Endeavour*	1/22/98	GR	Mir-08	Yarbrough (O1) Pettinger (O2) Lane (O3)		Snyder	Rickerl Guevara
STS-91 *Discovery*	6/2/98	GR	Mir-09 Spacehab	Pettinger (O1) O'Rourke (O2) Lane (O3)		Hamilton	Summa Moreno
STS-95 *Discovery*	10/29/98	D-R	Spartan 201-05	Lane (O1) Pettinger (O2) Bacher (O3)		Hamilton	Moreno Foti
STS-88 *Endeavour*	12/4/98	GR	ISS-2A Node 1 PMA-1&2	Yarbrough (O1) Lane (O2) Bacher (O3)		T. Snyder	Schrock Teymurian
STS-96 *Discovery*	5/27/99	GR	ISS-2A 1 Starshine	Yarbrough (O1) Bacher (O2) Pettinger (O3)		Miller	Summa Guevara
STS-93 *Columbia*	7/23/99	D	Chandra X-Ray Observatory				Brownfield
STS-103 *Discovery*	12/19/99	GR	HST SM-03A	Bacher (O1) Snow (O2) Stultz (O3)		Hallstrom	Moreno Foti
STS-101 *Atlantis*	5/19/00	GR	ISS-2A 2A	Pettinger (O1) Yarbrough (O2) Stuit (O3)		T. Snyder	Schrock Teymurian Pascucci
STS-106 *Atlantis*	9/8/00	GR	ISS-2A 2B	Pettinger (O1) Stuit (O2) Bacher (O3)		Hallstrom	Moreno Pascucci
STS-92 *Discovery*	10/11/00	GR	ISS-3A Z1 Truss PMA-3	Stultz (O1) Yarbrough (O2) Bacher (O3)		Snyder	Summa
STS-97 *Endeavour*	11/30/00	GR	ISS-4A P6 Truss	Stuit (O1) Pettinger (O2) Stultz (O3)		Bigonesse	Moreno
STS-98 *Atlantis*	2/7/01	GR	ISS-5A US Lab	Bacher (O1) Stultz (O2) Stuit (O3)		Hallstrom	Summa
STS-102 *Discovery*	3/8/01	GR	ISS-5A 1 MPLM-Leonardo	Pettinger (O1) Bacher (O2) Yarbrough (O3)		Hamilton	Schrock
STS-100 *Endeavour*	4/19/01	GR	ISS-6A SSRMS MPLM-Raffaello	Yarbrough (O1) Pettinger (O2) Stultz (O3)		T. Snyder	Summa

Mission	Launch Date	Rndz Type	Mission Payload	Profile Support	Maneuver and Rndz Specialist	Target	Prox Ops / Tools Checkout
STS-104 *Atlantis*	7/12/01	GR	ISS-7A Airlock	Stuit (O1) Yarbrough (O2) Pettinger (O3)		Hallstrom	Pascucci
STS-105 *Discovery*	8/10/01	GR	ISS-7A 1 EAS MPLM-Leonardo	Stultz (O1) Stuit (O2) Bacher (O3)		Hamilton	Schrock
STS-108 *Endeavour*	12/5/01	GR	ISS-UF1 MPLM-Raffaello Starshine 2	Pettinger (O1) Stultz (O2) Yarbrough (O3)		T. Snyder	Summa
STS-109 *Columbia*	3/1/02	GR	HST SM-03B	Bacher (O1) Pettinger (O2) Stuit (O3)		Hallstrom	Pascucci
STS-110 *Atlantis*	4/8/02	GR	ISS-8A S0 Truss	Yarbrough	Stultz (O1) Pettinger (O2) Stuit (O3)	Hamilton	Summa
STS-111 *Endeavour*	6/5/02	GR	ISS-UF2 MPLM-Leonardo	Bacher	Stuit (O1) Barrett (O2) Pettinger (O3)	T. Snyder	Schrock
STS-112 *Atlantis*	10/7/02	GR	ISS-9A S1 Truss	Bacher	Pettinger (O1) Stuit (O2) S. Snyder (O3)	Hallstrom	Summa
STS-113 *Endeavour*	11/23/02	GR	ISS-11A P1 Truss	Yarbrough	Pettinger (O1) Tracy (O2) S. Snyder (O3)	Meissen	LoPresti Schrock
STS-114 *Discovery*	7/26/05	GR	ISS-LF1 MPLM-Raffaello	Yarbrough (O1) Pettinger (O2) Steinmueller (O3)	S. Snyder (O1) Bacher (O2) Stuit (O3)	Hallstrom	Schrock LoPresti (C/O) Schrock Pascucci (C/O)
STS-121 *Discovery*	7/4/06	GR	ISS-ULF1 1 MPLM-Leonardo	Pettinger (O1) S. Synder (O2) Renshaw (O3)	Bacher (O1) Stuit (O2) Yarbrough (O3)	Hallstrom	Summa LoPresti (C/O) Summa Pascucci (C/O)
STS-115 *Atlantis*	9/9/06	GR	ISS-12A P3/P4 Truss	Bacher (O1) Renshaw (O3)	Stuit (O1) S. Snyder (O2) Pettinger (O3)	Hallstrom	Pascucci Schrock (C/O)
STS-116 *Discovery*	12/9/06	GR	ISS-12A 1 P5 Truss Spacehab	Yarbrough (DOL O1) Stuit (O1)	S. Snyder (O1) Stuit (DOL O1) Bacher (O2) Renshaw (O3)	Bartz	Pascucci Summa (C/O)
STS-117 *Atlantis*	6/22/07	GR	ISS-13A S3/S4 Truss	Renshaw (O1) Pascucci (O3)	Pettinger (O1) Tracy (O2) Stuit (FD2 O2) Yarbrough (O3)	Hallstrom	LoPresti Schrock (C/O)

Mission	Launch Date	Rndz Type	Mission Payload	Profile Support	Maneuver and Rndz Specialist	Target	Prox Ops / Tools Checkout
STS-118 *Endeavour*	8/8/07	GR	ISS-13A 1 S5 Truss Spacehab	Renshaw	Stuit (O1) S. Snyder (O2) Pettinger (O3)	Bartz	Schrock Summa (C/O)
STS-120 *Discovery*	10/23/07	GR	ISS-10A Node 2 P6 Relocate	Pascucci (O1) Sip (O3)	Yarbrough (O1) Pettinger (O2) S. Snyder (O3)	Hallstrom	Summa LoPresti-Bellock (C/O)
STS-122 *Atlantis*	2/7/08	GR	ISS-1E Columbus Lab	Pettinger (O1) Sip (O3)	Stuit (O1) Renshaw (O2) Yarbrough (O3)	Bartz	Schrock LoPresti-Bellock (C/O)
STS-123 *Endeavour*	3/11/08	GR	ISS-1J/A JEM-PS SPDM (Dexter)	Stuit (O1) Sip	Renshaw (O1) Yarbrough (O2) Pettinger (O3)	Hallstrom	Pascucci Summa (C/O)
STS-124 *Discovery*	5/31/08	GR	ISS-1J Kibo Module J-RMS	Sip	Pettinger (O1) S. Snyder (O2) Stuit (O3)	Bartz	LoPresti-Bellock Schrock (C/O)
STS-126 *Endeavour*	11/14/08	GR	ISS-ULF2 MPLM-Leonardo	Sip	S. Snyder (O1) Stuit (O2) Pascucci (O3)	Bartz	Summa Schrock (C/O) Summa Pascucci (C/O)
STS-119 *Discovery*	3/15/09	GR	ISS-15A S6 Truss	Pettinger (O1) Renshaw (O3)	Pascucci (O1) S. Snyder (O2) Sip (O3)	Hallstrom	Schrock Summa (C/O)
STS-125 *Atlantis*	5/11/09	GR	HST SM-04	Pettinger (O1) Sip (O2) Stuit (O3)	S. Snyder (O1) Yarbrough (O2) Renshaw (O3)	Hallstrom	Pascucci Summa (C/O) Pascucci LoPresti-Bellock (C/O)
STS-400 *Endeavour*	Not Flown	GR	*Atlantis*	Yarbrough (O1) Sip (O2) Renshaw (O3)	Stuit (O1) Snyder (O2) Pettinger (O3)	Hallstrom	Summa Pascucci (C/O)
STS-127 *Endeavour*	7/15/09	GR	ISS-2J/A JEM-ES ELM-ES	Stuit	Sip (O1) S. Snyder (O2) Pascucci (O2) Pettinger (O3)	Bartz	Summa Schrock (C/O)
STS-128 *Discovery*	8/28/09	GR	ISS-17A MPLM-Leonardo TriDAR	Yarbrough	Pettinger (O1) S. Snyder (O2) Stuit (O3)	Hallstrom	LoPresti-Bellock Pascucci (C/O)
STS-129 *Atlantis*	11/16/09	GR	ISS-ULF3 ELC 1&2	Renshaw	S. Snyder (O1) Sip (O2) Yarbrough (O3)	Bartz	Pascucci Schrock (C/O)

Mission	Launch Date	Rndz Type	Mission Payload	Profile Support	Maneuver and Rndz Specialist	Target	Prox Ops / Tools Checkout
STS-130 *Endeavour*	2/8/10	GR	ISS-20A Cupola/ Node 3	Sip	Pettinger (O1) Stuit (O2) S. Snyder (O3)	Hallstrom	LoPresti-Bellock Summa (C/O)
STS-131 *Discovery*	4/5/10	GR	ISS-19A MPLM	Pettinger	Yarbrough (O1) Sip (O2) Stuit (O3)	Hallstrom	Schrock Summa (C/O)
STS-132 *Atlantis*	5/14/10	GR	ULF4 MRM1	Pascucci	S. Snyder (O1) Stuit (O2) Pettinger (O3)	Bartz	LoPresti-Bellock Schrock (C/O)
STS-133 *Discovery*	2/24/11	GR	ULF5 PLM	S. Snyder	Sip (O1) Yarbrough (O2) Pettinger (O3)	Bartz	Pascucci Schrock (C/O)
STS-134 *Endeavour*	5/16/11	GR	ULF6 AMS STORRM DTO	Pettinger	Stuit (O1) S. Snyder (O2) Sip (O3)	Hallstrom	Summa LoPresti-Bellock (C/O)
STS-135 *Atlantis*	7/08/11	GR	ULF7 MPLM LMC	S. Snyder	Yarbrough (O1) Sip (O2) Stuit (O3)	LoPresti-Bellock	Summa (rndz) LoPresti-Bellock (undock C/O)

APPENDIX F – MISSION EVALUATION ROOM PERSONNEL

Starting with STS-51 (September 1993) NASA/JSC Engineering Directorate personnel and supporting contractor personnel supported rendezvous tools (Rendezvous and Proximity Operations Program, or RPOP, Hand Held Lidar, or HHL, and Trajectory Control Sensor, or TCS) in the Mission Evaluation Room (MER). These specialists worked closely with Mission Operations Rendezvous Guidance and Procedures Officer (RGPO) and crew training personnel before, during, and after flights. These names were supplied by Jack Brazzel and Jim Duron in February of 2010.

Rendezvous and Proximity Operations Program (RPOP)

Jim Barrett
Jack Brazzel
Fred Clark
Chris Foster
Heather Hinkel
Zoran Milenkovic
Pete Spehar
Scott Tamblyn

Hand Held Lidar (HHL)

Quinn Dunn
Bill Foster
Chris Hovanetz
Tiffany Biehl McFadden
Harold Nitschke
Chau Phan
Mark Schuette
Joe Victor

Trajectory Control Sensor (TCS)

Marty Barr
Mike Brieden
Tamara Cougar
Kent Dekome
Quinn Dunn
Jim Duron
Rodney Elmore
Tim Fisher
Bill Foster
John Handy
Hank Holt
Chris Hovanetz
Johnny Lewis
Tiffany Biehl McFadden
Ken Moreland
Joe Prather
Mark Schuette
Joe Victor

This page intentionally left blank.

APPENDIX G – RELATIVE FRAME

Relative motion is often depicted in a Local Vertical Local Horizontal (LVLH) or Local Vertical Curvilinear (LVC) frame (Figure G.1).[1]

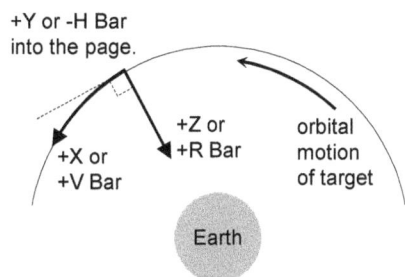

Figure G.1 Local Vertical Curvilinear reference frame.

The target position and velocity vectors are used to define the axes. Nomenclature for the axes follows the convention used within the Shuttle Program.

The +Z axis, also call the +R Bar axis, is defined as:

$$\mathbf{i}_Z = -\text{unit}[\mathbf{r}_T]$$

The +Y axis, also called the –H Bar axis, is defined as:

$$\mathbf{i}_Y = -\text{unit}[\mathbf{r}_T \times \mathbf{v}_T]$$

The +X axis, also called the +V Bar axis, is defined as:

$$\mathbf{i}_X = \text{unit}[(\mathbf{r}_T \times \mathbf{v}_T) \times \mathbf{r}_T]$$

In the LVC frame, the V Bar is curvilinear, rather than rectilinear.

Reference

1. Adamo, D. R., "A Meaningful Relative Motion Coordinate System For Generic Use," *2005 AAS/AIAA Astrodynamics Specialist Conference, Advances in the Astronautical Sciences*, Univelt, San Diego, CA, 2005.

This page intentionally left blank.

APPENDIX H – BURN NOMENCLATURE

The nomenclature for ground targeted burns (NC, NH, NPC, NSR) originated during rendezvous trajectory and Mission Control software development for the Gemini Program in the 1962-1964 time period. The naming convention has been retained through the Gemini, Apollo, Skylab, Apollo/Soyuz, Space Shuttle, and Orion Programs due to the large amount of software and documentation that use these burn names.

The NPC (Plane Change) maneuver controls out-of-plane motion relative to the target. The NH (Height) maneuver controls height relative to the target at a future time. NC (Catch-up) controls the phasing rate of the chaser spacecraft. NSR (Slow Rate) establishes a chaser coelliptic orbit with respect to the target orbit. If there is more than one burn of a type, such as NC, the burns are labeled as NC1, NC2, etc.

The burn names also served as counter variables in the Docking Initiation (DKI) program that was developed by late 1964 to compute rendezvous burns during pre-mission planning and during a flight in Mission Control. In reports and memos from that time the names were written as N_{SR}, N_C, N_H, and N_{PC}. NASA Mission Planning and Analysis Directorate (MPAD) personnel that developed the rendezvous trajectories and techniques also designed the burn targeting algorithms for DKI.

Various versions of DKI were used for pre-mission planning and burn targeting in Mission Control during the Gemini, Apollo, Skylab, and Apollo-Soyuz Programs. Some software from DKI was also included in the Orbital Maneuver Processor (OMP) used during the Shuttle Program.

The names N_C, N_H, N_{PC}, and N_{SR} represented the Nth crossing of the chaser line of apsides, the point where the burns were typically performed. The line of apsides was also called the maneuver line.* A numerical value for the burn enabled personnel and the DKI program to tell where in the mission plan a ground targeted maneuver would be executed. N=1 represented the first apogee, N=1.5 the first perigee, as so on. For example, if the second N_C maneuver was to be executed at the fifth apogee, the burn would be written as N_{C2}=5 in reports, memos, and presentations. Plane change burns (N_{PC}) typically occurred 90 degrees from an apsis crossing. An N_{PC} burn would be written as N_{PC}=2.25 or N_{PC}=3.75. However, over time, in discussions of burns in presentations and documents the variable names were simply written as NC, NH, NPC, and NSR, without the line of apsides crossing count number and subscripts.

In Gemini, Apollo, Skylab, and Apollo-Soyuz documentation the term M=x (where x is an integer) is occasionally encountered. M represented the number of the chaser spacecraft apogee nearest the rendezvous point. For example, M=17 meant that the final phase of rendezvous occurred near the 17th chaser spacecraft apogee.

The lowercase "n" in the below definitions represents the numerical sequence of burns in the mission.

CDH – Constant Delta Height established the LM in a coelliptic orbit with respect to the CSM (Apollo lunar missions).

CIRC – Establishes a circular orbit.

CSI – Coelliptic Sequence Initiation sets up the proper conditions for the CDH and TPI maneuvers (Apollo).

Insertion – Establishment of a safe orbit after powered flight ascent has been completed.

NCn – Phasing burns are used to control the relative catch-up rate, and are typically executed at apogee or perigee. However, an NC burn may be deliberately performed off the line of apsides in order to shift the line of apsides and eliminate the need for an NSR burn to control lighting during proximity operations. It may also be shifted off of an axis to lower the radial component of a subsequent Ti or NSR burn.

NCCn – A Corrective Combination maneuver is a combination of NC and NH burns, and can also include an out-of-plane component. This is a Lambert targeted burn.

NHn – A Height maneuver, executed at either apogee or perigee, controls the differential height (DH) between the chaser and target orbits. It may also be shifted off an axis for the same reasons as a NC burn.

NPC – A Plane Change maneuver corrects for planar dispersions, and is performed at the intersection (node) of the chaser and target phantom orbital planes. If a NC, NSR, or NH burn occurs near a nodal crossing an out-of-plane delta-velocity component may be added to it to avoid performing a separate NPC burn.

NSRn – Slow Rate maneuver places the chaser in a co-elliptic orbit with the target, aligning the lines of apsides of both vehicles. NSR burns can be used to meet lighting requirements on the day of rendezvous.

---

* Young, Kenneth A., and Catherine T. Osgood, *Preflight Orbital And Reentry Trajectory Data for Gemini VI*, MSC Internal Note No. 65-FM-125, Mission Planning and Analysis Division, NASA Manned Spacecraft Center, Houston, TX, September 17, 1965. See also JSC-35055 in the A Note on Sources chapter in this volume.

OMS-n – Shuttle Orbital Maneuvering System (OMS) burns conducted for orbit insertion following ascent.

Out-of-Plane Null – Performed manually during the on-board targeted phase to constrain chaser relative motion to the orbital plane of the target.

Sep – A separation burn establishes a safe departure trajectory after deployment of a target spacecraft or undocking from a space station.

SORn – nth Stable Orbit Rendezvous maneuver.

TPF – "Terminal Phase Finalization" was the total delta-velocity that had to be executed at intercept to null relative motion. In reality TPF was not executed. Braking gates at specified ranges were performed instead. This acronym was used in Gemini and Apollo documentation.

TPI – A "Terminal Phase Initiation" burn placed the chaser on an intercept trajectory with the target on Gemini and Apollo (lunar, Skylab, Apollo/Soyuz). For Orion it targeted the chaser for the Transition to Docking Axis (TDA) point. TPI is executed while the chaser is on an orbit that is coelliptic with the target spacecraft.

Ti – "Transition initiation" was the shuttle equivalent of TPI. On the stable orbit profile it placed the shuttle on an intercept trajectory with the target spacecraft. On an ORBT profile Ti targeted the shuttle for the MC-4 burn point. Immediately after MC-4 the manual piloting phase began. A lowercase "i" was used in "Ti" to avoid confusion with another Shuttle Program rendezvous acronym, T1. T1 and T2 were used in the on-board Lambert targeting software to denote the start and end times of a transfer.

TPMn, MCC-n or MC-n – Terminal Phase Midcourse or Mid-Course Correction burns are executed after TPI or Ti to adjust the intercept trajectory based on relative sensor measurements.

VBNn – nth V Bar Null maneuver that nulls relative motion on the V Bar or establishes coelliptic motion at a point near the V bar.

APPENDIX I – ACRONYMS

A = Assembly
AGS = Abort Guidance System
ALS = Advanced Logistics System
ASTP = Apollo Soyuz Test Project
ATDA = Augmented Target Docking Adapter
ATV = Automated Transfer Vehicle
AVGS = Advanced Video Guidance Sensor
CETA = Crew and Equipment Translation Aid
CIRC = circularization burn
CM = Command Module
COAS = Crew Optical Alignment Sight
CRISTA = Cryogenic Infrared Spectrometers and Telescopes for the Atmospheric
CMG = Control Moment Gyro
CSM = Command/Service Module
DART = Demonstration of Autonomous Rendezvous
DM = Docking Module
ESA = European Space Agency
ESOC = European Space Operations Center
ET = External Tank
EURECA = European Retrievable Carrier
EVA = Extra Vehicular Activity
FA = Fly Around
GATV = Gemini Agena Target Vehicle
GPS = Global Positioning System
GRO = Gamma Ray Observatory
HHL = Hand Held Laser
HST = Hubble Space Telescope
IAE = Inflatable Antenna Experiment
IBSS = Infrared Background Signature Survey
IFM = In Flight Maintenance
IMU = Inertial Measurement Unit
INS = Insertion
INTELSAT = International Telecommunications Satellite
IRT = Integrated Rendezvous Target
ISS = International Space Station
JSC = Johnson Space Center
KSC = Kennedy Space Center
LDEF = Long Duration Exposure Facility
LEO = Low Earth orbit
LF = Logistics Flight
LM = Lunar Module
LOR = Lunar Orbit Rendezvous
LVC = Local Vertical Curvilinear coordinate frame.
LVLH = Local Vertical Local Horizontal
M = docking on the mth revolution
MAHRSI = Middle Atmosphere High Resolution Spectograph Investigation
MC-n = n^{th} Mid-course Correction burn
MMU = Manned Maneuvering Unit
MORAD = Manned Orbital Rendezvous and Docking
MPAD = Mission Planning and Analysis Division
MPLM = Multi-Purpose Logistics Module
MSC = Manned Spacecraft Center

NC-n = n^{th} phasing burn
NCC = Corrective Combination burn
NH = Height burn
NPC = plane change burn
NSR = coelliptic burn
OAST = Office of Aeronautics and Space Technology
ODS = Orbiter Docking System
OEX DAP = Orbital Experiments Digital Autopilot
OFT = Orbital Flight Test
OMP = Orbit Maneuver Processor
OMS-n = n^{th} Orbital Maneuvering System burn
ORBT = Optimized R-Bar Targeted Rendezvous
ORFEUS = Orbiting and Retrievable Far and Extreme Ultraviolet Spectrometer
PAM = Payload Assist Module
PAMS-STU = Passive Aerodynamically Stabilized Magnetically Damped Satellite-Satellite Test Unit
PMA = Pressurized Mating Adapter
PDP = Plasma Diagnostics Package
RCS = Reaction Control System
REP = Radar Evaluation Pod
RMS = Remote Manipulator System
Rndz = Rendezvous
RPM = R Bar Pitch Maneuver
Rtrv = Retrieve
S-IVB = Second stage of the Saturn IB or third stage of the Saturn V
SAINT = Satellite Inspector
SEP = Separation
SFU = Space Flyer Unit
SLA = Spacecraft LM Adapter
Short = Direct Rendezvous
SK = Station-Keeping
SORn = n^{th} Stable Orbit Rendezvous burn
SPARTAN = Shuttle Pointed Autonomous Tool For Astronomy
SPAS = Shuttle Pallet Satellite
SPS = Service Propulsion System
STORRM DTO = Sensor Test for Orion RelNav Risk Mitigation Detailed Test Objective
SYNCOM = Synchronous Communication
TACAN = Tactical Air Navigation
TCS = Trajectory Control Sensor
TDRSS = Tracking and Data Relay Satellite System
TRS = Teleoperator Retrieval System
Ti = Transition initiation
TIG = Time of Ignition
TORRA = Twice Orbital Rate R Bar Approach
TORVA = Twice Orbital Rate V Bar Approach
TPI = Terminal Phase Initiation
TPM = Terminal Phase Mid-course correction
Transposition = Transposition and docking maneuver to extract the LM from the S-IVB
UF = Utilization Flight
VBNn = n^{th} V Bar Null burn

VGS = Video Guidance Sensor
VHF = Very High Frequency
WSF = Wake Shield Facility
ΔH = delta height
ΔV = delta velocity

Space Shuttle Rendezvous and Proximity Operations Missions to the ISS (See Inside Front Cover For Other Shuttle Missions)

Mission	Flight	Year	Profile	Target	Comments
ISS Assembly and Supply	88 (2A)	1998	Ground-Up	ISS	Captured Zarya with RMS, attached Unity Node with PMA 1 & 2.
	96 (2A.1)	1999	Ground-Up	ISS	First docking with ISS. ISS resupply and outfitting.
	101 (2A.2a)	2000	Ground-Up	ISS	ISS resupply and outfitting.
	106 (2A.2b)	2000	Ground-Up	ISS	ISS resupply and outfitting.
	92 (3A)	2000	Ground-Up	ISS	Radar failure. Z1 Truss, PMA 3, Ku comm, & CMGs installed.
	97 (4A)	2000	Ground-Up	ISS	Delivered P6 truss (with solar arrays & radiators).
	98 (5A)	2001	Ground-Up	ISS	Delivered Destiny lab.
	102 (5A.1)	2001	Ground-Up	ISS	Tail forward approach. MPLM resupply. Crew exchange.
	100 (6A)	2001	Ground-Up	ISS	Tail forward approach. Installed robotic arm. MPLM resupply.
	104 (7A)	2001	Ground-Up	ISS	Delivered Quest Airlock (installed with ISS robotic arm).
	105 (7A.1)	2001	Ground-Up	ISS	MPLM resupply. Crew exchange.
	108 (UF-1)	2001	Ground-Up	ISS	MPLM resupply. Crew exchange.
	110 (8A)	2002	Ground-Up	ISS	Delivered S0 truss and Mobile Transporter.
	111 (UF-2)	2002	Ground-Up	ISS	MPLM resupply. Mobile base installation. Crew exchange.
	112 (9A)	2002	Ground-Up	ISS	Delivered S1 truss, radiators & CETA cart A.
	113 (11A)	2002	Ground-Up	ISS	Delivered P1 truss, radiators & CETA cart B. Crew exchange.
	114 (LF-1)	2005	Ground-Up	ISS	MPLM Resupply. CMG replacement. First RPM.
	121 (ULF-1.1)	2006	Ground-Up	ISS	MPLM Resupply. Add third ISS crewmember.
	115 (12A)	2006	Ground-Up	ISS	P3/P4 truss.
	116 (12A.1)	2006	Ground-Up	ISS	P5 Truss, SPACEHAB
	117 (13A)	2007	Ground-Up	ISS	S3/S4 Truss
	118 (13A.1)	2007	Ground-Up	ISS	S5 Truss
	120 (10A)	2007	Ground-Up	ISS	U.S. Node 2, first flight of Lambert guidance upgrade.
	122 (1E)	2008	Ground-Up	ISS	Columbus Laboratory
	123 (1J/A)	2008	Ground-Up	ISS	Kibo Logistics Module, Dextre Robotics System
	124 (1J)	2008	Ground-Up	ISS	Kibo Pressurized Module, Japanese Remote Manipulator System
	126 (ULF2)	2008	Ground-Up	ISS	MPLM
	119 (15A)	2009	Ground-Up	ISS	S6 truss segment
	127 92J/A)	2009	Ground-Up	ISS	Kibo JEM EF, Kibo Japanese ELM-ES
	128 (17A)	2009	Ground-Up	ISS	Leonardo MPLM, LMPESSC, Vernier RCS failure.
	129 (ULF3)	2009	Ground-Up	ISS	ELC1, ELC2
	STS-130 (20A)	2010	Ground-Up	ISS	Tranquility Node 3, Cupola. TCS failure during approach.
	STS-131 (19A)	2010	Ground-Up	ISS	Leonardo MPLM, radar fail.
	STS-132 (ULF4)	2010	Ground-Up	ISS	ICC, MRM1, COAS bulb replacement.
	STS-133 (ULF5)	2011	Ground-Up	ISS	ELC4, Leonardo PMM
	STS-134 (ULF6)	2011	Ground-Up	ISS	ELC3, AMS-2, STORRM DTO during rndz & docking.
	STS-135 (ULF7)	2011	Ground-Up	ISS	Raffaello MPLM, LMC, return to Earth of failed ammonia pump. ISS yaw maneuver after orbiter undocking to facilitate engineering photos during orbiter half-lap fly-around.

A = Assembly, **AMS** = Alpha Magnetic Spectrometer, **ATV** = Automated Transfer Vehicle, **CETA** = Crew and Equipment Translation Aid, **CMG** = Control Moment Gyro, **ELC** = EXPRESS Logistics Carrier, **ELM-ES** = Experiment Logistics Module - Exposed Section, **EVA** = Extra Vehicular Activity, **ICC** = Integrated Cargo Carrier, **JEM EF** = Japanese Experiment Module Exposed Facility, **LF** = Logistics Flight, **LMC** = Lightweight Multi-purpose Carrier, **LMPESSC** = Lightweight Multi-Purpose Experiment Support Structure Carrier, **MPLM** = Multi-Purpose Logistics Module, **MRM** = Mini Research Module, **ORBT** = Optimized R-Bar Targeted Rendezvous, **PMA** = Pressurized Mating Adapter, **PMM** = Permanent Multi-Purpose Module, **Rndz** = Rendezvous, **RPM** = R Bar Pitch Maneuver, **STORRM DTO** = Sensor Test for Orion Relnav Risk Mitigation Detailed Test Objective, **TORRA** = Twice Orbital Rate R Bar Approach, **TORVA** = Twice Orbital Rate V Bar Approach, **UF** = Utilization Flight, **ULF** = Utilization & Logistics Flight

NASA/JSC

AK / Andrews, Cheryl
AK / Fontenot, Brent
DA111 / Hill, Paul
DA111 / Schaefer, Stan
DA335 / MOD Library (1)
DA712 / Barnett, Gail
DM / Banfield, Rob
DM / Gavin, Richard
DM / Fields, John
DM / Library
DM3 / Reichert, Christine
DM32 / Burley, Phillip
DM32 / González, Edward
DM32 / Rojas, Roger
DM32 / Stein, Robert
DM32 / Tracy, William
DM34 / Dannemiller, Dave
DM34 / Machula, Mike
DM34 / Ruiz, Sarah
DM34 / Spehar, Jeannette
DM34 / Walker, Steve
DM35 / Miller, Todd
DM35 / Tuxhorn, Jeff
DS21 / Dunham, Scott
EA3 / Parma, George
EA34 / McMahon, James
EG4 / Fitzgerald, Steve
EG5 / Cerimele, Chris
EG5 / Hoffman, David
EG5 / Merriam, Robert
EG5 / Whitley, Ryan
EG6 / Brazzel, Jack
EG6 / Chevray, Keiko
EG6 / Crain, Tim
EG6 / Cryan, Scott
EG6 / D'Souza, Chris
EG6 / Hart, Jeremy
EG6 / Hinkel, Heather
EG6 / Merkle, Scott
EG6 / Mitchell, Jennifer
EG6 / Mrozinski, Richard
EG6 / Ruiz, Jose
EG6 / Spehar, Pete
EG6 / Weeks, Michael
EV29 / Murray, Scott
EV813 / Duron, Jim
IM111 / Scroggins, Mark
IS20 / Ross-Nazzal, Jennifer
IS20 / Wright, Rebecca
IS23 / STI Center (Caballero) (4)
MA111 / Shannon, John P.
MG111 / Wooten, Peggy
MV6 / Fletcher, Malise
ZD / Pearson, Don

NASA Ames Research Center
Moffett Field, CA 94035
MS 202-3 / Library

NASA Dryden Flight Research Center
P.O. Box 273
Edwards, California 93523-0273
4839 / Gelzer, Dr. Christian

NASA Glenn Research Center
21000 Brookpark Road
Cleveland, OH 44135
105-3 / Burke, Laura
105-3 / Sjauw, Waldy
142-3 / Technical Library

NASA / Goddard Space Flight Center
Greenbelt, MD 20771 / USA
272 / Banholzer, Gordon
590 / Dennehy, Cornelius
595 / Carpenter, Dr. Russell
595 / Naasz, Bo
595 / Barbee, Brent
595 / Berry, Kevin
595 / Gaebler, John
595 / Getzandanner, Kenny
595 / Moreau, Michael

NASA Headquarters
Washington, DC 20546
BJ000 / Lengyel, David
KA0 / Bauer, Frank
NH / Barry, William
NH / Garber, Stephen

Jet Propulsion Laboratory
4800 Oak Grove Dr.
Pasadena, CA 91109
111-113 / Library

NASA Kennedy Space Center
Kennedy Space Center, Florida 32899
NE-A2 / Menendez, Alfred
Library-D

NASA Langley Research Center
Hampton, VA 23681
185 / Document (Miller, Sue)

NASA Marshall Space Flight Center
Huntsville, AL 35812
CS20 / Wright, Michael
ES32 / Bryan, Thomas
ES32 / Howard, Richard
EV42 / Heaton, Andrew
EV94 / Kulpa, Vyga

9781780398280